Nutrition, Metabolism, and Obesity: Novel Strategies and Molecular Mechanisms—Future Perspectives

Nutrition, Metabolism, and Obesity: Novel Strategies and Molecular Mechanisms—Future Perspectives

Guest Editor

Marion Korach-André

MDPI

Basel • Beijing • Wuhan • Barcelona • Belgrade • Novi Sad • Cluj • Manchester

Guest Editor
Marion Korach-André
Department of Research and
Innovation
University Marie and Louis
Pasteur
Besançon
France

Editorial Office
MDPI AG
Grosspeteranlage 5
4052 Basel, Switzerland

This is a reprint of the Special Issue, published open access by the journal *Biomolecules* (ISSN 2218-273X), freely accessible at: https://www.mdpi.com/journal/biomolecules/special_issues/FZ2VN9F16L.

For citation purposes, cite each article independently as indicated on the article page online and as indicated below:

Lastname, A.A.; Lastname, B.B. Article Title. *Journal Name* **Year**, *Volume Number*, Page Range.

ISBN 978-3-7258-7056-1 (Hbk)
ISBN 978-3-7258-7057-8 (PDF)
https://doi.org/10.3390/books978-3-7258-7057-8

Contents

About the Editor

Marion Korach-André

Marion Korach-André is a scientific expert with over 20 years of international experience in translational and biomedical research, specializing in nutrition, obesity, metabolic diseases, and metabolic programming. She holds a Ph.D. in Biochemistry, Human Nutrition and Physiology, and has led multidisciplinary projects across academia, healthcare, and industry.

Currently, she serves as a Health and Innovation Officer, Stars EU Alliance Officer and Head of Research and Innovation at the University Marie and Louis Pasteur, driving European collaboration and innovation. She also works as a Scientific Advisor for FoodFacts (Sweden), focusing on sustainable nutrition and public health.

Marion worked for 18 years at Karolinska Institutet where she led research on metabolic programming, obesity, and sex hormones, and as Research Coordinator at SciLifeLab in Sweden, where she coordinated projects on spatial transcriptomics and metabolic disorders.

Passionate about interdisciplinary collaboration, she bridges research, education, private sector, and societal impact, with a focus on innovative nutritional strategies and metabolic health.

Preface

Obesity is a global epidemic and a chronic, multifactorial disease with profound impacts on metabolic, cardiovascular, and overall health. This reprint of *Biomolecules* explores the latest scientific advancements in understanding and addressing obesity. Recent research has uncovered complex interactions between genetic, molecular, environmental, and behavioral factors, revealing new biomarkers and therapeutic targets[1-4]. The reprint covers innovative pharmacological approaches[5,6], the role of the immune system[7,8,9], gut–brain axis communication[7,10], molecular understanding11, the role of ultra-processed food[12], and the role of personalized nutrition13 and digital health technologies in obesity management[1-4].

The aim of the current work is to bridge the gap between groundbreaking research and clinical practice. By presenting cutting-edge insights into molecular mechanisms, this Reprint seeks to empower clinicians, researchers, and policymakers to develop evidence-based, multidisciplinary strategies. The purpose is to foster collaboration across fields—from endocrinology to public health—to create sustainable, equitable solutions for obesity and its comorbidities.

This work addresses the urgent need to address obesity as a complex disease, not merely a result of caloric excess. This Reprint aims to inspire further research and dialogue, ultimately contributing to a future where metabolic health is accessible to all.

1. Srivastava, G.; Campbell, S.L.; Hill, C.R.; Stanley, T.L.; A Lawson, E.; Apovian, C.M.; Almandoz, J.P.; Leggio, L.; Lakdawalla, D.N.; Dar, M.; et al. Novel strategies for medical management of obesity: mechanisms, clinical implications, and societal impacts—a report from the 25th Annual Harvard Nutrition Obesity Symposium. *Am. J. Clin. Nutr.* **2025**, *122*, 866–885.
2. Nicze, M.; Dec, A.; Borówka, M.; Krzyżak, D.; Bołdys, A.; Bułdak, Ł.; Okopień, B. Molecular Mechanisms behind Obesity and Their Potential Exploitation in Current and Future Therapy. *Int. J. Mol. Sci.* **2024**, *25*, 8202.
3. Młynarska, E.; Bojdo, K.; Bulicz, A.; Frankenstein, H.; Gąsior, M.; Kustosik, N.; Rysz, J.; Franczyk, B. Obesity as a Multifactorial Chronic Disease: Molecular Mechanisms, Systemic Impact, and Emerging Digital Interventions. *Curr. Issues Mol. Biol.* **2025**, *47*, 787.
4. Buldak, L. Special Issue: "Molecular Background of Obesity and Its Impact on Therapeutic Strategies". *Int. J. Mol. Sci.* **2025**, *27*, 40.
5. Melnyk, S.; Hakkak, R. Effect of Metformin Treatment on Serum Metabolic Profile Changes in Lean and Obese Zucker Rat Model for Fatty Liver Disease. *Biomolecules* **2023**, 13, 1234.
6. Kounatidis, D.; Vallianou, N.G.; Karampela, I.; Rebelos, E.; Kouveletsou, M.; Dalopoulos, V.; Koufopoulos, P.; Diakoumopoulou, E.; Tentolouris, N.; Dalamaga, M. Anti-Diabetic Therapies and Cancer: From Bench to Bedside. *Biomolecules* **2024**, 14, 1479.
7. Chen, X.; Hou, Y.; Liao, A.; Pan, L.; Yang, S.; Liu, Y.; Wang, J.; Xue, Y.; Zhang, M.; Zhu, Z.; et al. Integrated Analysis of Gut Microbiome and Adipose Transcriptome Reveals Beneficial Effects of Resistant Dextrin from Wheat Starch on Insulin Resistance in Kunming Mice. *Biomolecules* **2024**, 14, 186.
8. Al Mamun, A.; Rakib, A.; Mandal, M.; Kumar, S.; Singla, B.; Singh, U.P. Polyphenols: Role in Modulating Immune Function and Obesity. *Biomolecules* **2024**, 14, 221.
9. Barbosa, P.; Pinho, A.; Lázaro, A.; Paula, D.; Tralhão, J.G.; Paiva, A.; Pereira, M.J.; Carvalho, E.; Laranjeira, P. Bariatric Surgery Induces Alterations in the Immune Profile of Peripheral Blood T Cells. *Biomolecules* **2024**, 14, 219.
10. Chen, W.; Ma, Q.; Li, Y.; Wei, L.; Zhang, Z.; Khan, A.; Khan, M.Z.; Wang, C. Butyrate Supplementation Improves Intestinal Health and Growth Performance in Livestock: A Review. *Biomolecules* **2025**, 15, 85.

11. Vaganova, A.N.; Shemyakova, T.S.; Lenskaia, K.V.; Rodionov, R.N.; Steenblock, C.; Gainetdinov, R.R. Trace Amine-Associated Receptors and Monoamine-Mediated Regulation of Insulin Secretion in Pancreatic Islets. *Biomolecules* **2023**, *13*, 1618.
12. Abdisa, K.B.; Szerdahelyi, E.; Molnár, M.A.; Friedrich, L.; Lakner, Z.; Koris, A.; Toth, A.; Nath, A. Metabolic Syndrome and Biotherapeutic Activity of Dairy (Cow and Buffalo) Milk Proteins and Peptides: Fast Food-Induced Obesity Perspective—A Narrative Review. *Biomolecules* **2024**, *14*, 478.
13. Charles, S.; Liu, Y.; Bharmal, S.H.; Kimita, W.; Petrov, M.S. Effect of Acute Nutritional Ketosis on Circulating Levels of Growth Differentiation Factor 15: Findings from a Cross-Over Randomised Controlled Trial. *Biomolecules* **2024**, *14*, 665.

Marion Korach-André
Guest Editor

biomolecules

MDPI

Article

Effect of Acute Nutritional Ketosis on Circulating Levels of Growth Differentiation Factor 15: Findings from a Cross-Over Randomised Controlled Trial

Sanjali Charles, Yutong Liu, Sakina H. Bharmal, Wandia Kimita and Maxim S. Petrov *

School of Medicine, University of Auckland, Auckland 1010, New Zealand
* Correspondence: max.petrov@gmail.com; Tel.: +64-99232776

Abstract: Exogenous supplementation with ketone beverages has been shown to reduce plasma glucose levels during acute nutritional ketosis. It remains to be investigated whether growth differentiation factor 15 (GDF-15)—an anorexigenic hormone—is involved in this process. The aim was to investigate the effect of a ketone ester beverage delivering β-hydroxybutyrate (KEβHB) on plasma levels of GDF-15, as well as assess the influence of eating behaviour on it. The study was a randomised controlled trial (registered at clinicaltrials.gov as NCT03889210). Individuals were given a KEβHB beverage or placebo in a cross-over fashion. Blood samples were collected at baseline, 30, 60, 90, 120, and 150 min after ingestion. Eating behaviour was assessed using the three-factor eating questionnaire. GDF-15 levels were not significantly different ($p = 0.503$) after the KEβHB beverage compared with the placebo. This finding remained consistent across the cognitive restraint, emotional eating, and uncontrolled eating domains. Changes in the anorexigenic hormone GDF-15, irrespective of eating behaviour, do not appear to play a major role in the glucose-lowering effect of exogenous ketones.

Keywords: growth differentiation factor 15; appetite; eating behaviour; ketosis; β-hydroxybutyrate; nutrition; prediabetes

1. Introduction

Based on current trends, it is estimated that approximately 20% of the world's population will have obesity by 2025 [1,2]. The negative implications of high obesity rates cannot be overstated, especially because it is also a risk factor for non-communicable diseases such as diabetes mellitus [1]. Obesity is characterised by an excessive accumulation of body fat due to an energy imbalance where consumption outweighs expenditure [3]. Extensive research has been conducted to elucidate pathways in the brain that control satiety and hunger [3]. This knowledge has opened up new avenues for understanding individuals' eating behaviour and helping reduce the burden of obesity [3]. The area postrema—a circumventricular organ located in the hindbrain at the base of the fourth ventricle—plays a key role in integrating and transmitting chemical and neural signals to the brain regions that control feeding [4]. Receptors for various molecules are expressed on neurons located in the area postrema, including the glial cell-derived neurotrophic family receptor α-like (GFRAL) that binds specifically to growth differentiation factor 15 (GDF-15) [5]. GDF-15 is a member of the transforming growth factor-β superfamily [5]. Experimental murine studies have shown that food intake is reduced via GDF-15-GFRAL-RET signalling in the area postrema [5–7]. While GDF-15 levels have been significantly associated with obesity, it has been suggested that this increase in GDF-15 is likely not causing the disease but compensating for it by attempting to reduce food intake [6,8]. Native GDF-15 has a short half-life, which precludes it from being a viable therapeutic option [9]. Therefore, recombinant GDF-15 molecules have been developed with better pharmacokinetic properties and efficacy [9]. Recombinant GDF-15 molecules administered subcutaneously to obese mice and monkeys were shown to reduce food intake and glucose levels and improve insulin sensitivity [9,10]. Further, elevated GDF-15 levels stimulated an increase in hepatic triglyceride metabolism, leading to a decrease in plasma triglycerides [11]. These findings from preclinical studies suggest that targeting GDF-15 signalling pathways may have potential in the fields of obesity and metabolic health [6,8].

GDF-15 analogues for use in human studies are still under development [6,8]. Therefore, alternative methods that increase circulating levels of GDF-15 for better metabolic health outcomes may need to be explored. One possible way to increase the levels of GDF-15 is through a high-fat, low-carbohydrate ketogenic diet [12]. Ketogenic diets (compared with the normal chow diet) were shown to significantly increase serum GDF-15 levels in mice [12]. This diet induces the state of ketosis in the body, making ketone bodies (β-hydroxybutyrate, acetone, and acetoacetate) the primary source of energy [13]. The eating behaviour of people on a ketogenic diet can be assessed using the three-factor eating questionnaire (TFEQ), which measures cognitive restraint, emotional eating, and uncontrolled eating [14]. It was shown that uncontrolled eating and emotional eating decreased following a ketogenic diet administered for 12 months, suggesting that nutritional ketosis may change eating behaviour [14]. There is an unmet need to clarify whether GDF-15 plays a role in causing this change, given its role in reducing food intake. It can take weeks to reach moderate nutritional ketosis (0.5–5 nmol/L) on a ketogenic diet, and adherence to this diet can be difficult, resulting in the development of alternative methods to reach ketosis [15]. This includes acute nutritional ketosis, where exogenous ketones are ingested and the peak state of ketosis is reached within 30–60 min [15]. The Cross-over randomisEd Trial of β-hyroxybUtyrate in prediabeteS (CETUS) showed that the ingestion of a ketone ester beverage delivering β-hydroxybutyrate (KEβHB) reduced the levels of glucose and triglycerides in individuals with prediabetes [15,16]. However, there is a gap in the literature in regard to the effect exogenous ketones have on GDF-15 levels. Given that it is an anorexigenic hormone, investigating this effect may prove beneficial in further assessing the utility of acute nutritional ketosis in treating obesity and hyperglycaemia.

The primary aim of the present study was to investigate the effect of the KEβHB beverage (versus placebo) on GDF-15 levels. The secondary aim was to investigate whether eating behaviour impacted the levels of GDF-15 during nutritional ketosis. The tertiary aim was to assess the association between GDF-15 levels and changes in glucose and triglyceride levels during nutritional ketosis.

2. Materials and Methods

2.1. Study Design and Population

The study was run at the University of Auckland (New Zealand) and was part of the CETUS project. The project was a randomised controlled trial that primarily investigated the changes in the glucoregulatory response to a KEβHB beverage in individuals with prediabetes [15]. The Health and Disability Ethics Committee assessed and approved the project (18/NTB/161). It was conducted in accordance with the standards set by the Declaration of Helsinki and prospectively registered at clinicaltrials.gov (NCT03889210). The present study reports on a secondary outcome of the project, specifically plasma levels of GDF-15.

Adult individuals (≥18 years) were included in the study if they met the criteria for prediabetes diagnosis outlined by the American Diabetes Association (glycated haemoglobin (HbA1c) between 5.7% and 6.4% (39–47 mmol/mol) and/or fasting plasma glucose (FPG) between 100 and 125 mg/dL) [17]. Individuals were excluded if they took ketone supplements or followed a ketogenic diet, were diagnosed with diabetes, had received glucose-lowering medication, performed intensive physical training or participated in competitive sports, had malignancy, underwent bariatric or gastrointestinal surgery, had a cognitive disability, or had poor venous access for blood sampling at the time of the study. Participants were randomised into two sequences in a cross-over fashion. The research nurse and study participants were blinded to the sequence of allocation.

2.2. Study Protocol

For a total of 24 h prior to the clinic visit, participants were asked to start a food log (that aided them in consuming similar meals before both visits), undertake an overnight fast of 8–10 h, and refrain from exercise and alcohol consumption. Six blood samples were taken during each clinical visit. The first sample was collected prior to receiving the intervention, which was a 100 mL drink comprising water, a commercially available FDA-approved monoester of the ketone β-hydroxybutyrate (R-3-hydroxybutryl-1,3-hydroxybutyrate; ΔG®, TΔS Ltd., Oxford, UK; 4.4 mmol/kg of D-β-hydroxybutyrate—equivalent to 1.05 mL/kg and 1.9 kcal/kg of individual body weight). The placebo was a 100 mL beverage comprising water, malic acid (BioTrace®, Auckland, New Zealand), and arrowroot (McKenzie's®, Altona, Australia). Blood samples were taken 30, 60, 90, 120, and 150 min after ingestion. Participants returned for a second clinic visit after a washout period of 7–10 days and were provided with the alternate beverage.

2.3. Laboratory Measurements

β-hydroxybutyrate levels were measured on whole blood immediately after collection using a handheld ketone meter and FreeStyle Optimum β-ketone test strip (Abbott Laboratories, Chicago, IL, USA). The blood samples were collected in EDTA and lithium heparin tubes (BD Vacutainer®; Becton, Dickinson and Company, Franklin Lakes, NJ, USA). They were then centrifuged at $4000\times g$ for 5 min and separated into plasma aliquots to be stored at −80 °C. GDF-15 plasma levels were measured using a human enzyme-linked immunosorbent assay (ELISA) kit (intra-assay coefficient of variation = 4.3%, and inter-assay coefficient of variation = −1.6%) (Ansh Laboratories, Webster, TX, USA). Blood samples were also sent to LabPlus, a tertiary medical laboratory at Auckland City Hospital (New Zealand), to measure plasma glucose and triglyceride levels [15,16].

2.4. Anthropometric Measurements

To measure body mass index (BMI) (kg/m^2), a digital stadiometer (Health-o-meter®/ Pelstar LLC©, McCook, IL, USA) was used. Hip circumference (cm^3) and waist circumference (cm^3) were measured in the horizontal plane over the participant's clothing [18].

2.5. Assessment of Eating Behaviour

During a clinic visit, participants completed the TFEQ to assess their eating behaviour. The version used in this study was the TFEQ-R18, which consists of 18 different questions covering three domains (cognitive restraint, emotional eating, and uncontrolled eating) and was validated for use in the general population [19]. Cognitive restraint refers to the tendency to limit food intake regardless of hunger or satiety. Emotional eating refers to the tendency to increase food intake to deal with negative emotions. Uncontrolled eating refers to the tendency to increase food intake while feeling unable to regulate it [20]. The answers participants provided to each question were assigned a score from 1 to 4. There were six, three, and nine questions in the cognitive restraint, emotional eating, and uncontrolled eating domains, respectively [19].

2.6. Power Calculation

The sample size of the present study was determined based on the primary endpoint—change in plasma glucose [15]. Specifically, it was hypothesised that study participants would have a 15% decrease in glucose levels following ingestion of the KEβHB drink. It was determined that a sample size of 15 individuals is required to detect (with 80% statistical power) a difference in effect size of 0.88 at an α-level of 0.05. A total of 18 individuals were recruited to account for a 20% dropout or missing samples.

2.7. Statistical Analysis

Prism software version 9 (GraphPad Company, Boston, MA, USA) was used to conduct all statistical analyses. Baseline characteristics were reported as mean and standard deviation or median and interquartile range. Statistical significance was determined by a p-value of less than 0.05. Statistical analyses were conducted in four steps. First, to determine whether the difference in the levels of GDF-15 between the groups was statistically significant, a paired t-test was conducted. The levels of GDF-15 from 0 to 150 min for the KEβHB and placebo groups were log-transformed and plotted. The total area under the curve (AUC) was calculated using the trapezoidal rule. Q-Q plots were used to assess normality. Second, to determine whether there was an effect of time, treatment, or both on the levels of GDF-15, a repeated measure two-way ANOVA test was conducted. To meet the sphericity assumption, the Geisser-Greenhouse correction was used. Third, participants were stratified into low and high eating behaviour subgroups based on the median score of TFEQ for the three domains—cognitive restraint, emotional eating, and uncontrolled eating. Steps one and two of the analyses (described above) were repeated. Last, using data from the KEβHB group, a Spearman's correlation analysis was conducted, investigating the association between the levels of GDF-15 at 30, 60, 90, 120, and 150 min and the changes in the levels of glucose and triglycerides from baseline to 30, 60, 90, 120, and 150 min.

3. Results

3.1. GDF-15 in the Study Population

Of the 32 individuals considered, 18 met the eligibility criteria and were randomised (12 men and 6 women). All the participants completed the study. The characteristics of the 18 study participants are presented in Table 1. The median level of GDF-15 at baseline was 2.1 ng/mL, and the interquartile range was 1.1–2.5 ng/mL. Changes in the levels of glucose from baseline to 30–150 min were significantly ($p = 0.014$, $\rho = 0.273$) associated with the GDF-15 levels from 30 to 150 min. Changes in the levels of triglycerides from baseline to 30–150 min were not significantly ($p = 0.640$, $\rho = -0.053$) associated with the levels of GDF-15 from 30–150 min (Figure 1).

Table 1. Characteristics of study participants.

Characteristic	Participants (n = 18) [1]
Age, y	55 ± 14
Sex, n	
Men	12
Women	6
GDF-15, ng/mL	2.1 (1.1–2.5)
Systolic blood pressure, mmHg	135 ± 27
Diastolic blood pressure, mmHg	88 ± 13
BMI, kg/m^2	28.4 (24.5–30.9)
Hip circumference, cm	108.5 (99.8–112.3)
Waist circumference, cm	100.0 (90.3–107.5)

Footnotes: [1] values are presented as means ± standard deviation or median (interquartile range). Abbreviations: BMI = body mass index; GDF-15 = growth differentiation factor 15.

3.2. Effect of Acute Nutritional Ketosis on Levels of GDF-15

The difference between the total AUCs for the KEβHB (48.54 ± 35.21 ng/mL × min) and placebo (45.50 ± 30.72 ng/mL × min) drinks was not statistically significant ($p = 0.503$; d = 0.09) (Table 2). The effect of time alone ($p = 0.962$), treatment alone ($p = 0.687$), or interaction between time and treatment ($p = 0.812$) on the levels of GDF-15 was not statistically significant (Figure 2).

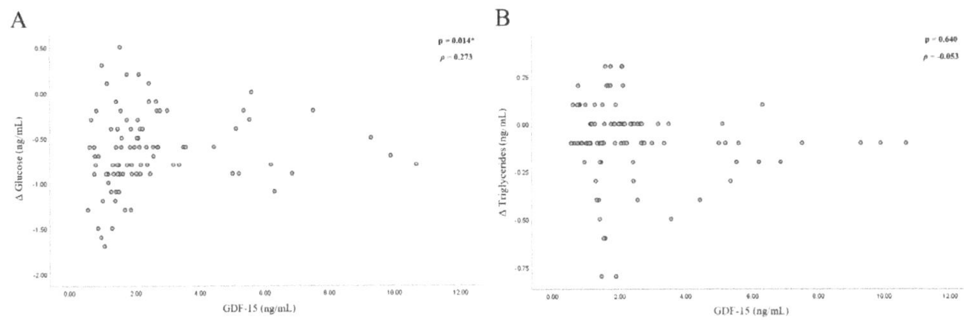

Figure 1. Associations between GDF-15 and changes in glucose and triglycerides. Footnotes: * indicates statistical significance ($p < 0.05$). Spearman's correlation analysis was conducted between two variables: 1. GDF-15 levels at 30, 60, 90, 120, and 150 min; 2. change in glucose (**A**) and triglycerides (**B**) from baseline to 30, 60, 90, 120, and 150 min. Analysis was conducted on data from the KEβHB group only. Abbreviations: GDF-15 = growth differentiation factor 15; KeβHB = ketone monoester β-hydroxybutyrate.

Table 2. Differences in the levels of GDF-15 after ingestion of the KEβHB and placebo beverages.

Molecule	KEβHB (n = 18)	Placebo (n = 18)	Mean Difference (95% CI)	d (Effect Size)	p-Value
GDF-15 (ng/mL × min)	48.54 ± 35.21	45.50 ± 30.72	−3.04 (−12.41, 6.33)	0.09	0.503

Footnotes: levels of GDF-15 from 0 to 150 min were log-transformed, and then AUCs were calculated. A paired t-test was run to determine the mean difference between the AUCs of both groups, effect size, and p-value. The values are presented as means ± standard deviation or median (IQR). Abbreviations: AUC = area under the curve; GDF-15 = growth differentiation factor 15; KeβHB = ketone monoester β-hydroxybutyrate; IQR = interquartile range.

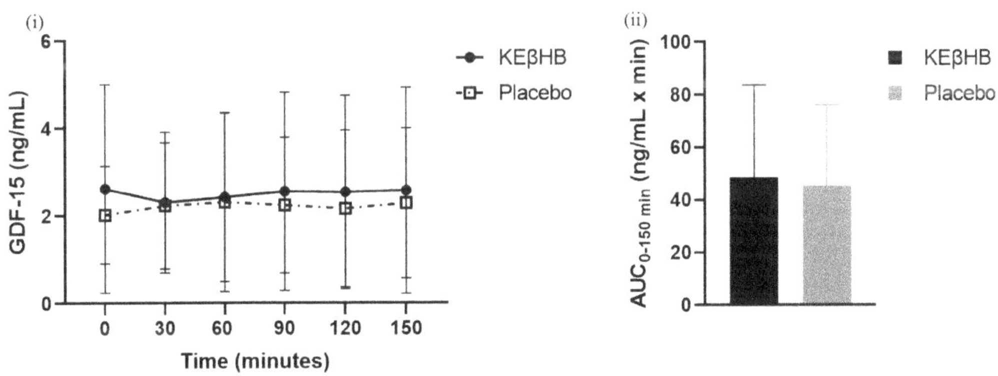

Figure 2. Change in the levels of GDF-15 after ingestion of the KEβHB and placebo beverages. Footnotes: mean absolute concentrations of GDF-15 from 0 to 150 min are presented in panel (**i**). Log-transformed AUCs of GDF-15 are presented in panel (**ii**). In both panels, error bars represent standard deviation. Abbreviations: AUC = area under the curve; GDF-15 = growth differentiation factor 15; KEβHB = ketone monoester β-hydroxybutyrate.

3.3. Role of Eating Behaviour

In participants with low cognitive restraint scores, the difference between the total AUCs for the KEβHB (52.51 ± 44.50 ng/mL × min) and placebo (53.48 ± 38.10 ng/mL × min) drinks was not statistically significant (p = 0.850; d = 0.02) (Table 3). The effect of time alone (p = 0.832), treatment alone (p = 0.996), or the interaction between time and treatment (p = 0.093) on the levels of GDF-15 was not statistically significant (Figure 3). In participants with high cognitive restraint scores, the difference between the total AUCs for the KEβHB

(44.56 ± 24.87 ng/mL × min) and placebo (37.52 ± 20.25 ng/mL × min) drinks was not statistically significant (p = 0.372; d = 0.31) (Table 3). The effect of time alone (p = 0.923), treatment alone (p = 0.454), or the interaction between time and treatment (p = 0.873) on the levels of GDF-15 was not statistically significant (Figure 3).

Table 3. Differences in the levels of GDF-15 after ingestion of the KEβHB and placebo beverages according to the eating behaviour domains.

Domain	Time	KEβHB (n = 9)	Placebo (n = 9)	Mean Difference (95% CI)	d (Effect Size)	p-Value
			Cognitive Restraint			
Low	Baseline	0.32 ± 0.40	0.28 ± 0.25	−0.04 (−0.24, 0.15)	0.13	0.624
	Total AUC	52.51 ± 44.50	53.48 ± 38.10	0.96 (14.82, 4.94)	0.02	0.850
High	Baseline	0.26 ± 0.26	0.20 ± 0.28	−0.06 (0.36, 0.12)	0.21	0.648
	Total AUC	44.56 ± 24.87	37.52 ± 20.25	−7.05 (−24.21, 10.12)	0.31	0.372
			Emotional Eating			
Low	Baseline	0.29 ± 0.39	0.28 ± 0.18	−0.01 (−0.23, 0.22)	0.02	0.941
	Total AUC	52.31 ± 43.85	50.22 ± 39.29	−2.09 (−12.03, 7.84)	0.05	0.640
High	Baseline	0.29 ± 0.29	0.20 ± 0.33	−0.09 (−0.34, 0.15)	0.30	0.410
	Total AUC	44.76 ± 26.07	40.78 ± 20.27	−3.99 (−22.59, 14.61)	0.17	0.634
			Uncontrolled Eating			
Low	Baseline	0.35 ± 0.35	0.19 ± 0.29	−0.16 (−0.41, 0.10)	0.48	0.197
	Total AUC	54.20 ± 43.32	49.81 ± 38.19	−4.39 (−12.75, 3.96)	0.11	0.260
High	Baseline	0.23 ± 0.31	0.28 ± 0.24	0.06 (−0.13, 0.24)	0.20	0.512
	Total AUC	42.87 ± 26.17	41.18 ± 22.47	−1.69 (−21.02, 17.64)	0.07	0.845

Footnotes: levels of GDF-15 (ng/mL) from 0 to 150 min were log-transformed, and then AUCs were calculated. Paired t-tests were run to determine the mean difference between the AUCs of both pairs, effect size, and p-values. The values are presented as means ± standard deviation or median (IQR). Abbreviations: AUC = area under the curve; GDF-15 = growth differentiation factor 15; KEβHB = ketone monoester β-hydroxybutyrate; IQR = interquartile range.

In participants with low emotional eating scores, the difference between the total AUCs for the KEβHB (52.31 ± 43.85 ng/mL × min) and placebo (50.22 ± 39.29 ng/mL × min) drinks was not statistically significant (p = 0.640; d = 0.05) (Table 3). The effect of time alone (p = 0.338), treatment alone (p = 0.891), or the interaction between time and treatment (p = 0.937) on the levels of GDF-15 was not statistically significant (Figure 3). In participants with high emotional eating scores, the difference between the total AUCs for the KEβHB (44.76 ± 26.07 ng/mL × min) and placebo (40.78 ± 20.27 ng/mL × min) drinks was not statistically significant (p = 0.634; d = 0.17) (Table 3). The effect of time alone (p = 0.644), treatment alone (p = 0.402), or the interaction between time and treatment (p = 0.538) on the levels of GDF-15 was not statistically significant (Figure 3).

In participants with low uncontrolled eating scores, the difference between the total AUCs for the KEβHB (54.20 ± 43.32 ng/mL × min) and placebo (49.81 ± 38.19 ng/mL × min) drinks was not statistically significant (p = 0.260; d = 0.11) (Table 3). The effect of time alone (p = 0.191), treatment alone (p = 0.740), or the interaction between time and treatment (p = 0.277) on the levels of GDF-15 was not statistically significant (Figure 3). In participants with high uncontrolled eating scores, the difference between the total AUCs for the KEβHB (42.87 ± 26.17 ng/mL × min) and placebo (41.18 ± 22.47 ng/mL × min) drinks was not statistically significant (p = 0.845; d = 0.07) (Table 3). The effect of time alone (p = 0.524), treatment alone (p = 0.817), or the interaction between time and treatment (p = 0.102) on the levels of GDF-15 was not statistically significant (Figure 3).

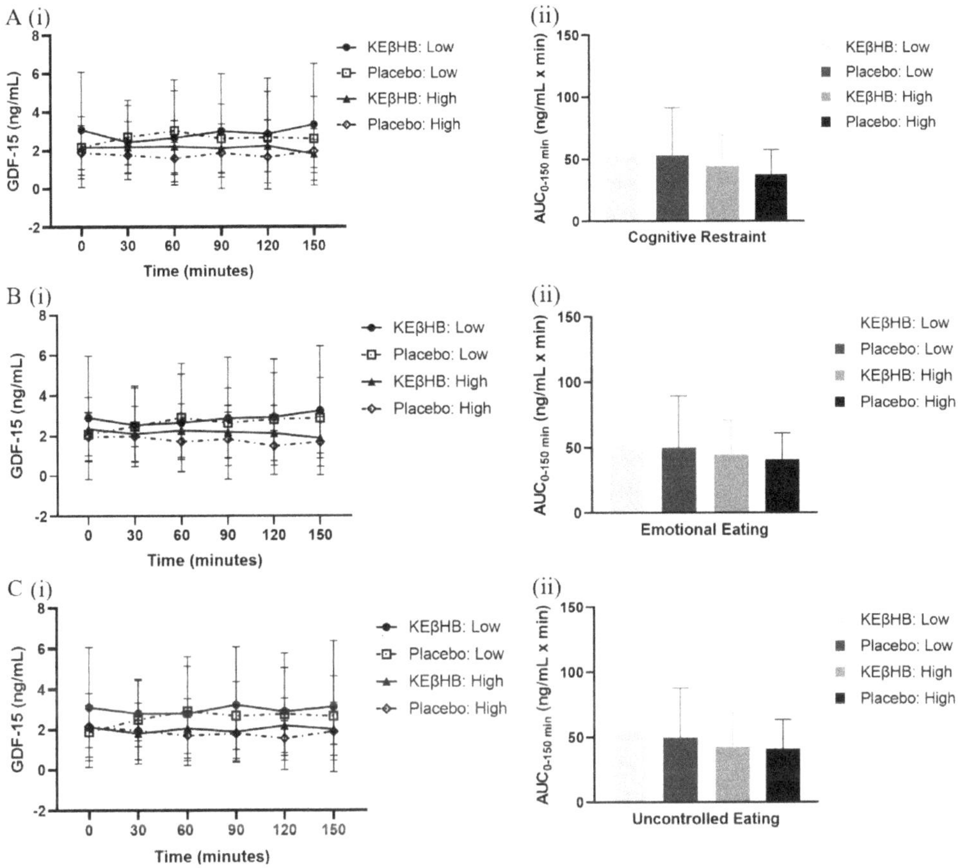

Figure 3. Change in the levels of GDF-15 after ingestion of KEβHB and placebo beverages according to the eating behaviour domains. Footnotes: mean absolute concentrations of GDF-15 from 0 to 150 min and log-transformed AUCs of GDF-15 grouped by high and low uncontrolled eating categories are presented in panels (**A**(**i**,**ii**)), respectively. Mean absolute concentrations of GDF-15 from 0 to 150 min and log-transformed AUCs of GDF-15 grouped by high and low cognitive restraint categories are presented in panels (**B**(**i**,**ii**)), respectively. Mean absolute concentrations of GDF-15 from 0 to 150 min and log-transformed AUCs of GDF-15 grouped by high and low EE categories are presented in panels (**C**(**i**,**ii**)), respectively. In all panels, error bars represent standard deviation. Abbreviations: AUC = area under the curve; GDF-15 = growth differentiation factor 15; KEβHB = ketone monoester β-hydroxybutyrate.

4. Discussion

The present study was the first randomised controlled trial that examined the effect of exogenous ketone supplementation on GDF-15 levels in participants with prediabetes. We found that the effect of the KEβHB beverage on the levels of GDF-15 was not statistically significant compared with the placebo beverage. This finding held true when participants were stratified according to their eating behaviour into cognitive restraint, emotional eating, and uncontrolled eating domains. Correlation analyses showed that the levels of GDF-15 were significantly associated with the changes in the levels of glucose from baseline and not significantly associated with the changes in the levels of triglycerides from baseline during ketosis.

Ketogenic diets have gained traction as an intervention for obesity. There is evidence that this diet not only reduces body weight but also improves glycaemic control. This is a result of the body entering a state of ketosis, where fat-derived molecules (i.e., ketone bodies) become the primary source of energy. Ketosis can also be effectively achieved by the ingestion of ketone supplements [15]. These easily consumable supplements have consistently been shown to readily reduce plasma glucose levels, which suggests that they may be a more accessible treatment for people with obesity and diabetes compared with a ketogenic diet. There are two important advantages that ketosis offers in comparison with other energy-generating mechanisms. First, it prevents the degradation of skeletal muscle. The body does not have protein stores, unlike the adipose tissue that stores fat. Most protein is stored in muscles. Therefore, when the non-carbohydrate substrates are amino acids (during gluconeogenesis), they are derived from the breakdown of skeletal muscle via proteolysis to generate fuel [21]. There is evidence that ketones may promote the synthesis of proteins in muscles [22,23]. The evolutionary adaptation of preferentially using fat stores over protein stores as a fuel source prevents the breakdown of protein stores [24]. Second, ketosis provides a fuel source that is used by the brain. Under most physiological conditions, the main source of energy for the brain is glucose. However, during starvation, while vital organs such as the liver are able to use free fatty acids as a fuel source, the brain preferentially metabolises ketones rather than free fatty acids for energy. Free fatty acids have to undergo β-oxidation to generate adenosine triphosphate—the molecule that provides energy. The process of β-oxidation requires high amounts of oxygen, which produces superoxide chemicals that can cause neuron injury due to oxidative stress. It is believed that the brain has evolved to lower the expression of proteins involved in β-oxidation to protect it from injury. Ketones do not require β-oxidation, which is why they are preferred for use by the brain [25]. Understanding the pathways that cause the glucose-lowering effect of exogenous ketone supplementation may provide deeper insights into improved metabolic health outcomes in people in the state of ketosis. Therefore, a key characteristic of obesity—problematic eating behaviour—was considered in the present study in an effort to elucidate the physiological mechanisms behind the established glucose-lowering effect of acute nutritional ketosis and inform the future application of exogenous ketone supplements.

An absence of statistically significant changes in GDF-15 levels during acute nutritional ketosis suggests that the main effects of the KEβHB drink are not achieved through alterations in food intake. Poffe and colleagues reported that exogenous ketone supplementation reduced the levels of GDF-15 compared with the control beverage when consumed after each endurance training session over three weeks [26]. The authors showed that the physiological stress response observed in the overtraining stages was significantly less prominent in the ketone supplementation group compared with the placebo group. Protection against the stress response is likely the reason that the levels of GDF-15 were lower after ketone supplementation in that study. Also, Poffe and colleagues observed that energy intake was increased in the ketone supplementation group compared with the control group. Since the control group had higher levels of GDF-15 (an anorexigenic hormone), their food intake would be expected to be lower [26]. In the present CETUS study, participants were sedentary, and this may explain why no difference in the levels of GDF-15 was observed (i.e., the magnitude of the stress response was much lower compared with that observed during intensive training). A study in recreational female athletes, who were placed on a diet that restricted calories by 30% for four weeks and consumed a ketone supplement before every meal, also reported no statistically significant changes in GDF-15 [27]. That study also reported no changes in the circulating levels of ghrelin and leptin [27]. It is pertinent to note here that another CETUS study also found no significant effect of acute ketone supplementation on the circulating levels of leptin [28]. The utility that exogenous ketone supplements may have during weight loss interventions is likely based on their ability to reduce stress markers during exercise and prolong the period of intense activity rather than modulate appetite [26,27]. This is further confirmed by an earlier CETUS study that reported no statistically significant changes in the appetite hormones ghrelin and peptide YY and no statistically significant differences in hunger when comparing the KEBHB group and the placebo group [29]. Prior to that study, there was a discrepancy in the literature about whether hunger is reduced during acute nutritional ketosis [29–31]. The earlier studies that reported on significant changes in appetite after ketone supplement consumption included controls that were isocaloric dextrose solutions and, therefore, contained carbohydrates (unlike the placebo used in the CETUS project) [29]. The carbohydrates likely caused a reduction in the levels of ghrelin, resulting in the observed effect of appetite modulation [29]. The present study takes the field further by grouping the participants according to high versus low scores in three eating behaviour categories before looking at the change in levels of GDF-15. The TFEQ has been validated in the general population and has been shown to successfully identify eating behaviour [19]. Even in this subgroup analysis, no statistically significant differences in the circulating levels of GDF-15 were observed. This finding is in line with the results reported by Liu and colleagues when participants were stratified into the same three subgroups and found no statistically significant differences in levels of ghrelin, PYY, and hunger [29]. The inference made in the earlier study that the regulation of appetite is not mediating the effects observed during acute ketosis is therefore strengthened by the findings presented in this study.

GDF-15 levels in the present study were associated with changes in the levels of glucose from baseline but not with changes in the levels of triglycerides from baseline. The statistically significant reductions in glucose and triglyceride levels during ketosis were documented in earlier CETUS studies [15,16]. It is possible that the significant correlation observed in the present study exists for glucose (but not triglycerides) because acute nutritional ketosis causes greater reductions in the levels of glucose compared with triglycerides. The changes in glucose were more evenly spread and ranged from approximately +0.50 to −1.75 ng/mL, whereas the changes in triglycerides ranged from approximately +0.25 to −1.75 ng/mL, and most data points were concentrated near the 0 to −0.10 ng/mL range (Figure 2). Therefore, the trend of higher GDF-15 levels being associated with a greater reduction is more prevalent in the glucose analysis. Our findings suggest that GDF-15 might not be involved directly (but rather indirectly) in the reduction in glucose levels during acute nutritional ketosis. An earlier CETUS study showed that this effect was likely mediated by a glucose-dependent insulinotropic peptide—an incretin secreted by enteroendocrine K cells in the gut [15]. It stimulates β cells in the pancreas with a view to releasing insulin, subsequently leading to a reduction in plasma blood glucose. Further studies are warranted to investigate the role of other gut and pancreatic hormones during acute nutritional ketosis.

The present study had several limitations. First, the present study might have been underpowered to detect changes in GDF-15 as the sample size calculation was based on the primary endpoint (i.e., change in plasma glucose). However, this study may help in the design of future adequately powered randomised controlled trials that investigate GDF-15. Second, the study participants included both men and women of varied ages. GDF-15 levels increase with age, and this increase is different in men and women [30]. Also, earlier studies that investigated the levels of GDF-15 during acute nutritional ketosis had study populations made up of either men alone or women alone [31,32]. However, the cross-over design mitigated this issue because every participant was paired with their own results when comparing the levels of GDF-15 after consumption of both beverages. Third, only the acute effect of nutritional ketosis was studied in the present study, and therefore, the effect of long-term consumption of the KEβHB beverage on the circulating levels of GDF-15 remains unknown. There have been at least three earlier studies that investigated the effects of chronic ingestion of exogenous ketone supplements on plasma glucose levels [33–35]. All studies published to date deemed exogenous ketone supplements to be safe, with the longest study period being 4 weeks [36]. In the future, an investigation into the long-term effect of ketone drinks on GDF-15 is expected.

5. Conclusions

The beneficial role KEBHB-supplemented beverages have for people with prediabetes does not include an acute change in plasma levels of GDF-15. Eating behaviour does not appear to affect the levels of GDF-15 during acute nutritional ketosis. While there is no effect of acute nutritional ketosis on the circulating levels of GDF-15 in the prediabetes human model, its effect on diabetes or obesity warrants purposefully designed clinical and basic science investigations.

Author Contributions: M.S.P. conceived and supervised the study. S.H.B. recruited study participants. S.C., Y.L. and W.K. contributed to data analysis and interpretation. S.C. drafted the manuscript. Y.L., W.K., S.H.B. and M.S.P. revised the manuscript for important intellectual content. All authors have read and agreed to the published version of the manuscript.

Funding: This study was part of the COSMOS programme. COSMOS is funded by the Royal Society of New Zealand (Rutherford Discovery Fellowship to Max Petrov).

Institutional Review Board Statement: The study was conducted in accordance with the Declaration of Helsinki, and approved by the Health and Disability Ethics Committee (18/NTB/161).

Informed Consent Statement: Informed consent was obtained from all subjects involved in the study.

Data Availability Statement: The original contributions presented in the study are included in the article, further inquiries can be directed to the corresponding author.

Conflicts of Interest: The authors declare no conflicts of interest.

References

1. Loos, R.J.; Yeo, G.S. The genetics of obesity: From discovery to biology. *Nat. Rev. Genet.* **2022**, *23*, 120–133. [CrossRef]
2. NCD Risk Factor Collaboration. Worldwide trends in body-mass index, underweight, overweight, and obesity from 1975 to 2016: A pooled analysis of 2416 population-based measurement studies in 128·9 million children, adolescents, and adults. *Lancet* **2017**, *390*, 2627–2642. [CrossRef]
3. Amin, T.; Mercer, J.G. Hunger and satiety mechanisms and their potential exploitation in the regulation of food intake. *Curr. Obes. Rep.* **2016**, *5*, 106–112. [CrossRef]
4. Price, C.J.; Hoyda, T.D.; Ferguson, A.V. The area postrema: A brain monitor and integrator of systemic autonomic state. *Neuroscientist* **2008**, *14*, 182–194. [CrossRef]
5. Yang, L.; Chang, C.C.; Sun, Z.; Madsen, D.; Zhu, H.; Padkjær, S.B.; Wu, X.; Huang, T.; Hultman, K.; Paulsen, S.J.; et al. GFRAL is the receptor for GDF15 and is required for the anti-obesity effects of the ligand. *Nat. Med.* **2017**, *23*, 1158–1166. [CrossRef]
6. Hale, C.; Véniant, M.M. Growth differentiation factor 15 as a potential therapeutic for treating obesity. *Mol. Metab.* **2021**, *46*, 101117. [CrossRef]
7. Tsai, V.W.; Manandhar, R.; Jrøgensen, S.B.; Lee-Ng, K.K.; Zhang, H.P.; Marquis, C.P.; Jiang, L.; Husaini, Y.; Lin, S.; Sainsbury, A.; et al. The anorectic actions of the TGFβ cytokine MIC-1/GDF15 require an intact brainstem area postrema and nucleus of the solitary tract. *PLoS ONE* **2014**, *9*, e100370. [CrossRef]
8. Wang, D.; Day, E.A.; Townsend, L.K.; Djordjevic, D.; Jørgensen, S.B.; Steinberg, G.R. GDF15: Emerging biology and therapeutic applications for obesity and cardiometabolic disease. *Nat. Rev. Endocrinol.* **2021**, *17*, 592–607. [CrossRef]
9. Xiong, Y.; Walker, K.; Min, X.; Hale, C.; Tran, T.; Komorowski, R.; Yang, J.; Davda, J.; Nuanmanee, N.; Kemp, D.; et al. Long-acting MIC-1/GDF15 molecules to treat obesity: Evidence from mice to monkeys. *Sci. Transl. Med.* **2017**, *9*, 412. [CrossRef] [PubMed]
10. Mullican, S.E.; Lin-Schmidt, X.; Chin, C.N.; Chavez, J.A.; Furman, J.L.; Armstrong, A.A.; Beck, S.C.; South, V.J.; Dinh, T.Q.; Cash-Mason, T.D.; et al. GFRAL is the receptor for GDF15 and the ligand promotes weight loss in mice and nonhuman primates. *Nat. Med.* **2017**, *23*, 1150–1157. [CrossRef] [PubMed]
11. Luan, H.H.; Wang, A.; Hilliard, B.K.; Carvalho, F.; Rosen, C.E.; Ahasic, A.M.; Herzog, E.L.; Kang, I.; Pisani, M.A.; Yu, S.; et al. GDF15 is an inflammation-induced central mediator of tissue tolerance. *Cell* **2019**, *178*, 1231–1244. [CrossRef]
12. Zhang, M.; Sun, W.; Qian, J.; Tang, Y. Fasting exacerbates hepatic growth differentiation factor 15 to promote fatty acid β-oxidation and ketogenesis via activating XBP1 signaling in liver. *Redox Biol.* **2018**, *16*, 87–96. [CrossRef]
13. Roekenes, J.; Martins, C. Ketogenic diets and appetite regulation. *Curr. Opin. Clin. Nutr. Metab. Care* **2021**, *24*, 359–363. [CrossRef]
14. Moriconi, E.; Camajani, E.; Fabbri, A.; Lenzi, A.; Massimiliano, C. Very-low-calorie ketogenic diet as a safe and valuable tool for long-term glycemic management in patients with obesity and type 2 diabetes. *Nutrients* **2021**, *13*, 758. [CrossRef]
15. Bharmal, S.H.; Cho, J.; CAlarcon Ramos, G.; Ko, J.; Cameron-Smith, D.; Petrov, M.S. Acute nutritional ketosis and its implications for plasma glucose and glucoregulatory peptides in adults with prediabetes: A crossover placebo-controlled randomized trial. *J. Nutr.* **2021**, *151*, 921–929. [CrossRef]

16. Liu, Y.; Bharmal, S.H.; Kimita, W.; Petrov, M.S. Effect of acute ketosis on lipid profile in prediabetes: Findings from a cross-over randomized controlled trial. *Cardiovasc. Diabetol.* **2022**, *21*, 138. [CrossRef]
17. American Diabetes Association. Classification and diagnosis of diabetes: Standards of medical care in diabetes-2020. *Diabetes Care* **2020**, *43*, S14–S31. [CrossRef]
18. Bharmal, S.H.; Alarcon Ramos, G.C.; Ko, J.; Petrov, M.S. Abdominal fat distribution modulates the metabolic effects of exogenous ketones in individuals with new-onset prediabetes after acute pancreatitis: Results from a randomized placebo-controlled trial. *Clin. Nutr. ESPEN* **2021**, *43*, 117–129. [CrossRef]
19. De Lauzon, B.; Romon, M.; Deschamps, V.; Lafay, L.; Borys, J.M.; Karlsson, J.; Ducimetière, P.; Charles, M.A.; Group, F.L. The Three-Factor Eating Questionnaire-R18 is able to distinguish among different eating patterns in a general population. *J. Nutr.* **2004**, *134*, 2372–2380. [CrossRef]
20. Anglé, S.; Engblom, J.; Eriksson, T.; Kautiainen, S.; Saha, M.-T.; Lindfors, P.; Lehtinen, M.; Rimpelä, A. Three factor eating questionnaire-R18 as a measure of cognitive restraint, uncontrolled eating and emotional eating in a sample of young Finnish females. *Int. J. Behav. Nutr. Phys. Act.* **2009**, *6*, 41. [CrossRef]
21. Mullins, G.; Hallam, C.L.; Broom, I. Ketosis, ketoacidosis and very-low-calorie diets: Putting the record straight. *Nutr. Bull.* **2011**, *36*, 397–402. [CrossRef]
22. Nair, S.N.; Welle, S.L.; Halliday, D.; Campbell, R.G. Effect of B-hydroxybutyrate on whole body leucine kinetics and fractional mixed skeletal muscle protein synthesis in humans. *Am. J. Physiol. Endocrinol. Metab.* **1988**, *254*, 198–205. [CrossRef]
23. Evans, M.; Cogan, K.E.; Egan, B. Metabolism of ketone bodies during exercise and training: Physiological basis for exogenous supplementation. *J. Physiol.* **2017**, *595*, 2857–2871. [CrossRef]
24. VanItallie, T.B.; Nufert, T.H. Ketones: Metabolism's ugly duckling. *Nutr. Rev.* **2003**, *61*, 327–341. [CrossRef]
25. Schönfeld, P.; Reiser, G. Why does brain metabolism not favor burning of fatty acids to provide energy-Reflections on disadvantages of the use of free fatty acids as fuel for brain. *J. Cereb. Blood Flow. Metab.* **2013**, *33*, 1493–1499. [CrossRef]
26. Poffé, C.; Ramaekers, M.; Van Thienen, R.; Hespel, P. Ketone ester supplementation blunts overreaching symptoms during endurance training overload. *J. Physiol.* **2019**, *597*, 3009–3027. [CrossRef]
27. Hiroux, C.; Schouten, M.; de Glisezinski, I.; Simon, C.; Crampes, F.; Hespel, P.; Koppo, K. Effect of increased protein intake and exogenous ketosis on body composition, energy expenditure and exercise capacity during a hypocaloric diet in recreational female athletes. *Front. Physiol.* **2023**, *13*, 1063956. [CrossRef]
28. Charles, S.; Liu, Y.; Kimita, W.; Ko, J.; Bharmal, S.H.; Petrov, M.S. Effect of D-β-hydroxybutyrate-(R)-1,3 butanediol on plasma levels of asprosin and leptin: Results from a randomised controlled trial. *Food Funct.* **2023**, *14*, 759–768. [CrossRef]
29. Liu, Y.; Bharmal, S.H.; Kimita, W.; Petrov, M.S. Effect of d-β-hydroxybutyrate-(R)-1,3 butanediol on appetite regulation in people with prediabetes. *Mol. Nutr. Food Res.* **2023**, *67*, e2200615. [CrossRef]
30. Stubbs, B.J.; Cox, P.J.; Evans, R.D.; Santer, P.; Miller, J.J.; Faull, O.K.; Magor-Elliott, S.; Hiyama, S.; Stirling, M.; Clarke, K. On the Metabolism of Exogenous Ketones in Humans. *Front. Physiol.* **2017**, *8*, 848. [CrossRef]
31. Vestergaard, E.T.; Zubanovic, N.B.; Rittig, N.; Møller, N.; Kuhre, R.E.; Holst, J.J.; Rehfeld, J.F.; Thomsen, H.H. Acute ketosis inhibits appetite and decreases plasma concentrations of acyl ghrelin in healthy young men. *Diabetes Obes. Metab.* **2021**, *23*, 1834–1842. [CrossRef]
32. Alcazar, J.; Frandsen, U.; Prokhorova, T.; Kamper, R.S.; Haddock, B.; Aagaard, P.; Suetta, C. Changes in systemic GDF15 across the adult lifespan and their impact on maximal muscle power: The Copenhagen Sarcopenia Study. *J. Cachex-Sarcopenia Muscle* **2021**, *12*, 1418–1427. [CrossRef]
33. Soto-Mota, A.; Norwitz, N.G.; Evans, R.; Clarke, K.; Barber, T.M. Exogenous ketosis in patients with type 2 diabetes: Safety, tolerability and effect on glycaemic control. *Endocrinol. Diabetes Metab.* **2021**, *4*, e00264. [CrossRef]
34. Soto-Mota, A.; Vansant, H.; Evans, R.D.; Clarke, K. Safety and tolerability of sustained exogenous ketosis using ketone monoester drinks for 28 days in healthy adults. *Regul. Toxicol. Pharmacol.* **2019**, *109*, 104506. [CrossRef]
35. Walsh, J.J.; Neudorf, H.; Little, J.P. 14-Day Ketone Supplementation Lowers Glucose and Improves Vascular Function in Obesity: A Randomized Crossover Trial. *J. Clin. Endocrinol. Metab.* **2021**, *106*, 1738–1754. [CrossRef]
36. Falkenhain, K.; Daraei, A.; Forbes, S.C.; Little, J.P. Effects of exogenous ketone supplemetation on blood glucose: A systematic review and meta-analysis. *Adv. Nutr.* **2022**, *13*, 1697–1714. [CrossRef]

biomolecules

MDPI

Article

Bariatric Surgery Induces Alterations in the Immune Profile of Peripheral Blood T Cells

Pedro Barbosa [1,2,3,4], Aryane Pinho [2,4,5], André Lázaro [6,7,8], Diogo Paula [6,7], José G. Tralhão [6,7,8,9], Artur Paiva [4,7,8,10,11], Maria J. Pereira [12], Eugenia Carvalho [2,3,4,*] and Paula Laranjeira [2,4,7,8,10,*]

1 University of Coimbra, Institute for Interdisciplinary Research, Doctoral Programme in Experimental Biology and Biomedicine (PDBEB), 3030-789 Coimbra, Portugal; pedrobarbosa@cnc.uc.pt
2 Center for Neuroscience and Cell Biology (CNC), University of Coimbra, 3004-504 Coimbra, Portugal; aryanepinho@cnc.uc.pt
3 Institute for Interdisciplinary Research (IIIUC), University of Coimbra, 3030-789 Coimbra, Portugal
4 Center for Innovative Biomedicine and Biotechnology (CIBB), University of Coimbra, 3000-504 Coimbra, Portugal; artur.paiva@chuc.min-saude.pt
5 Department of Life Science, University of Coimbra, 3000-456 Coimbra, Portugal
6 General Surgery Unit, Centro Hospitalar e Universitário de Coimbra, 3000-075 Coimbra, Portugal; uc42311@uc.pt (A.L.); 17681@chuc.min-saude.pt (D.P.); jgtralhao@chuc.min-saude.pt (J.G.T.)
7 Clinical Academic Center of Coimbra (CACC), 3004-061 Coimbra, Portugal
8 Coimbra Institute for Clinical and Biomedical Research (iCBR), Group of Environmental Genetics of Oncobiology (CIMAGO), Faculty of Medicine (FMUC), University of Coimbra, 3000-548 Coimbra, Portugal
9 Institute of Biophysics, Faculty of Medicine, University of Coimbra, 3000-548 Coimbra, Portugal
10 Flow Cytometry Unit, Department of Clinical Pathology, Centro Hospitalar e Universitário de Coimbra, 3000-076 Coimbra, Portugal
11 Instituto Politécnico de Coimbra, ESTESC-Coimbra Health School, Ciências Biomédicas Laboratoriais, 3046-854 Coimbra, Portugal
12 Department of Medical Sciences, Clinical Diabetology and Metabolism, Uppsala University, SE-75185 Uppsala, Sweden; maria.pereira@medsci.uu.se
* Correspondence: ecarvalh@cnc.uc.pt (E.C.); 1979paula@gmail.com (P.L.); Tel.: +351-304-502-923 (E.C.); +351-927-218-443 (P.L.); Fax: +351-239-822-776 (E.C.)

Abstract: Low-grade inflammation is closely linked to obesity and obesity-related comorbidities; therefore, immune cells have become an important topic in obesity research. Here, we performed a deep phenotypic characterization of circulating T cells in people with obesity, using flow cytometry. Forty-one individuals with obesity (OB) and clinical criteria for bariatric surgery were enrolled in this study. We identified and quantified 44 different circulating T cell subsets and assessed their activation status and the expression of immune-checkpoint molecules, immediately before (T1) and 7–18 months after (T2) the bariatric surgery. Twelve age- and sex-matched healthy individuals (nOB) were also recruited. The OB participants showed higher leukocyte counts and a higher percentage of neutrophils. The percentage of circulating Th1 cells were negatively correlated to HbA1c and insulin levels. OB Th1 cells displayed a higher activation status and lower PD-1 expression. The percentage of Th17 and Th1/17 cells were increased in OB, whereas the $CD4^+$ Tregs' percentage was decreased. Interestingly, a higher proportion of OB $CD4^+$ Tregs were polarized toward Th1- and Th1/17-like cells and expressed higher levels of CCR5. Bariatric surgery induced the recovery of $CD4^+$ Treg cell levels and the expansion and activation of Tfh and B cells. Our results show alterations in the distribution and phenotype of circulating T cells from OB people, including activation markers and immune-checkpoint proteins, demonstrating that different metabolic profiles are associated to distinct immune profiles, and both are modulated by bariatric surgery.

Keywords: T cells; Treg; obesity; bariatric surgery; insulin resistance; immune phenotype

1. Introduction

Obesity is a major health concern. It can trigger a number of diseases and conditions, including insulin resistance (IR), type 2 diabetes mellitus (T2D), and cardiovascular disease (CVD) [1,2]. Obesity-associated chronic low-grade inflammation is a common feature of these comorbidities [1]; therefore, the immune system has become a major focus of investigation of obesity [2,3].

Innate and adaptive immune cells play an important role in maintaining homeostasis and a perfect balance between anti-inflammatory and pro-inflammatory cell populations, including those infiltrated in the adipose tissue (AT) microenvironment. However, the adipokine-releasing patterns associated with dysfunctional adipocytes, present in obesity, prompts an imbalance within the AT-infiltrated immune cell populations. Increased production of interleukin (IL)-6, tumor necrosis factor (TNF)-α, and leptin by dysfunctional adipocytes not only modulates the local inflammation by induction of pro-inflammatory polarization of immune cells but also plays an important role in systemic inflammation [4–6]. Furthermore, a chronic pro-inflammatory profile of immune cells may have a deleterious impact on metabolism, contributing to insulin resistance and T2D development. Neutrophils and, to a higher extent, monocytes/macrophages have been described as major contributors to obesity-related low-grade inflammation and insulin resistance development [7–9].

Adaptive immune cells also play an important role in obesity-related inflammation. $CD4^+$ (Th) and $CD8^+$ (Tc) T lymphocytes have been described as displaying an altered phenotype and function in obesity and metabolic dysfunction [10,11]. In fact, studies suggest that T cells display a metabolic switch toward pro-inflammatory polarization, such as Th1 and Th17 [8,12,13]. Moreover, both the number and the percentage of circulating $CD4^+$ Treg cells within $CD4^+$ T cells are reduced in adults with morbid obesity [14]. In turn, important phenotypic alterations are present in $CD8^+$ Treg cells from individuals with obesity (our unpublished data). And changes in the functional compartments of $CD8^+$ T cells have also been described in people with morbid obesity and the metabolic syndrome [6,15]. Despite these findings, little is known about the $CD8^+$ T cell phenotype, either in obesity or T2D. Moreover, the information available about less-represented T cell subsets is even scarcer.

In fact, most of the available literature focuses on a restricted type and number of T cell subsets, such as Th1, Th17, or Treg [8,14]. Thus, there is lack of information on less-represented T cell subsets, which may play an important role in obesity-associated low-grade inflammation. In addition, a deep evaluation of the circulating immune cells in obesity, covering most T cell subsets and providing the big picture of T cell changes in obesity, is missing. Therefore, the present study describes to a large extent the circulating T cells in people with obesity at different metabolic states. The absolute counts and percentages of 44 different T cell subsets, which are further characterized in terms of activation status and expression of immunosuppressive molecules, are assessed in this study, in participants with obesity and in a non-obese control group. Additionally, the impact of bariatric surgery on each of the circulating T cell subsets analyzed has also been studied.

2. Materials and Methods

2.1. Participants

A group of 41 adults with obesity (OB), comprising 27 women and 14 men with a mean age of 44 ± 10.8 years old, undergoing bariatric surgery at the Bariatric Surgery Unit, Department of General Surgery, Centro Hospitalar e Universitário de Coimbra (CHUC), Portugal, were enrolled in the present study, after giving their written informed consent. This study was approved by the CHUC's Ethical Committee (protocol code: CHUC-115-20 and CHUC-202-20), and it was performed according to the Declaration of Helsinki (1964) guidelines. The participants were clinically indicated for bariatric surgery following specific criteria of BMI > 40 kg/m^2 or ≥35 kg/m^2 associated with dyslipidemia, hypertension, clinical diagnosis of T2D, or sleep apnea. Acute inflammatory conditions, cancer, and neurodegenerative and autoimmune diseases, as well as the use of immunosuppressive and anti-inflammatory drugs on a daily basis, were exclusion criteria. From the recruited participants, 35 were studied before bariatric surgery and 14 after bariatric surgery, as indicated in Figure 1 and Table 1. Eight of the participants were studied both before and after bariatric surgery, while six were studied exclusively after the bariatric surgery (Figure 1 and Supplementary Table S1). Only individuals with Class IV obesity undergo a second bariatric surgery. An age- and sex-matched group of healthy individuals (n = 12; 8 women and 4 men; mean age: 43 ± 11.9 years old) were also recruited to participate as a control group (non-obese group, nOB). A volume of 12 mL of EDTA-anticoagulated fasting peripheral blood (PB) was collected from all participants.

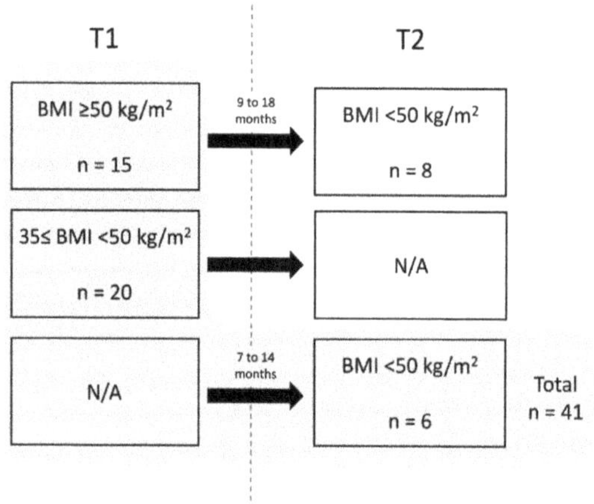

Figure 1. Distribution of the participants according to the time-point of sample collection. Thirty-five individuals were studied before bariatric surgery (T1). Eight of them, with BMI ≥50 kg/m^2 at T1, were also studied 9 to 18 months after the surgery (T2). Six individuals were only studied at T2. N/A: not analyzed.

Table 1. Anthropometric and biochemical characterization of participants, before (T1) and after (T2) bariatric surgery.

		nOB		OB at T1		OB at T2
	n		*n*		*n*	
Sex (female/male)	12	8/4	35	23/12	14	8/6
Age (years)	12	43 ± 12	35	45 ± 12	14	41 ± 9.0
BMI (kg/m^2)	12	24.8 ± 3.0	35	48.3 ± 7.6	14	41.6 ± 4.3 ^{A,B}
NC (cm)			25	43 ± 6	13	41 ± 5 ^{B}
WHR	6	0.82 ± 0.11	25	0.98 ± 0.10 ^{A}	13	0.96 ± 0.10 ^{A}
Obesity class (II/III/IV)			35	7/13/15	14	6/8/-
Type 2 Diabetes			35	14	14	2
HbA1c (%)	5	5.5 ± 0.48	35	6.0 ± 0.88	14	5.1 ± 0.21 ^{A,B}
Glucose (mg/dL)	7	88.6 ± 8.0	35	104.8 ± 25.5	14	83.2 ± 6.3 ^{B}
Insulin (µU/mL)	6	5.9 ± 2.3	35	17.1 ± 9.1 ^{A}	14	6.1 ± 2.0 ^{B}
HOMA-IR	6	1.3 ± 0.54	35	4.6 ± 3.2 ^{A}	14	1.5 ± 0.76 ^{B}
C-peptide (ng/mL)	3	1.34 ± 0.43	35	4.0 ± 1.5 ^{A}	13	2.1 ± 0.57 ^{B}
LDL (mg/dL)	7	101.7 ± 32.5	35	113.2 ± 41.2	14	118.7 ± 31.5
HDL (mg/dL)	7	67.0 ± 19.3	35	43.8 ± 10.9 ^{A}	14	44.9 ± 14.6 ^{A}
Triglycerides (mg/dL)	7	101.3 ± 52.7	35	124.8 ± 52.1	14	84.6 ± 26.5 ^{B}
Total cholesterol (mg/dL)	7	189.0 ± 53.2	35	182.5 ± 40.3	14	180.6 ± 36.4
Atherogenic index	6	3.0 ± 0.61	35	4.5 ± 1.4 ^{A}	14	4.4 ± 1.2 ^{A}
CRP (mg/dL)	3	0.20 ± 0.23	35	1.0 ± 0.77 ^{A}	14	0.64 ± 0.52 ^{B}

Table 1. *Cont.*

		nOB		OB at T1		OB at T2
	n		*n*		*n*	
Leptin (ng/mL)	6	2.6 ± 3.5	22	12.0 ± 6.0 [A]	10	5.9 ± 5.2 [B]
Adiponectin (μg/mL)	6	121.1 ± 57.9	22	86.2 ± 48.7	10	138.6 ± 121.9

nOB: healthy participants (without obesity); OB at T1: group of all individuals with obesity studied before surgery; OB at T2: group of all individuals studied 7 to 18 months after bariatric surgery; BMI: body mass index; NC: neck circumference; WHR: waist-to-hip ratio; HOMA-IR: homeostatic model assessment of insulin resistance; LDL: low-density lipoprotein cholesterol; HDL: high-density lipoprotein cholesterol; CRP: C-reactive protein. Statistical differences were considered when $p < 0.05$. [A] $p < 0.05$ vs. nOB; [B] $p < 0.05$ vs. OB before surgery, considering only participants with BMI ≥50 kg/m^2 (obesity Class IV).

2.1.1. Anthropometric and Biochemical Characterization of the Participants

The anthropometric and biochemical/metabolic characteristics of the participants are detailed in Table 1. A scale (seca 515 mBCA, seca, Hamburg, Germany) was used to measure the body weight (kg). The body mass index (BMI, kg/m^2) was calculated as previously described [16]: BMI = BW (kg)/height^2 (m^2). The waist (WC), hips (HC), and neck (NC) circumferences were measured using a flexible measuring tape (cm), while a digital sphygmomanometer (Philips IntelliVue MP20, Philips, Boebligen, Germany) was used to measure the systolic and diastolic blood pressure (mmHg) after 5 min of rest.

Glycated hemoglobin (HbA1C, %), fasting glucose (mg/dL) and insulin (μU/mL), C-peptide (μg/mL), and C-reactive protein (CRP, μg/mL), as well as triglycerides (TG, mg/dL), high-density lipoprotein cholesterol (HDL, mg/dL), and total cholesterol (mg/dL) were measured in PB after an 8 h fast, following the standard methods at the Department of Clinical Pathology at CHUC. The Friedewald formula [17], LDL = TC − HDL + (TG × 0.2), was used to calculate the low-density lipoprotein cholesterol (LDL, mg/dL), while the homeostatic model assessment insulin resistance index (HOMA-IR) was also calculated using the fasting glucose and insulin levels as previously described [16]: HOMA-IR = (insulin (μU/mL)) × glucose (mg/dL))/405 (Table 1).

Leptin and adiponectin levels were measured in plasma, by immunoassay, following the manufacturer's instructions (RayBio Tech Life Inc., Peachtree Corners, GA, USA) (Table 1).

2.1.2. Metabolic profile and obesity class of participants

Participants with obesity were divided per obesity class [18], into Class II (35 ≤ BMI < 40 kg/m^2), Class III (40 ≤ BMI < 50 kg/m^2), or Class IV (BMI ≥ 50 kg/m^2), as well as according to their metabolic status [19], based on HOMA-IR, into insulin sensitive (IS) or IR. The IR group was subdivided into either normoglycemic (IRn), with pre-diabetes (Pre-T2D), or with type 2 diabetes (T2D), according to their fasting glucose and HbA1c % (Table 2).

Table 2. Criteria used to classify people with obesity according to their metabolic profiles.

	Insulin Sensitive (IS)	Insulin Resistant (IR)		
HOMA-IR	<2		≥2	
		Insulin Resistant Normoglycemic (IRn)	Pre-Diabetes (Pre-T2D)	Type 2 Diabetes (T2D)
Fasting glucose (mg/dL)		<100	<125	≥125
HbA1c (%)		<5.7	≥5.70 and <6.49	≥6.5
		Non-Type 2 Diabetes (nT2D)		

Table 3 shows how participants from the different metabolic groups are distributed into obesity classes.

Table 3. Distribution of participants with obesity according to their metabolic group and obesity class.

		Metabolic Groups			
		IS	**IRn**	**Pre-T2**	**T2D**
Obesity Class	Class II	3	0	2	2
	Class III	2	4	3	4
	Class IV	0	5	2	8
	Total	5	9	7	14

IS: insulin sensitive; IRn: insulin resistant and normoglycemic; Pre-T2D: pre-diabetes; T2D: type 2 diabetes.

2.2. *Immunophenotyping*

PB immune cells were identified and characterized by flow cytometry, using a stain–lyse–wash procedure previously described [20–22]. In summary, 100 µL of PB was incubated with the monoclonal antibodies (mAbs) indicated in Table 4, in the presence of 50 µL of Brilliant stain buffer (Becton Dickinson Biosciences (BD), San Jose, CA, USA), for 30 min in the dark and at room temperature. Subsequently, erythrocytes were lysed using 2 mL of FACSLysing solution (BD) and a 10 min incubation period. After centrifugation at 540× *g* for 5 min, the FACSLysing solution was discarded, and the resulting cell pellet was washed with 2 mL of Dulbecco's phosphate-buffered saline (DPBS; Corning, Manassa, VA, USA). Lastly, the cell pellet was resuspended in 500 µL of DPBS, and the sample was acquired in a FACSLyric^{TM} (BD) flow cytometer, using the FACSuite acquisition software (v1.5.0.925; BD).

Table 4. Monoclonal antibody combination used to identify the different T cell subsets, presenting the information on each respective fluorochrome, clone, and commercial source.

	Monoclonal Antibodies										
	TIM-3	**CD25**	**CD4**	**CD127**	**CD3**	**CD20**	**CD8**	**CXCR5**	**CCR6**	**PD-1**	**CCR5**
Fluorochrome	FITC	PE	PerCP-Cy5.5	PE-Cy7	APC	APC-R700	APC-H7	BV421	BV510	BV605	BV711
Clone	7D3	M-A251	SK3	HIL-7R-M21	UCHT1	2H7	SK1	RF8B2	11A9	EH12.1	2D7/CCR5
Commercial source	BD Horizon	BD Pharmingen	BD Pharmingen	BD Pharmingen	BD Pharmingen	BD Horizon	BD Pharmingen	BD Horizon	BD Horizon	BD Horizon	BD Horizon

FITC: fluorescein isothiocyanate; PE: phycoerythrin; PerCP-Cy5.5: peridin chlorophyll protein–cyanine 5.5; PE-Cy7: phycoerythrin/cyanine 7; APC: allophycocyanin; APC-H7: allophycocyanin-Hilite7; APC-R700: allophycocyanin R700; BV: Brilliant Violet; BD Pharmingen (San Diego, CA, USA); BD Horizon (San Jose, CA, USA).

Identification and Characterization of the Major Immune Cell Populations and T Cell Subsets

Flow cytometry data were analyzed with the Infinicyt software (version 2.0.5; Cytognos SL, Salamanca, Spain). Supplementary Figure S1 describes the gate strategy used to identify the major immune cell populations (lymphocytes, monocytes, neutrophils, and eosinophils) and the distinct T cell subsets. Immune cell analysis was performed after excluding debris and cell doublets, considering their forward scatter (FSC)-A and FSC-H properties. Neutrophils were identified based on their FSC and side scatter (SSC) light dispersion characteristics, while monocytes were also identified based on CD4 expression (Figure 2A). T and B cells were identified by their positivity for CD3 (Supplementary Figure S1B) and CD20 (Supplementary Figure S1C), respectively. Our mAb combination (Table 4) was designed to perform a deep characterization of T lymphocytes, allowing us to identify 44 different subpopulations. Thus, T lymphocytes were primarily divided into $CD4^+$ (Th), $CD8^+$ (Tc), and $CD4^+CD8^+$ T cells (Supplementary Figure S1D). As $CD4^-CD8^-$ T cells comprised both TCRαβ $CD4^-CD8^-$ T cells and γδ T cells, our mAb combination did not allow for the distinction of these cell groups; therefore, they were not further analyzed. As demonstrated in Supplementary Figure S1, within each one of the three major T cell subpopulations (Th, Tc, and $CD4^+CD8^+$ T cells), we identified the following T cell subsets: $CD25^{high}CD127^{low/-}$ regulatory T (Treg) cells (Supplementary Figure S1E);

CXCR5$^+$ follicular T (Tf) cells (Supplementary Figure S1G); and, based on CCR5 (CD195) and CCR6 (CD196) expression (Supplementary Figure S1H), CCR5$^+$CCR6$^-$ T cells (T1), CCR5$^-$CCR6$^+$ T cells (T17), CCR5$^+$CCR6$^+$ T cells (T1/17), and CCR5$^-$CCR6$^-$ T cells. Then, Treg cells were subdivided into CXCR5$^+$ regulatory T (Tfr) cells (Supplementary Figure S1F). According to the expression of CCR5 and CCR6, T1-like, T17-like, T1/17-like, and CCR5$^-$CCR6$^-$ T cells were further identified within Treg, Tf, and Tfr cells. Finally, the percentage of early activated T cells (CD25$^+$, Supplementary Figure S1I) and the expression of the immune-checkpoint proteins PD-1 (CD279, Supplementary Figure S1J) and TIM-3 (CD366, Supplementary Figure S1K) were further identified within each one of the 44 T cell subsets identified in this study, whenever the number of events acquired was enough to enable this analysis.

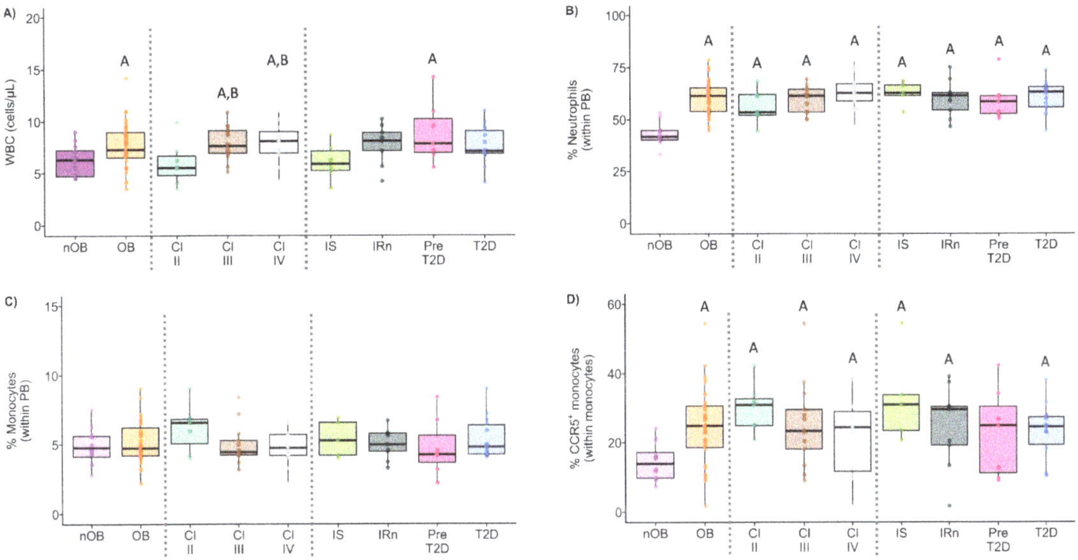

Figure 2. White blood cells count (WBC) in peripheral blood (**A**) and percentage of neutrophils (**B**) and monocytes within peripheral blood (PB) leukocytes (**C**) in nOB and OB, as well as among OB participants stratified by obesity class and metabolic profile. Percentage of monocytes expressing CCR5$^+$ within all the studied groups (**D**). nOB: healthy participants (without obesity); OB: participants with obesity; IS: insulin sensitive; IRn: insulin resistant and normoglycemic; Pre-T2D: pre-diabetes; T2D: type 2 diabetes. Statistical differences were considered when $p < 0.05$. ^{A} $p < 0.05$ vs. nOB; ^{B} $p < 0.05$ vs. Class II; ^{C} $p < 0.05$ vs. Class III; ^{D} $p < 0.05$ vs. IS; ^{E} $p < 0.05$ vs. IRn; ^{F} $p < 0.05$ vs. Pre-T2D.

2.3. *Statistical Analysis*

Data were presented using mean ± standard deviation or median and inter-quartile range (iqr) for all studied variables. The comparison between groups was performed using the Wilcoxon rank sum and Kruskal–Wallis test from package *coin* version 1.4.2 [23,24], followed by Dunn's test from package *rstatix* version 0.6.0 [25], as appropriate. Furthermore, for the participants studied both before and after bariatric surgery, paired-sample comparisons were performed using the Wilcoxon signed-rank test [26]. Spearman's correlations were performed between two continuous variables, using the rcorr() function from package *Hmisc* version 4.4-2 [27]. Statistically significant differences were considered when $p < 0.05$. The plots were performed using the package *ggplot2* [28]. All statistical analysis was performed using R (version 4.0.2; R Foundation for Statistical Computing, Vienna, Austria) [26,29].

3. Results

3.1. Obesity Induces Alterations in Peripheral Blood Neutrophils and Monocytes

People with obesity (OB) presented higher white blood cell (WBC) counts (7.6 cells/μL ± 2.2) in comparison to the nOB group (6.2×10^3 cells/μL ± 1.5, $p < 0.05$; Figure 2A and Supplementary Table S2). Additionally, the stratification of BMI into obesity classes indicated that Class II obesity (6.0×10^3 cells/μL ± 2.1) displayed a similar WBC count to the nOB group. On the other hand, Classes III (7.9 $\text{x}10^3$ cells/μL ± 1.7, $p < 0.05$ vs. nOB) and IV (8.2×10^3 cells/μL ± 2.4, IV vs. II and IV vs. nOB, $p < 0.05$) were significantly increased (Figure 2A and Supplementary Table S2). Furthermore, the BMI was positively correlated to WBC (rho = 0.39, $p < 0.05$), as shown in Figure 3. A 1.4-fold increase in the percentage of neutrophils in the OB group compared to nOB ($p < 0.05$, Figure 2B and Supplementary Table S2) was also detected, which translated into an increase of 1.7× the absolute number of neutrophils (Supplementary Table S3). Additionally, the BMI and neck circumference were highly correlated to the percentage of neutrophils (rho = 0.62 and rho = 0.58, respectively, $p < 0.05$, Figure 3). Monocytes from the OB group showed a higher percentage of $CCR5^+$ cells (Figure 2D and Supplementary Table S2) and a positive correlation between the percentage of monocytes and the HbA1c % for the OB group (rho = 0.36, $p < 0.05$; Figure 3). Considering the metabolic profile, the landscape was quite different. The IS group displayed a similar WBC count as the nOB group (nOB: 6.2 cells/μL ± 1.5; IS: 6.1 cells/μL ± 1.8; Figure 2A and Supplementary Table S2), while the non-IS OB groups showed higher WBC counts, with Pre-T2D (Pre: 8.8 cells/μL ± 3.0) displaying a significant difference vs. the nOB group (Figure 2A and Supplementary Table S2).

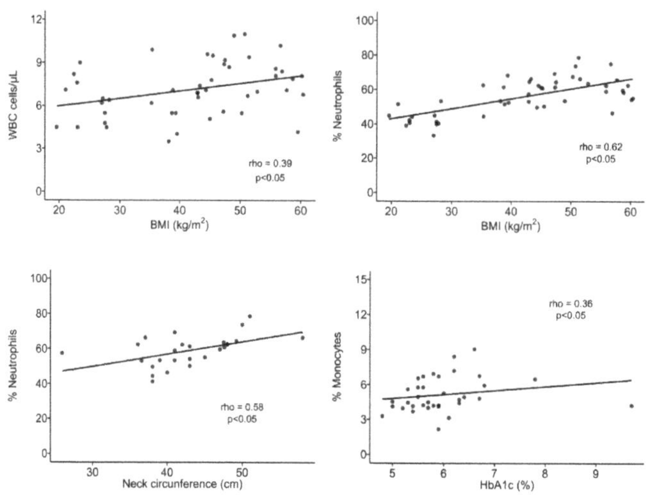

Figure 3. Spearman's correlation analysis between immune and anthropometric or metabolic parameters. BMI: body mass index; WBC: white blood cells. Statistical differences were considered when $p < 0.05$.

3.2. Obesity Alters the T Cell Percentage and Their Subsets, but Not That of B Cells, in Peripheral Blood

The percentage of lymphocytes in whole blood was decreased by 1.5-fold in the OB group when compared to nOB ($p < 0.05$; Figure 4A and Supplementary Table S2) and was negatively correlated to BMI and neck circumference (rho = −0.53 and rho = −0.44, respectively, $p < 0.05$), as displayed in Figure 5. Similarly, the percentage of T lymphocytes in whole blood was also decreased in the OB group (19% ± 6.8 vs. nOB: 26% ± 7.8, $p < 0.05$, Figure 4B and Supplementary Table S2) and negatively correlated to BMI and neck circumference (rho = −0.34 and rho = −0.46, respectively, $p < 0.05$; Figure 5). Importantly, the levels of plasma glucose showed a weak negative correlation to the percentage of lymphocytes and T lymphocytes (rho = −0.35 and rho = −0.31, respectively, $p < 0.05$; Figure 5). B lymphocytes did not show any differences among the studied groups (Figure 4C and Supplementary Table S2).

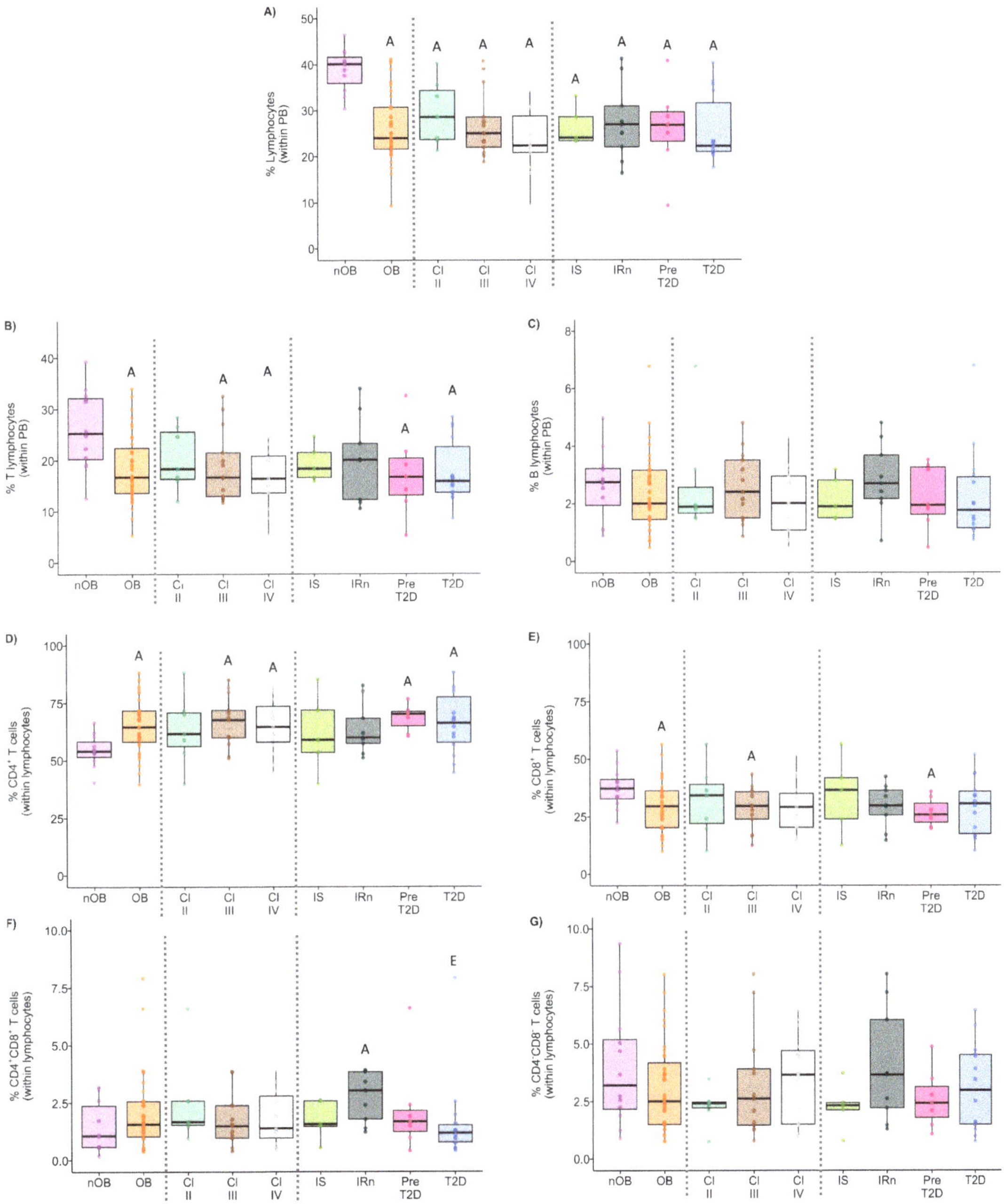

Figure 4. Percentage of lymphocytes (**A**), T lymphocytes (**B**), and B lymphocytes (**C**) within peripheral blood (PB) leukocytes and distribution of T lymphocytes into CD4$^+$ (**D**), CD8$^+$ (**E**), CD4$^+$CD8$^+$ (**F**), and CD4$^-$CD8$^-$ (**G**) compartments among all the studied groups. nOB: healthy participants (without

obesity); OB: participants with obesity; IS: insulin sensitive; IRn: insulin resistant and normoglycemic; Pre-T2D: pre-diabetes; T2D: type 2 diabetes. Statistical differences were considered when $p < 0.05$. ^{A} $p < 0.05$ vs. nOB; ^{B} $p < 0.05$ vs. Class II; ^{C} $p < 0.05$ vs. Class III; ^{D} $p < 0.05$ vs. IS; ^{E} $p < 0.05$ vs. IRn; ^{F} $p < 0.05$ vs. Pre-T2D.

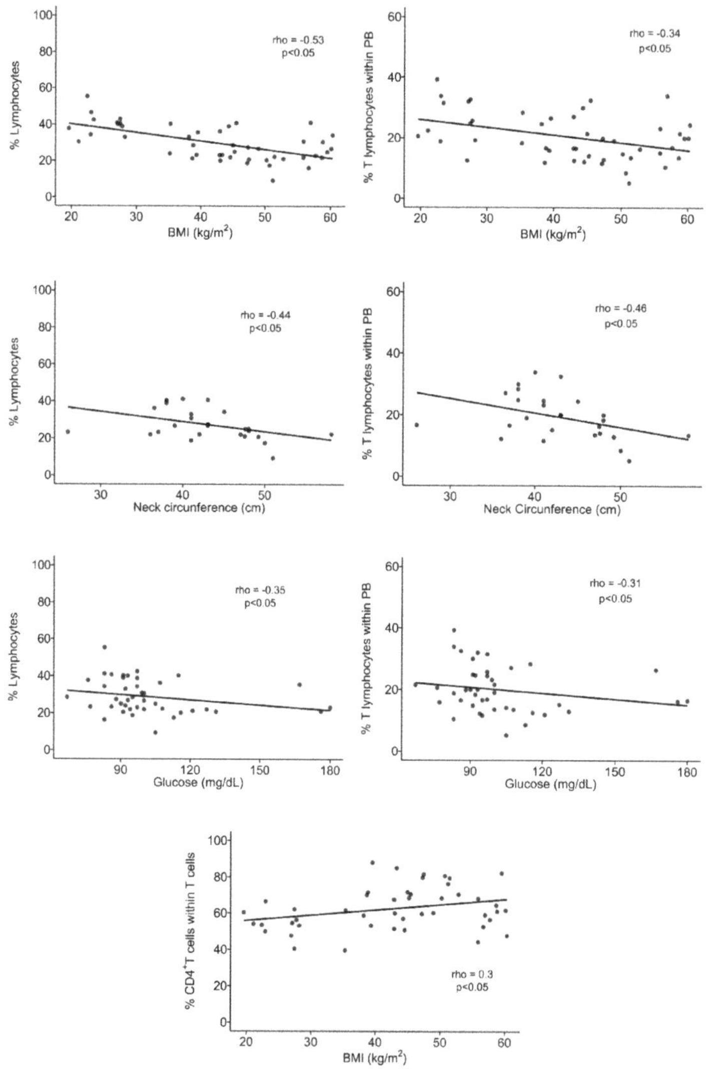

Figure 5. Spearman's correlation analysis between immune and anthropometric or metabolic parameters. BMI: body mass index; PB: peripheral blood. Statistical differences were considered when $p < 0.05$.

A significant increase in the percentage of $CD4^{+}$ T cells in the OB group (65% ± 12.0) vs. nOB (54% ± 7.1, $p < 0.05$; Figure 4D and Supplementary Table S2) was observed, while the percentage of $CD8^{+}$ T cells was significantly reduced (nOB: 38% ± 8.6 vs. OB: 30% ± 11.1, $p < 0.05$; Figure 4E and Supplementary Table S2). No differences were detected on $CD4^{+}CD8^{+}$ T cells (Figure 4F and Supplementary Table S2). BMI showed a moderate positive correlation with the percentage of $CD4^{+}$ T cells (rho = 0.30, $p < 0.05$) (Figure 5).

Looking at the metabolic profile, Pre-T2D (68% ± 5.8) and T2D groups (66% ±13.4) showed the highest percentage of CD4^{+} T cells in comparison to the nOB (54% ± 7.1, $p < 0.05$) group (Figure 4D and Supplementary Table S2). Moreover, the CD8^{+} T cells were decreased in insulin-resistant groups, with an emphasis in the Pre-T2D group (27% ± 6.1) when compared to nOB (38% ± 8.6, $p < 0.05$), as described in Supplementary Table S2 and Figure 4E.

3.3. CD4^{+} T Lymphocytes

The phenotype of the CD4^{+} T cells, as well as the percentage of activated T cells (identified as CD25^{+}), and the expression of immune regulatory molecules were summarized in Figures 6–9, and more details are provided in Supplementary Tables S4, S5 and S7.

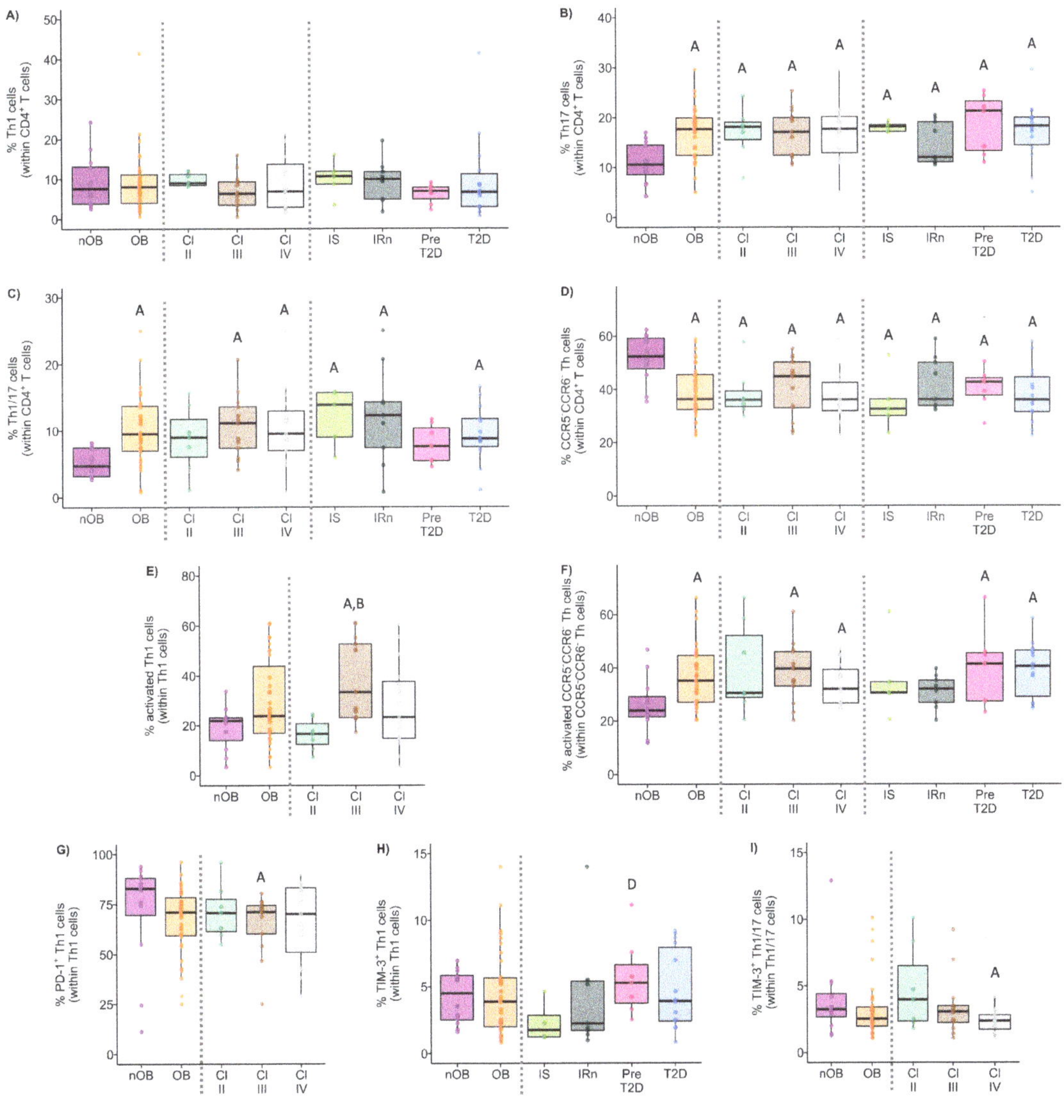

Figure 6. Percentage of Th1 (**A**), Th17 (**B**), Th1/17 (**C**), and CCR5^{-}CCR6^{-} (**D**) T cells within CD4^{+} T cells. Percentage of activated (CD25^{+}) Th1 (**E**) and CCR5^{+}CCR6^{+} Th cells (**F**). Percentage of PD-1^{+}

Th1 (**G**), TIM-3⁺ Th1 (**H**), and TIM-3⁺ Th1/17 cells (**I**) in nOB and OB, grouped according to obesity class and metabolic profile. nOB: healthy participants (without obesity); OB: participants with obesity; IS: insulin sensitive; IRn: insulin resistant and normoglycemic; Pre-T2D: pre-diabetes; T2D: type 2 diabetes; Th1—T helper 1 ($CCR5^{+}CCR6^{-}$); Th17—T helper 17 ($CCR5^{-}CCR6^{+}$); Th1/17—T helper 1/17 ($CCR5^{+}CCR6^{+}$). Statistical differences were considered when $p < 0.05$. ^{A} $p < 0.05$ vs. nOB; ^{B} $p < 0.05$ vs. Class II; ^{C} $p < 0.05$ vs. Class III; ^{D} $p < 0.05$ vs. IS; ^{E} $p < 0.05$ vs. IRn; ^{F} $p < 0.05$ vs. Pre-T2D.

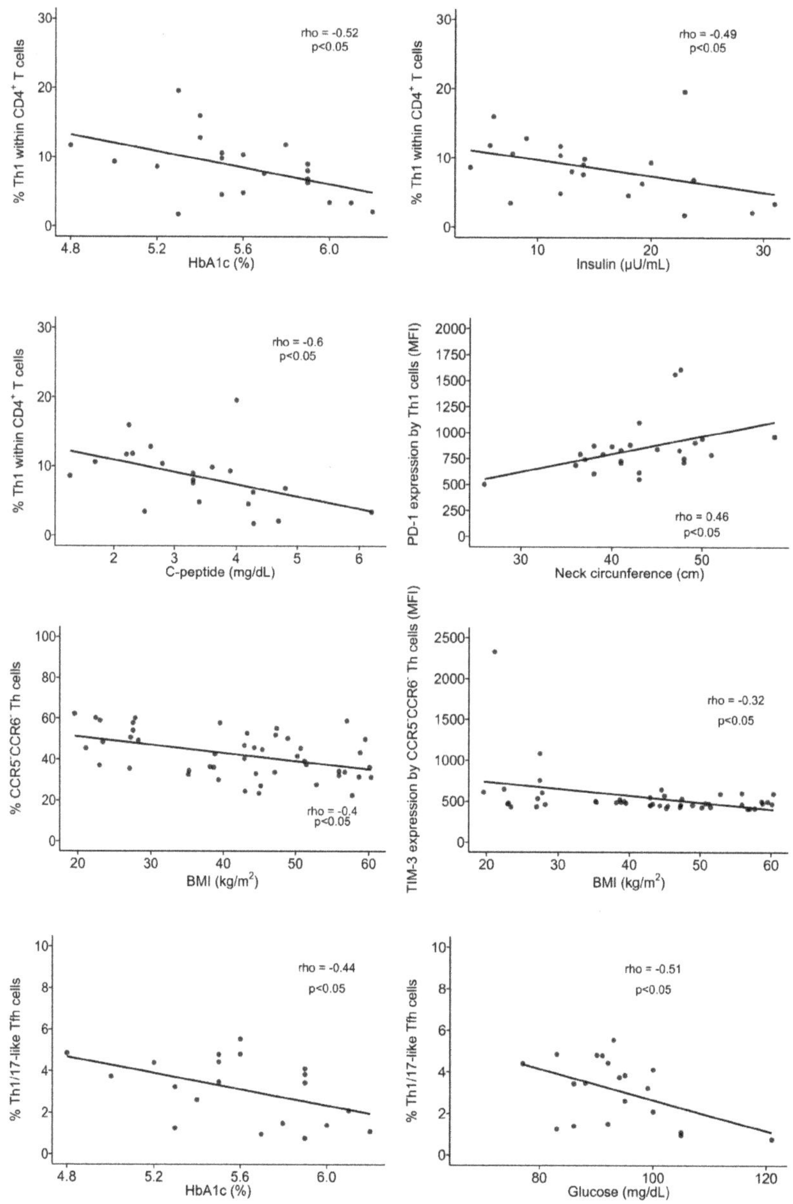

Figure 7. Spearman's correlation analysis between immune and anthropometric or metabolic parameters. BMI: body mass index; MFI: mean fluorescence intensity. Statistical differences were considered when $p < 0.05$.

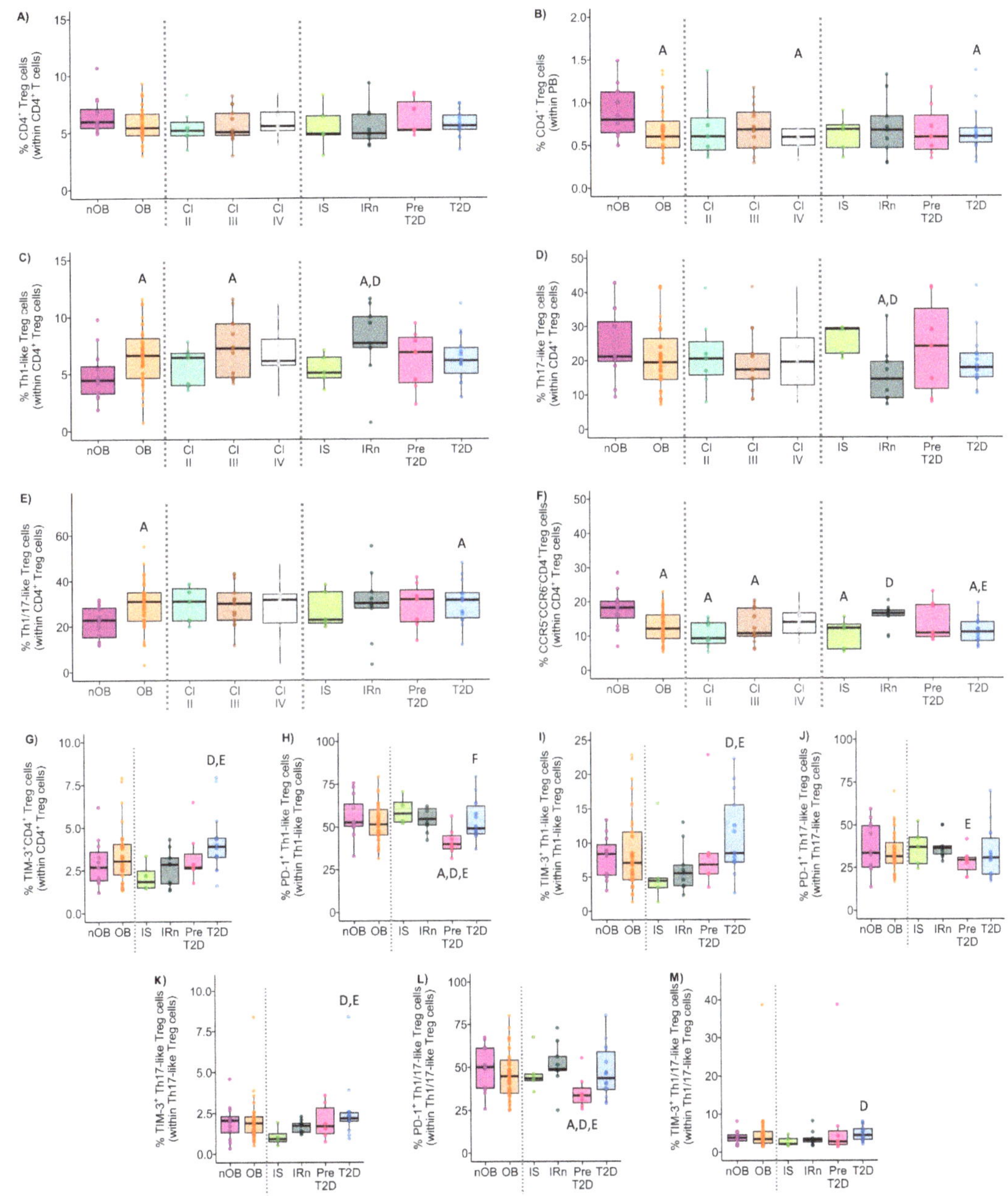

Figure 8. Percentage of CD4+ Tregs within CD4+ T cells (**A**) and within peripheral blood (PB) leukocytes (**B**). Polarization of CD4+ Tregs into Th1-like (**C**), Th17-like (**D**), Th1/17-like (**E**), and CCR5−CCR6− (**F**) CD4+ Tregs. Percentage of PD-1+ and TIM-3+ expressed by CD4+ Tregs (**G**) and by Th1-like (**H**,**I**), Th17-like (**J**,**L**), and Th1/17-like (**M**,**N**) CD4+ Tregs among the studied groups. nOB: healthy participants (without obesity); OB: participants with obesity; IS: insulin sensitive; IRn: insulin resistant and normoglycemic; Pre-T2D: pre-diabetes; T2D: type 2 diabetes. Statistical differences were considered when $p < 0.05$. ^{A} $p < 0.05$ vs. nOB; ^{B} $p < 0.05$ vs. Class II; ^{C} $p < 0.05$ vs. Class III; ^{D} $p < 0.05$ vs. IS; ^{E} $p < 0.05$ vs. IRn; ^{F} $p < 0.05$ vs. Pre-T2D.

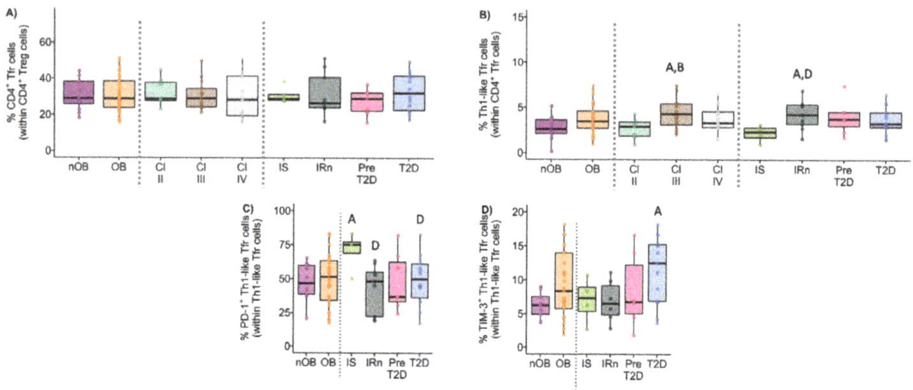

Figure 9. Treg cells with follicular phenotype and their polarization into Th1-like Tfr cells. Percentage of follicular CD4$^+$ Treg cells (Tfr) within CD4$^+$ Treg cells (**A**). Percentage of CD4$^+$ Tfr cells with Th1-like phenotype (**B**). Percentage of PD-1$^+$ (**C**) and TIM-3$^+$ (**D**) cells within Th1-like CD4$^+$ Tfr cells. nOB: healthy participants (without obesity); OB: participants with obesity; IS: insulin sensitive; IRn: insulin resistant and normoglycemic; Pre-T2D: pre-diabetes; T2D: type 2 diabetes. Statistical differences were considered when $p < 0.05$. ^{A} $p < 0.05$ vs. nOB; ^{B} $p < 0.05$ vs. Class II; ^{C} $p < 0.05$ vs. Class III; ^{D} $p < 0.05$ vs. IS; ^{E} $p < 0.05$ vs. IRn; ^{F} $p < 0.05$ vs. Pre-T2D.

3.3.1. Th1 Cells from OB Participants Present a Higher Activation Status and Lower Expression of PD-1

Though no differences in the percentage of Th1 cells between nOB and OB were found (Figure 6A and Supplementary Table S4), Th1 cells from OB displayed an increased percentage of activated cells (30% ± 17.2 vs. 19% ± 8.9 in nOB), more marked in Class III (39% ± 16.5, $p < 0.05$) and not verified in Class II (16% ± 6.4), as indicated in Figure 6E and Supplementary Table S4. In turn, a decrease in the percentage of PD-1$^+$ Th1 cells was found among OB subjects from Class III (65% ± 15.4, $p < 0.05$, Figure 6G and Supplementary Table S4)) and Class IV (66% ± 20.0, $p > 0.05$) in comparison to nOB (71% ± 27.1), as seen in Figure 6G and Supplementary Table S4. Interestingly, when evaluating the metabolic groups, a sequential increase in PD-1$^+$ Th1 cells from IS (54% ± 19.8) to IRn (62% ± 16.7), Pre-T2D (67% ± 14.6), and T2D (75% ± 14.9; Supplementary Table S4) was observed. In addition, T2D patients also presented an increased expression in PD-1 (measured as mean fluorescence intensity, MFI), as compared to the nOB ($p > 0.05$), IS, and IRn ($p < 0.05$) groups, as shown in Supplementary Table S4. Importantly, besides presenting a lower percentage of PD-1$^+$ Th1 cells, IS participants also showed a decreased percentage of TIM-3$^+$ Th1 cells (2.3% ± 1.6) compared to nOB (4.2% ± 2.0, $p > 0.05$) and the remaining groups (Figure 6H and Supplementary Table S4). Furthermore, a negative correlation between the percentage of Th1 and the percentage of HbA1c, insulin, or c-peptide levels were observed in IS, IRn, and Pre-T2D participants. Additionally, the neck circumference was positively correlated to the expression of PD-1 by Th1 cells (rho = 0.46, $p < 0.05$; Figure 7).

3.3.2. The OB Group Displays an Increased Percentage of Th17 and Th1/17 Cells That Is Associated with a Reduced Expression of TIM-3

The percentages of Th17 and Th1/17 cells from the OB group were increased (17% ± 5.2 and 10% ± 5.2, respectively) in comparison to the nOB (11% ± 4.1 and 5.2% ± 2.2, respectively, $p < 0.05$; Figure 6B,C and Supplementary Table S4). Interestingly, a decreased percentage of PD-1$^+$ cells within the Th17 subset was detected in the OB group (37% ± 9.9), with an emphasis on Class III obesity (34% ± 6.5), compared to nOB (44% ± 15.4, $p > 0.05$; Supplementary Table S4), in addition to the Th1/17 cells (nOB: 64% ± 17.6 vs. OB: 58% ± 11.5, $p > 0.05$; Supplementary Table S4). Furthermore, the expression of PD-1 in both Th17 and Th1/17 cells were reduced in Class III, in comparison to nOB ($p < 0.05$, Supplementary

Table S4). Similarly, there was a decreased percentage of TIM-3$^+$ Th1/17 cells in the OB group (3.2% ± 2.2) compared to nOB (4.0% ± 3.1), more pronounced in Class IV (2.3% ± 0.8, $p < 0.05$ vs. nOB; Figure 6I and Supplementary Table S4). The TIM-3 MFI in Th17 and Th1/17 cells, also showed a tendency to decrease in OB vs. nOB ($p > 0.05$; Supplementary Table S4).

3.3.3. CCR5$^-$CCR6$^-$ Helper T Cells Show Decreased Percentage and an Altered Phenotype in Insulin Resistant Groups

CCR5$^-$CCR6$^-$ Th cells comprised approximately 52% ± 9.0 of the CD4$^+$ T cells in the nOB group, while in OB groups they comprised 39% ± 9.8 of the CD4$^+$ T cell compartment ($p < 0.05$), even among obesity classes and metabolic groups (Figure 6D and Supplementary Table S4). It is noteworthy that, despite their reduced percentage, these cells were more activated in OB (37% ± 11.4 vs. nOB: 26% ± 10.2, $p < 0.05$), with Pre-T2D (40% ± 15.0) and T2D (39% ± 10.3) showing the highest percentages of activated cells (Figure 6F and Supplementary Table S4). This increased cell activation was accompanied by a reduction in the expression of TIM-3 on these cells in OB (487 ± 58 vs. nOB: 736 ± 534, $p < 0.05$), with Classes III and IV being the lowest ($p < 0.05$; Supplementary Table S4). Additionally, the IRn group also displayed a decrease in the expression of TIM-3 (471 ± 70.4) when compared to nOB and IS ($p < 0.05$), as shown in Supplementary Table S4. Moreover, there was a negative correlation between the BMI and the percentage of CCR5$^-$CCR6$^-$ Th cells (rho = 0.40, $p < 0.05$) and the expression of TIM-3 on CCR5$^-$CCR6$^-$ Th cells (rho = −0.32, $p < 0.05$; Figure 7).

3.3.4. Regulatory Helper T Cells

Participants with obesity presented a decrease in the percentage of CD4$^+$ regulatory T cells measured in whole blood (OB: 0.67% ± 0.3 vs. nOB: 0.89% ± 0.3, $p < 0.05$, Figure 8B). Interestingly, the higher the BMI, the lower the CD4$^+$ Treg percentage, with Class IV obesity showing the lowest value (nOB: 0.89% ± 0.3; II: 0.69 ± 0.36; III: 0.70 ± 0.27; IV: 0.63% ± 0.9; nOB vs. IV, $p < 0.05$; Figure 8B). Additionally, the percentage of regulatory T cells measured in whole blood was also reduced in T2D (IS: 0.62 ± 0.22; IRn: 0.70 ± 0.36; Pre-T2D: 0.67 ± 0.31; T2D: 0.66 ± 0.28; T2D vs. nOB $p < 0.05$; Figure 8B).

A deeper analysis of the CD4$^+$ Treg compartments showed a higher percentage of Th1-like and Th1/17-like Tregs in the OB group (6.5% ± 2.6 and 29% ± 10.9, respectively) vs. nOB (4.9% ± 2.3 and 22% ± 7.3, respectively, $p < 0.05$; Figure 8C,E and Supplementary Table S5). This was accompanied by a higher expression of CCR5 in both subpopulations ($p < 0.05$; Supplementary Table S5). These differences were more marked for Class III obesity (Th1-like: 7.3% ± 2.6; Th1/17-like: 29% ± 9.5, $p < 0.05$ vs. nOB) and IRn, which showed the highest percentage of Th1-like Treg (7.9% ± 3.3) (Figure 8C,E and Supplementary Table S5). An increased expression of CCR5 was observed in all IR groups, especially the T2D ($p < 0.05$, Supplementary Table S5). Moreover, the percentage of CCR5$^-$CCR6$^-$ Th-like Treg was significantly reduced in OB vs. nOB ($p < 0.05$; Figure 8F and Supplementary Table S5). Noteworthy, the percentage of PD-1$^+$ Th1-like Treg was decreased in Pre-T2D (41% ± 8.1 vs. nOB: 56% ± 12.3; Figure 8H and Supplementary Table S5), while the PD-1$^+$ Th1/17-like was reduced in the Pre-T2D group (36% ± 10.1, $p < 0.05$ vs. nOB; Figure 8M and Supplementary Table S5) and Class II obesity (39% ± 13.9), in comparison to the nOB (49% ± 13.3), with sequential increases in Classes III and IV (Supplementary Table S5). On the other hand, the percentages of TIM-3$^+$ Th1-like and TIM-3$^+$ Th1/17-like Tregs were higher in the T2D (Figure 8I,N and Supplementary Table S5).

No differences were found in the percentage of Th17-like Treg between OB and nOB groups. However, the IRn presented a reduction in the percentage of Th17-like Tregs (16% ± 8.2) when compared to the nOB group (25% ± 10.1, $p < 0.05$), as shown in Figure 8D and Supplementary Table S5.

Interestingly, when looking at the percentages of PD-1$^+$ Th17-like Tregs, these were reduced in Pre-T2D in comparison to the nOB and the remaining OB groups (Figure 8J and

Supplementary Table S5). On the other hand, the IRn showed a higher expression of PD-1 by Th17-like Treg as compared to nOB, Pre-T2D, and T2D ($p < 0.05$). The percentage of TIM-3$^+$ Th17-like Treg was increased in the Class II and Pre-T2D groups, when compared to nOB ($p > 0.05$), and it was decreased in the IS group ($p > 0.05$), as seen in Figure 8L and Supplementary Table S5. Of note, the IS group showed a similar distribution in the Treg phenotypes as the nOB group, except in the percentage of CCR5$^-$CCR6$^-$ Th-like Treg, which was significantly reduced (nOB: 18% ± 5.6 vs. IS: 10% ± 4.5, $p < 0.05$; Supplementary Table S5).

Importantly, almost 31% of the CD4$^+$ Tregs in circulation, from people with obesity, showed a Tfr phenotype, while 29% presented a Th1/Th17-like phenotype (Figure 9A and Supplementary Table S5). Similar to Th1-like Treg, CD4$^+$ Tfr showed a higher percentage of cells with a Th1-like phenotype within Class III obesity and IRn, when compared to nOB ($p < 0.05$; Figure 9B and Supplementary Table S5). Interestingly, the percentage of PD-1$^+$ Th1-like Tfr was increased in Class II (64% ± 19.4) and IS (71% ± 14.4) compared to nOB ($p < 0.05$; Figure 9C and Supplementary Table S5). Furthermore, obesity Class II patients also had higher percentage of TIM-3$^+$ Th1-like CD4$^+$Tfr cells (11% ± 4.6) compared to nOB ($p < 0.05$) (Figure 9D and Supplementary Table S5).

3.3.5. Follicular Helper T Cells

Tfh cells play an important role in B cell differentiation, and our results pointed to a slight increase in the percentage of these cells in people with obesity (nOB: 15% ± 4.3 vs. OB: 18% ± 4.0, $p < 0.05$), with no differences within the groups with obesity (Figure 10A and Supplementary Table S6). Interestingly, despite this increase in Tfh, no changes were observed in B lymphocytes. Furthermore, the expression of CXCR5 (Supplementary Table S6) was increased in the Tfh from OB participants (3224 ± 748 vs. nOB: 2766 ± 373, $p < 0.05$), especially in Pre-T2D (3501 ± 847, $p < 0.05$ vs. nOB). Interestingly, the percentage of Th1/17-like Tfh was higher in people with IRn (3.8% ± 1.3) compared to nOB (2.4% ± 1.5, $p < 0.05$; Supplementary Table S6). The percentage of PD-1$^+$ Tfh was reduced in people with Class III (44% ± 8.4) obesity compared to the remaining OB and nOB groups (54% ± 9.9, $p < 0.05$; Figure 11B and Supplementary Table S6). Likewise, the percentage of TIM-3$^+$ Tfh cells was decreased in Class IV (2.0% ± 0.73, $p > 0.05$) and IRn (1.9% ± 0.93, $p < 0.05$) compared to nOB (2.9% ± 1.2) (Figure 11C and Supplementary Table S6). The percentage of Th1/17-like Tfh in nT2D was negatively correlated to HbA1c (rho = −0.44, $p < 0.05$) and glucose (rho = −0.51, $p < 0.05$) (Figure 7).

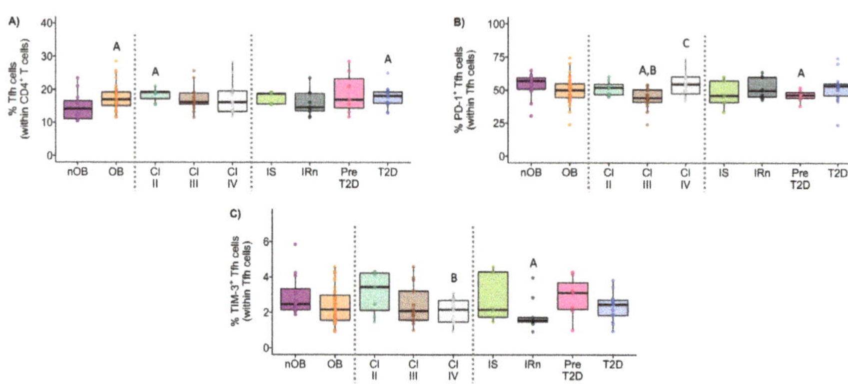

Figure 10. Frequency and expression of immune regulatory molecules by follicular CD4$^+$ T cells (Tfh). Percentage of follicular CD4$^+$ T cells (Tfh) within CD4$^+$ T cells (**A**). Percentage of PD-1$^+$ (**B**) and TIM-3$^+$ (**C**) cells within Tfh cells, in the different studied groups. nOB: healthy participants (without obesity); OB: participants with obesity; IS: insulin sensitive; IRn: insulin resistant and normoglycemic; Pre-T2D: pre-diabetes; T2D: type 2 diabetes. Statistical differences were considered when $p < 0.05$.

^{A} $p < 0.05$ vs. nOB; ^{B} $p < 0.05$ vs. Class II; ^{C} $p < 0.05$ vs. Class III; ^{D} $p < 0.05$ vs. IS; ^{E} $p < 0.05$ vs. IRn; ^{F} $p < 0.05$ vs. Pre-T2D.

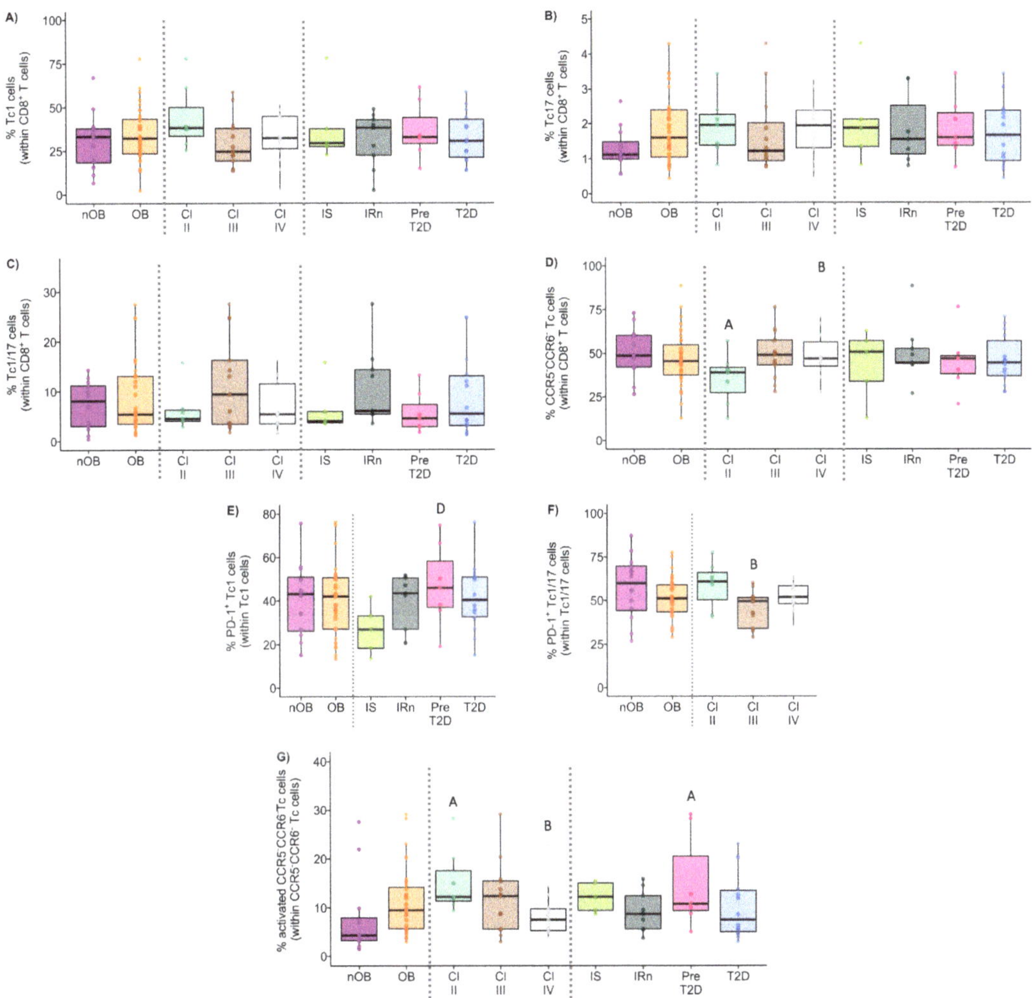

Figure 11. The Tc1 (**A**), Tc17 (**B**), Tc1/17 (**C**), and CCR5$^-$CCR6$^-$ (**D**) cell percentages within CD8$^+$ T cells, as well as the percentages of PD-1$^+$ Tc1 (**E**) and Tc1/17 (**F**) cells and of the activated CCR5$^-$CCR6$^-$ Tc cells (**G**) in nOB and OB, grouped according to obesity class and metabolic profile. nOB: healthy participants (without obesity); OB: participants with obesity; IS: insulin sensitive; IRn: insulin resistant and normoglycemic; Pre-T2D: pre-diabetes; T2D: type 2 diabetes; Tc1—T cytotoxic 1 (CCR5$^+$CCR6$^-$); Tc17—T cytotoxic 17 (CCR5$^-$CCR6$^+$); Tc1/17—T cytotoxic 1/17 (CCR5$^+$CCR6$^+$). Statistical differences were considered when $p < 0.05$. ^{A} $p < 0.05$ vs. nOB; ^{B} $p < 0.05$ vs. Class II; ^{C} $p < 0.05$ vs. Class III; ^{D} $p < 0.05$ vs. IS; ^{E} $p < 0.05$ vs. IRn; ^{F} $p < 0.05$ vs. Pre-T2D.

3.4. CD8$^+$ T Lymphocytes

The phenotypic features of CD8$^+$ T cells, the percentage of activated cells (CD25$^+$), and the expression of immune regulatory proteins were summarized in Figures 11–13, and more details are provided in Supplementary Tables S7 and S8.

3.4.1. Tc1 Cells Display Higher Expression of CCR5 in the OB Group and It Was Correlated to Insulin Levels

In similarity to the Th1 cells, there were no differences in the percentages of Tc1 among groups (Figure 11A and Supplementary Table S7). However, OB participants presented higher expression of CCR5 (measured as MFI) in Tc1 cells (3703 ± 1853), compared to nOB (2646 ± 1091, $p < 0.05$), particularly in the IS group (4164 ± 1671, $p < 0.05$) and T2D (3884 ± 2039, $p < 0.05$), as seen in Supplementary Table S7. Additionally, the percentage of PD-1$^+$ Tc1 cells was decreased in the IS (27% ± 11.4), while the remaining groups showed similar percentages to the nOB (41% ± 17.3, $p > 0.05$; Figure 11E and Supplementary Table S7). Interestingly, the percentage of PD-1$^+$ Tc1 cells was highly correlated to insulin levels and HOMA-IR (rho= 0.50 and rho = 0.51, $p < 0.05$, respectively; Figure 12) for the nOB and nT2D groups.

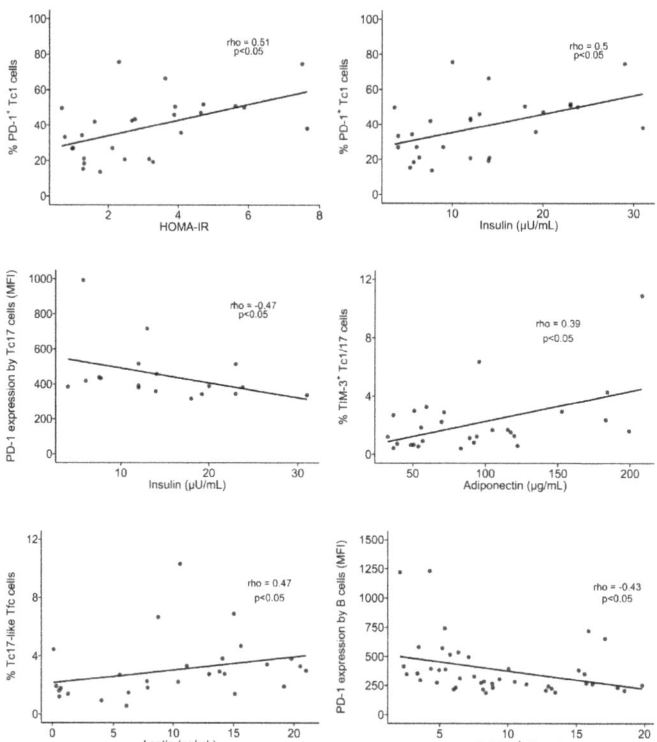

Figure 12. Spearman's correlation analysis between immune and anthropometric or metabolic parameters. BMI: body mass index; MFI: mean fluorescence intensity. Statistical differences were considered when $p < 0.05$.

3.4.2. In Tc17 Cells, the PD-1 Expression Correlates to Metabolic Parameters

An increase in the percentage of Tc17 was observed in the OB group (OB: 2.1% ± 1.4 vs. nOB: 1.3% ± 0.59, $p > 0.05$; Figure 12 and Supplementary Table S7). This was accompanied by a higher expression of CCR6 measured as MIF (OB: 983 ± 245 vs. nOB: 790 ± 206, $p < 0.05$) (Supplementary Table S7). On the other hand, nT2D showed a higher percentage of Tc17 (2.33% ± 1.6) compared to nOB (1.3 % ± 0.59, $p < 0.05$), with the Pre-T2D group presenting the highest expression CCR6 ($p < 0.05$ vs. nOB) (Figure 11B and Supplementary Table S7). Interestingly, Tc17 cells from nT2D participants presented a significant correlation between the expression of PD-1 (measured as MFI) and insulin (rho = −0.47), HOMA-IR (rho = −0.48), and C-peptide (rho = −0.49), as represented in Figure 6.

3.4.3. Increased Activation Status of Cytotoxic Tc1/17 and CCR5$^-$CCR6$^-$ T Cells in the OB Group

The percentage of Tc1/17 cells was similar among groups (Figure 11C and Supplementary Table S7); however, they displayed increased activation status in the OB (24% ± 14.1 vs. nOB: 17% ± 11.6, $p < 0.05$), which was more pronounced in Class II obesity (32% ± 16.9, $p < 0.05$ vs. nOB) (Supplementary Table S7). It is noteworthy that the percentage of PD-1$^+$ Tc1/17 cells tended to be decreased in OB (51% ± 11.7), especially in the Class III (45% ± 10.3), IS (49% ± 14.6) and IRn (48% ± 8.9) groups, compared to the nOB (57% ± 18.8, $p > 0.05$), (Figure 11F and Supplementary Table S7). Adiponectin levels were positively correlated to the percentage of TIM-3$^+$ Tc1/17 cells (rho = 0.39, $p < 0.05$; Figure 12). Interestingly, CCR5$^-$CCR6$^-$CD8$^+$ T cells also showed an activated pattern in the OB (11% ± 6.6) compared to the nOB (8.0% ± 8.3, $p > 0.05$), especially in Class II obesity (15% ± 6.7, $p < 0.05$) and Pre-T2D (15% ± 9.7, $p < 0.05$; Figure 11G and Supplementary Table S7).

3.4.4. CD8$^+$ Regulatory T Cells Are Expanded in Participants with Obesity

A deep characterization of CD8$^+$ Treg cells has been described in detail in another study [30]. Briefly, an increase in the percentage of CD8$^+$ Treg cells in the OB group (nOB: 0.14% ± 0.18 vs. OB: 0.27% ± 0.24, $p < 0.05$) was observed. This was more pronounced in pre-diabetes (Pre-T2D: 0.42% ± 0.31, $p < 0.05$ vs. nOB). Additionally, positive correlations between the percentage of circulating CD8$^+$ Treg cells and fasting insulin ($p < 0.05$) or CRP ($p < 0.05$) levels [30] were found. Moreover, the CD8$^+$ Treg phenotype from OB individuals were more prone to differentiate into Tc1- and Tc1/17-like cells and displayed increased expression of CCR5 [30].

3.4.5. Follicular Cytotoxic T Cells Present an Increased Activation Status in the OB Group

A slight increase in the percentage of Tfc was detected in the OB (2.6% ± 1.8 vs. nOB: 1.9% ± 0.88, $p > 0.05$), particularly in the T2D (3.1% ± 2.0, $p = 0.054$ vs. nOB) (Figure 13A and Supplementary Table S8). In similarity to the Tfh, the Tfc also display higher expression of CXCR5 in OB ($p < 0.05$; Supplementary Table S8), particularly in Class II obesity, as well as a higher percentage of activated Tfc cells (OB: 10% ± 5.3 vs. nOB: 6.6% ± 4.0, $p < 0.05$; Figure 13B and Supplementary Table S8).

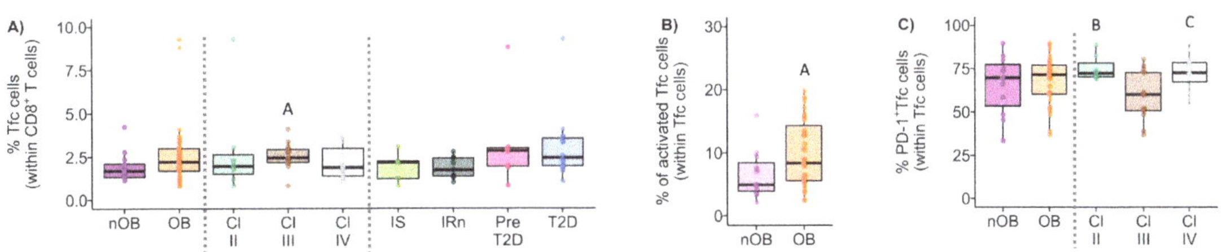

Figure 13. Frequency, activation profile, and immune regulatory markers expression by follicular CD8$^+$ T cells (Tfc). Percentage of follicular CD8$^+$ T cells (Tfc) within CD8$^+$ T cells (**A**). Percentage of activated Tfc (**B**) and PD-1$^+$ Tfc (**C**) cells in the different groups under study. nOB: healthy participants (without obesity); OB: participants with obesity; IS: insulin sensitive; Irn: insulin resistant and normoglycemic; Pre-T2D: pre-diabetes; T2D: type 2 diabetes. Statistical differences were considered when $p < 0.05$. ^{A} $p < 0.05$ vs. nOB; ^{B} $p < 0.05$ vs. Class II; ^{C} $p < 0.05$ vs. Class III; ^{D} $p < 0.05$ vs. IS; ^{E} $p < 0.05$ vs. IRn; ^{F} $p < 0.05$ vs. Pre-T2D.

Furthermore, the T2D group displayed an increase in Tc1-like Tfc in comparison to the nT2D (T2D: 69% ± 8.1 vs. nT2D: 60% ± 15.1, $p < 0.05$; nOB: 63% ± 12.8). On the other hand, the percentages of Tc1/17-like Tfc were decreased in Class III obesity (4.6% ± 2.3 vs. nOB: 8.2% ± 5.0, $p < 0.05$) and in the IRn (4.7% ± 3.4, $p > 0.05$ vs. nOB; Supplementary Table S8). Moreover, the Tfc from Class III obesity presented a reduction in the percentage and expression of PD-1, when compared to the nOB and the remaining obesity groups

(Figure 13C and Supplementary Table S8). Importantly, the percentage of Tc17-like Tfc cells was positively correlated to leptin levels (rho = 0.47, $p < 0.05$), while the expression of PD-1 (per MFI) in B cells was negatively correlated to the percentage of $CD25^+$ Tfc cells (rho = -0.43, $p < 0.05$) (Figure 12).

3.5. The Phenotype of $CD4^+CD8^+$ T Cells in Obesity

The $CD4^+CD8^+$ T cells comprised only 2% of the circulating T cells in both the nOB and OB groups (Figure 5 and Supplementary Table S2). In participants with OB, the $CD4^+CD8^+$ T cells did not exhibit any differences in their polarization compared to the nOB (Figure 14 and Supplementary Table S9). However, a decrease in T1/17-like $CD4^+CD8^+$ T cells was observed in Class II obesity ($p < 0.05$ vs. nOB; Figure 14C and Supplementary Table S9). Hence, the OB group displayed a higher percentage of activated T1/17-like $CD4^+CD8^+$ T cells (48% ± 17.2), especially in Class II (55% ± 15.6), compared to the nOB group (37% ± 15.6, $p < 0.05$) (Supplementary Table S9). A decreased percentage of $PD\text{-}1^+$ T1-like $CD4^+CD8^+$ T cells in OB (57% ± 24.9 vs. nOB: 68% ± 26.2, $p > 0.05$; Supplementary Table S9) was also observed, especially in Class III obesity (50% ± 24.3). Likewise, a reduction in $PD\text{-}1^+$ among T17-like and T1/17-like $CD4^+CD8^+$ T cells in the OB vs. nOB group ($p < 0.05$), more marked in Class III obesity, IRn, and Pre-T2D for the T17-like cells ($p < 0.05$) and in Class IV for the T1/17-like $CD4^+CD8^+$ T cells ($p < 0.05$), was observed (Supplementary Table S9). The OB participants had a lower percentage of $TIM\text{-}3^+$ T1-like $CD4^+CD8^+$ T cells, especially in Class III obesity, IS, and IRn, compared to nOB ($p < 0.05$) (Supplementary Table S9). Interestingly, the percentage of $TIM\text{-}3^+$ T1/17-like $CD4^+CD8^+$ T cells was negatively correlated to HbA1c levels in the nT2D group (rho = -0.54, $p < 0.05$; Figure 15).

Looking deeper into the $CCR5^-CCR6^-CD4^+CD8^+$ T cells, they were markedly more activated in individuals with obesity (nOB: 17% ± 15.5 vs. OB: 30% ± 19.1, $p < 0.05$), especially Pre-T2D and T2D ($p < 0.05$ vs. nOB; Supplementary Table S9), and displayed a decreased percentage of $PD\text{-}1^+$ cells (nOB: 48% ± 28.5 vs. OB: 31% ± 17.1, $p = 0.085$; Supplementary Table S9). This decrease was more pronounced in Class III obesity (22% ± 12.0, $p < 0.05$ vs. nOB) and Irn (27% ± 19.2, $p = 0.069$ vs. nOB; Supplementary Table S9). In addition, the percentage of activated $CCR5^-CCR6^-CD4^+CD8^+$ T was positively correlated to BMI (rho = 0.34, $p < 0.05$; Figure 15).

The $CD4^+CD8^+$ Treg cells were decreased only in Class II obesity (0.34% ± 0.17 vs. nOB: 1.1% ± 1.1, $p < 0.05$; Figure 14E and Supplementary Table S9).

Furthermore, the Tf compartment was increased in T2D (3.1% ± 1.7) in comparison to nOB (2.3% ± 0.96, $p > 0.05$), IS (1.3% ± 0.50, $p < 0.05$) and IRn (1.7% ± 1.0, $p < 0.05$; Figure 14F and Supplementary Table S9). Moreover, participants with Class II obesity displayed an increase in the percentage of T17-like $CD4^+CD8^+$ Tf cells compared to nOB, while in the remaining groups, T17-like cells tended to be decreased (nOB: 15% ± 14.5; OBII: 20% ± 12.3; OBIII: 11% ± 9.5; OBIV: 6.9% ± 3.8; OBII vs. OBIV, $p < 0.05$; Supplementary Table S9). In similarity to T17-like cells, T1/17-like $CD4^+CD8^+$ Tf cells were also reduced in participants with Class III obesity (nOB: 10 ± 9.1, OBII: 13% ± 7.7, OBIII: 6.9% ± 5.1, OBIV: 12% ± 5.4; OBIII vs. OBIV, $p < 0.05$) and in the IRn group (6.8% ± 5.7, $p > 0.05$ vs. nOB; Supplementary Table S9). Interestingly, the IRn and Pre-T2D also displayed a slight reduction in the percentage of the $PD\text{-}1^+$ $CD4^+CD8^+$ Tf cells, compared to the nOB group ($p > 0.05$; Supplementary Table S9).

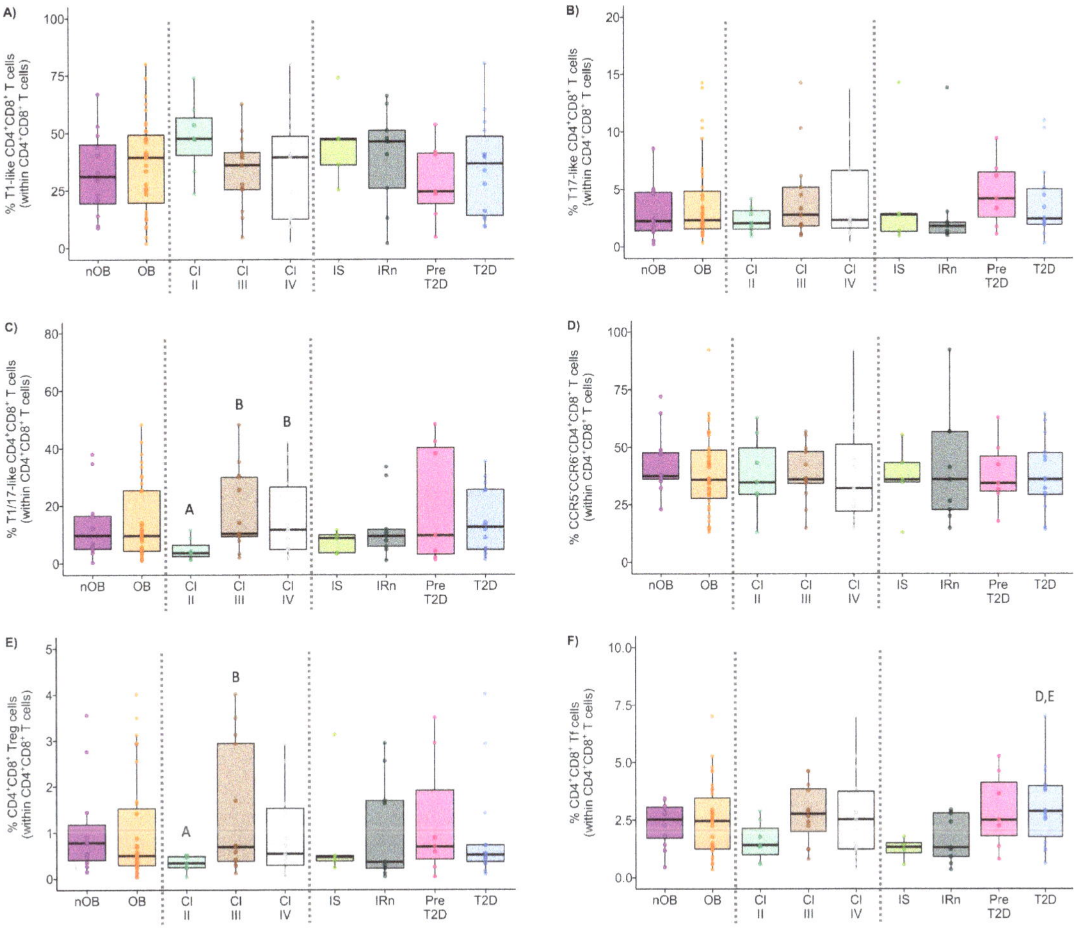

Figure 14. Percentage of CD4⁺CD8⁺ T cells presenting a T1-like (**A**), T17-like (**B**), T1/17-like (**C**), and CCR5⁻CCR6⁻ (**D**), Treg (**E**), and Tf (**F**) phenotype, within CD4⁺CD8⁺ T cells, in nOB and OB, grouped according to obesity class and metabolic profile. nOB: healthy participants (without obesity); OB: participants with obesity; IS: insulin sensitive; IRn: insulin resistant and normoglycemic; Pre-T2D: pre-diabetes; T2D: type 2 diabetes; T1-like—$CCR5^+CCR6^-$; T17-like—$CCR5^-CCR6^+$; T1/17-like—$CCR5^+CCR6^+$. Statistical differences were considered when $p < 0.05$. ^{A} $p < 0.05$ vs. nOB; ^{B} $p < 0.05$ vs. Class II; ^{C} $p < 0.05$ vs. Class III; ^{D} $p < 0.05$ vs. IS; ^{E} $p < 0.05$ vs. IRn; ^{F} $p < 0.05$ vs. Pre-T2D.

3.6. Differences within Metabolic Groups

Participants with obesity and insulin resistance, not taking medication for T2D, showed a higher percentage of $CD4^+CD8^+$ T cells within T cells (IRn: 2.8% ± 1.1; Pre-T2D: 2.2% ± 2.0) compared to T2D patients under treatment (1.6% ± 1.9, $p < 0.05$ for IRn vs. T2D; Figure 5 and Supplementary Table S2). Furthermore, insulin-resistant groups (including T2D) displayed a higher polarization pattern toward the Th1-like, namely in $CD4^+$ Treg cells (Figure 8 and Supplementary Table S5) and $CD4^+$ Tfr cells (Figure 9 and Supplementary Table S5), when compared to the IS group. Importantly, groups with higher percentages of HbA1c displayed higher percentages of $CD4^+CD8^+$ Tf cells, while the polarization of these cells presented different patterns among the groups, wherein Pre-T2D displayed the highest percentage of $CD4^+CD8^+$ Tf toward T17-like (Supplementary Table S9). Additionally, the Pre-T2D also displayed the highest percentage of $CD8^+$ Treg cells among groups [30].

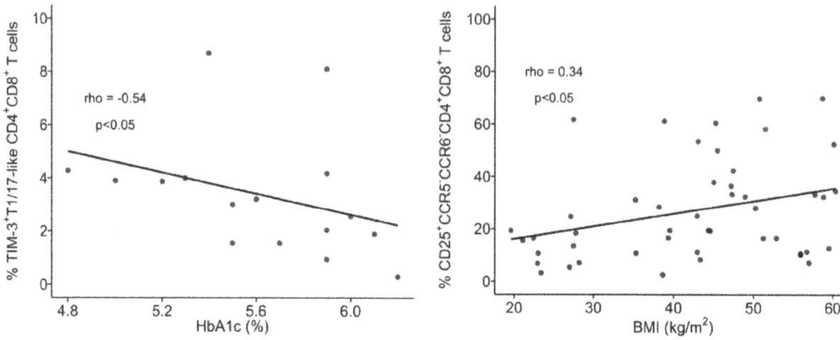

Figure 15. Spearman's correlation analysis between immune and anthropometric or metabolic parameters. BMI: body mass index. Statistical differences were considered when $p < 0.05$.

Interestingly, the IS group showed an overall reduction in the percentage of cells expressing PD-1, as well as in the amount of PD-1 protein expressed per cell (measured as MFI), while a sequential increase was observed with increasing IR (Supplementary Tables S4–S9). A similar pattern was observed regarding the percentage of cells expressing TIM-3. These variations were especially observed in Th1 (Figure 6H and Supplementary Table S4) and Tc1 (Supplementary Table S7), as well as $CD4^+$ Treg cells regarding TIM-3 (Figure 9G and Supplementary Table S5). On the other hand, the IS group had a higher percentage of $PD\text{-}1^+$ Th1-like Treg cells (60% ± 8.7), while the Pre-T2D had the lowest percentage (41% ± 8.1, $p < 0.05$; Figure 8H and Supplementary Table S5) of these cells. As previously observed, the Pre-T2D group had important alterations in the $CD8^+$ Treg cells, regarding their phenotype and the expression of immune checkpoint molecules [30].

3.7. Effect of Bariatric Surgery on Metabolic and Immune Profiles

To understand the effect of bariatric surgery on the obesity-associated immune profile, an unpaired analysis was performed, comparing 15 individuals with Class IV obesity before surgery (at baseline or T1) and a set of 14 participants, evaluated 7 to 18 months post-surgery (T2). Additionally, a follow-up study with paired comparisons including eight individuals with Class IV obesity, analyzed both at T1 and T2, was also performed (Figure 1). In this follow-up study, the time period between T1 and T2 ranged from 9 to 18 months (Figure 16 and Supplementary Table S10).

When comparing independent groups of patients at T1 and T2 (Figure 16 and Supplementary Table S10), a reduction in white blood cell count was detected after surgery (from 8.2 cells/µL ± 2.4 to 6.7 cells/µL ± 1.6, $p < 0.05$; Figure 16A and Supplementary Table S10), achieving values resembling those measured in the nOB group (6.2 cells/µL ± 1.5). Despite this reduction, the percentage of neutrophils remained similar before and after surgery (Supplementary Table S10). The percentage of lymphocytes was slightly increased at T2, despite remaining lower than in nOb (T1: 24% ± 7.8; T2: 29% ± 8.9; nOB: 40% ± 6.6; T2 vs. nOB, $p < 0.05$; Figure 16B and Supplementary Table S10). The percentage of T lymphocytes followed a similar pattern (Figure 16C and Supplementary Table S10). Interestingly, bariatric surgery impacted B lymphocytes, promoting increases in their percentage (T1: 2.1% ± 1.2; T2: 3.4% ± 1.7, $p < 0.05$; Figure 16E and Supplementary Table S10) and activation status ($p < 0.05$; Figure 16F), accompanied by an expansion of the Tfh cells (nOB: 15% ± 4.3; T1: 17% ± 4.9; T2: 20% ± 5.1; nOB vs. T2, $p < 0.05$) (Figure 16G and Supplementary Table S10).

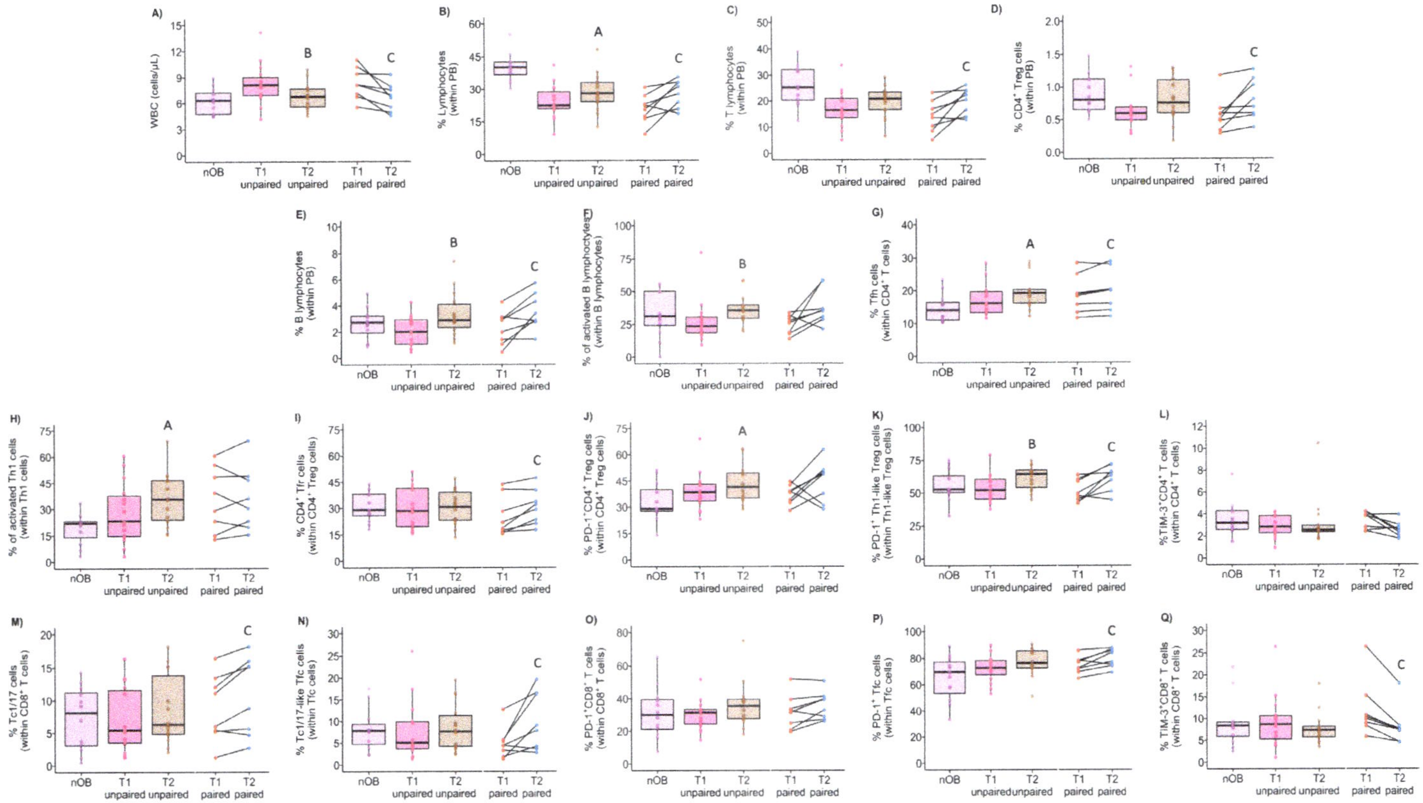

Figure 16. Phenotypical changes in T cells from individuals with obesity, before (T1) and after (T2) bariatric surgery. The analysis of unpaired samples comprised all individuals with Class IV obesity in T1 (n = 15) and 14 individuals after bariatric surgery (T2). The analysis of paired samples consisted in a follow-up of 8 individuals before (T1) and after (T2) bariatric surgery. The nOB group was used as reference. Distribution of white blood cells count (WBC) (cells/µL) before (T1)

and after (T2) surgery (**A**). Percentage of lymphocytes (**B**), T lymphocytes (**C**), $CD4^+$ Treg cells (**D**), and B lymphocytes (**E**), within peripheral blood leukocytes (PB), before (T1) and after (T2) surgery. Percentage of activated B cells (measured within B cells) (**F**) and Tfh cells (within $CD4^+$ T cells) (**G**), activated Th1 cells (within Th1 cells) (**H**), and $CD4^+$ Tfr cells (within $CD4^+$ Treg cells) (**I**) before (T1) and after (T2) surgery. Percentages of PD-1^+ $CD4^+$ Tregs (**J**) and Th1-like Tregs (**K**) and percentage of TIM-3^+CD4^+ T cells (**L**) before (T1) and after (T2) bariatric surgery. Percentage of Tc1/17, within $CD8^+$ T cells (**M**) and Tc1/17-like Tfc cells, within Tfc (**N**), at T1 and T2. Percentages of PD-1^+ $CD8^+$ T cells (**O**) and Tfc cells (**P**) and percentage of TIM-3^+CD8^+ T cells (**Q**) before (T1) and after (T2) surgery. nOB: healthy participants (without obesity). Statistical differences were considered when $p < 0.05$. ^{A} $p < 0.05$ vs. nOB; ^{B} $p < 0.05$ vs. T1 unpaired; ^{C} $p < 0.05$ vs. T1 paired.

Regarding $CD4^+$ T cells, the group participants evaluated post-surgery displayed an increased percentage of $CD4^+$ Treg cells in whole blood (0.80% ± 0.34) compared to the group evaluated at baseline (0.63% ± 0.29, $p > 0.05$; Figure 16D). This increase was accompanied by an increased percentage of $CD4^+$ Treg expressing PD-1 post-surgery (T2: 44% ± 11.0 vs. T1: 39% ± 11.02, $p > 0.05$; and vs. nOB: 33% ± 11.0, $p < 0.05$; Figure 16J and Supplementary Table S10), especially marked in Th1-like Tregs (T2: 62% ± 9.4 vs. T1: 54% ± 11.0, $p < 0.05$) (Figure 16K and Supplementary Table S10). An increased percentage of activated Th1 cells was detected post-surgery (36% ± 15.4) when compared to nOB (19% ± 8.9, $p < 0.05$) and at baseline (28% ± 17.4, $p > 0.05$; Figure 16H and Supplementary Table S10), and a similar activation pattern occurred in $CCR5^-CCR6^-CD4^+$ T cells (Supplementary Table S10).

Bariatric surgery also resulted in a slight increase in the percentage of PD-1^+ $CD8^+$ T cells (Figure 16O and Supplementary Table S10). In turn, T17-like $CD4^+CD8^+$ Tf cells were significantly increased at T2 (18% ± 18.6) vs. T1 (6.9% ± 3.8, $p < 0.05$) (Supplementary Table S10).

In paired comparisons of participants before (T1) and after (T2) bariatric surgery, an amelioration of the metabolic profile was observed, with the normalization of metabolic parameters, including HbA1c, glucose, insulin, HOMA-IR, and triglycerides. When the immune profile of these participants was analyzed (Table S10), bariatric surgery induced a reduction in the WBC count (T1: 8.3 cells/µL ± 1.7 vs. T2: 6.7 cells/µL ± 1.5, $p < 0.05$; Figure 16A and Supplementary Table S10) and an increase in the percentage of lymphocytes ($p < 0.05$; Figure 16B and Supplementary Table S10), due to the increase in T ($p < 0.05$; Figure 16C and Supplementary Table S10) and B cells ($p < 0.05$; Figure 16E and Supplementary Table S10). Although no differences were found in the proportion of $CD4^+$ Treg cells within $CD4^+$ T cells (Supplementary Table S10), the percentage of $CD4^+$ Treg measured in whole blood was increased post-surgery (T1: 0.58% ±0.27 vs. T2: 0.81% ± 0.31, $p < 0.05$; Figure 16D and Supplementary Table S10), along with an increased percentage of PD-1^+ Th1-like Tregs ($p < 0.05$; Figure 16K and Supplementary Table S10) and the expression of CCR6 in Th17-like Tregs ($p < 0.05$; Supplementary Table S10). Similarly, $CD4^+$ Tfr and Tfh cells were increased post-surgery ($p < 0.05$; Figure 16I and Supplementary Table S10), while the percentage of $CCR5^-CCR6^-$ Th cells was reduced ($p < 0.05$). Within $CD8^+$ T cells, an increased percentage of Tc1/17 cells (Figure 16M and Supplementary Table S10), Tc1/17-like Tfc cells (Figure 16N and Supplementary Table S10), and PD-1^+ Tfc cells was observed post-surgery ($p < 0.05$; Figure 16P and Supplementary Table S10). Conversely, the percentage of TIM-3^+ cells was decreased among $CD4^+$ T ($p > 0.05$) and $CD8^+$ T cells ($p < 0.05$; Figure 16Q and Supplementary Table S10). Importantly, these decreases affected specific T cell subsets, namely the Tfh, Th1-like Treg, $CCR5^-CCR6^-$ Th, and $CCR5^-CCR6^-$ Tc cells (Supplementary Table S10).

4. Discussion

Obesity is a potential trigger for different comorbidities, with chronic low-grade inflammation perpetuating and underlying these conditions [1,2]. The immune system is paramount in stimulating and maintaining inflammatory processes. It plays an important role in the onset of obesity-related comorbidities [2]. In addition, the interaction of immune cells and the extracellular matrix (ECM) also plays an important role in the onset of low-grade inflammation in obesity. This is driven by the release of different molecules, such as small fragments of hyaluronan, which induce the activation of innate immune cells via toll-like receptors. Moreover, the involvement of these molecules in the development of insulin resistance has previously been suggested [31].

The present study brings new light in regard to the phenotype and function of circulating immune cells in people with obesity, with obesity-associated insulin resistance and T2D. The study aimed to obtain a deep characterization of circulating T cells. The impact of obesity on the imbalance of immune homeostasis is evident in both innate and adaptive immune systems. Higher white blood cell counts were observed in people with obesity, except for those in the insulin-sensitive group who displayed similar WBC counts as the nOB. Increased percentages of neutrophils and a reduction in T lymphocytes, with consequent reduction in total lymphocytes, in people with obesity were also observed. In previous studies, the important role of neutrophils was reflected in the onset of the inflammatory cascade in obesity [32] and in the onset of insulin resistance by the release of elastase, promoting the degradation of insulin receptors [7,33]. Moreover, different authors indicated the impact of obesity on the adaptive immune cells, with emphasis on T cells, either in circulation or infiltrated in adipose tissue [11]. In our cohort, increases in $CD4^+$ T cells, accompanied by reductions in $CD8^+$ T cells, were observed within circulating immune cells.

In 2015, Łuczyński and collaborators found an increased count of circulating $CD4^+$ T cells with the Th17 phenotype in children with obesity (10–18 years old) and suggested these cells played a key role in low-grade inflammation and established the link between obesity and diabetes [34]. In addition, they also observed a higher percentage of Th17 cells expressing IFN-γ [34] that have been further identified as Th1/17 cells [35]. Importantly, our data showed that both populations were increased in the peripheral blood of people with obesity, accompanied by a reduction in $CCR5^-CCR6^-$ Th cells. This profile was observed independently of their metabolic status. However, despite no difference being observed between the Th1 cell percentages, the IS and IRn participants presented lower expression of PD-1 by Th1 cells, as compared to T2D participants. The reduction in the expression of PD-1 could be indicative of an increased function of these cells in those participants. According to previous studies [reviewed [36]], the expression of PD-1 by Th cells was related to loss of function [36] and exhaustion [37]. In fact, studies indicated that the blockade of PD-1, as well as the blockade of TIM-3, induced an increase in IFN-γ production by $PD\text{-}1^+CD4^+$ T cells [36]. On the other hand, $CD8^+$ T and $CD4^+CD8^+$ T cells from the OB group showed a different polarization tendency, with preferential Tc1- and T1-like polarization, respectively. Noteworthy, the Tc1 cells from individuals with obesity showed higher expression of CCR5, allowing these cells to migrate into inflammatory sites [38].

Some controversy remains regarding the circulating $CD4^+$ Treg cells in people with obesity. Some studies indicate that $CD4^+$ Treg cells are reduced in circulation in people with obesity [14], while others do not [39]. In the present study, no differences were identified in the percentage of $CD4^+$ Treg cells within $CD4^+$ T cells, although a significant reduction was observed in the percentage of $CD4^+$ Treg cells measured within whole blood. It is important to mention that this reduction can be attributed to the increased white blood cell count in people with obesity, rather than to an actual reduction in the number of $CD4^+$ Tregs in the PB. Therefore, we evaluated the absolute number of $CD4^+$ Tregs in whole blood and found that the absolute count was not significantly different between the nOB and the OB groups, which displayed similar counts (Supplementary Table S3). Interestingly, $CD8^+$ Treg cells follow the opposite direction, displaying an increase in their percentage in people with obesity—more marked in people with pre-diabetes [30]. It is noteworthy that this cell population was previously described as being positively correlated to BMI [40]. The expression of chemokine receptors by Treg cells allows them to migrate into inflammatory locations [38]. As previously reported, $CD4^+$ T cells infiltrated into adipose tissue display a higher tendency to polarize toward Th1 cells [13] (and our unpublished data). It is important to acknowledge that the increase in the percentage of circulating Th1- and Th1/17-like Treg cells in people with obesity described in our study may indicate active migration of these Treg cells into adipose tissue as an attempt to mitigate the inflammation in the tissue microenvironment. Additionally, this increase was accompanied by a worsened metabolic condition. Furthermore, regarding the expression of the immune checkpoint proteins, PD-1 has commonly been associated with Treg homeostasis and immunosuppressive capability [reviewed [41]], as well as TIM-3 [42]. In fact, people with obesity and pre-diabetes showed a reduced percentage of $PD\text{-}1^+$Th1-like Treg in circulation but also a reduction in $PD\text{-}1^+CD8^+$ Treg cells and $TIM\text{-}3^+CD8^+$ Treg cells [30]. This fact supports a previous study reporting the impaired immune suppressive function of Tregs during insulin resistance and T2D [43].

Follicular T cells play an important role in the maintenance and differentiation of B cells [44]. Guo and collaborators previously reported that people with T2D display higher percentages of Tfh cells in circulation, in agreement with our study. It is important to mention that the research conducted by Guo and collaborators (2020) was performed in people without obesity (BMI of 25.90 $\pm$.99) [44]. Regardless of obesity, T2D can induce alterations in the percentage of Tfh cells. Interestingly, Tfc displayed a similar tendency, despite the expression of CXCR5 being reduced in T2D compared to the IS group.

People with Pre-T2D and T2D displayed a higher expression of TIM-3 in some $CD4^+$ T and $CD8^+$ T cell subsets, compared to the remaining OB groups. However, Sun and collaborators (2020) reported a reduction in the expression of TIM-3 within $CD4^+$ and $CD8^+$ T cells, when comparing people with obesity and people with obesity and T2D [45]. In fact, the authors attributed the difference in the expression of TIM-3 to a transient upregulation of plasma glucose levels during the early stages of T2D, with a restoration of the T cells function in long-term T2D. Moreover, an increase in TIM-3 has also been previously described in Tfh cells [44].

Importantly, bariatric surgery has been used as a gold standard in the treatment of obesity and, consequently, obesity-related comorbidities [15]. The impact of this procedure has been well described regarding its positive effects on metabolic profiles. However, the effects of bariatric surgery have not been well studied regarding immunometabolism and the modulation of the immune system. Here, we present a deep characterization of T cells before and post-surgery. Our results show that bariatric surgery has important modulatory effects on the circulatory immune profile of people with obesity. Wijngaarden et al. (2022) postulated a recovery of the $CD4^+$ T cell compartment in patients three months after bariatric surgery toward the lean control phenotype and found no changes in $CD8^+$ T cells [15]. Our study analyzes the circulating immune profile of people with obesity, 7 to 18 months post-bariatric surgery. A higher percentage of $CD4^+$ Tregs and B cells within whole blood was observed, with an increased activation profile, also accompanied by an expansion of Tfh.

A large heterogeneity found among participants and the medication used to treat/mitigate the obesity comorbidities may interfere directly or indirectly with the immune system. The most prescribed anti-diabetic, metformin, greatly impacts immune modulation, as reviewed in [46]. Furthermore, statins also play an important role in immune regulation by inhibiting T cell proliferation [47].

Despite these confounding factors, we demonstrated important new alterations on several T cell subsets associated with obesity. How distinct immune profiles impact the development of comorbidities and the obesity itself emerges as an interesting research area. Similarly, it is important to understand whether the outcome of bariatric surgery differs for patients with different immune profiles. Finally, our study demonstrated that obesity modulates differently distinct T cell subsets; therefore, it is mandatory to accurately identify these T cell subsets to understand which ones are altered. Otherwise, some differences observed in obesity would go unnoticed.

5. Conclusions

We describe significant and novel changes in circulating T cells associated with obesity. Remarkably, $CD4^+$ and $CD8^+$ T cells exhibited distinct alterations, even though both subsets exhibited increased CCR5 expression, potentially serving as a common link between the two populations. Notably, T cells from individuals with obesity displayed a pro-inflammatory profile and presented changes in the expression of immune regulatory molecules, which may serve as a counterbalance to mitigate the inflammation exacerbation. The precise impact of the immune alterations reported in this study on the development of insulin resistance and type 2 diabetes remains uncertain. The effects of bariatric surgery on the immune system were observed around 12 months after surgery, but it is mandatory to follow the participants for a longer period to understand the long-term effects of bariatric surgery on the different immune cells.

Supplementary Materials: The following supporting information can be downloaded at https://www.mdpi.com/article/10.3390/biom14020219/s1: Supplementary Figure S1: Dotplot histograms illustrating the gating strategy to identify T cells and B cells, as wells as the different T cell subpopulations. The scheme is specifically representing the $CD4^+$ T cell subpopulations (G–K), but the same strategy was applied to the $CD8^+$ and $CD4^+CD8^+$ T cells. (A) FSC and SSC light-dispersion properties allow the identification of the major immune cell populations. (B) T cells were identified by the expression of CD3, and (C) B cells were identified by the expression of CD20. (D) Identification of $CD4^+$, $CD8^+$, $CD4^+CD8^+$, and $CD4^-CD8^-$ T cells. (E) Identification of $CD4^+$ Treg cells within $CD4^+$ T cells, according to their typical $CD25^{++}CD127^{low/-}$ phenotype. (F) Identification of $CD4^+$ Tfr and $CXCR5^-$ Treg cells. (G) Identification of Tfh cells ($CXCR5^+$) within non-Treg $CD4^+$ T cells. (H) Identification of Th1, Th17, Th1/17, and $CCR5^-CCR6^-$ Th cells, based on CCR5 and CCR6 expression. These last four subsets were also identified within the Treg, Tfr, and Tfh cell subsets. (I) Identification of cells expressing the early activation marker CD25. (J) and (K) Strategy for the identification of cells expressing the immune checkpoint proteins PD-1 and TIM-3, respectively. Supplementary Table S1: Anthropometric and biochemical characterization of the participants who were evaluated both before (T1) and after (T2) the bariatric surgery. Supplementary Table S2: Absolute number of white blood cells in circulation and percentage of each immune cell type in nOB and OB, as well as among OB participants stratified by obesity class and metabolic profile. Supplementary Table S3: Absolute number (in cells/μL) of all peripheral blood immune cell populations evaluated in healthy (non-obese) controls (nOb) and individuals with obesity (Ob). NA: not analyzed due to low numbers of events. Mann–Whitney and Kruskal–Wallis tests were applied as appropriate. Statistical differences were considered when $p < 0.05$. Supplementary Table S4: The Th1, Th17, Th1/17, and $CCR5^-CCR6^-$ cell percentages within $CD4^+$ T cells and expression of activation and immune regulatory markers in nOB and OB, grouped according to obesity class and metabolic profile. Supplementary Table S5: Percentage and phenotype of circulating $CD4^+$ Treg cells from nOB and OB individuals. Supplementary Table S6: Percentage and phenotype of circulating $CD4^+$ follicular T cells from nOB and OB individuals. Supplementary Table S7: Percentage and phenotype of circulating $CD8^+$ T cells in nOB and OB groups. Supplementary Table S8: Percentage and phenotype of circulating $CD8^+$ T follicular cells from nOB and OB groups. Supplementary Table S9: Percentage and phenotype of circulating $CD4^+CD8^+$ T cells within each group of participants. Supplementary Table S10: Absolute number of white blood cells and percentage of the major peripheral blood immune cell populations before (T1) and after (T2) bariatric surgery. Distribution of circulating T cells among their different subsets and characterization of their activation status and immune checkpoint protein expression before (T1) and after (T2) bariatric surgery. The phenotypic analyses were performed in two independent groups of individuals (unpaired) before (T1) and 7 to 18 months after (T2) bariatric surgery. A follow-up study was carried out in a set of 8 individuals (paired), analyzed both before (T1) and 9 to 18 months after (T2) bariatric surgery.

Author Contributions: Conceptualization, P.B., E.C. and P.L.; methodology, P.B. and A.P. (Aryane Pinho); formal analysis, P.B.; investigation, P.B., A.P. (Aryane Pinho), A.L. and D.P.; resources, A.L., D.P., J.G.T., M.J.P. and A.P. (Artur Paiva); data curation, P.B.; writing—original draft preparation, P.B.; writing—review and editing, E.C. and P.L.; visualization, P.B.; supervision, E.C. and P.L.; funding acquisition, M.J.P. and E.C. All authors have read and agreed to the published version of the manuscript.

Funding: This work was financed by the European Regional Development Fund (ERDF), through the Centro 2020 Regional Operational Programme under the project Healthy Aging 2020—CENTRO-01-0145-FEDER-000012 and through the COMPETE 2020—Operational Programme for Competitiveness and Internationalisation and Portuguese national funds via FCT—Fundação para a Ciência e a Tecnologia, under projects POCI-01-0145-FEDER-007440, UIDB/04539/2020, UIDP/04539/2020, and LA/P/0058/2020, and also Pedro Barbosa's (SFRH/BD/143849/2019) and Aryane Pinho's (SFRH/BD/145054/2019) PhD grants. Funds were also obtained from EXODIAB (Excellence of Diabetes Research in Sweden) and the Rudberg Agnes and Mac Foundation.

Institutional Review Board Statement: This study was conducted in accordance with the Declaration of Helsinki and approved by the Institutional Review Board (or Ethics Committee) of Centro Hospitalar e Universitário de Coimbra (CHUC), Portugal (protocol code CHUC-115-20 and CHUC-202-20).

Informed Consent Statement: Informed consent was obtained from all subjects involved in the study.

Data Availability Statement: Dataset available on request from the authors.

Conflicts of Interest: The authors declare that there are no conflicts of interest.

References

1. Khanna, D.; Khanna, S.; Khanna, P.; Kahar, P.; Patel, B.M. Obesity: A Chronic Low-Grade Inflammation and Its Markers. *Cureus* **2022**, *14*, e22711. [CrossRef]
2. SantaCruz-Calvo, S.; Bharath, L.; Pugh, G.; SantaCruz-Calvo, L.; Lenin, R.R.; Lutshumba, J.; Liu, R.; Bachstetter, A.D.; Zhu, B.; Nikolajczyk, B.S. Adaptive Immune Cells Shape Obesity-Associated Type 2 Diabetes Mellitus and Less Prominent Comorbidities. *Nat. Rev. Endocrinol.* **2022**, *18*, 23–42. [CrossRef]
3. Viardot, A.; Heilbronn, L.K.; Samocha-Bonet, D.; Mackay, F.; Campbell, L.V.; Samaras, K. Obesity Is Associated with Activated and Insulin Resistant Immune Cells. *Diabetes/Metab. Res. Rev.* **2012**, *28*, 447–454. [CrossRef]
4. Reggio, S.; Rouault, C.; Poitou, C.; Bichet, J.C.; Prifti, E.; Bouillot, J.L.; Rizkalla, S.; Lacasa, D.; Tordjman, J.; Clément, K. Increased Basement Membrane Components in Adipose Tissue during Obesity: Links with TGFβ and Metabolic Phenotypes. *J. Clin. Endocrinol. Metab.* **2016**, *101*, 2578–2587. [CrossRef]
5. Muir, L.A.; Neeley, C.K.; Meyer, K.A.; Baker, N.A.; Brosius, A.M.; Washabaugh, A.R.; Varban, O.A.; Finks, J.F.; Zamarron, B.F.; Flesher, C.G.; et al. Adipose Tissue Fibrosis, Hypertrophy, and Hyperplasia: Correlations with Diabetes in Human Obesity. *Obesity* **2016**, *24*, 597–605. [CrossRef]
6. Wijngaarden, L.H.; van der Harst, E.; Klaassen, R.A.; Dunkelgrun, M.; Kuijper, T.M.; Klepper, M.; Ambagtsheer, G.; IJzermans, J.N.M.; de Bruin, R.W.F.; Litjens, N.H.R. Effects of Morbid Obesity and Metabolic Syndrome on the Composition of Circulating Immune Subsets. *Front. Immunol.* **2021**, *12*, 675018. [CrossRef] [PubMed]
7. Uribe-Querol, E.; Rosales, C. Neutrophils Actively Contribute to Obesity-Associated Inflammation and Pathological Complications. *Cells* **2022**, *11*, 1883. [CrossRef] [PubMed]
8. Rizk, N.M.; Fadel, A.; Alshammari, W.; Younes, N.; Bashah, M. The Immunophenotyping Changes of Peripheral CD4+ T Lymphocytes and Inflammatory Markers of Class III Obesity Subjects after Laparoscopic Gastric Sleeve Surgery—A Follow-up Study. *J. Inflamm. Res.* **2021**, *14*, 1743. [CrossRef]
9. Krinninger, P.; Ensenauer, R.; Ehlers, K.; Rauh, K.; Stoll, J.; Krauss-Etschmann, S.; Hauner, H.; Laumen, H. Peripheral Monocytes of Obese Women Display Increased Chemokine Receptor Expression and Migration Capacity. *J. Clin. Endocrinol. Metab.* **2014**, *99*, 2500–2509. [CrossRef]
10. Park, C.S.; Shastri, N. The Role of T Cells in Obesity-Associated Inflammation and Metabolic Disease. *Immune Netw.* **2022**, *22*, e13. [CrossRef] [PubMed]
11. Sbierski-Kind, J.; Goldeck, D.; Buchmann, N.; Spranger, J.; Volk, H.D.; Steinhagen-Thiessen, E.; Pawelec, G.; Demuth, I.; Spira, D. T Cell Phenotypes Associated with Insulin Resistance: Results from the Berlin Aging Study II. *Immun. Ageing* **2020**, *17*, 40. [CrossRef] [PubMed]
12. Artemniak-Wojtowicz, D.; Kucharska, A.M.; Stelmaszczyk-Emmel, A.; Majcher, A.; Pyrżak, B. Changes of Peripheral Th17 Cells Subset in Overweight and Obese Children after Body Weight Reduction. *Front. Endocrinol.* **2022**, *13*, 917402. [CrossRef] [PubMed]
13. Bradley, D.; Smith, A.J.; Blaszczak, A.; Shantaram, D.; Bergin, S.M.; Jalilvand, A.; Wright, V.; Wyne, K.L.; Dewal, R.S.; Baer, L.A.; et al. Interferon Gamma Mediates the Reduction of Adipose Tissue Regulatory T Cells in Human Obesity. *Nat. Commun.* **2022**, *13*, 5606. [CrossRef] [PubMed]
14. Agabiti-Rosei, C.; Trapletti, V.; Piantoni, S.; Airo, P.; Tincani, A.; De Ciuceis, C.; Rossini, C.; Mittempergher, F.; Titi, A.; Portolani, N.; et al. Decreased Circulating T Regulatory Lymphocytes in Obese Patients Undergoing Bariatric Surgery. *PLoS ONE* **2018**, *13*, e0197178. [CrossRef] [PubMed]
15. Wijngaarden, L.H.; Taselaar, A.E.; Nuijten, F.; Harst, E.V.d.; Klaassen, R.A.; Kuijper, T.M.; Jongbloed, F.; Ambagtsheer, G.; Klepper, M.; IJzermans, J.N.M.; et al. T and B Cell Composition and Cytokine Producing Capacity before and after Bariatric Surgery. *Front. Immunol.* **2022**, *13*, 888278. [CrossRef] [PubMed]
16. Barbosa, P.; Melnyk, S.; Bennuri, S.C.; Delhey, L.; Reis, A.; Moura, G.R.; Børsheim, E.; Rose, S.; Carvalho, E. Redox Imbalance and Methylation Disturbances in Early Childhood Obesity. *Oxidative Med. Cell. Longev.* **2021**, *2021*, 2207125. [CrossRef] [PubMed]
17. Knopfholz, J.; Disserol, C.C.D.; Pierin, A.J.; Schirr, F.L.; Streisky, L.; Takito, L.L.; Massucheto Ledesma, P.; Faria-Neto, J.R.; Olandoski, M.; Da Cunha, C.L.P.; et al. Validation of the Friedewald Formula in Patients with Metabolic Syndrome. *Cholesterol* **2014**, *2014*, 261878. [CrossRef]
18. Romanski, P.A.; Farland, L.V.; Tsen, L.C.; Ginsburg, E.S.; Lewis, E.I. Effect of Class III and Class IV Obesity on Oocyte Retrieval Complications and Outcomes. *Fertil. Steril.* **2019**, *111*, 294–301.e1. [CrossRef]

19. Tavares, G.; Rosendo-Silva, D.; Simões, F.; Eickhoff, H.; Marques, D.; Sacramento, J.F.; Capucho, A.M.; Seiça, R.; Conde, S.V.; Matafome, P. Circulating Dopamine Is Regulated by Dietary Glucose and Controls Glucagon-like 1 Peptide Action in White Adipose Tissue. *Int. J. Mol. Sci.* **2023**, *24*, 2464. [CrossRef]
20. Laranjeira, P.; dos Santos, F.; Salvador, M.J.; Simões, I.N.; Cardoso, C.M.P.; Silva, B.M.; Henriques-Antunes, H.; Corte-Real, L.; Couceiro, S.; Monteiro, F.; et al. Umbilical-Cord-Derived Mesenchymal Stromal Cells Modulate 26 Out of 41 T Cell Subsets from Systemic Sclerosis Patients. *Biomedicines* **2023**, *11*, 1329. [CrossRef]
21. Laranjeira, P.; Rodrigues, T.; Silva, A.; Barbosa, P.; Reis, T.; Lopes, C.; Pinho, A.; Silva, S.; Silva, I.; Carvalho, A.; et al. A Single Dose of COVID-19 Vaccine Induces a Strong T Cell and B Cell Response in Healthcare Professionals Recovered from SARS-CoV-2 Infection. *Clin. Exp. Med.* **2022**, *23*, 529–537. [CrossRef]
22. Laranjeira, P.; Rodrigues, R.; Carvalheiro, T.; Constanço, C.; Vitória, H.; Matarraz, S.; Trindade, H.; Órfão, A.; Paiva, A. Expression of CD44 and CD35 during Normal and Myelodysplastic Erythropoiesis. *Leuk. Res.* **2015**, *39*, 361–370. [CrossRef] [PubMed]
23. Hothorn, T.; Hornik, K.; van de Wiel, M.A.; Zeileis, A. A Lego System for Conditional Inference. *Am. Stat.* **2006**, *60*, 257–263. [CrossRef]
24. Hothorn, T.; Hornik, K.; van de Wiel, M.A.; Zeileis, A. Implementing a Class of Permutation Tests: The Coin Package. *J. Stat. Softw.* **2008**, *28*, 1–23. [CrossRef]
25. Kassambara, A. Rstatix: Pipe-Friendly Framework for Basic Statistical Tests; R Package Version 0.6.0. Available online: https://cran.r-project.org/web/packages/rstatix/index.html (accessed on 16 July 2020).
26. R Core Team. R: A Language and Environment for Statistical Computing. 2019. Available online: https://search.r-project.org/CRAN/refmans/bnpa/html/create.cluster.html (accessed on 16 July 2020).
27. Harrell, F.E.; Dupont, C. Hmisc: Harrell Miscellaneous. R Package Version 4.4-2. Available online: https://CRAN.R-project.org/package=Hmisc (accessed on 16 July 2020).
28. Wickham, H. *Ggplot2*; Springer: New York, NY, 2009; ISBN 978-0-387-98140-6.
29. RStudio Team, A. RStudio: Integrrated Development Environment for R 2021. Available online: https://www.r-project.org/conferences/useR-2011/abstracts/180111-allairejj.pdf (accessed on 16 July 2020).
30. Barbosa, P.; Pinho, A.; Lázaro, A.; Rosendo-Silva, D.; Paula, D.; Campos, J.; Tralhão, J.G.; Pereira, M.J.; Paiva, A.; Laranjeira, P.; et al. CD8+ Treg Cells Play a Role in the Obesity-Associated Insulin Resistance. *Life Sci.* **2024**, *336*, 122306. [CrossRef]
31. Romo, M.; López-Vicario, C.; Pérez-Romero, N.; Casulleras, M.; Martínez-Puchol, A.I.; Sánchez, B.; Flores-Costa, R.; Alcaraz-Quiles, J.; Duran-Güell, M.; Ibarzábal, A.; et al. Small Fragments of Hyaluronan Are Increased in Individuals with Obesity and Contribute to Low-Grade Inflammation through TLR-Mediated Activation of Innate Immune Cells. *Int. J. Obes.* **2022**, *46*, 1960–1969. [CrossRef]
32. Talukdar, S.; Oh, D.Y.; Bandyopadhyay, G.; Li, D.; Xu, J.; McNelis, J.; Lu, M.; Li, P.; Yan, Q.; Zhu, Y.; et al. Neutrophils Mediate Insulin Resistance in Mice Fed a High-Fat Diet through Secreted Elastase. *Nat. Med.* **2012**, *18*, 1407–1412. [CrossRef] [PubMed]
33. Yoshimura, A.; Ohnishi, S.; Orito, C.; Kawahara, Y.; Takasaki, H.; Takeda, H.; Sakamoto, N.; Hashino, S. Association of Peripheral Total and Differential Leukocyte Counts with Obesity-Related Complications in Young Adults. *Obes. Facts* **2015**, *8*, 1–16. [CrossRef]
34. Łuczyński, W.; Grubczak, K.; Moniuszko, M.; Głowińska-Olszewska, B.; Bossowski, A. Elevated Levels of Th17 Cells in Children with Central Obesity. *Scand. J. Clin. Lab. Investig.* **2015**, *75*, 595–601. [CrossRef]
35. Tsanaktsi, A.; Solomou, E.E.; Liossis, S.N.C. Th1/17 Cells, a Subset of Th17 Cells, Are Expanded in Patients with Active Systemic Lupus Erythematosus. *Clin. Immunol.* **2018**, *195*, 101–106. [CrossRef]
36. Laba, S.; Mallett, G.; Amarnath, S. The Depths of PD-1 Function within the Tumor Microenvironment beyond CD8+ T Cells. *Semin. Cancer Biol.* **2022**, *86*, 1045–1055. [CrossRef]
37. He, X.; Xu, C. PD-1: A Driver or Passenger of T Cell Exhaustion? *Mol. Cell* **2020**, *77*, 930–931. [CrossRef]
38. Zhang, R.; Miao, J.; Zhang, K.; Zhang, B.; Luo, X.; Sun, H.; Zheng, Z.; Zhu, P. Th1-like Treg Cells Are Increased but Deficient in Function in Rheumatoid Arthritis. *Front. Immunol.* **2022**, *13*, 863753. [CrossRef]
39. Donninelli, G.; Cornò, M.D.; Pierdominici, M.; Scazzocchio, B.; Varì, R.; Varano, B.; Pacella, I.; Piconese, S.; Barnaba, V.; D'Archivio, M.; et al. Distinct Blood and Visceral Adipose Tissue Regulatory T Cell and Innate Lymphocyte Profiles Characterize Obesity and Colorectal Cancer. *Front. Immunol.* **2017**, *8*, 643. [CrossRef] [PubMed]
40. Gonzalez-Espinosa, L.O.; Montiel-Cervantes, L.A.; Guerra-Márquez, A.; Peñaflor-Juárez, K.; Reyes-Maldonado, E.; Vela-Ojeda, J. Maternal Obesity Associated with Increase in Natural Killer T Cells and CD8+ Regulatory T Cells in Cord Blood Units. *Transfusion* **2016**, *56*, 1075–1081. [CrossRef] [PubMed]
41. Gianchecchi, E.; Fierabracci, A. Inhibitory Receptors and Pathways of Lymphocytes: The Role of PD-1 in Treg Development and Their Involvement in Autoimmunity Onset and Cancer Progression. *Front. Immunol.* **2018**, *9*, 2374. [CrossRef] [PubMed]
42. Banerjee, H.; Nieves-Rosado, H.; Kulkarni, A.; Murter, B.; McGrath, K.V.; Chandran, U.R.; Chang, A.; Szymczak-Workman, A.L.; Vujanovic, L.; Delgoffe, G.M.; et al. Expression of Tim-3 Drives Phenotypic and Functional Changes in Treg Cells in Secondary Lymphoid Organs and the Tumor Microenvironment. *Cell Rep.* **2021**, *36*, 109699. [CrossRef] [PubMed]
43. Kartika, R.; Wibowo, H. Impaired Function of Regulatory T Cells in Type 2 Diabetes Mellitus. *Mol. Cell. Biomed. Sci.* **2020**, *4*, 1–9. [CrossRef]
44. Guo, S.; Yu, X.; Wang, L.; Jing, J.; Sun, Y.; Li, N.; Kuang, J.; Zhao, D.; Yu, X.; Yang, J.; et al. The Frequency of Tim-3 on Circulating Tfh Cells Was Increased in Type 2 Diabetes Mellitus Patients. *Eur. J. Inflamm.* **2020**, *18*, 1–8. [CrossRef]

45. Sun, L.; Zou, S.; Ding, S.; Du, X.; Shen, Y.; Liu, C.; Shi, B.; Zhang, X. Circulating T Cells Exhibit Different TIM3/Galectin-9 Expression in Patients with Obesity and Obesity-Related Diabetes. *J. Diabetes Res.* **2020**, *2020*, 2583257. [CrossRef] [PubMed]
46. Marcucci, F.; Romeo, E.; Caserta, C.A.; Rumio, C.; Lefoulon, F. Context-Dependent Pharmacological Effects of Metformin on the Immune System. *Trends Pharmacol. Sci.* **2020**, *41*, 162–171. [CrossRef] [PubMed]
47. Zeiser, R. Immune Modulatory Effects of Statins. *Immunology* **2018**, *154*, 69–75. [CrossRef] [PubMed]

biomolecules

MDPI

Article

Integrated Analysis of Gut Microbiome and Adipose Transcriptome Reveals Beneficial Effects of Resistant Dextrin from Wheat Starch on Insulin Resistance in Kunming Mice

Xinyang Chen [1,2], Yinchen Hou [1,3], Aimei Liao [1,2], Long Pan [1,2], Shengru Yang [3], Yingying Liu [1,2], Jingjing Wang [1,2], Yingchun Xue [1,2], Mingyi Zhang [1,2], Zhitong Zhu [1,2] and Jihong Huang [1,2,4,5,*]

[1] Food Laboratory of Zhongyuan, Luohe 462300, China; 2021930593@stu.haut.edu.cn (X.C.); 80891@hnuahe.edu.cn (Y.H.); aimeiliao@haut.edu.cn (A.L.); panlong@haut.edu.cn (L.P.); 2021920157@stu.haut.edu.cn (Y.L.); 2021930602@stu.haut.edu.cn (J.W.); 2021930597@stu.haut.edu.cn (Y.X.); 2022930656@stu.haut.edu.cn (M.Z.); zzt@stu.haut.edu.cn (Z.Z.)

[2] School of Biological Engineering, Henan University of Technology, Zhengzhou 450001, China

[3] College of Food and Biological Engineering, Henan University of Animal Husbandry and Economy, Zhengzhou 450046, China; 81692@hnuahe.edu.cn

[4] State Key Laboratory of Crop Stress Adaptation and Improvement, College of Agriculture, Henan University, Kaifeng 475004, China

[5] School of Food and Pharmacy, Xuchang University, Xuchang 461000, China

* Correspondence: huangjihong@haut.edu.cn

Abstract: Systemic chronic inflammation is recognized as a significant contributor to the development of obesity-related insulin resistance. Previous studies have revealed the physiological benefits of resistant dextrin (RD), including obesity reduction, lower fasting glucose levels, and anti-inflammation. The present study investigated the effects of RD intervention on insulin resistance (IR) in Kunming mice, expounding the mechanisms through the gut microbiome and transcriptome of white adipose. In this eight-week study, we investigated changes in tissue weight, glucose–lipid metabolism levels, serum inflammation levels, and lesions of epididymal white adipose tissue (eWAT) evaluated via Hematoxylin and Eosin (H&E) staining. Moreover, we analyzed the gut microbiota composition and transcriptome of eWAT to assess the potential protective effects of RD intervention. Compared with a high-fat, high-sugar diet (HFHSD) group, the RD intervention significantly enhanced glucose homeostasis (e.g., AUC-OGTT, HOMA-IR, $p < 0.001$), and reduced lipid metabolism (e.g., TG, LDL-C, $p < 0.001$) and serum inflammation levels (e.g., IL-1β, IL-6, $p < 0.001$). The RD intervention also led to changes in the gut microbiota composition, with an increase in the abundance of probiotics (e.g., *Parabacteroides*, *Faecalibaculum*, and *Muribaculum*, $p < 0.05$) and a decrease in harmful bacteria (*Colidextribacter*, $p < 0.05$). Moreover, the RD intervention had a noticeable effect on the gene transcription profile of eWAT, and KEGG enrichment analysis revealed that differential genes were enriched in PI3K/AKT, AMPK, in glucose-lipid metabolism, and in the regulation of lipolysis in adipocytes signaling pathways. The findings demonstrated that RD not only ameliorated IR, but also remodeled the gut microbiota and modified the transcriptome profile of eWAT.

Keywords: resistant dextrin; insulin resistance; inflammation; gut microbiota; transcriptome

1. Introduction

Insulin resistance (IR) plays a crucial role in metabolic syndrome (MetS) [1]. It typically manifests alongside hyperglycemia, hyperlipidemia, hypertension, and abdominal obesity, serving as a foundational factor in the pathogenesis of various metabolic diseases, including obesity, nonalcoholic fatty liver disease, type 2 diabetes mellitus, atherosclerosis and Alzheimer's disease [2–5]. Consequently, alleviating IR has emerged as a pivotal strategy for early intervention in MetS.

The three primary target organs for insulin metabolism are the liver, adipose tissue, and muscle [6]. White adipose tissue, a significant regulator of energy metabolism in the body, is particularly noteworthy. Inflammation within white adipose tissue is considered an initiating factor in the development of IR [7,8]. In adipose tissue, the main actions of insulin are to inhibit lipolysis by enhancing lipid synthesis and maintaining free fatty acid levels [9]. Furthermore, insulin facilitates glucose transport in adipose tissue through the PI3K/AKT/GLUT-4 pathway [10]. Previous studies have demonstrated that the byproducts of metabolism generated by gut microbiota can modulate adipose-tissue energy metabolism and insulin sensitivity [11,12].

The gut microbiota, a delicate ecosystem, plays a pivotal role in both health and disease [13]. Recent studies have shown a significant alteration in the composition of the gut microbiota in insulin-resistant patients, implicating the gut microbiota in carbohydrate and lipid metabolism as well as inflammatory responses, thereby mediating the development of metabolic diseases [3,14–16]. Moreover, numerous studies underscore the influence of diet on gut microbiota dysbiosis [17–19], and this behavior can be sustained. Therefore, it would be effective to alleviate IR through gut microbiota homeostasis. Dietary fiber, resistant to digestion and absorption in the small intestine, undergoes fermentation by gut microbiota, eliciting a specific response in the gut microbiota [20]. Meanwhile, dietary fiber exerts a profound influence on the concentration of bioactive compounds, such as short-chain fatty acids, bile acids, and branched-chain amino acids, originating from bacterial metabolites, which facilitate metabolic interactions between host and microbiota [21,22].

Metformin is widely recognized as a hypoglycemic agent, known for its ability to reduce blood glucose levels by inhibiting hepatic glucose production [23]. Recent studies have presented compelling evidence that metformin exerts notable effects on the gut microbiota, influencing both the small and large intestines [24,25]. In a rat model subjected to a high-fat diet, metformin demonstrated its capacity to modulate glucose-sensing pathways in the upper small intestine. This modulation was achieved through alterations in the gut microbiota, showing the intricate relationship between metformin, the gut microbiota, and the regulation of glucose-sensing pathways [26]. Furthermore, in a mouse model concentrating on colorectal cancer, metformin exhibited its ability to influence the relative abundance of genera associated with colorectal cancer. In particular, it played a role in inhibiting the formation of colorectal tumors induced by *Clostridium nucleatum* [27]. This underscores the fact that metformin modulates the microbiota and plays multiple roles. However, administering metformin in conjunction with a high-fat, high-sugar diet or at the onset of abnormal glucose metabolism is suboptimal [28], as it may result in gastrointestinal distress. Therefore, investigating functional foods as an alternative approach would be a more prudent and sensible course of action.

RD is generally produced by starch dextrination using specific processes, including thermal-acid, dry heat, and microwave methods [29]. RD, a soluble dietary fiber, exhibits diverse nutritional functions, including the potential to improve obesity, type 2 diabetes, sleep disorders in obese patients, non-alcoholic fatty liver disease, and polycystic ovary syndrome [30–34]. Moreover, RDs act mainly by modifying the composition of the gut microbiota, including the proliferation of probiotics and the inhibition of harmful bacteria [32,35,36]. However, previous studies reveal that the influence of RD on the gut microbiota is inconsistent, and possibly correlated with factors such as raw material source, molecular weight and processing methods [19,35,37–39]. In addition, there are fewer studies on whether RD affects the transcriptome profile of adipose tissue, and the effects of RD from wheat starch on the gut microbiota and its potential involvement in improving IR remain unclear.

In this study, our objective was to investigate glucose–lipid metabolic parameters and gut microbial changes on HFHSD Kunming mice following the intervention of RD from wheat starch, and to explore its potential effects on eWAT. Specifically, we employed 16S rRNA sequencing of cecum contents to discern the effects of RD intervention on gut microbial communities. Additionally, we conducted RNA sequencing analysis of eWAT to identify significantly enriched metabolic pathways and the underlying molecular mechanisms. This study provides new insights into the potential of RD intervention for improving IR.

2. Materials and Methods

2.1. Materials

Wheat starch was purchased from Liangrun Food Co., Ltd. (Xinxiang, China). Thermostable α-amylase (40,000 U/g) and amyloglucosidase (100,000 U/g) were purchased from Macklin Biochemical Technology Co., Ltd. (Shanghai, China). Biochemical kits for TC, TG, LDL-C, HDL-C were purchased from Beijing Solarbio Science & Technology Co., Ltd. (Beijing, China). ELISA kits for INS, IL-1β, and IL-6 were procured from Meimian Industrial Co., Ltd. (Yancheng, China). Metformin Hydrochloride Tablets were obtained from Jingfeng Pharmaceutical Co., Ltd. (Beijing, China). Adipose tissue Fix Solution was obtained from Servicebio Biochemical Technology Co., Ltd. (Wuhan, China).

2.2. Animal Experimental Design

Kunming mice (male; aged 4 weeks; classed specific-pathogen free) were obtained from Huaxing Experimental Animal Co., Ltd. (Zhengzhou, China, SCXY(Yu) 2019-0002). The mice were raised under standard specific conditions, with a room temperature of 22–26 °C, relative humidity ranging between 40 and 60%, a 12 h artificial light cycle, and proper ventilation. After one week of acclimatization, the mice were divided into four groups based on body weight (Figure 1). The average body weights did not differ significantly between four groups ($n = 10$ per group): normal diet (ND), high-fat, high-sugar diet group (HFHSD), HFHSD with RD intervention group (HFHSD+RD), and HFHSD with metformin control group (HFHSD+MC). The basal feed (mouse growth and breeding feed) was purchased from Beijing Keao Xieli Feed Co., Ltd. The feed formula for the basic diet (ND) is corn, soybean meal, fish meal, flour, yeast meal, vegetable oil, salt, multivitamins and minerals. The high-fat and high-sugar diet, consisting of 10% lard, 20% sucrose, 2.5% cholesterol, 1% sodium cholate, and 66.5% basic diet, was prepared by Huaxing Experimental Animal Farm Co., Ltd. (Zhengzhou, China). RD, administered via gavage at a dose of 6.2 g/kg/day, is equivalent to the recommended adult (60 kg) intake of about 0.5 g/kg/day. Similarly, metformin was administered by gavage at a dose of 200 mg/kg/day, translating to an equivalent adult dose of 16.2 mg/kg/day. During the intervention, the mice had unrestricted access to food and water, and fasting blood glucose levels were assessed every two weeks. The intervention lasted until the 8th week. All mice were fasted overnight for 12 h. All mice were anesthetized with isoflurane, sacrificed, and blood and tissue samples were collected. Blood samples were incubated at room temperature (25–27 °C) for 1 h, and then centrifuged at 4 °C and 2500 rpm for 25 min to separate the serum. Simultaneously, the remaining samples were rapidly frozen in liquid nitrogen and stored in a freezer at −80 °C. Additionally, the general health and welfare of the mice were monitored throughout the entire study duration. All animal experimentation protocols were conducted in accordance with the guidelines of the EU Directive (2010/63/EU) on the treatment and usage of laboratory animals. All experiments were approved by the Henan University of Technology Ethics Committee.

2.3. Preparation of Resitant Dextrin (RD)

The combination of thermal acidolysis treatment and enzymatic methods was used for the preparation of RD [40]. Wheat starch was suspended with HCl solution of 0.075 $mol^{-1} \cdot L$ (1:1.5, *w/v*). After stirring for 30 min, the precipitate was dried in an oven at 40 °C for 24 h. The dried mixture was heated (180 °C, 90 min) to obtain crude dextrin. The crude dextrin was dispersed (1:3, *w/v*), with pH adjusting to 6.0 using 0.1 M NaOH. Subsequently, thermostable α-amylase (1%, *w/w*) was added and stirred at 95 °C for 120 min, then with pH adjusting to 6.0 utilizing 0.1 M HCl, amyloglucosidase (0.5%, *v/w*) was added and stirred at 60 °C for 60 min. The enzymes were then inactivated by heating the mixture at

100 °C for 10 min. The mixture was added to 4-times the volume of liquid ethanol (95%). The sample was the left, and was dried to obtain the RD [29]. The total fiber content of RD was determined using GB/T 22224-2008 (2), which revealed a dietary fiber content of 82.69%.

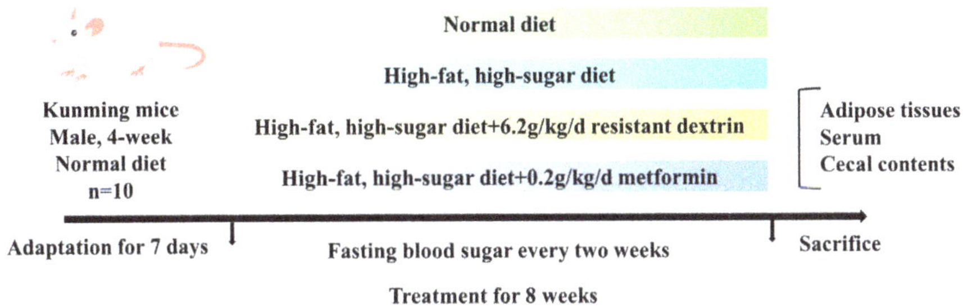

Figure 1. The animal experimental design.

2.4. Blood Glucose and HOMA-IR Tests

Fasting blood glucose (FBG) was measured after an overnight fast lasting 12 h, using a glucometer (Roche, Shanghai, China). Fasting insulin levels (FINS) were determined employing Elisa kits (Meimian, Yancheng, China). For the oral glucose tolerance test (OGTT), the mice, after fasting for 16 h, received a glucose load (20%, m/m) of 2 g/kg body weight. Blood glucose levels were assessed at specific time periods (0, 30, 60, 90, and 120 min). The area under the curve (AUC) of OGTT and HOMA-IR index were calculated based on previous methodologies [41], with the following formula: HOMA-IR = FINS $\times$ FBG/22.5.

2.5. Histology Examination

Briefly, the eWAT were fixed and then embedded in paraffin. Subsequently, the tissue samples were transversely sectioned into slices measuring 4–5 μm in thickness. These slices were stained with H&E to evaluate the pathological status of the eWAT. Finally, the images were analyzed with the CaseViewer 2.4 (3DHISTECH, Budapest, Hungary).

2.6. RNA-Seq Analysis

Total RNA extracted from eWAT was assessed for purity and integrity using both the Nanodrop 2000 (Thermo Fisher, Waltham, MA, USA) and the Agilent 2100 (Agilent, Santa Clara, CA, USA). We chose the number of samples per group based on the RNA QC results and chose to retain 2 samples per group, following comprehensive analysis. After library construction, preliminary quantification and insert size calculation of the libraries were performed. Subsequently, RNA was sequenced via the Illumina Novaseq 6000 (Illumina, San Diego, CA, USA). The low quality reads and adaptor sequences were trimmed with Trimmomatic. The analysis pipeline involved aligning paired-end clean reads to the reference genome using HISAT2 v2.1.0 and quantifying read counts per gene with featureCounts v2.0.3. Transcripts per million (TPM) were then calculated for each gene, based on its length. Differential expression analysis was conducted using the edgeR software (v3.40.2), with p-values adjusted using the Benjamini & Hochberg method. Significantly differentially expressed genes were identified based on corrected p-values and absolute $|\log_2 FC|$ thresholds ($|\log_2 FC| \geq 2$, FDR < 0.01). Subsequently, clusterProfiler R software (v4.6.2) was utilized for statistical enrichment analysis of these genes in KEGG pathways. This service was conducted by Chi-Biotech Co., Ltd. (Shenzhen, China).

2.7. Gut Microbiome 16S rRNA Sequencing Analysis

Total genomic DNA of the gut microbiome was extracted from cecum content samples using previously described methods [24]. The hypervariable region V3-V4 of the bacterial

16S rRNA gene was amplified with primer pairs 338F (5′-ACTCCTACGGGAGGCAGCAG-3′) and 806R (5′-GGACTACHVGGGTWTCTAAT-3′), employing an ABI GeneAmp® 9700 PCR thermocycler (ABI, Foster, CA, USA). The PCR amplification of 16S rRNA gene was performed as follows: initial denaturation at 95 °C for 3 min, followed by 30 cycles of denaturing at 95 °C for 30 s, annealing at 55 °C for 30 s and extension at 72 °C for 45 s, single extension at 72 °C for 10 min, and end at 4 °C. The PCR product was extracted from 2% agarose gel, purified with the AxyPrep DNA Gel Extraction Kit (Axygen Biosci-ences, Union City, CA, USA), following the manufacturer's instructions, and quantified using the Quantus™ Fluorometer (Promega, Madison, WI, USA). Purified amplicons were pooled equimolarly and subjected to paired-end sequencing on an Illumina MiSeq PE300 platform (Illumina, San Diego, CA, USA). Operational taxonomic units (OTUs) with 97% similarity cutoff were clustered using UPARSE version 7.1. This service was performed by Majorbio Co., Ltd. (Shanghai, China).

2.8. Statistical Analysis

All experimental results were presented as mean ± SDs. Multiple comparisons were conducted using one-way ANOVA with Tukey's *post hoc* test. GraphPad Prism 7.0 (GraphPad Inc., San Diego, CA, USA) was used for the statistical analyses. A p-value < 0.05 was regarded as statistically significant.

3. Results and Discussion

3.1. Resistant Dextrin Improves Metabolic Disturbance in HFHSD Kunming Mice

To investigate the effects of RD intervention on mice with glucose–lipid metabolism disorders induced by HFHSD, we examined various parameters including eWAT weight, glucose–lipid metabolism indicators, and serum inflammatory factor levels. By the eighth week of treatment, FBG, HOMA-IR index, AUC-OGTT, TC, and LDL-C in HFHSD mice were significantly elevated compared to those in ND mice ($p < 0.01$, one-way ANOVA with Tukey's *post hoc* test). These data confirmed the successful establishment of the IR model in Kunming mice in this study. In contrast, RD intervention led to a significant reduction in FBG, HOMA-IR index, AUC-OGTT, TG, and LDL-C in HFHSD-fed mice ($p < 0.001$, one-way ANOVA with Tukey's *post hoc* test, Figure 2A,C,D,F,H). The metformin intervention exhibited a similar effect. Changes in body weight (Figure S1), diet intake (Figure S2), FBG of OGTT-0min (Figure S3), and FINS (Figure S4) of Kunming mice during the intervention are shown in the Supplementary File.

Compared to the HFHSD group, RD or metformin interventions resulted in a significant reduction in eWAT weight ($p < 0.01$, one-way ANOVA with Tukey's *post hoc* test, Figure 3A). And the levels of serum IL-1β and IL-6 were elevated in HFHSD mice compared to those in ND, whereas they were significantly reduced in the HFHSD+RD and HFHSD+MC group ($p < 0.001$, one-way ANOVA with Tukey's *post hoc* test, Figure 3C,D). H&E staining of eWAT showed cell enlargement in HFHSD mice, but RD or metformin intervention alleviated these effects (Figure 3B). The size of adipocytes could affect IR, and metabolic disorders have been associated with larger adipocytes [42]. This implies that RD possesses the capability to impact adipocyte size and potentially alleviate insulin resistance through specific pathways or actions.

3.2. Resistant Dextrin Modifes Transcriptome of EWAT in HFHSD Kunming Mice

Typically, white adipose tissue (WAT) is considered detrimental to health because of the presence of white adipocytes containing substantial lipid droplets. These cells play a role in influencing insulin sensitivity, lipid metabolism, and low-grade inflammation, ultimately contributing to the development of metabolic disorders [43]. However, the influence of RD on white adipose tissue remains uncertain. RNA-seq analysis of eWAT was conducted to further investigate the potential mechanisms underlying the improvement of IR in HFHSD-induced Kunming mice by RD intervention. Principal component analysis (PCA) of the transcriptome data revealed distinct clusters between the ND, HFHSD and HFHSD+RD groups (Figure 4A), indicating that mice were categorized based on dietary treatment and showed diet-specific transcriptional responses in all mice. In the comparison between ND and HFHSD groups, there were differentially expressed genes (DEGs) ($|\log_2 FC| \geq 2$, FDR < 0.01) (n = 1758), whereas HFHSD+RD versus HFHSD showed differences in more genes (n = 5364) (Figure 4B–D).

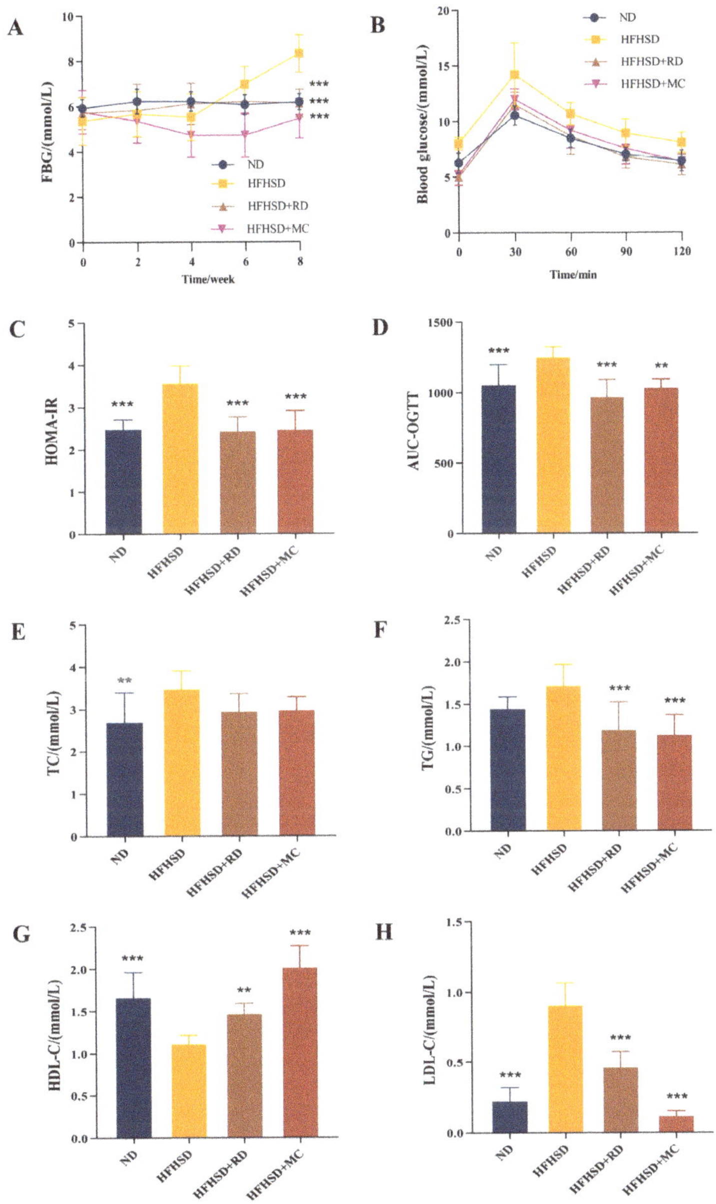

Figure 2. RD ameliorates glucose–lipid metabolism disorders in HFHSD Kunming mice (n = 10). (**A**) FBG every two weeks. (**B**) OGTT at the end of intervention. (**C**) HOMA-IR at the end of intervention. (**D**) Area under the curve (AUC) of OGTT. (**E**–**H**) Serum lipid profiles. ** $p < 0.01$, and *** $p < 0.001$ vs. HFHSD group.

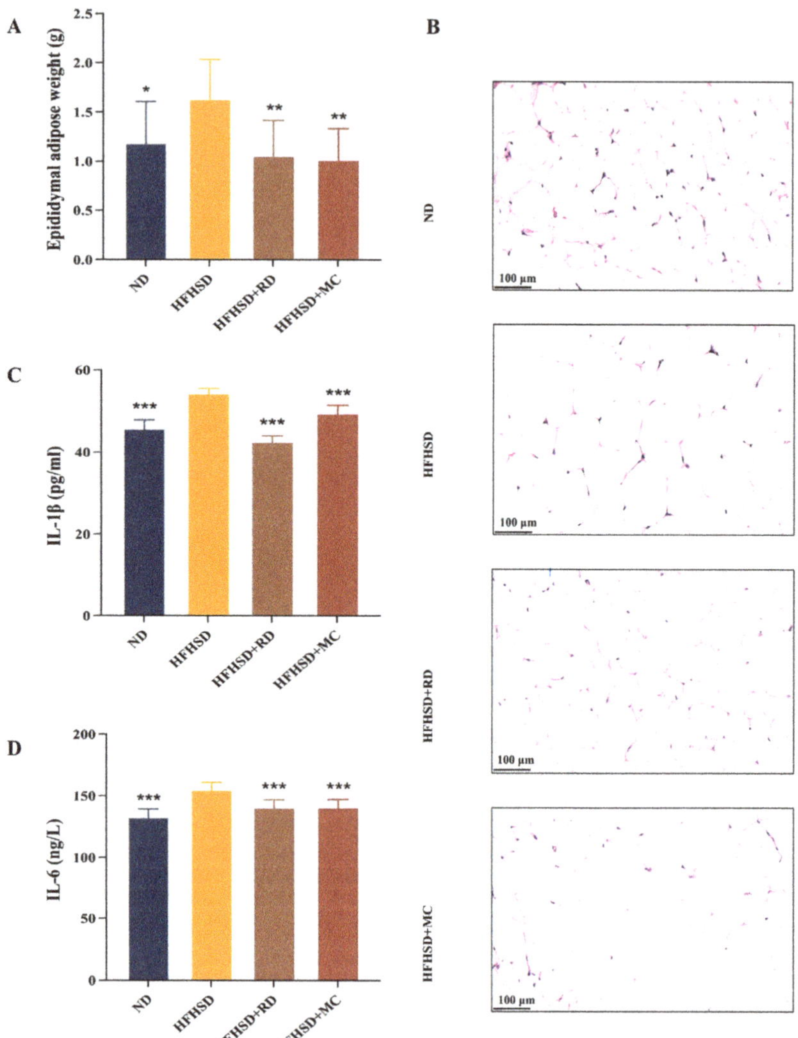

Figure 3. RD ameliorates eWAT weight, cellular hypertrophy and serum inflammation levels in HFHSD Kunming mice (n = 10). (**A**) The eWAT weight. (**B**) H&E staining. (**C**,**D**) Serum IL-1β and IL-6 levels. * $p < 0.05$, ** $p < 0.01$ and *** $p < 0.001$ vs. HFHSD group.

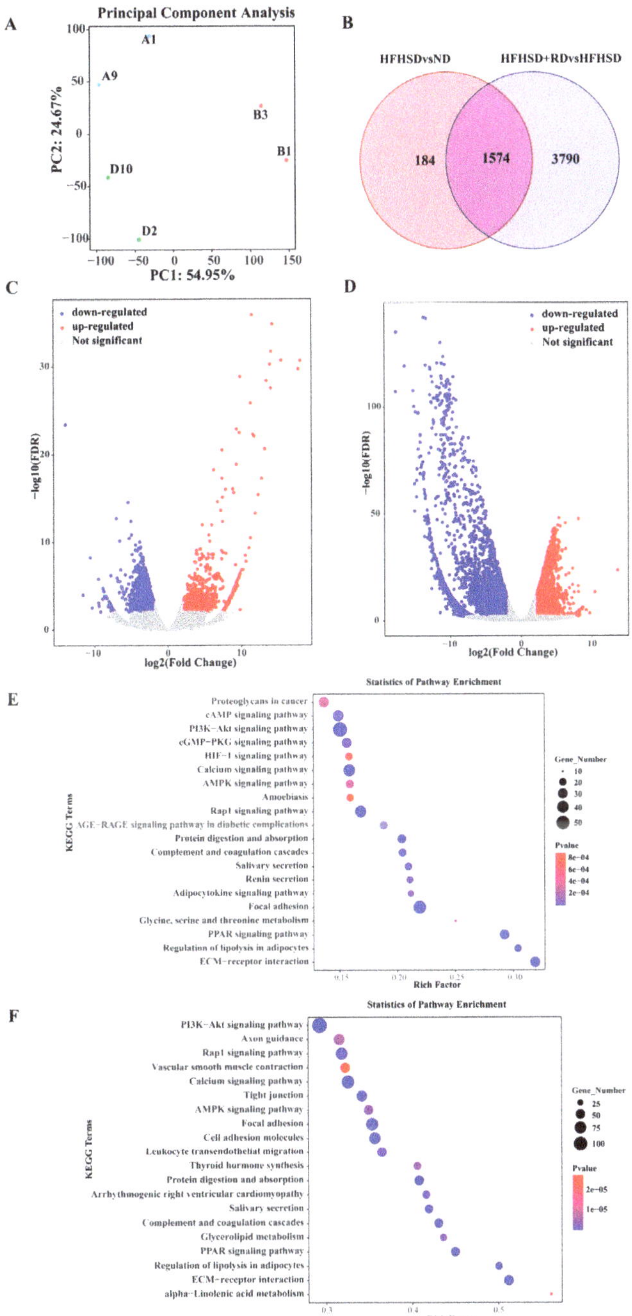

Figure 4. RD intervention modifies the eWAT transcriptome profiles in HFHSD Kunming mice ($n = 2$). (**A**) Principal component analysis, ND group: A1 and A9; HFHSD group: B1 and B3; HFHSD+RD group: D2 and D10. (**B**) Venn diagram showing the total number of DEGs in the HFHSD vs. ND and HFHSD+RD vs. HFHSD comparisons. (**C**) Volcano plot showing the significant DEGs for the HFHSD and ND groups. (**D**) Volcano plot showing the significant DEGs for HFHSD+RD and HFHSD groups.

(**E**) KEGG enrichment analysis showing the top 20 signaling pathways affected by DEGs between HFHSD and ND groups (FDR < 0.01 $|\log_2FC| \geq 2$). (**F**) KEGG enrichment analysis showing the top 20 signaling pathways affected by DEGs between HFHSD+RD and HFHSD groups (FDR < 0.01 $|\log_2FC| \geq 2$).

Through KEGG signaling pathway enrichment analysis, we observed a significant enrichment of the PI3K/AKT and AGE-RAGE signaling pathway in the diabetic complications, indicating substantial differences in gene expression associated with glucose metabolism between the ND and HFHSD groups (Figure 4E). We identified enrichment of DEGs in metabolic pathways including PI3K/AKT, PPAR, AMPK, glycerolipid metabolism and regulation of lipolysis in adipocytes when comparing HFHSD+RD with HFHSD groups (Figure 4F). The PI3K/AKT signaling pathway was the most significantly enriched, and it was found that the genes on the PI3K/AKT signaling pathway were significantly different between the HFHSD and HFHSD+RD group ($|\log_2FC| \geq 2$, FDR < 0.01). Moreover, the genes were up-regulated following the RD intervention (Table 1), suggesting an improvement in IR and maintenance of the normal level of glucose–lipid metabolism in the body. The gene expression of the insulin metabolism (PI3K/AKT) signaling pathway between the HFHSD+RD and the ND group is described in the Supplementary Material (Table S1). Previous studies have reported that the PI3K/AKT signaling pathway was important for insulin in maintaining glucose–lipid metabolic homeostasis [44]. And when IR occurred in adipose tissue, the release of numerous free fatty acids into the circulation aggravated dyslipidemia and IR, mainly due to the blockade of insulin metabolism signaling pathways [45].

Table 1. Gene expression of insulin metabolism (PI3K/AKT) signaling pathway.

Symbol	logFC	FDR	HFHSD	HFHSD+RD	Phenotype
Pik3r1	2.43	1.58×10^{-10}	13.55	101.23	Disruptions can lead to impaired insulin signaling.
Pik3cb	2.26	2.34×10^{-12}	6.48	42.90	Impaired PI3K activity may affect downstream signaling.
Pik3r6	2.20	2.49×10^{-5}	1.33	8.29	Plays a role in insulin signaling and may impact glucose metabolism.
Akt2	2.69	1.42×10^{-18}	28.54	251.00	Akt2 dysregulation is associated with insulin resistance.
Irs1	4.54	5.80×10^{-20}	0.54	17.00	Impaired IRS1 can lead to insulin resistance.
Igf1	4.35	1.14×10^{-39}	3.97	111.66	Deficiency levels can impact metabolism.
Slc2a4	3.78	5.19×10^{-34}	11.18	207.59	Dysregulation can lead to impaired glucose uptake and insulin resistance.
Foxo1	2.17	1.16×10^{-13}	4.57	28.26	Activation can enhance energy balance.

Meanwhile, another data analysis revealed a significant upregulation of genes associated with lipid metabolism, namely, *CIDEA*, *OTOP1*, *CMKLR1*, *C/EBPα*, *PPARγ* and *ADIPOQ* between the HFHSD+RD group and the HFHSD group ($|\log_2FC| \geq 2$, FDR < 0.01). These gene expression levels are shown in the Supplementary Material (Table S2). The activation of *OTOP1* is believed to yield metabolic benefits for obesity, contributing to the maintenance of adipose immune homeostasis in obese patients [46]. *CIDEA1*, identified as a key gene for white adipose tissue "browning" [47], suggests that RD activates pathways promoting "browning", potentially improving IR in adipose tissue. Regarding *CMKLR1*, its close association with inflammatory responses is documented. However, these findings were somewhat counterintuitive. Some studies suggest that *CMKLR1* deficiency may exacerbate glucose homeostasis and IR [48,49]. Conversely, other studies indicate that *CMKLR1* overexpression in adipose tissue significantly ameliorated liver histopathological changes in non-alcoholic steatohepatitis in rats [50]. These findings suggest that adipose tissue, with its distinct and specific role among tissues, requires investigation of the enigmatic role of eWAT in IR, especially after the RD intervention. The Gene Set Enrichment Analysis (GSEA) among the ND, HFHSD, and HFHSD+RD groups is presented in the Supplementary File (Figures S5 and S6). Additionally, a heat map depicting the gene expression on the enriched pathway is included.

3.3. Resistant Dextrin Alters Gut Microbiota Composition in HFHSD Kunming Mice

To investigate the effects of RD on the gut microbiota, we conducted the 16s rRNA sequencing analysis of the cecal contents. The data showed 782 common OTUs among the four groups (Figure 5A). Principal coordinate analyses (PCoA) of the weighted unifrac for these mice demonstrated good dispersion at the OTU levels, based on the within-group difference test ANOSIM (Figure 5B). There was no significant difference in the alpha diversity indices for OTU levels in the gut microbiota between ND, HFHSD, and HFHSD+RD. However, the intervention with metformin significantly reduced the alpha diversity of the gut microbiota. Detailed data can be found in the Supplementary File (Figure S7). At the phylum level, the HFHSD+RD group showed higher relative abundances of *Bacteroidota*, *Proteobacteria*, and *unclassified_k_norank_d_Bacteria* compared to the HFHSD group. Conversely, *Desulfobacterota*, *Actinobacteriota*, *Patescibacteria*, *Verrucomicrobiota*, and *Deferribacterota* were decreased in the HFHSD+RD group (Figure 5C). This is further illustrated by circos plots of phylum-level flora differences between the four groups, and this is shown in the Supplementary Files (Figure S8). Interestingly, *Chloroflexi* was exclusively identified in the HFHSD+RD group, but absent in the other three groups, and it was considered a potentially anti-inflammatory bacterium [51]. At the genus level, the relative abundances of *Bacteroides*, *Faecalibaculum*, *Parabacteroides*, *norank_f_Ruminococcaceae*, *GCA-900066575*, *Alloprevotella*, *Muribaculum*, and *Parasutterella* were significantly increased in the HFHSD+RD group compared to the HFHSD group (Wilcoxon rank-sum test, $p < 0.05$, Figure 5D). RD intervention led to a reduction in the relative abundances of *unclassified_f_Lachnospiraceae*, *norank_f_Lachnospiraceae*, *Enterorhabdus*, *Candidatus_Saccharimonas*, and *Colidextribacter* (Wilcoxon rank-sum test, $p < 0.05$). Linear discriminant analysis effect size (LEfSe) was used to evaluate statistical significance and biologically relevant differences in the gut microbiota composition among the three groups of mice (Figure 5E). At the genus level, the LEfSe analyses revealed that *Bacteroides*, *Parabacteroides*, *Alistipes*, *GCA-900066575*, *norank_f_Ruminococcaceae*, and *Muribaculum* were highly enriched in the HFHSD+RD group compared to the HFHSD group. On the other hand, *Lactobacillus*, *Bifidobacterium*, *Enterorhabdus*, *unclassified_f_Lachnospiraceae* and *unclassified_f_Erysipelotrichaceae* were enriched in the HFHSD+MC group. It is worth noting that the HFHSD group showed an enrichment of *Colidextribacter*, which is typically considered to be a harmful bacterium.

RD intervention modulates the homeostasis of the microbiota and acts by altering the structure of the gut microbiota, which is consistent with previous reports [34]. However, the specific changes in gut microbiota may vary, possibly due to differences in the methods of preparation or the raw materials of RD [52].The abundances of these prominent bacteria enriched by either the RD or metformin-supplemented groups have been associated with the improvement in glucose–lipid metabolism disorders, as well as the well-known ability of *Bacteroides*, *Muribaculum*, and *Bifidobacterium* to produce short-chain fatty acids. Some articles suggest that metformin has a limited impact on the regulation of gut microbiota in the cecum of high-fat diet-induced diabetic mice, primarily influencing the small intestine's flora composition [24]. However, numerous studies increasingly emphasize the role of cecum flora, which can directly impact metabolism [53–55]. There is a possibility that the role of gut microbiota varies at different stages of glucose and lipid metabolism abnormalities. Therefore, it is imperative that subsequent experiments focus on detecting alterations in gut microbiota during high-fat and high-sugar dietary conditions, with particular emphasis on investigating the correlation between the changes and factors influencing disturbed glucose metabolism.

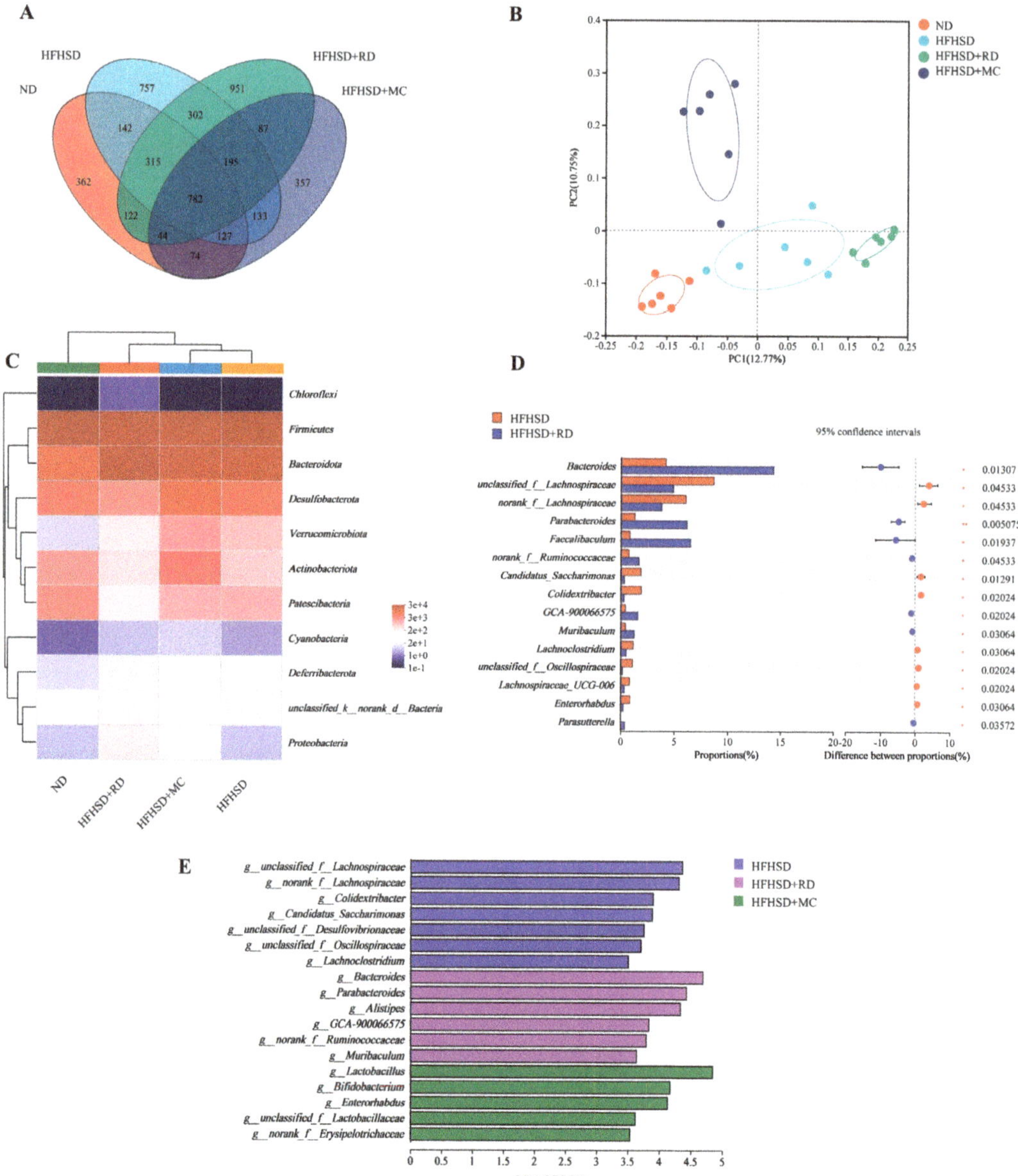

Figure 5. RD alters gut microbiota composition in HFHSD Kunming mice ($n = 6$). (**A**) Venn diagram depicting the number of bacteria at the OTU level between ND, HFHSD, HFHSD+RD, and HFHSD+MC groups. (**B**) Principal coordinate analyses based on OTUs of the microbial community in the four groups. (**C**) Gut microbiota composition at the genus level between the four groups of mice (abundance > 20). (**D**) Variances in the genus-level flora between HFHSD and HFHSD+RD mice employing the Wilcoxon rank-sum test. (**E**) LDA scores of differentially abundant taxa among HFHSD, HFHSD+RD, and HFHSD+MC mice, using the LEfSE method (LDA score > 3.5).

3.4. Correlation of Gut Microbiota with Parameters of Glucose–Lipid Metabolism

The relationships between eWAT weight, serum lipid metabolism parameters (TC, TG, HDL-C, LDL-C), inflammatory cytokines (IL-1β, IL-6), glucose metabolism indicators (AUC-OGTT, HOMA-IR), and gut microbiota composition were examined using Spearman's rank correlation analysis. Our results showed positive correlations between *Lachnoclostridium*, *Candidatus_Saccharimonas*, *norank_f__Oscillospiraceae* and *Colidextribacter* with markers related to glucose–lipid metabolism. Interestingly, the abundances of these bacteria exhibited a significant decrease following RD intervention, suggesting an improvement in glucose–lipid metabolism indicators. On the other hand, *Faecalibaculum*, *Bacteroides*, *Parabacteroides*, and *norank_f_Ruminococcaceae* showed negative correlations with glucose–lipid metabolism and inflammatory markers (Figure 6). In conclusion, these findings suggested that the RD intervention effectively abrogated the glucose–lipid metabolism disorders in HFHSD Kunming mice. This was evidenced by a decrease in TC, IL-1β, IL-6, LDL-C, HOMA-IR, AUC-OGTT and eWAT weight, accompanied by an increase in HDL-C. These improvements were primarily attributed to alterations in gut microbiota composition.

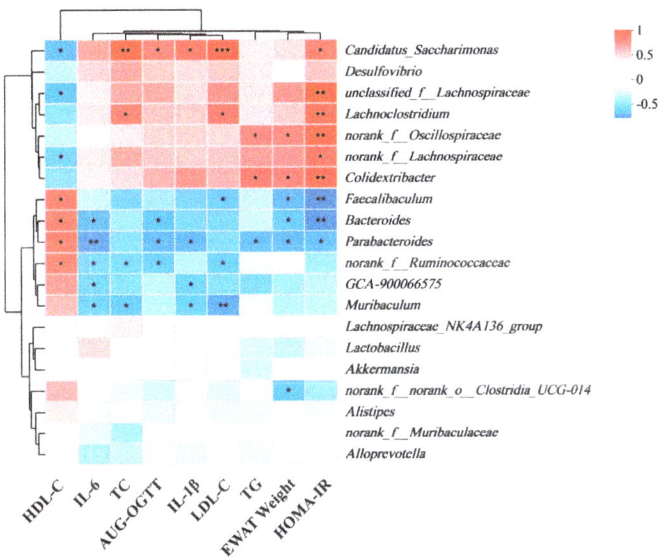

Figure 6. Correlation analysis of gut microbiota and glucose–lipid metabolism factors. ($n = 6$, * $p < 0.05$, ** $p < 0.01$, *** $p < 0.001$ indicates a significant difference).

4. General Discussion

The importance of IR in the development of MetS is evident, given its role in hastening the onset of diseases like obesity and type 2 diabetes [56]. Previous studies have shown that RD improves glucose homeostasis and lipid parameters, modifies the composition of the gut microbiota, and ameliorates IR via liver signaling pathways [34,36], but they have tended to ignore the role of white adipose tissue. In this study, we assessed the effects of RD and metformin on HFHSD Kunming mice. Analysis of TC, TG, HDL-C, LDL-C, FBG, AUC-OGTT, HOMA-IR, IL-1β, and IL-6 revealed the effectiveness of both RD and metformin in improving IR and maintaining glucose–lipid metabolism homeostasis. Additionally, RD intervention resulted in a significant reduction in eWAT weight and prevented further enlargement of adipose cells. These beneficial effects suggest a positive influence on the early development of MetS.

Short-chain fatty acids are primarily produced by gut microbiota through the fermentation of dietary fiber. They play beneficial roles, including reducing obesity, lowering blood glucose, and improving IR [57,58]. The abundances of short-chain fatty acid producers, namely *Bacteroides*, *Muribaculum*, *norank_f__Ruminococcaceae*, and *Parabacteroides*, increased in the HFHSD+RD group. Notably, RD intervention emerged as a significant contributor to preventing IR by mediating the gut microbiota, and the relative abundances of *Bacteroides*, *Faecalibaculum*, *Parabacteroides*, *norank_f_Ruminococcaceae*, *GCA-900066575*, *Alloprevotella* and *Muribaculum* were significantly increased. These results were consistent with previous studies identifying these bacterial genera as primarily associated with improving glucose–lipid metabolism disorders [59–61]. *Faecalibaculum*, classified as lactic acid-producing bacteria, is believed to exert anti-obesity effects through the production of lactic acid [60]. Furthermore, a noteworthy negative association was observed between *Bacteroides* and IR-linked carbohydrate metabolites [14]. *Lactobacillus*, *Bifidobacterium*, and *unclassified_f_Lachnospiraceae* were enriched in the HFHSD+MC group and are considered probiotics. These findings suggest that both RD and metformin may contribute to improving Kunming mice IR by altering gut microbiota composition. The distinct variations in the effects of RD and metformin interventions on gut microbiota are evident, and are likely attributable to the secondary action of metformin on the gut microbiota.

Among the various tissues constituting the WAT in male mice, eWAT is particularly susceptible to obesity development [62]. Previous studies suggested that RD supplementation could potentially alleviate adipose tissue inflammation by reducing macrophage infiltration, modulating macrophage polarization, and inhibiting NF-κB signaling, specifically in the eWAT [63]. Macrophage hypertrophy and chronic inflammation in adipose tissue are important factors in developing systemic IR [64]. The PI3K/AKT pathway, a classical pathway critical for insulin metabolism, exhibited significantly upregulated genes, suggesting that the RD intervention activated the ISR/PI3K/AKT pathway, and provided a new rationale for RD to ameliorate IR. Additionally, the PPAR signaling pathway influenced gene expression related to energy metabolism, cell development and differentiation. Our sequencing data also showed significantly increased expression levels of genes, including *C/EBPα*, *PPARγ*, *GLUT4*, and *ADIPOQ* in both the ND and HFHSD+RD groups, in contrast to the HFHSD group. In the terminal differentiation stage, which is characterized by the cessation of growth in targeted preadipocytes, *C/EBPα* and *PPARγ* play a pivotal role in inducing and sustaining the expression of key adipogenic genes (*GLUT4*, *AP2*, and *ADIPOQ*), crucial for the formation of functional adipocytes [65]. These alterations appear closely connected to the gut microbiota composition, implying that metabolites from the gut microbiota may mediate the role of the gut-adipose axis. This study provides novel insights into the mechanism of RD as a means to enhance insulin sensitivity. Certainly, there are some limitations in our study. Notably, there were noteworthy observations during the modeling of mice on a high-fat, high-sugar diet, particularly concerning the differences between females and males in glucose and lipid metabolism [66,67]. It is crucial to emphasize that, despite exposure to an identical diet, males demonstrated a heightened susceptibility to metabolic disorders. Therefore, validating the effectiveness of dietary interventions on female mice is necessary. This will lead to a more comprehensive understanding of the variability of metabolic abnormalities.

Subsequently, we consider investigating potential microbial changes in eWAT and biomarkers of the gut-adipose axis following RD intervention, aiming to further elucidate the link between gut microbiota and alterations in the gene expression. We propose conducting animal experiments to retrospectively investigate glycolipid metabolic parameters, changes in lipid transcriptome profiles, and alterations in gut flora throughout the process. Employing gut flora metabolomics and lipidomics would aid in exploring the effects of high-fat and high-sugar diets on stage-specific glycolipid metabolism changes in healthy mice, elucidating linkages, and uncovering underlying mechanistic processes. To ensure comprehensive insights, variations in animal sexes and feed formulations, particularly varying fat content (e.g., 15%, 30%, 45%, 60%), should be incorporated to investigate generalized mechanisms.

5. Conclusions

In summary, carrying out RD intervention (6.2 g/kg/day) on HFHSD Kunming mice can lead to a reduction in FBG, TG, LDL-C, AUC-OGTT, HOMA-IR, IL-1β, IL-6, and eWAT weight. These findings demonstrate the effectiveness of RD in alleviating IR in HFHSD-induced mice by reshaping the composition of gut microbiota. Moreover, RD intervention alters the transcriptomic profiles of eWAT in HFHSD mice. Our study provides new insights on the mechanisms through which RD intervention exerts beneficial effects in alleviating glucose–lipid metabolism disorders in HFHSD Kunming mice. The central role of gut microbiota and its complex interactions with the adipose tissue should be emphasized. These findings necessitate additional research to validate their clinical significance and explore potential therapeutic applications.

Supplementary Materials: The following supporting information can be downloaded at: https://www.mdpi.com/article/10.3390/biom14020186/s1, Figure S1: Change in body weight of mice in each group; Figure S2: Dietary changes in four groups of mice; Figure S3: Fasting blood glucose in each group of mice during oral glucose tolerance test (0 min); Figure S4: Serum fasting insulin levels in mice at week 8; Figure S5: Gene expression changes in insulin resistance signaling pathway (GSEA); Figure S6: Gene expression changes in regulation of lipolysis in adipocytes and AGE-RAGE in diabetic complications signaling pathway (GSEA); Figure S7: Alpha diversity of gut microbiota in various groups of mice; Figure S8: Circos plot of gut microbiota in mice at the phylum level. * $p < 0.05$, ** $p < 0.01$, and *** $p < 0.001$ vs. HFHSD group. one-way ANOVA with Tukey's post hoc test. Table S1: PI3K-AKT signaling pathway genes expressions between ND and HFHSD+RD groups; Table S2: Expression of selected genes between HFHSD and HFHSD+RD groups.

Author Contributions: X.C.: conceptualization, methodology and writing—original draft; Y.H.: supervision and writing—review and editing; A.L., L.P. and S.Y.: methodology and formal analysis; J.W. and Y.L.: data curation; M.Z., Y.X. and Z.Z.: methodology and validation; J.H.: supervision and writing—review and editing. All authors have read and agreed to the published version of the manuscript.

Funding: Major Science and Technology Projects for Public Welfare of Henan Province: 201300110300; Zhongyuan Scholars of Henan Province in China: 192101510004; Zhongyuan Scholar Workstation Funded Project: ZYGZZ2021056, 224400510026; Central Government Guides the Local Science and Technology Development Special Fund: Z20221341069; Henan Provincial Key Science & Technology Special Project: 232102111060; Key Research and Development Project of Henan Province: 231111310700.

Institutional Review Board Statement: The study followed ethical guidelines approved by the Institutional Animal Care and Use Committee of Henan University of Technology (protocol code: P20230315-1).

Informed Consent Statement: Not applicable.

Data Availability Statement: The datasets generated for this study are available on request to the corresponding author.

Conflicts of Interest: The authors declare no conflicts of interest.

References

1. Gluvic, Z.; Zaric, B.; Resanovic, I.; Obradovic, M.; Mitrovic, A.; Radak, D.; Isenovic, E.R. Link between Metabolic Syndrome and Insulin Resistance. *Curr. Vasc. Pharmacol.* **2017**, *15*, 30–39. [CrossRef]
2. Kim, B.; Feldman, E.L. Insulin resistance as a key link for the increased risk of cognitive impairment in the metabolic syndrome. *Exp. Mol. Med.* **2015**, *47*, e149. [CrossRef] [PubMed]
3. Wang, T.; Lu, J.; Shi, L.; Chen, G.; Xu, M.; Xu, Y.; Su, Q.; Mu, Y.; Chen, L.; Hu, R.; et al. Association of insulin resistance and β-cell dysfunction with incident diabetes among adults in China: A nationwide, population-based, prospective cohort study. *Lancet Diabetes Endocrinol.* **2020**, *8*, 115–124. [CrossRef] [PubMed]
4. Reardon, C.A.; Lingaraju, A.; Schoenfelt, K.Q.; Zhou, G.; Cui, C.; Jacobs-El, H.; Babenko, I.; Hoofnagle, A.; Czyz, D.; Shuman, H.; et al. Obesity and Insulin Resistance Promote Atherosclerosis through an IFNγ-Regulated Macrophage Protein Network. *Cell Rep.* **2018**, *23*, 3021–3030. [CrossRef] [PubMed]

5. Minowa, K.; Rodriguez-Agudo, D.; Suzuki, M.; Muto, Y.; Hirai, S.; Wang, Y.; Su, L.; Zhou, H.; Chen, Q.; Lesnefsky, E.J.; et al. Insulin dysregulation drives mitochondrial cholesterol metabolite accumulation: Initiating hepatic toxicity in nonalcoholic fatty liver disease. *J. Lipid Res.* **2023**, *64*, 100363. [CrossRef] [PubMed]
6. Qi, M.; Peng, H.; Song, J.; Zhang, J.; Wang, S.; Ma, C. Effects of extrusion and enzymatic hydrolysis on the in vitro starch digestibility, protein structure andrheological properties of pea flour. *Food Sci.* **2022**, *43*, 76–82.
7. Zheng, X.; Huang, W.; Li, Q.; Chen, Y.; Wu, L.; Dong, Y.; Huang, X.; He, X.; Ou, Z.; Peng, Y. Membrane Protein Amuc_1100 Derived from Akkermansia muciniphila Facilitates Lipolysis and Browning via Activating the AC3/PKA/HSL Pathway. *Microbiol. Spectr.* **2023**, *11*, e0432322. [CrossRef] [PubMed]
8. Emont, M.P.; Jacobs, C.; Essene, A.L.; Pant, D.; Tenen, D.; Colleluori, G.; Di Vincenzo, A.; Jørgensen, A.M.; Dashti, H.; Stefek, A.; et al. A single-cell atlas of human and mouse white adipose tissue. *Nature* **2022**, *603*, 926–933. [CrossRef] [PubMed]
9. Choi, W.G.; Choi, W.; Oh, T.J.; Cha, H.N.; Hwang, I.; Lee, Y.K.; Lee, S.Y.; Shin, H.; Lim, A.; Ryu, D.; et al. Inhibiting serotonin signaling through HTR2B in visceral adipose tissue improves obesity-related insulin resistance. *J. Clin. Investig.* **2021**, *23*, 131. [CrossRef]
10. Kuai, M.; Li, Y.; Sun, X.; Ma, Z.; Lin, C.; Jing, Y.; Lu, Y.; Chen, Q.; Wu, X.; Kong, X.; et al. A novel formula Sang-Tong-Jian improves glycometabolism and ameliorates insulin resistance by activating PI3K/AKT pathway in type 2 diabetic KKAy mice. *Biomed. Pharmacother.* **2016**, *84*, 1585–1594. [CrossRef]
11. Virtue, A.T.; McCright, S.J.; Wright, J.M.; Jimenez, M.T.; Mowel, W.K.; Kotzin, J.J.; Joannas, L.; Basavappa, M.G.; Spencer, S.P.; Clark, M.L.; et al. The gut microbiota regulates white adipose tissue inflammation and obesity via a family of microRNAs. *Sci. Transl. Med.* **2019**, *11*, eaav1892. [CrossRef]
12. Moreno-Navarrete, J.M.; Fernandez-Real, J.M. The gut microbiota modulates both browning of white adipose tissue and the activity of brown adipose tissue. *Rev. Endocr. Metab. Disord.* **2019**, *20*, 387–397. [CrossRef] [PubMed]
13. Maifeld, A.; Bartolomaeus, H.; Löber, U.; Avery, E.G.; Steckhan, N.; Markó, L.; Wilck, N.; Hamad, I.; Šušnjar, U.; Mähler, A.; et al. Fasting alters the gut microbiome reducing blood pressure and body weight in metabolic syndrome patients. *Nat. Commun.* **2021**, *12*, 1970. [CrossRef] [PubMed]
14. Takeuchi, T.; Kubota, T.; Nakanishi, Y.; Tsugawa, H.; Suda, W.; Kwon, A.T.; Yazaki, J.; Ikeda, K.; Nemoto, S.; Mochizuki, Y.; et al. Gut microbial carbohydrate metabolism contributes to insulin resistance. *Nature* **2023**, *621*, 389–395. [CrossRef] [PubMed]
15. Wu, J.; Wang, K.; Wang, X.; Pang, Y.; Jiang, C. The role of the gut microbiome and its metabolites in metabolic diseases. *Protein Cell* **2021**, *12*, 360–373. [CrossRef] [PubMed]
16. Biemann, R.; Buß, E.; Benndorf, D.; Lehmann, T.; Schallert, K.; Püttker, S.; Reichl, U.; Isermann, B.; Schneider, J.G.; Saake, G.; et al. Fecal Metaproteomics Reveals Reduced Gut Inflammation and Changed Microbial Metabolism Following Lifestyle-Induced Weight Loss. *Biomolecules* **2021**, *11*, 726. [CrossRef] [PubMed]
17. Liu, Y.; Wang, C.; Li, J.; Li, T.; Zhang, Y.; Liang, Y.; Mei, Y. Phellinus linteus polysaccharide extract improves insulin resistance by regulating gut microbiota composition. *FASEB J.* **2020**, *34*, 1065–1078. [CrossRef] [PubMed]
18. Sartor, R.B. Gut microbiota: Diet promotes dysbiosis and colitis in susceptible hosts. *Nat. Rev. Gastroenterol. Hepatol.* **2012**, *9*, 561–562. [CrossRef] [PubMed]
19. Cao, Y.; Chen, X.; Sun, Y.; Shi, J.; Xu, X.; Shi, Y.C. Hypoglycemic Effects of Pyrodextrins with Different Molecular Weights and Digestibilities in Mice with Diet-Induced Obesity. *J. Agric. Food. Chem.* **2018**, *66*, 2988–2995. [CrossRef]
20. Nie, Q.; Sun, Y.; Li, M.; Zuo, S.; Chen, C.; Lin, Q.; Nie, S. Targeted modification of gut microbiota and related metabolites via dietary fiber. *Carbohydr. Polym.* **2023**, *316*, 120986. [CrossRef]
21. Murga-Garrido, S.M.; Hong, Q.; Cross, T.L.; Hutchison, E.R.; Han, J.; Thomas, S.P.; Vivas, E.I.; Denu, J.; Ceschin, D.G.; Tang, Z.Z.; et al. Gut microbiome variation modulates the effects of dietary fiber on host metabolism. *Microbiome* **2021**, *9*, 117. [CrossRef] [PubMed]
22. Makki, K.; Brolin, H.; Petersen, N.; Henricsson, M.; Christensen, D.P.; Khan, M.T.; Wahlström, A.; Bergh, P.O.; Tremaroli, V.; Schoonjans, K.; et al. 6α-hydroxylated bile acids mediate TGR5 signalling to improve glucose metabolism upon dietary fiber supplementation in mice. *Gut* **2023**, *72*, 314–324. [CrossRef] [PubMed]
23. Hunter, R.W.; Hughey, C.C.; Lantier, L.; Sundelin, E.I.; Peggie, M.; Zeqiraj, E.; Sicheri, F.; Jessen, N.; Wasserman, D.H.; Sakamoto, K. Metformin reduces liver glucose production by inhibition of fructose-1-6-bisphosphatase. *Nat. Med.* **2018**, *24*, 1395–1406. [CrossRef] [PubMed]
24. Silamiķele, L.; Saksis, R.; Silamiķelis, I.; Kotoviča, P.P.; Brīvība, M.; Kalniņa, I.; Kalniņa, Z.; Fridmanis, D.; Kloviņš, J. Spatial variation of the gut microbiome in response to long-term metformin treatment in high-fat diet-induced type 2 diabetes mouse model of both sexes. *Gut Microbes* **2023**, *15*, 2188663. [CrossRef] [PubMed]
25. Yan, Y.; Li, Q.; Shen, L.; Guo, K.; Zhou, X. Chlorogenic acid improves glucose tolerance, lipid metabolism, inflammation and microbiota composition in diabetic db/db mice. *Front. Endocrinol.* **2022**, *13*, 1042044. [CrossRef]
26. Bauer, P.V.; Duca, F.A.; Waise, T.M.Z.; Rasmussen, B.A.; Abraham, M.A.; Dranse, H.J.; Puri, A.; O'Brien, C.A.; Lam, T.K.T. Metformin Alters Upper Small Intestinal Microbiota that Impact a Glucose-SGLT1-Sensing Glucoregulatory Pathway. *Cell Metab.* **2018**, *27*, 101–117.e105. [CrossRef]
27. Huang, X.; Hong, X.; Wang, J.; Sun, T.; Yu, T.; Yu, Y.; Fang, J.; Xiong, H. Metformin elicits antitumour effect by modulation of the gut microbiota and rescues Fusobacterium nucleatum-induced colorectal tumourigenesis. *EBioMedicine* **2020**, *61*, 103037. [CrossRef]

28. Huang, Z.; Wang, C.; Chen, J.; He, X.; Chen, K.; Jiang, X.; Zhao, C.; Liu, B. Ameliorating Effect on Glycolipid Metabolism of Spirulina Functional Formulation Combination from Traditional Chinese Medicine. *Oxid. Med. Cell. Longev.* **2022**, *2022*, 3910116. [CrossRef]
29. Chen, X.; Hou, Y.; Wang, Z.; Liao, A.; Pan, L.; Zhang, M.; Xue, Y.; Wang, J.; Liu, Y.; Huang, J. A Comparative Study of Resistant Dextrins and Resistant Maltodextrins from Different Tuber Crop Starches. *Polymers* **2023**, *15*, 4545. [CrossRef]
30. Farhangi, M.A.; Javid, A.Z.; Sarmadi, B.; Karimi, P.; Dehghan, P. A randomized controlled trial on the efficacy of resistant dextrin, as functional food, in women with type 2 diabetes: Targeting the hypothalamic-pituitary-adrenal axis and immune system. *Clin. Nutr.* **2018**, *37*, 1216–1223. [CrossRef]
31. Mateo-Gallego, R.; Moreno-Indias, I.; Bea, A.M.; Sánchez-Alcoholado, L.; Fumanal, A.J.; Quesada-Molina, M.; Prieto-Martín, A.; Gutiérrez-Repiso, C.; Civeira, F.; Tinahones, F.J. An alcohol-free beer enriched with isomaltulose and a resistant dextrin modulates gut microbiome in subjects with type 2 diabetes mellitus and overweight or obesity: A pilot study. *Food Funct.* **2021**, *12*, 3635–3646. [CrossRef]
32. Hobden, M.R.; Guérin-Deremaux, L.; Rowland, I.; Gibson, G.R.; Kennedy, O.B. Potential anti-obesogenic properties of non-digestible carbohydrates: Specific focus on resistant dextrin. *Proc. Nutr. Soc.* **2015**, *74*, 258–267. [CrossRef]
33. Gholizadeh Shamasbi, S.; Dehgan, P.; Mohammad-Alizadeh Charandabi, S.; Aliasgarzadeh, A.; Mirghafourvand, M. The effect of resistant dextrin as a prebiotic on metabolic parameters and androgen level in women with polycystic ovarian syndrome: A randomized, triple-blind, controlled, clinical trial. *Eur. J. Nutr.* **2019**, *58*, 629–640. [CrossRef]
34. Zhang, Z.; Chen, X.; Cui, B. Modulation of the fecal microbiome and metabolome by resistant dextrin ameliorates hepatic steatosis and mitochondrial abnormalities in mice. *Food Funct.* **2021**, *12*, 4504–4518. [CrossRef]
35. Włodarczyk, M.; Śliżewska, K.; Barczyńska, R.; Kapuśniak, J. Effects of Resistant Dextrin from Potato Starch on the Growth Dynamics of Selected Co-Cultured Strains of Gastrointestinal Bacteria and the Activity of Fecal Enzymes. *Nutrients* **2022**, *14*, 2158. [CrossRef]
36. Hu, F.; Niu, Y.; Xu, X.; Hu, Q.; Su, Q.; Zhang, H. Resistant dextrin improves high-fat-high-fructose diet induced insulin resistance. *Nutr. Metab.* **2020**, *17*, 36. [CrossRef]
37. Chen, W.; Zhang, T.; Ma, Q.; Zhu, Y.; Shen, R. Structure Characterization and Potential Probiotic Effects of Sorghum and Oat Resistant Dextrins. *Foods* **2022**, *11*, 1877. [CrossRef]
38. Barber, C.; Sabater, C.; Ávila-Gálvez, M.; Vallejo, F.; Bendezu, R.A.; Guérin-Deremaux, L.; Guarner, F.; Espín, J.C.; Margolles, A.; Azpiroz, F. Effect of Resistant Dextrin on Intestinal Gas Homeostasis and Microbiota. *Nutrients* **2022**, *14*, 4611. [CrossRef]
39. Laurentin, A.; Edwards, C.A. Differential fermentation of glucose-based carbohydrates in vitro by human faecal bacteria--a study of pyrodextrinised starches from different sources. *Eur. J. Nutr.* **2004**, *43*, 183–189. [CrossRef]
40. Trithavisup, K.; Shi, Y.C.; Krusong, K.; Tananuwong, K. Molecular structure and properties of cassava-based resistant maltodextrins. *Food Chem.* **2022**, *369*, 130876. [CrossRef]
41. Su, Z.; Li, H.; Ye, Z.; Zhu, Y.; Feng, B.; Tang, L.; Zheng, G. Qidan Tiaozhi capsule attenuates metabolic syndrome via activating AMPK/PINK1-Parkin-mediated mitophagy. *J. Ethnopharmacol.* **2023**, *307*, 116091. [CrossRef]
42. Nelson, M.E.; Madsen, S.; Cooke, K.C.; Fritzen, A.M.; Thorius, I.H.; Masson, S.W.C.; Carroll, L.; Weiss, F.C.; Seldin, M.M.; Potter, M.; et al. Systems-level analysis of insulin action in mouse strains provides insight into tissue- and pathway-specific interactions that drive insulin resistance. *Cell Metab.* **2022**, *34*, 227–239.e226. [CrossRef]
43. Jaganathan, R.; Ravindran, R.; Dhanasekaran, S. Emerging Role of Adipocytokines in Type 2 Diabetes as Mediators of Insulin Resistance and Cardiovascular Disease. *Can. J. Diabetes* **2018**, *42*, 446–456.e441. [CrossRef]
44. Zhang, Y.; Hai, J.; Cao, M.; Zhang, Y.; Pei, S.; Wang, J.; Zhang, Q. Silibinin ameliorates steatosis and insulin resistance during non-alcoholic fatty liver disease development partly through targeting IRS-1/PI3K/Akt pathway. *Int. Immunopharmacol.* **2013**, *17*, 714–720. [CrossRef]
45. Ahmed, B.; Sultana, R.; Greene, M.W. Adipose tissue and insulin resistance in obese. *Biomed. Pharmacother.* **2021**, *137*, 111315. [CrossRef]
46. Wang, G.X.; Cho, K.W.; Uhm, M.; Hu, C.R.; Li, S.; Cozacov, Z.; Xu, A.E.; Cheng, J.X.; Saltiel, A.R.; Lumeng, C.N.; et al. Otopetrin 1 protects mice from obesity-associated metabolic dysfunction through attenuating adipose tissue inflammation. *Diabetes* **2014**, *63*, 1340–1352. [CrossRef]
47. Carson, C.; Macias-Velasco, J.F.; Gunawardana, S.; Miranda, M.A.; Oyama, S.; St Pierre, C.L.; Schmidt, H.; Wayhart, J.P.; Lawson, H.A. Brown Adipose Expansion and Remission of Glycemic Dysfunction in Obese SM/J Mice. *Cell Rep.* **2020**, *33*, 108237. [CrossRef]
48. Ernst, M.C.; Haidl, I.D.; Zúñiga, L.A.; Dranse, H.J.; Rourke, J.L.; Zabel, B.A.; Butcher, E.C.; Sinal, C.J. Disruption of the chemokine-like receptor-1 (CMKLR1) gene is associated with reduced adiposity and glucose intolerance. *Endocrinology* **2012**, *153*, 672–682. [CrossRef]
49. Huang, C.; Wang, M.; Ren, L.; Xiang, L.; Chen, J.; Li, M.; Xiao, T.; Ren, P.; Xiong, L.; Zhang, J.V. CMKLR1 deficiency influences glucose tolerance and thermogenesis in mice on high fat diet. *Biochem. Biophys. Res. Commun.* **2016**, *473*, 435–441. [CrossRef]
50. Liu, J. The role and significance of CMKLR1 expression inadipose tissue of rats with non-alcoholicsteatohepatitis. *Chin. J. Pathophysiol.* **2015**, *31*, 1877.

51. Liu, Q.; Hao, H.; Li, J.; Zheng, T.; Yao, Y.; Tian, X.; Zhang, Z.; Yi, H. Oral Administration of Bovine Milk-Derived Extracellular Vesicles Attenuates Cartilage Degeneration via Modulating Gut Microbiota in DMM-Induced Mice. *Nutrients* **2023**, *15*, 747. [CrossRef]
52. Sasaki, H.; Hayashi, K.; Imamura, M.; Hirota, Y.; Hosoki, H.; Nitta, L.; Furutani, A.; Shibata, S. Combined resistant dextrin and low-dose Mg oxide administration increases short-chain fatty acid and lactic acid production by gut microbiota. *J. Nutr. Biochem.* **2023**, *120*, 109420. [CrossRef]
53. Fu, T.; Huan, T.; Rahman, G.; Zhi, H.; Xu, Z.; Oh, T.G.; Guo, J.; Coulter, S.; Tripathi, A.; Martino, C.; et al. Paired microbiome and metabolome analyses associate bile acid changes with colorectal cancer progression. *Cell Rep.* **2023**, *42*, 112997. [CrossRef]
54. Krisko, T.I.; Nicholls, H.T.; Bare, C.J.; Holman, C.D.; Putzel, G.G.; Jansen, R.S.; Sun, N.; Rhee, K.Y.; Banks, A.S.; Cohen, D.E. Dissociation of Adaptive Thermogenesis from Glucose Homeostasis in Microbiome-Deficient Mice. *Cell Metab.* **2020**, *31*, 592–604.e599. [CrossRef]
55. Tanca, A.; Manghina, V.; Fraumene, C.; Palomba, A.; Abbondio, M.; Deligios, M.; Silverman, M.; Uzzau, S. Metaproteogenomics Reveals Taxonomic and Functional Changes between Cecal and Fecal Microbiota in Mouse. *Front. Microbiol.* **2017**, *8*, 391. [CrossRef]
56. Zhang, S.; Xu, M.; Zhang, W.; Liu, C.; Chen, S. Natural Polyphenols in Metabolic Syndrome: Protective Mechanisms and Clinical Applications. *Int. J. Mol. Sci.* **2021**, *22*, 6110. [CrossRef]
57. Wu, R.; Tang, X.; Kang, X.; Luo, Y.; Wang, L.; Li, J.; Wu, X.; Liu, D. Effect of a Chinese medical nutrition therapy diet on gut microbiota and short chain fatty acids in the simulator of the human intestinal microbial ecosystem (SHIME). *J. Funct. Foods* **2019**, *62*, 103555. [CrossRef]
58. Shi, H.; Li, X.; Hou, C.; Chen, L.; Zhang, Y.; Li, J. Effects of Pomegranate Peel Polyphenols Combined with Inulin on Gut Microbiota and Serum Metabolites of High-Fat-Induced Obesity Rats. *J. Agric. Food Chem.* **2023**, *71*, 5733–5744. [CrossRef]
59. Zhuang, P.; Li, H.; Jia, W.; Shou, Q.; Zhu, Y.; Mao, L.; Wang, W.; Wu, F.; Chen, X.; Wan, X.; et al. Eicosapentaenoic and docosahexaenoic acids attenuate hyperglycemia through the microbiome-gut-organs axis in db/db mice. *Microbiome* **2021**, *9*, 185. [CrossRef]
60. Liao, J.; Cao, Y.; Zhao, J.; Yu, B.; Wang, Y.; Li, W.; Li, H.; Lv, S.; Wen, W.; Cui, H.; et al. Aqueous extract of Polygala japonica Houtt. ameliorated nonalcoholic steatohepatitis in mice through restoring the gut microbiota disorders and affecting the metabolites in feces and liver. *Phytomedicine* **2023**, *118*, 154937. [CrossRef]
61. Xie, J.; Tian, S.; Liu, J.; Huang, S.; Yang, M.; Yang, X.; Xu, R.; Lin, J.; Han, L.; Zhang, D. Combination Therapy with Indigo and Indirubin for Ulcerative Colitis via Reinforcing Intestinal Barrier Function. *Oxid. Med. Cell. Longev.* **2023**, *2023*, 2894695. [CrossRef]
62. Sárvári, A.K.; Van Hauwaert, E.L.; Markussen, L.K.; Gammelmark, E.; Marcher, A.B.; Ebbesen, M.F.; Nielsen, R.; Brewer, J.R.; Madsen, J.G.S.; Mandrup, S. Plasticity of Epididymal Adipose Tissue in Response to Diet-Induced Obesity at Single-Nucleus Resolution. *Cell Metab.* **2021**, *33*, 437–453.e435. [CrossRef]
63. Hu, Q.; Lu, Y.; Hu, F.; He, S.; Xu, X.; Niu, Y.; Zhang, H.; Li, X.; Su, Q. Resistant dextrin reduces obesity and attenuates adipose tissue inflammation in high-fat diet-fed mice. *Int. J. Med. Sci.* **2020**, *17*, 2611–2621. [CrossRef]
64. Duan, H.; Jing, L.; Xiang, J.; Ju, C.; Wu, Z.; Liu, J.; Ma, X.; Chen, X.; Liu, Z.; Feng, J.; et al. CD146 Associates with Gp130 to Control a Macrophage Pro-inflammatory Program That Regulates the Metabolic Response to Obesity. *Adv. Sci.* **2022**, *9*, e2103719. [CrossRef]
65. Longo, M.; Zatterale, F.; Naderi, J.; Parrillo, L.; Formisano, P.; Raciti, G.A.; Beguinot, F.; Miele, C. Adipose Tissue Dysfunction as Determinant of Obesity-Associated Metabolic Complications. *Int. J. Mol. Sci.* **2019**, *20*, 2358. [CrossRef]
66. Arnetz, L.; Ekberg, N.R.; Alvarsson, M. Sex differences in type 2 diabetes: Focus on disease course and outcomes. *Diabetes. Metab. Syndr. Obes.* **2014**, *7*, 409–420. [CrossRef]
67. Pettersson, U.S.; Waldén, T.B.; Carlsson, P.O.; Jansson, L.; Phillipson, M. Female mice are protected against high-fat diet induced metabolic syndrome and increase the regulatory T cell population in adipose tissue. *PLoS ONE* **2012**, *7*, e46057. [CrossRef]

biomolecules

MDPI

Article

Effect of Metformin Treatment on Serum Metabolic Profile Changes in Lean and Obese Zucker Rat Model for Fatty Liver Disease

Stepan Melnyk [1] **and Reza Hakkak** [1,2,3,*]

[1] Arkansas Children's Research Institute, Little Rock, AR 72202, USA
[2] Department of Dietetics and Nutrition, University of Arkansas for Medical Sciences, Little Rock, AR 72205, USA
[3] Department of Pediatrics, University of Arkansas for Medical Sciences, Little Rock, AR 72205, USA
* Correspondence: rhakkak@uams.edu

Abstract: Excessive weight and obesity are the leading risk factors for the development of chronic diseases, including diabetes. Metformin is capable of significantly improving coexisting complications of diabetes. We used a metabolomics approach to examine the effects of metformin administration on lean and obese (fa/fa) Zucker rats. After 1 week of acclimation, twenty-eight 5-week-old female lean and obese rats were randomly assigned to and maintained in the following four groups (seven rats/group) for 10 weeks: (1) lean control (LC); (2) obese control (OC); (3) lean metformin (LM); and (4) obese metformin (OM). At the end of 10 weeks, serum was collected and analyzed using HPLC with electrochemical detection, HPLC with UV detection, and liquid chromatography mass spectrometry. We selected 50 metabolites' peaks that were shared by all four groups of rats. Peak heights, as a defining factor, generally decreased in metformin-treated lean rats vs. untreated lean controls (3 LM:16 LC). Peak heights generally increased in metformin-treated obese rats vs. untreated obese controls (14 OM:5 OC). Overall, individual peaks were distributed as 11 that represented only lean rats, 11 that represented only obese rats, and 8 that were common among both lean and obese rats. In future studies, we will use a targeted metabolomics approach to identify those metabolites, map them to biochemical pathways and create a list of biomarkers. In summary, the current study contributed to a better understanding of the basic metabolic changes of lean and obese rats and demonstrated that both obesity and metformin make a significant impact on the metabolome of Zucker rats.

Keywords: metformin; serum metabolic; obese rats

1. Introduction

Obesity is a multifactorial and complex chronic, relapsing, and even life-threatening condition characterized by an excessive accumulation of body fat. Since the 1970s, obesity prevalence in adults and children has increased globally, and currently, the number of obese or overweight people in the world is greater than the number of those who are underweight [1,2]. Excessive weight and obesity are the leading risk factors for the development of diabetes and other chronic diseases. Chronic sterile meta-inflammation accompanying obesity is the main contributor to both insulin resistance and the initiation of type 2 diabetes [3–5]. Adult type 2 diabetes has been considered a serious threat as it is frequently accompanied by several major complications, such as a higher prevalence of cancers, cardiovascular diseases, chronic kidney disease, osteoporosis, and neurodegenerative diseases. Thus, the increase in obesity prevalence has led to increases in diabetes and other comorbid conditions, along with a related increase in community health care costs and patient mortality [6,7].

One of the oldest and most widely used oral antidiabetic drugs is metformin, which is capable of significantly improving coexisting complications of diabetes [8]. Metformin is used as a first-line treatment for type 2 diabetes and is a natural cationic biguanide capable of utilizing a broad spectrum of molecular mechanisms to control blood glucose. These mechanisms range from decreasing intestinal glucose absorption and inhibition of hepatic gluconeogenesis to decreasing insulin resistance, although the exact details of these mechanisms are unknown [9]. The mechanism action of metformin is shown in Figure 1.

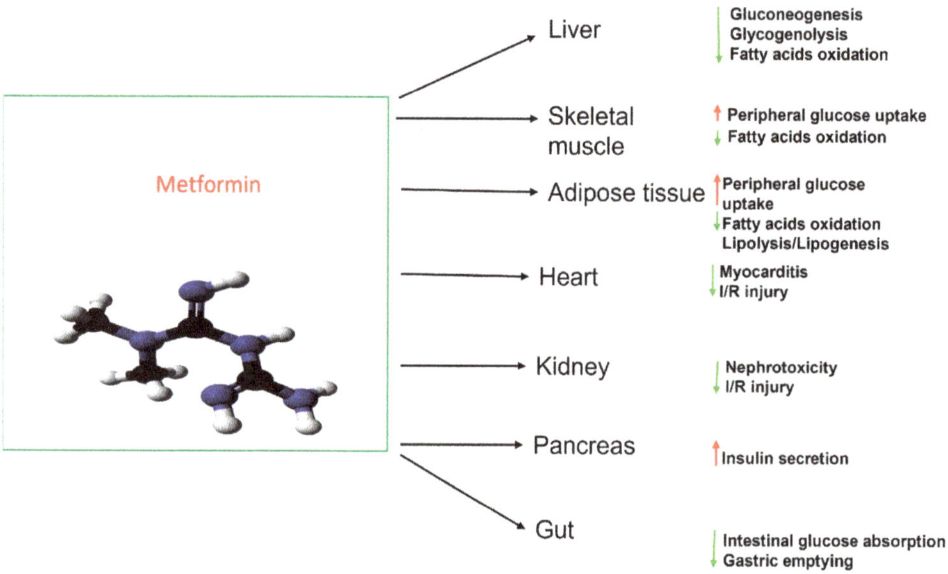

Figure 1. Metformin mechanism of action. (green color means decreasing risk and red is increasing risk).

Small molecules and their metabolites create unique and distinctive metabolic profiles for a biological system known as metabolome. The large-scale study of these small biomolecules are called metabolomics. Utilizing advanced technological tools, metabolomics are capable of not only performing a deep metabolic analysis of biological systems (human, animal, cell) at certain time points and under certain conditions (diseases, treatment), but are also able to provide a personalized approach. In a recent review by Costanzo et al. [10], the authors analyzed numerous human studies and concluded that metabolic differences in age and gender need to be adequately considered by investigators in future studies.

The use of a metabolomics approach could be extremely beneficial in the search for explanations on the multiple positive effects of metformin treatment. Metabolomic analyses have been applied to investigate metabolic alterations in response to metformin treatment and have shown significant systemic metabolome changes in a variety of biofluids, tissues, and cells [11]. The meta-analysis review by Xie et al. [12] systematized metabolic data from numerous studies that used a Zucker rat model to explain the development of obesity. The effect of metformin treatment on Zucker rats and the systemic changes of metabolism still need to be studied. Treatment with metformin in patients with non-alcoholic fatty liver disease (NAFLD) has been reported with contradictory findings regarding its effects on blood lipids and liver histology [13,14]. Obesity increases non-alcoholic fatty liver disease (NAFLD), and we recently reported that metformin treatment could reduce NAFLD using the obese Zucker rat model [15]. Using the same model, it was also observed that metformin treatment could change the gut microbiome [16].

Zucker rats (*fa/fa*) are the best-known, most widely used rat model for obesity and have been used to study non-insulin-dependent diabetes mellitus. The *fa* mutation, discovered by Zucker and Zucker in 1961, is a leptin receptor mutation that impairs satiety, leading to hyperphagia and obesity [17–20]. The Zucker rat is used as a model of human early onset hyperplastic–hypertrophic obesity. Obese Zucker rats develop hyperleptinemia, obesity, hyperinsulinemia, hyperlipidemia, insulin resistance, and hyperglycemia—classic signs of human metabolic syndrome [21]. As such, the obese Zucker rat may represent a useful tool to further understand the etiopathology of metabolic syndrome in humans.

We recently reported that obesity significantly changed the metabolic profile of 62% of the selected metabolites in obese Zucker rats. In general, we observed metabolite increases in obese rats, compared to lean rats, that were randomly distributed across the serum metabolic profile, except for one segment where the level of metabolites in obese rats was consistently lower [22].

In the present study, we used metabolomics tools to better understand the fundamental metabolic consequences of metformin treatment on lean and obese Zucker rats. An evidence-based scientific approach to modulate and study this process will be helpful in further clinical studies.

2. Materials and Methods

Most of the details regarding experimental design, sample preparation, data collection, and analysis have been described previously [22]. We will present an abbreviated description of these steps.

2.1. Experimental Design

Twenty-eight 5-week-old female lean and obese (*fa/fa*) Zucker rats (Envigo, Indianapolis, IN, USA) were used in this study. All animal care and procedures were approved by the University of Arkansas for Medical Sciences/Arkansas Children's Research Institute Institutional Animal Care and Use Committee (protocol # 3882) and adhered to the institutional policies and procedures.

Animals were maintained according to USDA and NIH guidelines. After a 1-week acclimation, lean and obese rats were randomly assigned to the following four groups (7 rats/group) and maintained for 10 weeks: (1) lean control (LC), (2) obese control (OC), (3) lean administered metformin (LM), and (4) obese administered metformin (OM). After 10 weeks, all rats were euthanized and were sacrificed using CO_2 (30%) prior to decapitation; serum was collected and stored at −80 °C for further analysis.

2.2. Sample Preparation for Measurement of Serum Metabolites

Serum samples for chromatographic analyses were prepared by adding 100 µL of ice-cold 10% meta-phosphoric acid to 100 µL of serum followed by a 10 min incubation on ice, centrifugation (18,000× *g* for 15 min at 4 °C), and filtration (0.2 µm nylon filter). Aliquots of the resultant extract (20 µL) were analyzed with HPLC with electrochemical detection (ECD) and HPLC with ultraviolet detection (UV) (see below).

For liquid chromatography mass spectrometry analyses (LC-MS), 50 µL of serum was mixed with 300 µL of HPLC-grade methanol, followed by centrifugation (18,000× *g* for 15 min at 4 °C). Supernatants were dried under nitrogen flow and then dissolved in 50 µL of 50% methanol/0.2% formic acid, after which 5 µL samples were injected and analyzed with LC-MS (see below).

2.3. HPLC and LC-MS Methods

2.3.1. HPLC-ECD

Unique methodological details for metabolite analysis with HPLC-ECD have been published previously [23,24]. Serum extract samples (20 µL) were analyzed using an HPLC-ECD system (Model 580, ESA Inc., Chelmsford, MA, USA) with a Phenomenex (Phenomenex Inc., Torrance, CA, USA) Capcell Pak reverse phase C_{18} column (4.6 × 150 mm,

3 μm particle size). The isocratic mobile phase was composed of 50 mM sodium phosphate, 1.0 mM octanesulfonic acid, and 2% acetonitrile (v/v), adjusted to pH 2.7, at a flow rate of 1 mL/min and a column compartment thermostat temperature of 25 °C. Metabolite identification (based on retention time) and quantification (based on peak areas of the standard and sample) were performed using HPLC software (Dionex Corp., Sunnyvale, CA, USA, 7.3).

2.3.2. HPLC-UV

Serum extract samples (20 μL) were also analyzed using an HPLC-UV system (Thermo Scientific UltiMate 3000, Thermo Fisher Scientific Inc., Waltham, MA, USA) with a Raptor AR C_{18} column (150 × 4.6 mm, 2.7 μm particle size) purchased from Restek Co., (Bellefonte, PA, USA). The isocratic mobile phase was composed of 0.05 mM KH_2PO_4 and 10% acetonitrile (v/v) at a flow rate of 1 mL/min and UV detection at 220, 240, 260, and 280 nm. Peak identification and plasma concentration of metabolites, calculated from peak areas and standard calibration curves, were determined using HPLC software (Dionex Corp., Sunnyvale, CA, USA, 7.3).

2.3.3. LC-MS

LC-MS was performed with a Thermo Fisher Scientific LTQ XL mass spectrometer coupled with a Thermo Scientific UltiMate 3000 HPLC system with a +3.5 kV spray voltage, a 275 °C capillary temperature, a 300 °C heater temperature, a sheath gas flow of 25 L/min, an auxiliary gas flow of 10 L/min, and a mass scan range of 80–900 m/z. The mobile phase was a 30 min gradient flow of (A) 10 mM ammonium formate in 95:5 acetonitrile/water with 0.1% formic acid and (B) 10 mM ammonium formate in 50:50 acetonitrile/water with 0.1% formic acid at a flow rate of 300 μL/min. Extract samples (5 μL) were injected into a Thermo Fisher Scientific Accurore C_{18} column (100 × 2.1 mm, 2.6 μm particle size) with a column temperature of 40 °C. MS data collection and analysis were performed using Thermo Fisher Scientific Xcalibur™ software v 2.2 SP1.48. Metabolites and their concentrations were determined based on retention time and molecular weight (m/z) using Thermo Fisher Scientific Mass Frontier 7.0 software.

2.4. Data Processing and Statistical Analysis

Mass Frontier 7.0 (Thermo Fisher Scientific software, San Jose, CA, USA) was used for the management, evaluation, and interpretation of mass spectra. For statistical analyses, one-way ANOVA and SigmaPlot 13.0 software (Systat Software, Inc., San Jose, CA, USA) were used, and the results are presented as mean ± standard deviation (SD). Student's two-tailed t test was used for independent comparisons of the heights of common peaks between the groups of Zucker rats. A p-value < 0.05 was considered a significant difference.

3. Results

3.1. Body Weight

The mean ± SD body weight (g) of lean and obese rats at the beginning of the experiment were as follows: lean (n = 14), 98 ± 7 g; obese (n = 14), 155 ± 13 g. After 10 weeks, the body weight of untreated lean and obese rats and those treated with metformin were LC, 269 ± 26 g; OC, 590 ± 41; LM, 278 ± 17 g; and OM, 573 ± 48 g. Our results showed that obese rats gained significantly (<0.001) more weight than lean rats after 8 weeks. Also, after 10 weeks of metformin treatment, obese rats gained body weight significantly higher in both the control and metformin treatment groups (p < 0.001) compared to lean rats. However, there were no differences (p = 0.20) between OC and OM groups.

3.2. Untargeted Metabolomics Data

To generate the data for this study, we used an untargeted metabolomics approach using different analytical liquid chromatographic platforms (HPLC-ECD, HPLC-UV, and LC-MS). We were able to collect over 400 high-quality (clean and distinguished peaks on the chromatograms) and high-intensity (the minimum noise-to-peak ratio was 1:10) peaks. We selected the 50 highest peaks that were shared between the four groups of rats. Each peak represented a single metabolite.

Analysis of all 50 individual peaks and average peak heights of seven LC or OC and seven LM or OM rats are presented in Figure 2. Student's *t* test was used to compare the average heights between the groups. We divided these peaks into two different categories: (1) groups not significantly different ($p > 0.05$) with subcategory of slight difference between the groups with *p*-values between 0.1 and 0.05; and (2) groups significantly different ($p < 0.05$).

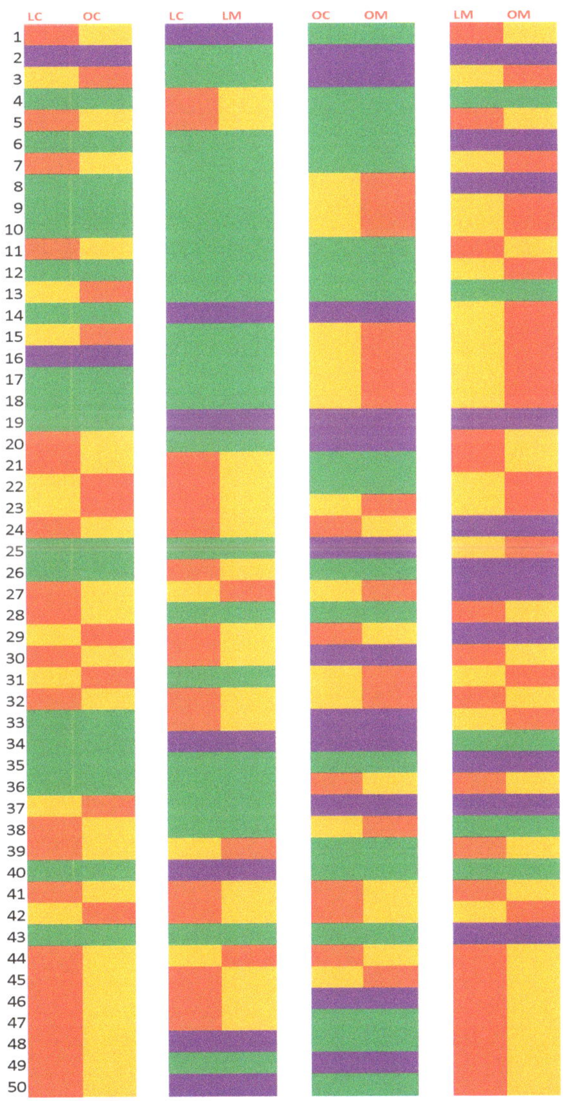

Figure 2. Heat map of metabolic differences in serum between lean and obese, control and metformin-treated Zucker rats. ■ Statistically higher ($p < 0.05$); ■ Statistically lower ($p < 0.05$); ■ Marginal ($0.1 > p > 0.05$); ■ No difference ($p > 0.05$).

Based on our data in Figure 2, we compared the metabolic profiles of the LC and OC rats and found that 30 peaks (Figure 3A) demonstrated statistically different heights, 18 peaks demonstrated no differences, and 2 peaks demonstrated marginal differences. Statistically different groups were further divided (Figure 3B) into two subgroups: 21 peaks were higher in LC rats, and only 9 peaks were higher in OC rats.

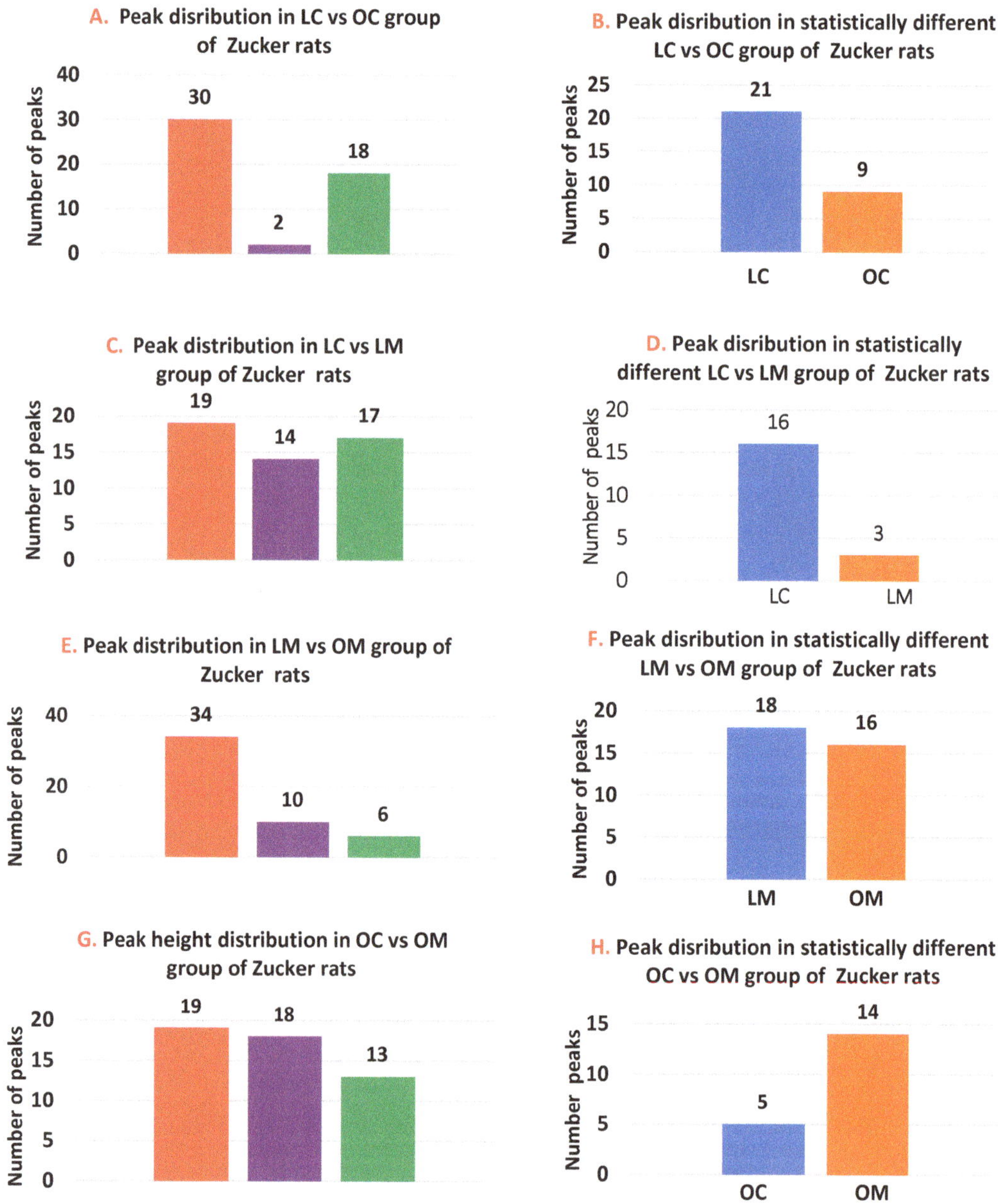

Figure 3. Peaks distribution in lean control (LC), lean metformin (LM), obese control (OC), and obese metformin (OM) rats. ■ $p < 0.05$, ■ $0.1 > p > 0.05$, ■ $p > 0.05$.

By comparing the peak heights of the LC and LM groups (Figure 3C), we found that 19 peaks demonstrated statistical differences, 17 peaks demonstrated no differences, and 14 peaks demonstrated marginal differences. Statistically different groups were further divided (Figure 3D) into two subgroups: 16 peaks were higher in LC rats, and only 3 peaks were higher in LM rats.

Upon comparing the peak heights of the LM and OM groups (Figure 3E), we found that 34 peaks demonstrated statistical differences, 6 peaks demonstrated no differences, and 10 peaks demonstrated marginal differences. Statistically different groups were further divided (Figure 3F) into two subgroups: 18 peaks were higher in LM rats, and 16 peaks were higher in OM rats.

Finally, by comparing the peak heights of the OC and OM groups (Figure 3G), we found that 19 peaks demonstrated statistical difference, 13 peaks demonstrated no difference, and 18 peaks demonstrated marginal differences between animals. Statistically different groups were further divided (Figure 3H) into two subgroups: only 5 peaks were higher in OC rats, and 14 peaks were higher in OM rats.

After calculating the total number of peaks in the statistically higher groups of Zucker rats, we analyzed their distribution according to the individual number (ID) of peaks between groups of animals. We observed that the metformin treatment of lean rats (LC vs. LM) (Figure 4A) demonstrated a predominant (16:3) decrease in the height of the peaks, and the metformin treatment of obese animals (OC vs. OM) (Figure 4B) demonstrated a predominant (14:5) increase in the height of the peaks. All individual peaks could be divided into three subcategories: (1) individual higher peak IDs that represent only lean animals (LC vs. LM), (2) individual higher peak IDs that represent only obese animals (OC vs. OM), and (3) individual higher peak IDs that are common between the two groups (Figure 4C).

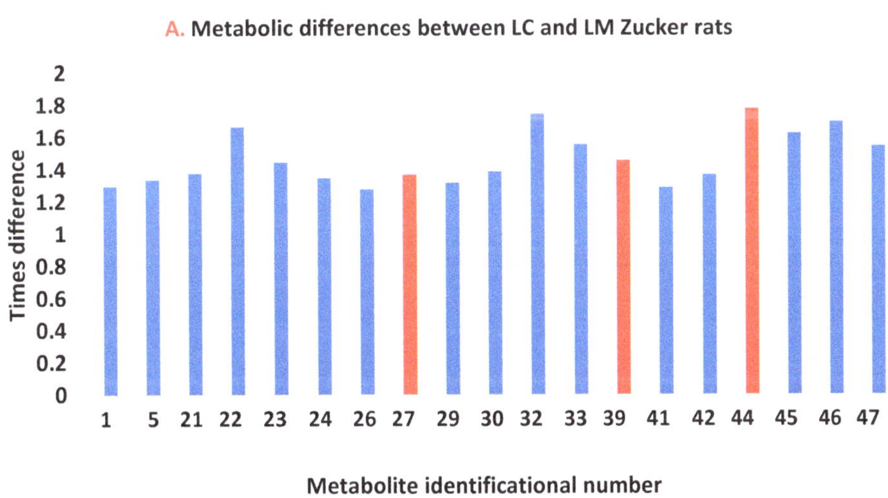

Figure 4. *Cont.*

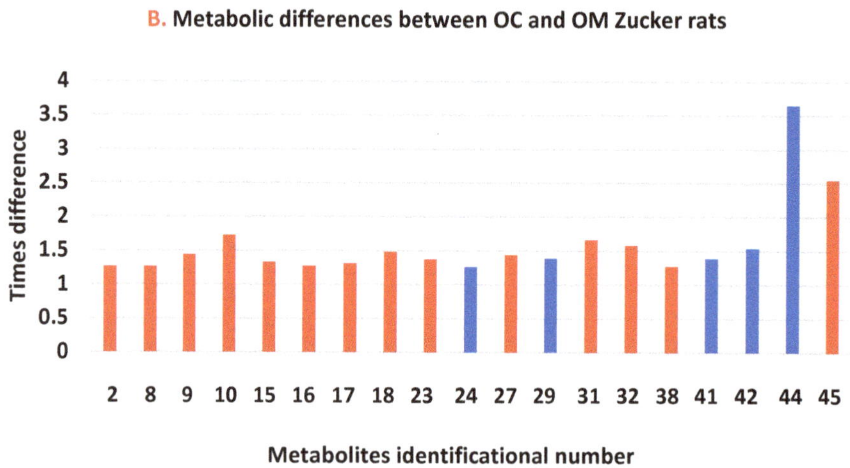

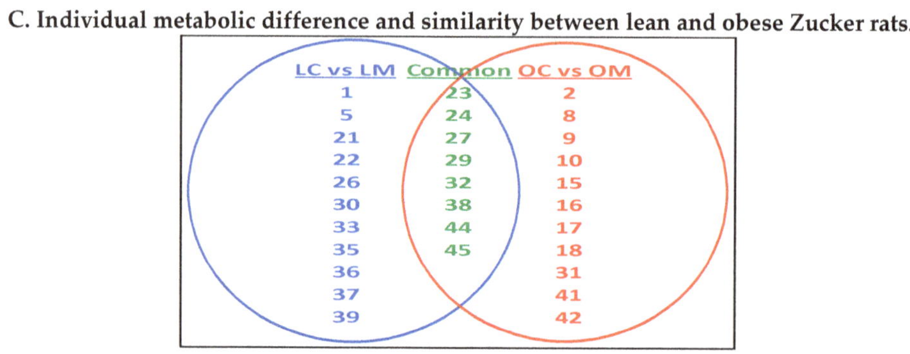

Figure 4. Metabolic differences in individual peak heights between lean and obese Zucker rats. ■ Peaks more pronounced in obese animals. ■ Peaks more pronounced in lean animals. ■ Peak that are similar among lean and obese animals.

4. Discussion

We used a lean and obese Zucker rat model to investigate the effects of metformin treatment on serum metabolic changes using an untargeted metabolomics approach. The studies of metabolic reactions in Zucker rats are very important for the generation of basic systemic knowledge of metabolism and to direct further steps in experimentation with this rat model. Few investigators have reported [25] data regarding metabolic changes in Zucker rats that were collected using a metabolomics approach [26]. The majority of research focuses on a better understanding of metformin's protective mechanisms related to certain pathological conditions, including diabetes, but does not focus on fundamental metabolic changes [27–31] or even extend applications of metformin [32]. The general metabolic studies of the effect of metformin on Zucker rats that have used a metabolomics approach have been extremely limited.

In the present study, we introduced a combination of analytical platforms from those that are complicated and technically demanding, such as LC-MS, to the relatively simple and affordable ones, such as HPLC-UV and HPLC-ECD. Our results demonstrated excellent practical potential and advantages whether we used all platforms together or separately in metabolomics studies.

Previously, we reported that the basic metabolic status of untreated lean and obese Zucker rats (LC and OC in this study) demonstrated a difference in 60% of the metabolites we analyzed and also demonstrated significant differences in peak heights (amplitude) and direction (shift of the peaks toward lean or obese group) [22]. These differences were present in both lean and obese animals, but the height of the selected peaks' domination (42% vs. 18%) was present in lean animals compared to obese. Such basic phenotypic differences could also predict metabolism. These metabolites could be critical in the use of targeted metabolomics and identification in future experiments in Zucker rats and other models. Our data demonstrated that the treatment of Zucker rats with metformin led to a variety of changes in metabolic status in both lean and obese animals. These metabolic changes/modifications, however, developed a different amplitude and direction in lean and obese animals. We found that peak heights (which represent metabolite concentrations) in lean control and lean metformin-treated rats were affected in approximately one-third of metabolites of interest with definite domination in the lean control group compared to the metformin-treated group (84% in LC vs. 16% in LM). On the other hand, metformin treatment resulted in decreased serum concentrations of certain metabolites in lean rats. In obese rats, we observed similar total metabolic changes (38%) in peak amplitudes but in the opposite direction (31% in OC vs. 69% in OM). The metformin treatment of obese rats stimulated an increase in some metabolite concentrations.

The different directions in the reaction of metformin-treated lean and obese rats make us question how each individual metabolite (from the 50 selected) will react in lean and obese rats. We found that lean and obese rats have a pool of eight common metabolites that react in the same direction in both groups and a pool of 11 distinct metabolites for each group. These 11 metabolites in obese rats could potentially be good candidates for further study by targeted metabolomics to identify, quantify, and tie them to specific metabolic pathways. They could also be good candidates for the determination of treatment biomarkers. In general, upon comparing peak heights of lean and obese metformin-treated (LM vs. OM) rat groups (Figure 5), we observed that 68% of peaks/metabolites affected were statistically different. They were distributed equally (17 in LM and 17 in OM), and each animal group had its own pool of peaks/metabolites that independently reacted to the metformin treatment.

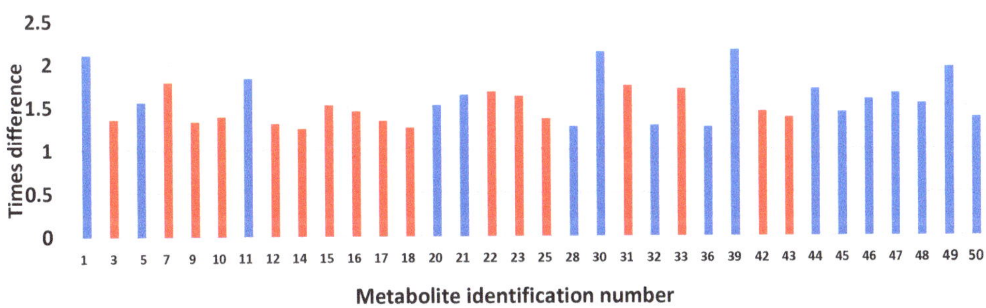

Figure 5. Changes of individual peak heights of metformin-treated lean and obese Zucker rats. ■ Peaks higher in the lean group. ■ Peaks higher in the obese group.

The protective effect of metformin alone or in combination with other medical treatment of diabetes have been reported in human and animal model studies. This effect is very complex, multifactorial, and most importantly, not fully understood. In our current study, we applied an untargeted metabolomics approach to maximize the data collection from our analytical installation of an obese Zucker rat model. These data will contribute to better understanding the metabolic bases of metformin effects on obese animals with promise for a future better manipulation through diet/treatment to optimize the protective effect of metformin. We found (Figure 4B) that 19 out of 50 (38%) of the metabolites in obese Zucker rats react to the treatment of metformin by increasing/decreasing their levels in serum. The identification of these metabolites is a critically important future step of our research.

Previously, as was reported in [22], there is a significant difference in serum metabolic profile between lean and obese Zucker rats. We reported that obese rats had a significantly lower methionine, free cysteine, tryptophan, kynurenic acid, and higher levels of cystine and tyrosine compared to lean rats. Cysteine/cystine ratio, also known as an oxidative ratio [33], was lower in obese rats. This couple plays an important reductive/oxidative role in support of oxidative homeostasis, and along with the decrease in the ratio, obese rats are becoming more vulnerable to oxidative stress changes and damages. In the present study, we selected the same metabolic row for a comparison in obese control and obese metformin-treated Zucker rats (Table 1). We noticed that the metformin treatment of obese rats is also capable of changing a metabolic profile in its own unique way. It was observed to increase the levels of free cysteine, tryptophan, and kynurenic acid in obese metformin-treated animals compared to control animals. Such changes can positively affect resistance to oxidative stress through the further involvement of cysteine in the synthesis of glutathione as well as affect the gut–brain axis through the involvement of tryptophan and kynurenic acid. The question about a full scale of metabolic perturbations, including the interconnection between these two pathways mentioned above, are open for further study.

Table 1. Serum metabolite concentrations of obese control (OC) and obese metformin (OM)-treated rats.

	OC	OM	*p*-Value
Methionine, nmol/mL	36.4 ± 5.72	34.4 ± 10.18	0.34
Cysteine, nmol/mL	10.1 ± 1.34	12.74 ± 2.61	0.024
Cystine, nmol/mL	14.7 ± 2.71	18.69 ± 5.75	0.071
Cysteine/Cystine	0.71 ± 0.168	0.74 ± 0.317	0.42
Tryptophan, nmol/mL	39.1 ± 9.14	48.2 ± 6.28	0.029
Kynurenic acid, nmol/mL	1.66 ± 0.332	2.27 ± 0.576	0.023
Tryptophane/Kynurenic acid	24.6 ± 8.02	22.7 ± 7.04	0.32
Tyrosine, nmol/mL	62.2 ± 10.83	66.5 ± 8.21	0.21

5. Conclusions

The current study helped us to better understand the basic metabolic changes in lean and obese Zucker rats with and without metformin treatment. Our data showed that both obesity and metformin treatment make significant impacts on the metabolome of Zucker rats, including oxidative stress and the gut–brain axis metabolites and their mutual interactions. These observed changes need to be further explored with a much more detailed targeted metabolomics approach to discover metabolic biomarkers.

Such biomarkers could be extremely helpful in the clarification of our options for choosing adequate tools for assessing the metabolic status of experimental obesity models using an untargeted metabolomics approach. In addition, biomarkers discovered in the future will help to identify the involvement of biochemical pathways in the treatment and management of obesity and glucose-lowering actions.

Author Contributions: S.M. participated in the study design, performed the experiments, and participated in the collection of data, interpretation of the study results, and writing of the manuscript. R.H. designed the study, funded this project, and participated in the collection of data and writing of the manuscript. All authors have read and agreed to the published version of the manuscript.

Funding: The present study was supported by funding from the Arkansas Bioscience Institute and Arkansas Children's Research Institute (RH).

Institutional Review Board Statement: The animal protocols used in the present study were approved by the Institutional Animal Care and Use Committee at the University of Arkansas for Medical Sciences in 2018 (approval no. 3882).

Data Availability Statement: The data presented in this study are available in this article.

Acknowledgments: We would like to thank David Irby for helping with animal experiments. This study was supported by the Arkansas Biosciences Institute.

Conflicts of Interest: The authors declare no conflict of interest.

References

1. Lustig, R.H.; Collier, D.; Kassotis, C.; Roepke, T.A.; Kim, M.J.; Blanc, E.; Barouki, R.; Bansal, A.; Cave, M.C.; Chatterjee, S.; et al. Obesity I: Overview and molecular and biochemical mechanisms. *Biochem. Pharmacol.* **2022**, *199*, 115012. [CrossRef] [PubMed]
2. Engin, A. The Definition and Prevalence of Obesity and Metabolic Syndrome. *Adv. Exp. Med. Biol.* **2017**, *960*, 1–17. [CrossRef]
3. Ala, M.; Ala, M. Metformin for Cardiovascular Protection, Inflammatory Bowel Disease, Osteoporosis, Periodontitis, Polycystic Ovarian Syndrome, Neurodegeneration, Cancer, Inflammation and Senescence: What Is Next? *ACS Pharmacol. Transl. Sci.* **2021**, *4*, 1747–1770. [CrossRef] [PubMed]
4. Guh, D.P.; Zhang, W.; Bansback, N.; Amarsi, Z.; Birmingham, C.L.; Anis, A.H. The incidence of co-morbidities related to obesity and overweight: A systematic review and meta-analysis. *BMC Public Health* **2009**, *9*, 88. [CrossRef]
5. Solis-Herrera, C.; Triplitt, C.; Cersosimo, E.; DeFronzo, R.A. Pathogenesis of Type 2 Diabetes Mellitus. In *Endotext*; Internet; MDText.com, Inc.: South Dartmouth, MA, USA, 2021.
6. Aras, M.; Tchang, B.G.; Pape, J. Obesity and Diabetes. *Nurs. Clin. N. Am.* **2021**, *56*, 527–541. [CrossRef]
7. Apovian, C.M. Obesity: Definition, comorbidities, causes, and burden. *Am. J. Manag. Care* **2016**, *22* (Suppl. S7), S176–S185. [PubMed]
8. Khodadadi, M.; Jafari-Gharabaghlou, D.; Zarghami, N. An update on mode of ac-tion of metformin in modulation of meta-inflammation and inflammaging. *Pharmacol. Rep.* **2022**, *74*, 310–322. [CrossRef]
9. Clish, C.B. Metabolomics: An emerging but powerful tool for precision medicine. *Cold Spring Harb. Mol. Case Stud.* **2015**, *1*, a000588. [CrossRef]
10. Costanzo, M.; Caterino, M.; Sotgiu, G.; Ruoppolo, M.; Franconi, F.; Campesi, I. Sex differences in the human metabolome. *Biol. Sex Differ.* **2022**, *13*, 30. [CrossRef]
11. Kim, H.W. Metabolomic Approaches to Investigate the Effect of Metformin: An Overview. *Int. J. Mol. Sci.* **2021**, *22*, 10275. [CrossRef]
12. Xie, B.; Waters, M.J.; Schirra, H.J. Investigating Potential Mechanisms of Obesity by Metabolomics. *J. Biomed. Biotechnol.* **2012**, 805683. [CrossRef] [PubMed]
13. Mazza, A.; Fruci, B.; Garinis, G.A.; Giuliano, S.; Malaguarnera, R.; Belfiore, A. The role of metformin in the management of NAFLD. *Exp. Diabetes Res.* **2012**, *2012*, 716404. [CrossRef] [PubMed]
14. Huang, K.H.; Lee, C.H.; Cheng, Y.D.; Gau, S.Y.; Tsai, T.H.; Chung, N.J.; Lee, C.Y. Correlation between long-term use of metformin and incidence of NAFLD among patients with type 2 diabetes mellitus: A real-world cohort study. *Front. Endocrinol.* **2022**, *13*, 1027484. [CrossRef] [PubMed]
15. Hakkak, R.; Rose, S.; Spray, B.; Kozaczek, M.; Korourian, S. Effects of Obesity and 10 Weeks of Metformin Treatment on Liver Steatosis. *J. Biomed. Rep.* **2021**, *14*, 49. [CrossRef]
16. Robeson, M.; Manna, K.; Randolph, C.; Byrum, S.; Hakkak, R. Short-term Metformin Treatment En-riches Bacteroides dorei in Obese Liver Steatosis Zucker Rat Model. *Front. Microbiol.* **2021**, *13*, 834776. [CrossRef]
17. Zucker, T.F.; Zucker, L.M. Fat accretion and growth in the rat. *J. Nutr.* **1963**, *80*, 6–19. [CrossRef]
18. Zucker, L.M.; Zucker, T.F. Fatty, a new mutation in the rat. *J. Hered.* **1961**, *52*, 275–278. [CrossRef]
19. Chua, S.C.; Chung, W.K.; Wu-Peng, X.S.; Zhang, Y.; Liu, S.M.; Tartaglia, L.; Leibel, R.L. Phenotypes of mouse 'diabetes' and rat 'fatty' due to mutations in the OB (leptin) receptor. *Science* **1996**, *271*, 994–996. [CrossRef]
20. Tartaglia, L.A.; Dembski, M.; Weng, X.; Deng, N.; Culpepper, J.; Devos, R.; Richards, G.J.; Campfield, L.A.; Clark, F.T.; Deeds, J.; et al. Identification and expression cloning of a leptin receptor, OB-R. *Cell* **1995**, *83*, 1263–1271. [CrossRef]
21. Zucker, L.M.; Antoniades, H.N. Insulin and obesity in the Zucker genetically obese rat "fatty". *Endocrinology* **1971**, *90*, 1320–1330. [CrossRef]
22. Melnyk, S.; Hakkak, R. Metabolic Status of Lean and Obese Zucker Rats Based on Untarget-ed and Targeted Metabolomics Analysis of Serum. *Biomedicines* **2022**, *10*, 153. [CrossRef] [PubMed]
23. Melnyk, S.; Pogribna, M.; Pogribny, I.; Hine, R.J.; James, S.J. A new HPLC method for the simultaneous determination of oxidized and reduced plasma aminothiols using coulometric electrochemical detection. *J. Nutr. Biochem.* **1999**, *10*, 490–497. [CrossRef] [PubMed]
24. Melnyk, S.; Pogribna, M.; Pogribny, I.P.; Yi, P.; James, S.J. Measurement of plasma and intracellular S-adenosylmethionine and S-adenosylhomocysteine utilizing coulometric electrochemical detection: Alter-ations with plasma homocysteine and pyridoxal 5′-phosphate concentrations. *Clin. Chem.* **2000**, *46*, 265–272. [CrossRef] [PubMed]
25. Dong, Y.; Chen, Y.T.; Yang, Y.X.; Shou, D.; Li, C.Y. Urinary Metabolomic Profiling in Zucker Diabetic Fatty Rats with Type 2 Diabetes Mellitus Treated with Glimepiride, Metformin, and Their Combination. *Molecules* **2016**, *21*, 1446. [CrossRef]
26. Wang, H.; Teng, Y.; Li, S.; Li, Y.; Li, H.; Jiao, L.; Wu, W. UHPLC-MS-Based Serum and Urine Metabolomics Reveals the Anti-Diabetic Mechanism of Ginsenoside Re in Type 2 Diabetic Rats. *Molecules* **2021**, *26*, 6657. [CrossRef]

27. Sambe, T.; Mason, R.P.; Dawoud, H.; Bhatt, D.L.; Malinski, T. Metformin treatment decreases nitroxidative stress, restores nitric oxide bioavailability and endothelial function beyond glucose control. *Biomed. Pharmacother.* **2018**, *98*, 149–156. [CrossRef]
28. Shoghi, K.I.; Finck, B.N.; Schechtman, K.B.; Sharp, T.; Herrero, P.; Gropler, R.J.; Welch, M.J. In vivo metabolic phenotyping of myocardial substrate metabolism in rodents: Differential efficacy of metformin and rosig-litazone monotherapy. *Circ. Cardiovasc. Imaging* **2009**, *2*, 373–381. [CrossRef]
29. Triggle, C.R.; Mohammed, I.; Bshesh, K.; Marei, I.; Ye, K.; Ding, H.; MacDonald, R.; Hollenberg, M.D.; Hill, M.A. Metformin: Is it a drug for all reasons and diseases? *Metabolism* **2022**, *133*, 155223. [CrossRef]
30. Sui, Y.; Kong, X.; Fan, R.; Ye, Y.; Mai, H.; Zhuo, S.; Lu, W.; Ruan, P.; Fang, S.; Yang, T. Long-term treatment with metformin in the prevention of fatty liver in Zucker diabetic fatty rats. *Diabetol. Metab. Syndr.* **2019**, *11*, 94. [CrossRef]
31. Cwynar-Zając, Ł. Metformin—A new approach. *Pediatr. Endocrinol. Diabetes Metab.* **2021**, *27*, 134–140. [CrossRef]
32. Zhou, J.; Massey, S.; Story, D.; Li, L. Metformin: An Old Drug with New Applications. *Int. J. Mol. Sci.* **2018**, *19*, 2863. [CrossRef] [PubMed]
33. Jones, D.P.; Go, Y.-M.; Anderson, C.L.; Ziegler, T.R.; Kinkade, J.M., Jr.; Kirlin, W.G. Cysteine/cystine couple is a newly recognized node in the circuitry for biologic redox signaling and control. *FASEB J.* **2004**, *18*, 1246–1248. [CrossRef] [PubMed]

biomolecules

MDPI

Review

Butyrate Supplementation Improves Intestinal Health and Growth Performance in Livestock: A Review

Wenting Chen [1,†], Qingshan Ma [1,†], Yan Li [1], Lin Wei [1], Zhenwei Zhang [1], Adnan Khan [2], Muhammad Zahoor Khan [1,*] and Changfa Wang [1,*]

[1] School of Agricultural Science and Engineering, Liaocheng University, Liaocheng 252000, China
[2] Animal Genomics Laboratory, School of Agriculture and Food Science, University College Dublin, D04 V1W8 Dublin, Ireland; adnan.khan@ucd.ie
* Correspondence: zahoorkhan@lcu.edu.cn (M.Z.K.); wangchangfa@lcu.edu.cn (C.W.)
† These authors contributed equally to this work.

Abstract: Butyrate supplementation has gained considerable attention for its potential benefits in livestock, particularly concerning intestinal health and growth performance. This review synthesizes recent research on the diverse roles of butyrate, across various livestock species. As a short-chain fatty acid, butyrate is known for enhancing intestinal development, improving immune function, and modulating microbial diversity. Studies indicate that butyrate supports gut barrier integrity, reduces inflammation, and optimizes feed efficiency, especially during the critical weaning and post-weaning periods in calves, piglets, and lambs. Supplementation with butyrate in livestock has been shown to increase average daily gain (ADG), improve gut microbiota balance, promote growth, enhance gut health, boost antioxidant capacity, and reduce diarrhea. Additionally, butyrate plays a role in the epigenetic regulation of gene expression through histone acetylation, influencing tissue development and immune modulation. Its anti-inflammatory and antioxidant effects have been demonstrated across various species, positioning butyrate as a potential therapeutic agent in animal nutrition. This review suggests that optimizing butyrate supplementation strategies to meet the specific needs of each species may yield additional benefits, establishing butyrate as an important dietary additive for enhancing growth performance and health in livestock.

Keywords: butyrate; livestock; microbiota; intestinal health; inflammation; antioxidants; growth performance

1. Introduction

Butyrate plays a critical role in animal health, serving not only as an important energy source for colonocytes but also as a cellular mediator that regulates various functions in intestinal cells and other tissues [1,2]. Additionally, it is a key supplement that effectively promotes intestinal development, enhances overall health, and improves growth performance [3–7]. Recent studies have emphasized its role in strengthening the integrity of the gut barrier and supporting a balanced gut microbiota. For example, sodium butyrate supplementation has been shown to significantly increase the average daily gain (ADG) and enhance microbiota diversity, notably impacting *Firmicutes*, *Bacteroidetes*, and *Lactobacillus*, which are crucial for calf development [8–10]. Moreover, butyrate supplementation supports improvements in gastrointestinal structures, such as the development of rumen papillae and duodenal villi, which are essential for nutrient absorption and overall gut health [11]. Besides these structural

https://doi.org/10.3390/biom15010085

benefits, continuous supplementation of butyrate can reduce diarrhea incidence, enhance calf health, and improve postnatal growth and immune responses [11,12].

One of the most notable advantages of butyrate supplementation is its ability to promote healthy gut microbiota, which are vital for optimal growth and health. Specifically, butyrate increases microbial diversity and supports the growth of beneficial bacteria such as *Firmicutes*, *Bacteroides*, *Prevotella*, and *Ruminococcus*, while simultaneously reducing harmful pathogens like *Proteobacteria* and *Escherichia/Shigella* [13,14]. This beneficial modulation of the gut microbiota is particularly significant during early intestinal development in animals, when microbial diversity is critical for establishing a robust digestive system.

In addition to modulating the gut microbiota, butyrate also plays a crucial role in immune regulation and inflammation control. It reduces the production of pro-inflammatory cytokines, inhibits the activation of nuclear factor kappa B (NF-κB), and protects the integrity of tight junctions in epithelial cells [15]. Notably, sodium butyrate supplementation has been linked to decreased levels of inflammatory markers such as interleukin-1β (IL-1β), IL-6, and tumor necrosis factor-α (TNF-α), leading to improved immune responses and reduced stress during critical growth phases [16,17]. Furthermore, these immune-regulatory effects extend to enhancing resistance to infections, as demonstrated by studies where butyrate improved host defenses against *Mycobacterium bovis*, the causative agent of bovine tuberculosis [18].

At the cellular level, butyrate influences gene expression through epigenetic mechanisms, particularly by inhibiting histone deacetylases (HDACs), which results in increased histone acetylation [19]. This process promotes the transcription of anti-inflammatory genes while suppressing the expression of pro-inflammatory cytokines, thereby enhancing anti-inflammatory responses [20,21]. Additionally, butyrate regulates key signaling pathways, such as AMP-activated protein kinase (AMPK) and mitogen-activated protein kinase (MAPK), which are essential for tissue development and metabolic regulation [15,22]. These epigenetic effects are particularly beneficial for improving growth performance and supporting animal health during stressful periods such as weaning.

This review aims to synthesize and assess the current research on the effects of butyrate supplementation on intestinal health and growth performance in livestock, including calves, piglets, sheep, and goats. As a short-chain fatty acid, butyrate plays a pivotal role in enhancing gut barrier function, modulating immunity, maintaining microbial balance, and influencing epigenetic regulation. This comprehensive overview highlights the multifaceted benefits of butyrate, including growth promotion, increased microbial diversity, inflammation control, and improved cellular functions (Figure 1). It offers valuable insights for researchers and industry professionals aiming to optimize livestock health and productivity.

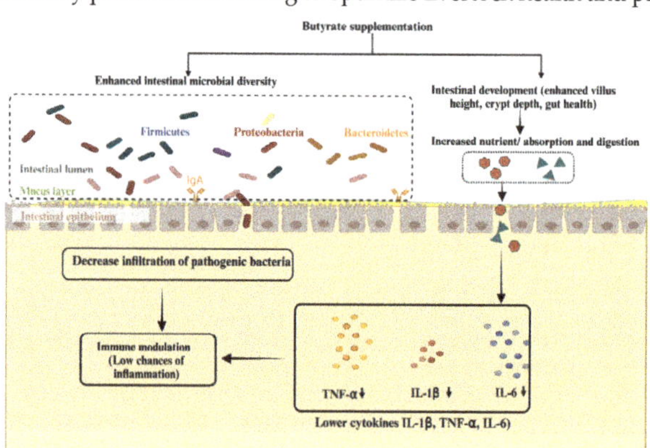

Figure 1. Butyrate's role in intestinal development, microbiota balance, immune response, and promotion of antiinflammation. Butyrate plays a pivotal role in maintaining microbiota balance, promoting

intestinal development, and enhancing growth performance. It supports anti-inflammatory mechanisms by reducing levels of pro-inflammatory cytokines such as IL-6, IL-1β, and TNF-α. Additionally, butyrate contributes to immune regulation by modulating immune cell function, thereby supporting intestinal health and sustaining overall immune homeostasis.

2. Methodology for Literature Search

This review synthesized data from published articles ranging from 2017 to the present, along with five additional articles published between 2011 and 2016, to ensure a comprehensive analysis of the content. The literature search employed the following keywords: "butyrate", "intestinal health", "growth", "animal production performance", "inflammation", "immunity", "intestinal microbiota", "sodium butyrate", "tributyrin", "livestock", "cows", "pigs", "sheep", and "goats". We exclusively considered articles published in peer-reviewed journals indexed in the Science Citation Index (SCI) and written in the English language. Additionally, we excluded data from books, book chapters, conference papers, and unpublished materials to maintain the rigor and relevance of the review. Finally, all the figures used in this study were created using https://BioRender.com.

3. Butyrate Role in Livestock Intestinal Health and Growth Performance

Butyrate is naturally produced in the large intestine and the foregut of ruminants through bacterial fermentation, a process primarily mediated by the enzyme butyryl-CoA/acetate-CoA transferase. Importantly, a balanced diet rich in fermentable fiber and non-structural carbohydrates can optimize butyrate production. Butyrate plays a crucial role in livestock nutrition, with numerous studies consistently demonstrating its significant positive effects on growth, intestinal health, and modulation of the gut microbiota [23–26]. Notably, the multifunctional properties of butyrate extend beyond its role in growth promotion; it also enhances intestinal integrity and shows promise as a treatment for various gastrointestinal disorders [27–33]. Furthermore, this short-chain fatty acid regulates critical gastrointestinal functions by modulating epithelial cell proliferation, apoptosis, and differentiation in response to microbial growth and signaling mechanisms [10,34]. In the following sections, we have discussed the essential role of butyrate in promoting livestock health, with a focus on its anti-inflammatory, antioxidant, and immune-regulating properties, as well as its impact on growth performance, including microbiota balance, daily weight gain, and enhanced intestinal development.

3.1. *Butyrate's Role in Calf Intestinal Health and Growth Promotion*

Butyrate, as a key feed additive, plays a significant role in promoting intestinal health and growth in calves, particularly during the vulnerable weaning and post-weaning periods. Recent studies have shown that butyrate has multiple effects, including supporting intestinal development, regulating immune responses, maintaining gut barrier integrity, and promoting gut microbiota balance in calves [8,35,36]. The supplementation with 8.78 g of sodium butyrate per day significantly increased ADG and positively affected microbial diversity and composition, notably enhancing the populations of *Lactobacillus*, *Ruminococcus*, *Romboutsia*, and *Erysipelotrichaceae_UCG-003* [9]. Furthermore, a quadratic pattern was observed in the improvements in growth performance and gut health, suggesting that adjusting the sodium butyrate dosage provides targeted benefits [9]. Consistently, research on supplementing 45 g/day of sodium butyrate showed significant improvements in bacterial diversity. Further, it has been shown to promote the proliferation of beneficial bacterial genera, including *Firmicutes*, *Bacteroides*, *Prevotella*, and *Ruminococcus*, while simultaneously decreasing the abundance of potentially pathogenic bacteria such as *Proteobacteria* and *Escherichia-Shigella*. These findings further confirm the critical role of sodium butyrate in promoting a balanced and robust microbial environment during the early stages of calf development [13]. Notably, the effects of sodium butyrate are not limited to microbiota modulation. A study investigating the continuous addition of sodium butyrate to the liquid diet of dairy calves demonstrated a significant reduction in the incidence of diarrhea and overall morbidity. Furthermore, calves receiving 4 g of sodium butyrate per day showed improvements in growth, gastrointestinal development, and immune responses, underscoring the importance of butyrate in enhancing the structural and functional aspects of the digestive system during the critical post-weaning period [11]. Similarly, another study examined the impact of sodium butyrate as a milk replacer additive in Holstein's calves. Twenty-four calves were divided into three groups receiving 0, 5, or 10 g of sodium butyrate per day for 49 days. Their results indicated that sodium butyrate supplementation improved feed conversion ratios and ADG, with the 5 g per day group showing the most pronounced benefits. Additionally, sodium butyrate increased total blood protein, enhanced rumen bacterial populations, and lowered ruminal pH, ultimately promoting rumen microbial development and energy metabolism [24]. In a parallel study, researchers explored the effects of sodium butyrate on the transcriptomic pathways of the gastrointestinal epithelium and the rumen microbiome in calves fed a high-fiber diet [3]. Their results showed that sodium butyrate supplementation improved ADG, enhanced jejunum and rumen papillae development, and downregulated inflammatory pathways such as IL-17 and NF-κB. Additionally, sodium butyrate increased the populations of beneficial bacteria, including *Bacillus subtilis* and *Eubacterium limosum*, further emphasizing its role in promoting calf health and nutrition [3].

Similarly, another study investigated the effects of sodium butyrate on pre-weaning calves, focusing on growth performance, diarrhea incidence, gastrointestinal tract (GIT) development, and tight junction protein expression. Sodium butyrate treatment effectively reduced diarrhea rates and increased rumen weight and papillae surface area. Additionally, sodium butyrate increased mucosal thickness in the abomasum and duodenum, improved ileal villus height and villus height-to-crypt depth ratios, and upregulated claudin-1 expression in the rumen, which was negatively correlated with diarrhea incidence. Furthermore, adding 4 g/day of sodium butyrate to milk replacer from day 4 to day 60 significantly affected starter feed intake, total dry matter intake, and hematological parameters such as glucose, urea, insulin-like growth factor-1 (IGF-1), and differential blood cell counts [24,30]. Furthermore, extending these findings, the research evaluated the impact of sodium butyrate supplementation on glucagon-like peptide (GLP)-1 and GLP-2 levels in beef cattle during late pregnancy and early postpartum periods. Twelve Japanese black cows were divided into a treatment group receiving sodium butyrate supplementation and a control group [37]. They further showed that the treatment group had significantly increased plasma total cholesterol and GLP-1 concentrations three days after calving. Notably, GLP-1 concentrations were higher in colostrum compared to plasma, although the treatment did not affect the composition of colostrum and transition milk [37]. In another significant study, the effects of sodium butyrate on pancreatic development in dairy calves over a 70-day period were examined [31]. Fourteen male Holstein calves were divided into two groups and fed either standard milk or milk supplemented with sodium butyrate. The results revealed that sodium butyrate significantly stimulated pancreatic growth, as evidenced by increased organ size, enhanced protein content, and elevated cellular proliferation. Notably, exocrine pancreatic function showed marked improvements, including higher enzyme activity and upregulated gene expression. Endocrine benefits were also observed, with increased insulin secretion and β-cell proliferation. Proteomic analysis identified G protein subunit alpha-15 (GNA15) as a key regulator of these processes, underscoring the role of sodium butyrate in promoting both exocrine and endocrine pancreatic functions in calves [31].

Effects of encapsulated butyric acid and zinc (BZ) with varying doses have been shown to be associated with growth, rumen morphology, and small intestine histology of feedlot steers [38]. Furthermore, it has been revealed that BZ supplementation (2 g BZ/kg dietary dry matter) improved dressed yield and growth performance during the receiving period and reduced liver abscess incidence. The researchers proposed that BZ supplementation could enhance early growth and decrease liver abscess incidence, though further research is needed to confirm these findings [38]. Additionally, magnesium butyrate (MgB) supplementation at a rate of 105 g/cow/day significantly increased colostrum yield and immunoglobulin G (IgG) levels, enhanced milk yield during the first week of lactation, and improved body condition scores from 3 to 9 weeks postpartum [12]. Interestingly, MgB also delayed the decline in milk production, improved milk synthesis efficiency, increased milk fat content, and reduced pro-inflammatory cytokines and somatic cell counts by inhibiting bacterial cell wall components [39]. Furthermore, the effects of sodium butyrate and sodium acetate supplementation on diarrheic yak calves in Tibet have been evaluated [40]. They used nineteen calves (divided into control and treatment groups), with the treatment group receiving sodium butyrate (10 g/kg) and sodium acetate (5 g/kg) for 28 days. Notably, their analysis showed increased short-chain fatty acids (SCFAs) and beneficial shifts in gut microbiota composition in the supplemented group. Moreover, the supplemented group experienced reduced diarrhea incidence, improved weight gain, enhanced total antioxidant capacity (T-AOC), superoxide dismutase (SOD), and glutathione peroxidase (GSH-Px) levels, as well as reduced inflammatory responses, including decreases in TNF-α and IL-1β levels [40]. Building upon these findings, a notable study explored the effects of sodium butyrate supplementation in calf starters with varying starch levels. Conducted over a 65-day period, the study focused on the inflammatory response and growth performance of post-weaned dairy calves [30]. Remarkably, sodium butyrate supplementation (5 g/kg of dietary dry matter) significantly increased ADG, particularly in calves consuming a low-starch starter (270 g/kg of dietary dry matter). Additionally, key inflammatory markers, such as toll-like receptor 4 (TLR4), nuclear factor kappa B (p65), and several pro-inflammatory cytokines were affected by dietary intervention, underscoring the importance of tailored sodium butyrate strategies in optimizing post-weaning calf growth and health [30]. This perspective is consistent with numerous studies demonstrating the challenges faced during the weaning process, which not only requires rumen development [41] but also the regulation of inflammatory responses [42,43]. In this context, current research increasingly focuses on identifying key inflammatory biomarkers that can predict disease susceptibility and growth outcomes in calves [44]. Complementing these findings, another study showed that supplementing high-starch calf starters (365 g of starch/kg of starter) with rumen-protected butyrate (10 g/kg butyrate/starter) during the critical two-week weaning transition increased starter intake, which may reflect a reduction in inflammation and stress during this period [16]. Furthermore, research has explored the protective effects of sodium butyrate against lipopolysaccharide (LPS)-induced inflammation in bovine rumen epithelial cells (BRECs), focusing on the role of G protein-coupled receptor 41 (GPR41) activation [45]. Pretreatment with sodium butyrate successfully reduced the expression of pro-inflammatory genes and increased the levels of tight junction proteins in LPS-stimulated BRECs. Additionally, sodium butyrate inhibited the activation of NF-κB and IκBα, decreased the expression of apoptosis markers such as Bax, caspase 3, and caspase 9, and favorably altered cell cycle distribution. Sodium butyrate also promoted the expression of genes involved in volatile fatty acid metabolism (*ACAT1* and *BDH1*) and upregulated *GPR41* expression. Notably, the protective effects of short-chain fatty acids, including sodium butyrate, were not observed in GPR41 knockdown BRECs, indicating that GPR41 activation is crucial for sodium butyrate's anti-inflammatory effects [46].

The antioxidant potential of butyrate has also been documented. Consistently, a study investigated the protective effects of sodium butyrate against LPS-induced mastitis in bovine mammary epithelial cells (BMECs) [47]. Based on their findings, sodium butyrate significantly reduced LPS-induced cell death in a concentration-dependent manner, lowered oxidative stress (as indicated by reduced intracellular reactive oxygen species (ROS) and malondialdehyde (MDA) levels), and enhanced the activity of antioxidant enzymes including SOD, catalase (CAT), and GPX. Moreover, it suppressed inflammatory markers (IL-6, IL-1β, TNF-α) and apoptosis-related proteins (caspases, Bax, Bcl-2), suggesting that it regulates oxidative stress, inflammation, and apoptosis in BMECs through the NF-κB and caspase/Bax pathways [47]. The research also showed that sodium butyrate exhibited anti-inflammatory effects on bovine embryo tracheal cells (EBTr), reversing LPS-induced inflammatory responses by reducing the expression of TLR4, phosphorylated NF-κB, IκBα, and IL1α, and inhibiting the nuclear translocation of NF-κB p65 through modulation of the TLR4 and NF-κB signaling pathways [48]. Interestingly, sodium butyrate also played a key role in inhibiting neutrophil extracellular trap (NET) formation induced by Staphylococcus aureus. Notably, sodium butyrate reduced the expression of NET markers, such as DNA, histones, myeloperoxidase, and neutrophil elastase [49]. Additionally, sodium butyrate activated the mammalian target of the rapamycin (mTOR) pathway, providing further mechanistic insights into its NET-suppressing properties [49]. Sodium butyrate also exhibited significant anticoccidial potential by reducing the viability of *Eimeria bovis* sporozoites and protecting Madin-Darby bovine kidney (MDBK) cells from damage, as evidenced by decreased lactate dehydrogenase (LDH) release. Furthermore, sodium butyrate supplementation (45 g/day) improved growth rates, ADG, feed conversion efficiency, and serum antioxidant markers, including increased GSH-Px activity and decreased MDA levels [50–52].

Tributyrin is a triglyceride formed by three butyrate molecules bound to a glycerol backbone. When ingested or administered, tributyrin is broken down by digestive enzymes (such as lipases), releasing butyrate. This process makes tributyrin an effective delivery form of butyrate, particularly in situations where direct butyrate supplementation may be difficult or inefficient. As a precursor of butyric acid, tributyrin has demonstrated significant benefits in reducing diarrhea and enhancing intestinal barrier function in calves. Studies have reported increased ADG, higher weaning weights, and reduced oxidative stress, evidenced by lower ROS and MDA levels, along with higher SOD and serum immunoglobulin M (IgM) levels. Additionally, tributyrin supplementation has been associated with an increase in beneficial gut bacteria, such as *Rikenellaceae*, *Bacteroides*, and *Ruminococcaceae*, as well as improvements in ileal morphology, plasma GLP-2 concentration, and post-weaning body weight [53–57]. Furthermore, tributyrin supplementation significantly reduced diarrhea frequency in calves, correlating with improved intestinal development and barrier function, as indicated by increased villus height and claudin-4 expression [53]. For example, as a derivative of butyrate, tributyrin effectively regulated inflammation, specifically IL-1β, during the pre-weaning phase, providing further evidence of butyrate's essential role in managing early-life inflammation [53]. Consistently, another study reported that supplementing milk with 2 g/L tributyrin significantly reduced diarrhea frequency between days 29 and 56, along with notable reductions in the expression of IL-1β, TLR2, and serum amyloid A, while simultaneously increasing claudin-4 and GPR41 levels in the jejunum and ileum. Furthermore, tributyrin supplementation promoted intestinal development, as shown by increased villus height and improved villus height-to-crypt depth ratios in the jejunum. Although no significant changes were observed in microbiota abundance, tributyrin remarkably increased the presence of beneficial SCFA-producing bacteria such as *Ruminococcaceae*, *Lachnospiraceae*, *Prevotella*, and *Rikenellaceae*, which correlated negatively with inflammatory gene expression and positively with intestinal barrier function and morphology [53].

Butyrate modulates gene expression by inhibiting HDACs, leading to increased histone acetylation, which enhances anti-inflammatory gene transcription and suppresses pro-inflammatory cytokines (Figure 2). A study demonstrated that sodium butyrate prevents LPS-induced inflammation in bovine mammary epithelial cells (MAC-T) by reducing apoptosis, lowering pro-inflammatory cytokines (TNF-α, IL6, IL-1β) in a dose-dependent manner, and preserving histone H3 acetylation. Sodium butyrate also inhibited NF-κB signaling, further supporting its potential as a protective agent against mastitis-related pathogens [58]. Additionally, sodium butyrate has shown therapeutic potential in enhancing host defenses against *Mycobacterium bovis*, the causative agent of bovine tuberculosis. In in vitro studies, sodium butyrate upregulated the expression of the antimicrobial peptide cathelicidin (LL37) in infected cells, while downregulating HDACs and inhibiting NF-κB signaling pathways. Furthermore, in in vivo mouse models, sodium butyrate reduced the production of inflammatory cytokines and enhanced resistance to *M. bovis* infection [18]. In a related study, the effects of butyrate on inflammation and tight junction (TJ) integrity were explored in BMECs [15]. Butyrate treatment significantly lowered HDAC3 levels, inhibited NF-κB activation, and reduced inflammatory cytokine production in response to γ-D-glutamyl-meso-diaminopimelic acid (iE-DAP) stimulation. Additionally, butyrate suppressed the expression of myosin light chain kinase (MLCK), increased tight junction protein levels, and improved epithelial barrier function. Importantly, the overexpression of HDAC3 counteracted these protective effects, demonstrating that butyrate alleviates inflammatory responses and maintains TJ integrity through an HDAC3-dependent mechanism [15]. Building on this, another study found that HDAC activity increased linearly with higher doses of LPS, accompanied by a significant reduction in histone H3 acetylation levels [22]. Beyond its well-established gastrointestinal benefits, sodium butyrate also regulates cellular function and tissue development within the rumen through epigenetic mechanisms. A recent study revealed that butyrate regulates gene expression via genome-wide H3K27ac acetylation, profoundly affecting pathways such as AMPK and MAPK signaling, and further underscoring its roles in cholesterol metabolism, cell adhesion, and overall cellular function in bovine rumen epithelial cells [19]. These findings provide deep insights into the epigenetic regulation of inflammatory responses, particularly highlighting the role of HDACs in modulating histone acetylation and inflammation. Overall, research on sodium butyrate emphasizes its multifaceted role in promoting calf health and development. Studies have shown that sodium butyrate enhances intestinal growth, modulates immune responses, improves microbial diversity, and reduces inflammation, collectively contributing to improved growth performance, feed efficiency, and reduced gastrointestinal disturbances such as diarrhea during critical weaning and post-weaning periods in calves, as summarized in Table 1.

Table 1. Summary of research on butyrate's role in enhancing intestinal health and growth performance in calves.

Biological Effect of Butyrate Treatment	Experimental Model	References
Sodium butyrate (1 mM for 15 h), followed by LPS treatment (5 μg/mL for 3 h), attenuated LPS-induced inflammatory responses by downregulating the expression of IL-6, IL-1β, COX-2, NLRP3, and iNOS, and by inhibiting the phosphorylation of NF-κB, IκB, and p65 induced by LPS in bovine macrophages.	Bovine Macrophages	[59]
Sodium butyrate (5 mM, 18 h) followed by LPS treatment (1 μg/mL, 6 h) inhibited the LPS-induced upregulation of TNF-α, IL6, and IL1B, and suppressed NF-κB p65 phosphorylation (5 μg/mL) in BMECs.	BMECs	[58]

Table 1. *Cont.*

Biological Effect of Butyrate Treatment	Experimental Model	References
Sodium butyrate (2 mM for 12 h) enhanced the antioxidant response in H_2O_2 treated (600 μM for 12 h) BMECs by activating G protein-coupled receptor 109A (GPR109A) and nuclear factor erythroid 2-related factor 2 (Nrf2), while promoting H3K9/14 acetylation. This treatment also facilitated Nrf2 nuclear accumulation through the AMPK signaling pathway, thereby promoting antioxidant and anti-apoptotic activities via the GPR109A/AMPK/Nrf2 signaling axis. Additionally, increased H3K9/14 acetylation further stimulated Nrf2 transcription, contributing to an enhanced antioxidant capacity in BMECs	BMECs	[60]
LPS (100 EU/mL) and LPS (100 EU/mL) plus sodium butyrate (10 mmol/L) treatments for 48 h increased the expression of lactation-related genes, including ACACA and ribosomal protein S6 kinase 1 (S6K1), which had been downregulated by LPS treatment in MAC-T cells. Additionally, sodium butyrate treatment suppressed HDAC activity, leading to a decline in histone H3 acetylation.	MAC-T cells	[22]
Sodium butyrate (5 mM) treatment for 2 h prior to LPS (1 μg/mL) inhibited HDACs, leading to enhanced acetylation of H3K9/14, H3K18, and H3K27 in MAC-Ts. However, despite the changes in HDAC activity and histone acetylation, sodium butyrate did not significantly reduce overall inflammatory gene expression.	MAC-Ts	[61]
Pretreatment with 0.5 mmol/L sodium butyrate for 18 h followed by a 6 h LPS challenge (4 μg/mL) in bovine hepatocytes showed that LPS exposure increased TNF-α and IL-6 production, but sodium butyrate pretreatment significantly reduced these levels in bovine hepatocytes. Sodium butyrate also inhibited the expression and nuclear translocation of phospho-p65 and phospho-IκBα, key proteins involved in inflammatory signaling stimulated by LPS. Additionally, genes and proteins associated with fatty acid metabolism, such as SREBP1c, SCD1, and DGAT1, were upregulated in cells pretreated with sodium butyrate. The ratios of phospho-AMPKα to AMPKα and phospho-ACCα to ACCα were significantly lower in the sodium butyrate-pretreated group, indicating altered fatty acid oxidation. Furthermore, sodium butyrate reversed the histone H3 deacetylation induced by LPS, suggesting a protective epigenetic effect.	Bovine Hepatocytes	[62]
The combination of milk replacer (MR) with 0.24% butyrate supplementation has shown greater MR intake and body weight in calves, exhibiting higher plasma IGF-I and IGFBP-3 levels, and lower IGFBP-2 levels. After weaning, plasma IGF-I and IGFBP-4 levels decreased in calves, while IGFBP-2 levels increased. At day 50, the combination of sodium butyrate and milk replacers resulted in higher mRNA levels of *GHR1A* and *IGF1*, along with lower levels of *IGFBP2*.	Dairy calves	[63]
Calves fed MR with 0.24% butyrate supplementation ad libitum had higher MR intake and body weight, with butyrate decreasing the sucking rate but increasing MR intake per meal. All calves showed increased immunoglobulin levels after colostrum intake. Body temperature was higher in ad libitum-fed calves during the first two weeks, suggesting better welfare due to fewer signs of hunger. However, neither feeding strategy nor butyrate supplementation consistently impacted health or immune status.	Dairy calves	[64]
Sodium butyrate, included in milk replacer at 4 g/kg of dry matter (DM) from days 7 to 56 of life, improved the concentrations of propionate, acetate, and total SCFAs in the colon of calves. This supplementation decreased the cecal abundance of butyrate-producing bacteria, *Butyrivibrio* and *Shuttleworthia* while enhancing the level of the propionate producer *Phascolarctobacterium*. Concurrently, it decreased the cecal concentration of *Mogibacterium*, which was associated with impaired gut health.	Dairy calves	[65]
Sodium butyrate supplementation along with a high-starch diet for a 21-day period significantly improved plasma GLP-2 concentration, increased n-butyrate concentration in the ruminal fluid, and tended to enhance the digestibility of dry matter and organic matter in Holstein cows.	Holstein cows	[66]
Tributyrin supplementation in milk replacers has been shown to reduce the occurrence of diarrhea in preweaning Holstein calves. This dietary intervention enhances gut health by promoting the development of the intestinal epithelium and improving gut barrier function, which in turn helps mitigate gastrointestinal disturbances commonly observed in young calves. The addition of tributyrin plays a crucial role in maintaining intestinal integrity and supporting the overall health and well-being of preweaning Holstein calves.	Holstein's calves	[67]
Sodium butyrate supplementation (4.4 g/day for 6 weeks) has been found to promote the ruminal microbiota, enhancing microbial diversity and stability within the rumen. This intervention positively influences the composition and activity of ruminal microorganisms, supporting improved fermentation processes and nutrient utilization. By fostering a more balanced microbial environment, sodium butyrate contributes to better overall ruminal health and function, which can enhance the digestive efficiency and performance of Holstein calves.	Holstein's calves	[68]

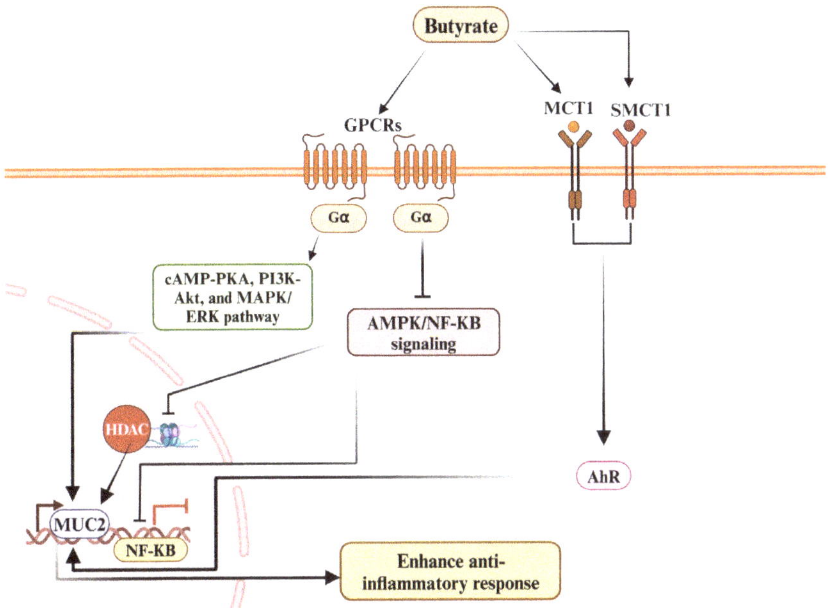

Figure 2. Butyrate's role in promoting anti-inflammatory response. Butyrate via monocarboxylate transporter 1 (MCT1) and sodium-coupled monocarboxylate transporter 1 (SMCT1) activate aryl hydrocarbon receptor (AHR) to promote anti-inflammatory response.

3.2. The Role of Butyrate in Piglet Intestinal Health and Growth Promotion

Butyrate plays a crucial role in promoting intestinal health and supporting piglet growth, particularly during the vulnerable post-weaning period, as widely documented in the literature. Chemically protected sodium butyrate has demonstrated significant effects in enhancing gut health, especially in weaned piglets. Notably, supplementation with a high dose of sodium butyrate (2000 mg/kg) led to an 18.5% improvement in ADG, while a lower dose (1000 mg/kg) still achieved a 6.9% increase [69]. Beyond promoting growth, high doses of sodium butyrate significantly upregulated anti-inflammatory cytokines (e.g., IL-10) and key antioxidant markers, including Keap1, Nrf-2, CAT, SOD, and GSH-Px. Additionally, this treatment promoted the proliferation of beneficial SCFA-producing bacteria, suggesting a multifaceted role in enhancing gut health through the modulation of microbiota composition [69]. Furthermore, butyrate serves as a vital energy source for colonocytes and helps alleviate stress in weaned piglets, although the precise mechanisms remain incompletely understood [70]. In the LPS-induced inflammatory model, butyrate supplementation improved gut morphology and reduced inflammation, evidenced by increased villus height and villus height-to-crypt depth ratio, while lowering inflammatory markers such as NF-κB and IL-1β, thereby contributing to intestinal integrity. Additionally, it altered the microbial composition, increasing beneficial bacteria while reducing harmful ones, indicating broad protective effects on gut health during stress and infection [70]. Similarly, adding sodium butyrate to low-protein diets improved the expression of antioxidant and anti-inflammatory genes, further demonstrating the compound's capability to balance immune responses and oxidative stress [71].

In line with these findings, other studies have shown that butyrate is essential during critical developmental stages such as weaning, where it helps maintain jejunal homeostasis and supports intestinal integrity. Supplementation with sodium butyrate (0.45 g/kg) has been proven to suppress inflammatory mediators such as IL-1β, IL-6, and TNF-α while enhancing intestinal barrier function [72]. Moreover, another form of butyrate, tributyrin, effectively reduced diarrhea incidence and improved gut morphology in nursery piglets [73].

Recent research has further expanded on butyrate's role beyond gut health, highlighting its potential to reduce liver injury caused by mycotoxins such as deoxynivalenol (DON). Sodium butyrate, through the modulation of oxidative stress pathways and histone acetylation mechanisms, successfully reversed liver damage and inflammation associated with DON exposure, underscoring its therapeutic potential [74]. This finding aligns with previous studies that repeatedly emphasize the ability of sodium butyrate to enhance growth performance and support immune function in piglets [75]. They reported that sodium butyrate significantly improved ADG and feed conversion efficiency, increased ileal villus height, and reduced the number of Peyer's patches in the ileum. Sodium butyrate treatment also upregulated the expression of several key genes, including nutrient transporters like sodium/glucose cotransporter 1 (*SGLT1*) and monocarboxylate transporter 1 (*MCT1*), tight junction proteins (*OCL* and zonula occludens-1), and antioxidants (*GPX* and *SOD*), which are critical for maintaining intestinal integrity and function [75]. Furthermore, sodium butyrate supplementation in low-protein diets over a four-week period significantly improved gut microbiota composition and markedly reduced inflammatory markers such as IL-8, TLR4, and IKKα in both plasma and colonic mucosa. Moreover, sodium butyrate enhanced antioxidant activity by upregulating GSH-Px expression and increasing Nrf2 activity, thereby effectively mitigating oxidative stress [71].

Additionally, enterotoxigenic *Escherichia coli* (ETEC) strains, including ETEC F4 and ETEC F18, are known to be major contributors to post-weaning diarrhea in piglets. Consistently, Kovanda et al. [76] found that butyrate treatment significantly alleviated diarrhea in piglets co-infected with ETEC F4 and ETEC F18 by reducing neutrophil infiltration and lowering serum TNF-α levels. Similarly, López-Colom et al. [77] highlighted the protective effects of medium-chain fatty acid-protected butyrate in maintaining intestinal health in pigs challenged with ETEC F4. Specifically, butyrate acts as a primary energy source for colonocytes, enhances intestinal barrier integrity, and exhibits strong anti-inflammatory properties, thereby mitigating damage caused by infections. By preserving tight junctions between intestinal cells, butyrate reduces gut permeability, prevents bacterial toxin translocation, and modulates immune responses to reduce inflammation and promote intestinal healing [77].

A study reported that 0.5% glyceryl butyrate supplementation in piglets exhibited significantly lower levels of pro-inflammatory cytokines (IL-1β, IL-6, and TNF-α) in the jejunum and ileum compared to controls [78]. Glyceryl butyrate supplementation also enhanced the expression of porcine beta-defensins (pBDs) and tight junction-associated proteins (Claudin 1, Occludin, and ZO-1), thereby improving gut barrier function. These improvements were attributed to the reduced activation of the NF-κB/MAPK signaling pathways and promoted the growth of beneficial gut microbiota, particularly Lactobacillus species [78]. The dietary supplementation of sows with sodium butyrate, medium-chain fatty acids, and omega-3 polyunsaturated fatty acids has shown a profound positive impact on both reproductive performance and the intestinal health of their piglets [79]. This supplementation regimen resulted in shorter weaning-to-estrus intervals, improved oxidative status, reduced diarrhea incidence, and enhanced colostrum quality [79]. Additionally, Zhong et al. [80] emphasized the critical role of microbial-derived butyrate in regulating cell apoptosis and proliferation, as well as modulating immune responses in the jejunum during the crucial first three weeks post-weaning. Notably, during the first week, a reduction in butyrate production, driven by an increase in specific bacterial families (*Erysipelotrichaceae* and *Lachnospiraceae*), led to inflammation, cell apoptosis, and impaired intestinal development. However, by the second week, an increase in other beneficial bacterial families (*Lactobacillaceae* and *Ruminococcaceae*) promoted butyrate production, which subsequently reduced inflammation, stimulated cell proliferation, and supported intestinal health and development [80]. In summary, these findings underscore the pivotal role of sodium butyrate in promoting growth, enhancing gut health, and alleviating inflammation and oxidative stress, particularly during the vulnerable post-weaning period for piglets, as presented in Table 2.

Table 2. Summary of research on butyrate's role in enhancing intestinal health and growth performance in pigs.

Biological Effect of Butyrate Treatment	Experimental Model	References
Sodium butyrate (3.00 g/kg coated butyrate for 21 days), followed by LPS treatment (100 μg/kg body weight), significantly improved intestinal health by enhancing villus height, improving the villus height to crypt depth ratios, and reducing tissue damage in piglets. Butyrate also reversed the LPS-induced activation of inflammatory pathways, including NF-κB p65 and PPARα, decreased pro-inflammatory cytokines such as IL-1β, IL-6, TNF-α, and TLR4, and promoted anti-inflammatory responses by increasing IL-10 and IL-13 levels. Additionally, butyrate modulated the piglets' gut microbiota by increasing beneficial bacteria, including *Firmicutes*, *Bacteroidetes*, *Clostridiaceae*, *Lactobacillus*, and *Prevotella*, while reducing harmful bacteria like *Proteobacteria*, *Enterobacteriaceae*, and *Escherichia/Shigella*. It also reduced apoptosis by regulating mitochondrial pathways and enhanced enterocyte energy metabolism.	Piglets	[70]
The inclusion of 3 kg of protected sodium butyrate/ton of feed enhanced the pig intestinal microbiota, promoting the growth of beneficial families such as *Prevotellaceae*, *Lachnospiraceae*, *Peptostreptococcaceae*, *Peptococcaceae*, and *Terrisporobacter*. This supplementation also promoted *Deinococcus-Thermus*, a group associated with the production of carotenoids that possess antioxidant, anti-apoptotic, and anti-inflammatory properties. Furthermore, the increased presence of *Clostridium butyricum* was linked to overall positive gut effects in weanling pigs, including improvements in villus height, body weight, and a reduction in diarrhea.	Piglets	[81]
The inclusion of 3 kg of protected sodium butyrate/ton of feed reduced the clinical and histological severity of colitis while decreasing pro-inflammatory cytokines such as IL-1β, IL-6, IL-8, and TNF-α in piglets. It promoted the growth of beneficial gut bacteria, including Faecalibacterium and Lactobacillus, and enhanced the tricarboxylic acid (TCA) cycle in colonocytes, thereby improving metabolism in the gut-liver axis. Additionally, the supplementation increased butyric acid levels in the colon and portal vein, facilitated the utilization of amino acids and vitamin B, and reversed LPS-induced fatty acid synthesis. Furthermore, it enhanced the production of anti-inflammatory cytokines, including IL-10 and TGF-β, while suppressing inflammatory mediators such as hypoxia-inducible factor 1α.	Piglets	[82]
The supplementation of a basal diet with 0.2% sodium butyrate demonstrated significant improvements in growth performance, as evidenced by enhanced final body weight, daily weight gain, and daily feed intake, alongside better feed efficiency and increased carcass weight in piglets. Additionally, it positively influenced the gut microbiota by promoting the abundance of *Bacteroidetes* while reducing the levels of *Firmicutes* and *Proteobacteria* in the caecal environment.	Piglets	[83]
The addition of 0.2% sodium butyrate in a basal diet counteracted the detrimental effects of 4 mg/kg deoxynivalenol (DON) exposure, which typically led to decreased performance, impaired intestinal barrier integrity, reduced expression of host defense peptides (HDPs), and disrupted gut microbiota in pigs. Sodium butyrate supplementation not only mitigated these adverse outcomes but also improved liver health and reduced levels of harmful enzymes. In vitro studies further demonstrated its protective role against DON-induced damage. Mechanistically, sodium butyrate activated the NOD2/caspase-12 pathway, subsequently enhancing HDP expression, which played a critical role in preventing intestinal barrier dysfunction caused by DON.	piglets	[84]
The supplementation of coated sodium butyrate at a dosage of 450 mg/kg significantly enhanced ADG, improved the *Lactobacilli* to *E. coli* ratio in the jejunum, and effectively reduced the incidence of post-weaning diarrhea and intestinal permeability in piglets.	Piglets	[85]
The supplementation of 5 g sodium butyrate per kilogram of feed during the first three weeks significantly enhanced growth performance in piglets, as evidenced by increased final body weight, ADG, and feed efficiency. Additionally, it improved dry matter digestibility, reduced coliform counts, and promoted the growth of lactic acid bacteria in the intestine. Furthermore, fecal gas emissions, particularly hydrogen sulfide (H_2S), were reduced, and villi length in the small intestine showed notable improvement, contributing to overall intestinal health and functionality.	Piglets	[86]
The administration of 450 mg of sodium butyrate/kg of feed over a two-week period significantly enhanced growth performance and improved intestinal structure in piglets, as indicated by increased villus height and strengthened intestinal barrier function. Additionally, it reduced mast cell degranulation and decreased levels of inflammatory mediators, including histamine, tryptase, TNF-α, and IL-6, in the intestines. The suppression of the JNK signaling pathway activation further demonstrated that sodium butyrate protects intestinal integrity by inhibiting mast cell activation and inflammation through this mechanism, highlighting its potential role in promoting intestinal health and reducing inflammatory responses.	Piglets	[72]

Table 2. *Cont.*

Biological Effect of Butyrate Treatment	Experimental Model	References
The intravenous administration of 40 mg of sodium butyrate/kg of body weight over seven days resulted in the upregulation of *MCT1* in the colon and a reduction in the expression of *HDAC1* and pro-inflammatory cytokine genes, including *IL-6*, *IL-18*, *IL-12p40*, and *TNF-α*. This treatment also bolstered anti-inflammatory responses and enhanced key markers of intestinal development in piglets, such as ZO-1, occludin, and epidermal growth factor (EGF), demonstrating its potential to support intestinal integrity and modulate inflammatory pathways effectively.	Piglets	[87]
The supplementation of 2000 mg of sodium butyrate/ kg of feed for 21 days significantly reduced diarrhea incidence and enhanced intestinal barrier function in piglets by upregulating key tight junction proteins, including Claudin-3, Occludin, and ZO-1, in the piglet colon. These beneficial effects were mediated through the activation of the GPR109A and the Akt signaling pathway. In vitro studies further supported these findings, demonstrating that blocking either GPR109A or Akt led to a reduction in Claudin-3 expression, confirming that sodium butyrate promotes intestinal integrity and barrier function through these specific molecular mechanisms.	Piglets	[88]
The inclusion of 1 g sodium butyrate/kg in the basal diet from day 85 of gestation through day 22 of lactation resulted in a shortened weaning-to-estrus interval and a reduced incidence of diarrhea in piglets. It also enhanced the nutritional and immunological quality of colostrum, as evidenced by increased concentrations of fat, protein, IgA, IgG, and IgM. Furthermore, sodium butyrate upregulated mRNA expression of key intestinal health markers, including claudin-1, zona occludens 1, and interleukin-10 while downregulating *TLR4* expression. Additionally, it influenced microbial diversity by reducing the abundance of *Firmicutes*, *Actinobacteria*, and *Synergistetes*, highlighting its role in improving reproductive performance, colostrum quality, and intestinal health.	Piglets	[89]

3.3. The Role of Butyrate in Intestinal Health and Growth Promotion in Sheep and Goats

The significant role of butyrate in enhancing intestinal health and promoting growth performance in sheep and goats has been widely documented. Consistent with these findings, supplementation with sodium butyrate (3 g/kg of starter feed for 73 days) significantly improved starter feed intake and increased weaning weight in lambs. The combination of forage and sodium butyrate resulted in positive synergistic effects on lamb weight gain, wither height, and body barrel size [90]. Sodium butyrate also increased the concentrations of SCFA and butyrate, while forage increased ruminal pH and acetate levels, but decreased butyrate and propionate concentrations [90]. A study reported that sodium butyrate supplementation (36 g/kg of dry matter for 14 days) not only increased the final body weight but also raised the butyrate concentrations in both reticuloruminal fluid and abomasal digesta [91]. Notably, significant structural improvements were observed in the omasum and abomasum, including increased epithelial thickness and enhanced hydrolytic activity in the rumen [91,92]. Similarly, Sun et al. [93] reported that administering sodium butyrate (1.8 mL/kg body weight) from days 7 to 35 significantly increased the average daily gain, improved jejunal villus height, and enhanced the villus-to-crypt ratio in the ileum of lambs. Moreover, the treatment promoted the expression of intestinal barrier proteins (Claudin-1 and Occludin) and modulated the gut microbiota, reducing harmful bacteria (e.g., *Clostridia_UCG-014* and *Romboutsia*) while increasing beneficial bacteria (e.g., *Succiniclasticum*) [93]. Furthermore, Świerk et al. [94] demonstrated that sodium butyrate supplementation, combined with higher concentrate intake, affected epithelial thickness in specific rumen regions and upregulated the expression of genes and proteins related to rumen barrier function, such as MCT1 and Claudin-4 [95,96]. Beyond intestinal health, butyrate supplementation also exhibits broader physiological effects. For instance, Mohamed et al. [97] reported that feeding lambs with a milk replacer containing sodium butyrate significantly increased total and daily weight gain, elevated testosterone levels, and improved testicular development. Additionally, the combination of sodium butyrate and milk replacer effectively reduced oxidative stress markers and enhanced antioxidant enzyme activity [97]. Another study investigated the application of the butyrate-producing strain *Clostridium beijerinckii R8*, isolated from sheep rumen, in goat kids [98]. Remarkably, the use of this strain significantly reduced diarrhea incidence and fecal scores, while promoting intestinal health by inhibiting inflammation, enhancing antioxidant capacity, and modulating the gut microbiota. Furthermore, it strengthened the intestinal barrier and improved jejunal morphology, offering a comprehensive approach to enhancing gut health in goat kids [98]. Consistently, Zhao et al. [99] reported that supplementation with coated sodium butyrate (3 g/kg feed for 21 days) in lambs significantly improved growth traits such as ADG and average daily feed intake (ADFI), primarily by promoting the colonic microbiota. This was accompanied by improvements in antioxidant markers, including increased levels of T-AOC, SOD, GSH-Px, and catalase (CAT), while MDA levels were reduced [99]. Furthermore, another study indicated that infection with Cryptosporidium parvum exacerbated intestinal inflammation in goats by suppressing butyrate-producing bacteria [100]. In lambs, supplementation with sodium butyrate (3 g/kg feed for 28 days) significantly reduced inflammatory responses by decreasing levels of pro-inflammatory cytokines such as TNF-α, IL-1β, and IL-8, and by inhibiting the LPS-induced TLR4/NF-κB signaling pathway at both mRNA and protein levels. Simultaneously, the treatment upregulated the expression of tight junction proteins (e.g., Claudin-3, ZO-1, and Occludin), thereby improving gut barrier function and enhancing the ruminal microbiota [101]. Similarly, Chang et al. [102] revealed that sodium butyrate supplementation significantly increased the pH of both the rumen and cecum, which was associated with the downregulation of GPR41/43 gene expression, inflammatory cytokines (IL-1β, IL-6, and IL-8), and chemokines (CCL5, CXCL13). At the same time, tight junction proteins were upregulated, further strengthening gut barrier integrity. Interestingly, the study also suggested that sodium butyrate induced epigenetic modifications, such as DNA methylation and chromatin compaction, in the promoter regions of GPR41/43 genes, indicating a potential epigenetic mechanism behind the observed changes [102]. In summary, these studies highlight the diverse benefits of sodium butyrate in enhancing intestinal health, growth performance, and the overall physiological well-being of sheep and goats, as illustrated in Table 3.

Table 3. Summary of research on butyrate's role in enhancing intestinal health and growth performance in sheep and goats.

Biological Effect of Butyrate Treatment	Experimental Model	References
The administration of coated sodium butyrate at a dosage of 3 g/kg over a 21-day period significantly enhanced the colonic microbiota in weaned lambs, promoting the growth of beneficial bacterial families such as *Lachnospiraceae*, *Verrucomicrobiota*, *Akkermansia*, *Roseburia*, and *Sinobacteraceae*. This intervention also improved the ADG and ADFI of the lambs. Additionally, it enhanced the antioxidant capacity, as evidenced by increased levels of T-AOC, T-SOD, GSH-Px, and CAT, while reducing MDA levels, thereby mitigating weaning stress.	Lamb	[99]
The inclusion of 6.25 g of calcium butyrate/kg of diet dry matter significantly enhanced feed intake and ADG while improving liver health in lamb, as indicated by reduced levels of alanine aminotransferase (ALT) and aspartate aminotransferase (AST).	Lamb	[103]
Sodium butyrate demonstrated the ability to enhance antimicrobial responses in goat mammary gland cells by significantly increasing the concentrations of β-defensin-1 and S100A7 in milk, thereby bolstering the innate immune defense mechanisms.	Goat MECs	[104]
The supplementation of tributyrin at 4.0 g/kg of DM significantly improved ADG and increased dry matter intake (DMI), while enhancing nutrient retention by reducing nitrogen, calcium, and phosphorus losses through feces and urine in lamb. It also promoted gastrointestinal development, evidenced by increased papillae length and width in the dorsal and ventral sacs and caudodorsal blind sac, as well as greater duodenal and ileal thickness and increased villus height in the duodenum, ileum, jejunum, and cecum. Additionally, reduced crypt depths in the duodenum and cecum suggested improved epithelial turnover and absorption efficiency. Tributyrin further enhanced rumen fermentation, as indicated by a significant decrease in rumen pH and an increase in total volatile fatty acid (VFA) concentration, while enriching the rumen microbial community by significantly boosting populations of *Clostridium*, *Butyrivibrio*, *Streptococcus*, *Prevotella*, *Ruminobacter*, and *Fibrobacter*, which are key bacteria associated with VFA production and fiber digestion.	Lambs	[105]
Sodium butyrate effectively mitigated oxidative stress induced by grain-induced sub-acute ruminal acidosis (SARA) in dairy goats by increasing mean ruminal pH and reducing LPS levels in the ruminal, portal, and hepatic regions. It enhanced the mRNA expression of antioxidant-related genes (*SOD1*, *SOD2*, *SOD3*, *GPX1*, and *CAT*), elevated TSOD and CAT enzyme activities, improved T-AOC, and reduced MDA levels in both the liver and plasma. Additionally, sodium butyrate upregulated the mRNA of expression of *UGT1A1*, *NQO1*, *MGST3*, and *Nrf2*, along with increasing total Nrf2 protein levels in goats fed a high-grain diet supplemented with sodium butyrate (HG + NaB).	Goat	[106]
The inclusion of 1% sodium butyrate effectively reversed damage to the rumen epithelium tight junction by inhibiting the protein kinase C (PKC) and MAPK signaling pathways, thereby providing protective effects on the rumen epithelium during subacute rumen acidosis induced by a high-concentrate diet.	Goat	[107]

Table 3. *Cont.*

Biological Effect of Butyrate Treatment	Experimental Model	References
The supplementation of 3 g of sodium butyrate/kg of starter DM over 73 days significantly improved starter intake and increased weaning weight in lambs. While forage provision alone did not affect intake or weight gain, the combination of sodium butyrate and forage demonstrated a positive interaction, enhancing lamb weight gain. Lambs in the forage-sodium butyrate group exhibited the greatest wither height and a larger body barrel. Additionally, this treatment led to increased concentrations of SCFAs and butyrate, further supporting its beneficial effects on growth and development.	Lamb	[90]
Oral infusion of sodium butyrate at a dose of 0.36 g/kg body weight from 10 to 49 days of age significantly increased ADFI, ADG, and BW of lambs at 5 and 6 weeks. The treatment resulted in higher concentrations of acetate, butyrate, and total volatile fatty acids (VFA) in rumen fluid, along with elevated levels of β-hydroxybutyrate (BHBA), insulin-like growth factor-1 (IGF-1), and insulin in plasma. Sodium butyrate promoted rumen papillae growth, as evidenced by increased rumen weight, papillae length, width, surface area, and thickness of the stratum corneum and epithelium. It upregulated the mRNA expression of genes related to cell cycle regulation (*cyclin A2, cyclin D1, CDK6*) and downregulated pro-apoptotic genes (*caspase-3, Bax*). Additionally, sodium butyrate elevated the expression of IGF-1 signaling genes (*IGF-1R, IGFBP-5*) while downregulating *IGFBP-3*. It also increased the mRNA expression of the VFA transporter *MCT1* gene and decreased the expression of genes involved in ion transport (*NHE2*) and fatty acid synthesis (*HMGCS2, HMGCL*).	Lamb	[108]
Intraruminal infusion of sodium butyrate at a dose of 0.3 g/kg of body weight, administered in 50 mL of 0.1 mol/L potassium phosphate buffer for 28 days, was associated with elevated ruminal butyrate concentration, larger papillae, and a higher number of cell layers in the epithelial strata of lamb. The treatment increased the mRNA expression of cell cycle regulators (*cyclin A, cyclin D1, cyclin E1, CDK2, CDK4,* and *CDK6*) and upregulated apoptosis-linked genes (*caspase 3, caspase 9,* and *Bax*).	Lamb	[109]

4. Discussion

One of the most important functions of butyrate is its ability to regulate inflammatory responses. Its anti-inflammatory effects are mainly achieved by inhibiting NF-κB, a key transcription factor responsible for the expression of pro-inflammatory cytokines such as IL-1β and TNF-α [15,110]. By suppressing NF-κB activation, butyrate effectively reduces the secretion of cytokines, creating a less inflammatory environment. This mechanism is particularly crucial during key developmental stages when organisms are more sensitive to stress and inflammatory conditions [16,17]. Moreover, butyrate's role as an HDAC inhibitor is central to its anti-inflammatory function [110]. Through HDAC inhibition, butyrate increases histone acetylation, promoting the transcription of anti-inflammatory genes and downregulating pro-inflammatory ones (Figure 2) [19]. Further research indicates that butyrate activates the aryl hydrocarbon receptor (AHR) via MCT1 and sodium-coupled monocarboxylate transporter 1 (SMCT1), contributing to its anti-inflammatory effects. Supporting this, a study showed that HDAC activity increased linearly under higher doses of LPS, while histone H3 acetylation levels significantly decreased [22]. This dual mechanism not only fosters an anti-inflammatory environment but also aids in the differentiation of regulatory T cells (Tregs), which are critical for maintaining immune balance and controlling excessive inflammatory responses [111,112]. Numerous studies have demonstrated that the inhibition of HDAC activity significantly upregulates the proliferation and functional efficacy of regulatory T-cells. These cells play a pivotal role in modulating immune responses by attenuating inflammation through the secretion of anti-inflammatory cytokines, including interleukin-10 (IL-10) and transforming growth factor-beta (TGF-β) [113,114]. Other studies have also reported that butyrate promotes Treg cell differentiation and the secretion of anti-inflammatory mediators such as IL-10 and secretory IgA while inhibiting Th17 cell differentiation, thereby enhancing immune and anti-inflammatory responses (Figure 3) [12,111,115]. Consistently, another study reported that Butyrate treatment significantly downregulated the level of natural killer (NK) cell function during chronic inflammation [116]. NK cells are crucial for defending against infections and tumors by targeting infected or malignant cells. However, in chronic inflammation, overactive or dysregulated NK cells can cause tissue damage by attacking stressed but healthy cells, potentially leading to autoimmune responses. Prolonged activation of NK cells can worsen chronic diseases by triggering tissue inflammation or fibrosis. To prevent this, the immune system must regulate NK cell activity to avoid excessive responses that may harm healthy tissues. In chronic inflammation, dysregulation can lead NK cells to mistakenly target the body's own tissues, further promoting inflammation and hindering recovery. Research by Shin et al. [117] demonstrated that sodium butyrate (NaB) promotes gene transcription in bovine cells by acetylating histones H3K9 and H3K27. Similarly, another study confirmed that sodium butyrate exerts antioxidant effects through GPR109A, Nrf2, and H3K9/14 acetylation [60]. The study further revealed that butyrate enhances Nrf2 nuclear accumulation and H3K9/14 acetylation via the AMPK signaling pathway, showcasing butyrate's antioxidant and anti-apoptotic roles through the GPR109A/AMPK/Nrf2 signaling axis. Acetylation of H3K9/14 promotes Nrf2 transcription, thereby boosting the cell's antioxidant defenses. In addition to its anti-inflammatory and gut-protective functions, butyrate also exhibits significant antioxidant properties (Figure 4). It increases the expression of antioxidant enzymes such as GSH-Px and activates Nrf2, a key regulator of the body's antioxidant response [71]. By activating Nrf2 and upregulating the expression of antioxidant genes, butyrate helps mitigate oxidative stress, preventing cellular damage and exacerbation of inflammatory processes. Further studies by Correia et al. [75] showed that sodium butyrate enhances ADG and feed conversion efficiency, increases ileal villus height, and reduces the number of Peyer's patches in the ileum. Sodium butyrate also upregulates the expression of several key genes related to nutrient transport (SGLT1, MCT1), tight junction integrity (OCL, ZO-1), and antioxidant defense (GPX, SOD), all of which are essential for maintaining intestinal function and integrity [75].

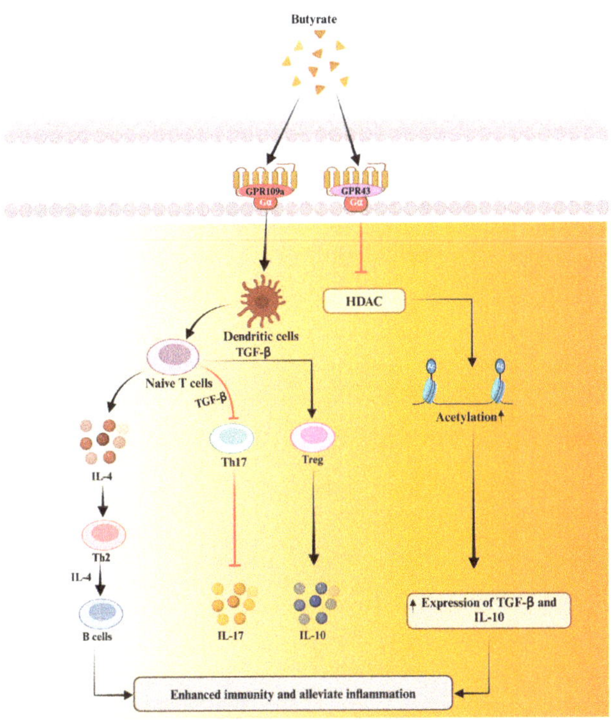

Figure 3. The schematic overview illustrates the role of butyrate in enhancing immunity. Butyrate is crucial for maintaining intestinal immune homeostasis. Butyrate inhibiting histone deacetylases (HDAC) activity by promoting its acetylation. HDAC acetylation significantly upregulates the proliferation and functional efficacy of regulatory T-cells. These cells play a pivotal role in modulating immune responses by attenuating inflammation through the secretion of anti-inflammatory cytokines, including interleu-kin-10 (IL-10) and transforming growth factor-beta (TGF-β) [111–114].

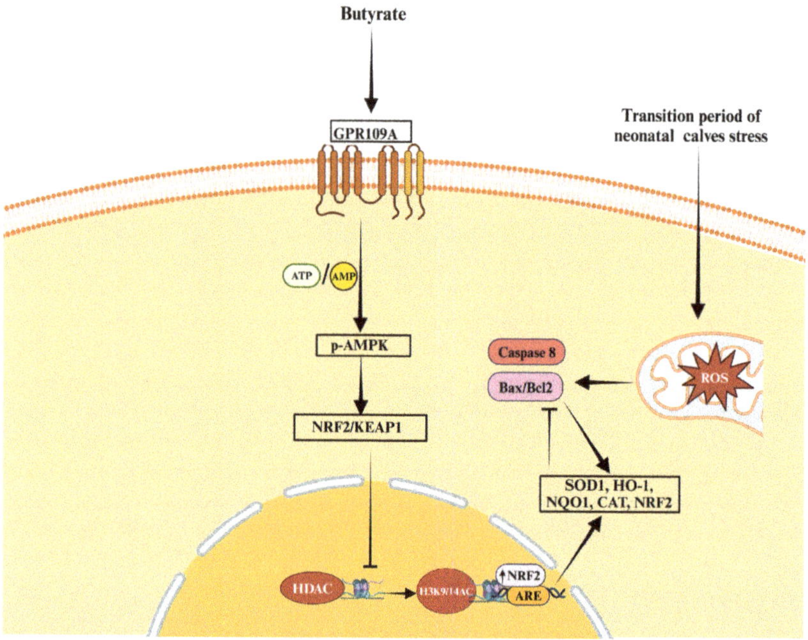

Figure 4. The antioxidant role of butyrate in promoting animal health. Butyrate alleviates stress in high-yielding dairy cows through the GPR109A/AMPK/NRF2 signaling pathway and H3K9/14 acetylation. Furthermore, NRF2 nuclear accumulation and H3K9/14 acetylation enhance antioxidant and anti-apoptotic effects in bovine mammary epithelial cells.

5. Conclusions

In conclusion, butyrate supplementation has emerged as a promising strategy for enhancing intestinal health and growth performance in livestock. Its multifaceted benefits include improving gut barrier integrity, modulating immune responses, optimizing microbial diversity, and reducing inflammation and oxidative stress. Butyrate's role in gene expression and epigenetic regulation further underscores its potential to promote tissue development and metabolic efficiency. Studies across calves, piglets, sheep, and goats consistently demonstrate its ability to enhance average daily gain, feed efficiency, and overall health, particularly during critical developmental stages like weaning. Future opportunities in this area include optimizing butyrate supplementation protocols to maximize benefits across various livestock species, developmental stages, and environmental conditions. Research should focus on precise dosing, long-term health impacts, and the mechanisms of butyrate's interaction with the gut microbiome and immune system. Exploring synergistic effects with other feed additives and their potential in disease mitigation could lead to new therapeutic applications. Future studies should also explore the potential of butyrate as an alternative to antibiotics as a growth promoter, emphasizing its role in mitigating the issue of antibiotic resistance. By tailoring strategies to species-specific needs, butyrate can be optimized as a valuable dietary additive, enhancing livestock productivity and welfare sustainably.

Author Contributions: W.C., M.Z.K., C.W. and Q.M.: writing—original draft. M.Z.K., W.C., Y.L., Z.Z., L.W. and A.K.: writing—review and editing, and literature search. M.Z.K. and C.W.: proofreading and supervision, resources, and funding: C.W. All authors have read and agreed to the published version of the manuscript.

Funding: This research was funded by the National Key R&D Program of China (grant numbers 2022YFD1600103; 2023YFD1302004), the Shandong Province Modern Agricultural Technology System Donkey Industrial Innovation Team (grant no. SDAIT-27), Livestock and Poultry Breeding Industry Project of the Ministry of Agriculture and Rural Affairs (grant number 19211162), the National Natural Science Foundation of China (grant no. 31671287), The Open Project of Liaocheng University Animal Husbandry Discipline (grant no. 319312101-14), the Open Project of Shandong Collaborative Innovation Center for Donkey Industry Technology (grant no. 3193308), Research on Donkey Pregnancy Improvement (grant no. K20LC0901) and Liaocheng University Scientific Research Fund (grant no. 318052025), and Shandong Province Agricultural Major Technology Collaborative Promotion Plan (SDNYXTTG-2024-13).

Institutional Review Board Statement: Not applicable.

Informed Consent Statement: Not applicable.

Data Availability Statement: Not applicable.

Conflicts of Interest: The authors declare no conflicts of interest. We declare that we have no financial and personal relationships with other people or organizations that can inappropriately influence our work, and there is no professional or other personal interest of any nature or kind in any product, service, and/or company that could be construed as influencing the content of this review paper.

References

1. Fagundes, R.R.; Belt, S.C.; Bakker, B.M.; Dijkstra, G.; Harmsen, H.J.; Faber, K.N. Beyond butyrate: Microbial fiber metabolism supporting colonic epithelial homeostasis. *Trends Microbiol.* **2024**, *32*, 178–189. [CrossRef] [PubMed]
2. Park, B.; Kim, J.Y.; Riffey, O.F.; Walsh, T.J.; Johnson, J.; Donohoe, D.R. Crosstalk between butyrate oxidation in colonocyte and butyrate-producing bacteria. *iScience* **2024**, *27*, 9. [CrossRef] [PubMed]
3. Zhong, H.; Yu, W.; Wang, M.; Lin, B.; Sun, X.; Zheng, N.; Wang, J.; Zhao, S. Sodium butyrate promotes gastrointestinal development of preweaning bull calves via inhibiting inflammation, balancing nutrient metabolism, and optimizing microbial community functions. *Anim. Nutr.* **2023**, *14*, 88–100. [CrossRef]
4. Bedford, A.; Gong, J. Implications of butyrate and its derivatives for gut health and animal production. *Anim. Nutr.* **2018**, *4*, 151–159. [CrossRef]
5. Górka, P.; Kowalski, Z.M.; Zabielski, R.; Guilloteau, P. Invited review: Use of butyrate to promote gastrointestinal tract development in calves. *J. Dairy Sci.* **2018**, *101*, 4785–4800. [CrossRef]
6. Herrick, K.J.; Hippen, A.R.; Kalscheur, K.F.; Schingoethe, D.J.; Ranathunga, S.D.; Anderson, J.L.; Moreland, S.C.; Van Eys, J.E. Infusion of butyrate affects plasma glucose, butyrate, and β-hydroxybutyrate but not plasma insulin in lactating dairy cows. *J. Dairy Sci.* **2018**, *101*, 3524–3536. [CrossRef]
7. Carretta, M.D.; Hidalgo, A.I.; Burgos, J.; Opazo, L.; Castro, L.; Hidalgo, M.A.; Figueroa, C.D.; Taubert, A.; Hermosilla, C.; Burgos, R.A. Butyric acid stimulates bovine neutrophil functions and potentiates the effect of platelet activating factor. *Vet. Immunol. Immunopathol.* **2016**, *176*, 18–27. [CrossRef]
8. He, Z.; Dong, H. The roles of short-chain fatty acids derived from colonic bacteria fermentation of non-digestible carbohydrates and exogenous forms in ameliorating intestinal mucosal immunity of young ruminants. *Front. Immunol.* **2023**, *14*, 1291846. [CrossRef]
9. Wu, D.; Zhang, Z.; Shao, K.; Wang, X.; Huang, F.; Qi, J.; Duan, Y.; Jia, Y.; Xu, M. Effects of sodium butyrate supplementation in milk on the growth performance and intestinal microbiota of preweaning Holstein calves. *Animals* **2023**, *13*, 2069. [CrossRef]
10. Plöger, S.; Stumpff, F.; Penner, G.B.; Schulzke, J.D.; Gäbel, G.; Martens, H.; Shen, Z.; Günzel, D.; Aschenbach, J.R. Microbial butyrate and its role for barrier function in the gastrointestinal tract. *Ann. N. Y. Acad. Sci.* **2012**, *1258*, 52–59. [CrossRef]
11. Nicola, M.S.; Kalb, A.L.; Barbosa, A.A.; Velasquez, B.E.; Rincon, J.A.; Feijó, J.O.; Dellagostin, E.N.; Martins, A.W.; Blödorn, E.B.; Domingues, W.B.; et al. Butyrate supplementation in the liquid diet of dairy calves leads to a rapid recovery from diarrhea and reduces its occurrence and relapses in the preweaning period. *J. Dairy Sci.* **2023**, *106*, 7908–7923. [CrossRef] [PubMed]
12. Kovács, L.; Pajor, F.; Bakony, M.; Fébel, H.; Edwards, J.E. Prepartum magnesium butyrate supplementation of dairy cows improves colostrum yield, calving ease, fertility, early lactation performance and neonatal vitality. *Animals* **2023**, *13*, 1319. [CrossRef] [PubMed]
13. Ma, L.; Yang, Y.; Liu, W.; Bu, D. Sodium butyrate supplementation impacts the gastrointestinal bacteria of dairy calves before weaning. *Appl. Microbiol. Biotechnol.* **2023**, *107*, 3291–3304. [CrossRef]

14. Wu, D.; Zhang, Z.; Wang, Q.; Shao, K.; Wang, D.; Wang, X.; Qi, J.; Jia, Y.; Xu, M. Interactions of starter starch and sodium butyrate for preweaned dairy calves: Growth performance, rumen development, and microbiota. *Anim. Feed Sci. Technol.* **2023**, *305*, 115780. [CrossRef]
15. Wang, Y.; Xu, Q.; Meng, M.; Chang, G.; Ma, N.; Shen, X. Butyrate protects against γ-d-glutamyl-meso-diaminopimelic acid-induced inflammatory response and tight junction disruption through histone deacetylase 3 inhibition in bovine mammary epithelial cells. *J. Agric. Food Chem.* **2023**, *71*, 14638–14648. [CrossRef]
16. McCurdy, D.E.; Wilkins, K.R.; Hiltz, R.L.; Moreland, S.; Klanderman, K.; Laarman, A.H. Effects of supplemental butyrate and weaning on rumen fermentation in Holstein calves. *J. Dairy Sci.* **2019**, *102*, 8874–8882. [CrossRef]
17. Dai, H.; Liu, X.; Yan, J.; Aabdin, Z.U.; Bilal, M.S.; Shen, X. Sodium butyrate ameliorates high-concentrate diet-induced inflammation in the rumen epithelium of dairy goats. *J. Agric. Food Chem.* **2017**, *65*, 596–604. [CrossRef]
18. Zhang, K.; Hussain, T.; Wang, J.; Li, M.; Wang, W.; Ma, X.; Liao, Y.; Yao, J.; Song, Y.; Liang, Z.; et al. Sodium butyrate abrogates the growth and pathogenesis of *Mycobacterium bovis* via regulation of cathelicidin (LL37) expression and NF-κB signaling. *Front. Microbiol.* **2020**, *11*, 433. [CrossRef]
19. Kang, X.; Li, C.; Liu, S.; Baldwin, R.L.; Liu, G.E.; Li, C.J. Genome-wide acetylation modification of H3K27ac in bovine rumen cell following butyrate exposure. *Biomolecules* **2023**, *13*, 1137. [CrossRef]
20. Kang, H.; Park, Y.K.; Lee, J.Y.; Bae, M. Roles of histone deacetylase 4 in the inflammatory and metabolic processes. *Diabetes Metab. J.* **2024**, *48*, 340–353. [CrossRef]
21. Park, O.J.; Ha, Y.E.; Sim, J.R.; Lee, D.; Lee, E.H.; Kim, S.Y.; Yun, C.H.; Han, S.H. Butyrate potentiates *Enterococcus faecalis* lipoteichoic acid-induced inflammasome activation via histone deacetylase inhibition. *Cell Death Discov.* **2023**, *9*, 107. [CrossRef] [PubMed]
22. Chen, J.; Wu, Y.; Sun, Y.; Dong, X.; Wang, Z.; Zhang, Z.; Xiao, Y.; Dong, G. Bacterial endotoxin decreased histone H3 acetylation of bovine mammary epithelial cells and the adverse effect was suppressed by sodium butyrate. *BMC Vet. Res.* **2019**, *15*, 267. [CrossRef] [PubMed]
23. Watanabe, D.H.; Doelman, J.; Steele, M.A.; Guan, L.L.; Seymour, D.J.; Penner, G.B. A comparison of post-ruminal provision of Ca-gluconate and Ca-butyrate on growth performance, gastrointestinal barrier function, short-chain fatty acid absorption, intestinal histology, and brush-border enzyme activity in beef heifers. *J. Anim. Sci.* **2023**, *101*, skad050. [CrossRef]
24. Zhang, J.; Bu, L.; Liu, Y.; Huo, W.; Xia, C.; Pei, C.; Liu, Q. Dietary supplementation of sodium butyrate enhances lactation performance by promoting nutrient digestion and mammary gland development in dairy cows. *Anim. Nutr.* **2023**, *15*, 137–148. [CrossRef]
25. Dengler, F.; Kraetzig, A.; Gäbel, G. Butyrate protects porcine colon epithelium from hypoxia-induced damage on a functional level. *Nutrients* **2021**, *13*, 305. [CrossRef]
26. Dengler, F.; Rackwitz, R.; Benesch, F.; Pfannkuche, H.; Gäbel, G. Both butyrate incubation and hypoxia upregulate genes involved in the ruminal transport of SCFA and their metabolites. *J. Anim. Physiol. Anim. Nutr.* **2015**, *99*, 379–390. [CrossRef]
27. Belli, A.L.; Coelho, S.G.; Campolina, J.P.; Neves, L.F.; Neto, H.C.; Silva, C.S.; Machado, F.S.; Pereira, L.G.; Tomich, T.R.; Carvalho, W.A.; et al. Effects of supplementing milk replacer with sodium butyrate on dairy calves. *Animals* **2024**, *14*, 277. [CrossRef]
28. Rodríguez, G.B.; Carmona, D.A.; Elghandour, M.M.; Salem, A.Z.; Soto, H.R.; Sánchez, R.R.; Adegbeye, M.J. Sustainable use of sodium butyrate as a source of bioactive additive: Impact on calf growth performance, rumen fermentation characteristics, and microbial count. *Biomass Convers. Biorefin.* **2024**, *14*, 6229–6235. [CrossRef]
29. Wu, D.; Meng, Q.S.; Wang, Y.D.; Wang, M.Y.; Xu, E.H.; Xiao, L.; Xu, M. Dietary supplementation of free or two fat-coated sodium butyrate with varying release times on gastrointestinal development and tight junctions in preweaning Holstein calves. *Anim. Feed Sci. Technol.* **2022**, *285*, 115224. [CrossRef]
30. Wu, D.; Wu, C.; Shao, K.; Zhang, Z.; Wang, D.; Qi, J.; Jia, Y.; Xu, M. Interactions of starter starch and sodium butyrate for postweaned dairy calves: Growth performance, blood indices, and inflammation. *Anim. Feed Sci. Technol.* **2024**, *309*, 115898. [CrossRef]
31. Wu, D.; Zhang, Z.; Wang, X.; Harmon, D.L.; Jia, Y.; Qi, J.; Li, X.; Jia, H.; Xu, M. Exploring the Role of G Protein Expression in Sodium Butyrate-Enhanced Pancreas Development of Dairy Calves: A Proteomic Perspective. *J. Agric. Food Chem.* **2024**, *72*, 5645–5658. [CrossRef] [PubMed]
32. Zhuang, Y.; Abdelsattar, M.M.; Fu, Y.; Zhang, N.; Chai, J. Butyrate metabolism in rumen epithelium affected by host and diet regime through regulating microbiota in a goat model. *Anim. Nutr.* **2024**, *19*, 41–55. [CrossRef] [PubMed]
33. Boschiero, C.; Gao, Y.; Baldwin, R.L.; Ma, L.; Li, C.J.; Liu, G.E. Differentially CTCF-binding sites in cattle rumen tissue during weaning. *Int. J. Mol. Sci.* **2022**, *23*, 9070. [CrossRef] [PubMed]
34. Guilloteau, P.; Savary, G.; Jaguelin-Peyrault, Y.; Romé, V.; Le Normand, L.; Zabielski, R. Dietary sodium butyrate supplementation increases digestibility and pancreatic secretion in young milk-fed calves. *J. Dairy Sci.* **2010**, *93*, 5842–5850. [CrossRef]
35. Amin, A.B.; Trabi, E.B.; Zhu, C.; Mao, S. Role of butyrate as part of milk replacer and starter diet on intestinal development in pre-weaned calves: A systematic review. *Anim. Feed Sci. Technol.* **2022**, *292*, 115423. [CrossRef]

36. Niwińska, B.; Hanczakowska, E.; Arciszewski, M.B.; Klebaniuk, R. Exogenous butyrate: Implications for the functional development of ruminal epithelium and calf performance. *Anim.* **2017**, *11*, 1522–1530. [CrossRef]
37. Inabu, Y.; Horike, H.; Yamano, H.; Taguchi, Y.; Okada, S.; Etoh, T.; Shiotsuka, Y.; Fujino, R.; Takahashi, H. Effect of feeding sodium butyrate to beef female cows during pre- and post-partum period on concentrations of glucagon-like peptides in plasma and colostrum. *Anim. Sci. J.* **2024**, *95*, e13961. [CrossRef]
38. Francis, F.L.; Rusche, W.C.; LaFleur, D.; Hergenreder, J.E.; Smith, Z.K. Dose effects of encapsulated butyric acid and zinc on beef feedlot steer growth performance, dietary net energy utilization, rumen morphometrics, small intestine histology, and carcass characteristics. *J. Anim. Sci.* **2024**, *102*, skae240. [CrossRef]
39. Wu, Y.; Sun, Y.; Zhang, R.; He, T.; Huang, G.; Tian, K.; Liu, J.; Chen, J.; Dong, G. Sodium butyrate more effectively mitigates the negative effects of high-concentrate diet in dairy cows than sodium β-hydroxybutyrate via reducing free bacterial cell wall components in rumen fluid and plasma. *Toxins* **2021**, *13*, 352. [CrossRef]
40. Kong, Q.; Chen, X.; Liu, Y.; Ali, F.; Idrees, A.; Ataya, F.S.; Shang, Z.; Li, K. Sodium acetate and sodium butyrate attenuate diarrhea in yak calves by regulating gut microbiota and metabolites. *Heliyon* **2024**, *10*, e12796. [CrossRef]
41. Meale, S.J.; Chaucheyras-Durand, F.; Berends, H.; Guan, L.L.; Steele, M.A. From pre- to postweaning: Transformation of the young calf's gastrointestinal tract. *J. Dairy Sci.* **2017**, *100*, 5984–5995. [CrossRef] [PubMed]
42. Steele, M.A.; Penner, G.B.; Chaucheyras-Durand, F.; Guan, L.L. Development and physiology of the rumen and the lower gut: Targets for improving gut health. *J. Dairy Sci.* **2016**, *99*, 4955–4966. [CrossRef]
43. Kim, M.H.; Yang, J.Y.; Upadhaya, S.D.; Lee, H.J.; Yun, C.H.; Ha, J.K. The stress of weaning influences serum levels of acute-phase proteins, iron-binding proteins, inflammatory cytokines, cortisol, and leukocyte subsets in Holstein calves. *J. Vet. Sci.* **2011**, *12*, 151–157. [CrossRef]
44. Machado, V.S.; Ballou, M.A. Overview of common practices in calf raising facilities. *Transl. Anim. Sci.* **2022**, *6*, txab234. [CrossRef]
45. Yang, T.; Datsomor, O.; Jiang, M.; Ma, X.; Zhao, G.; Zhan, K. Protective roles of sodium butyrate in lipopolysaccharide-induced bovine ruminal epithelial cells by activating G protein-coupled receptors 41. *Front. Nutr.* **2022**, *9*, 842634. [CrossRef]
46. Zhan, K.; Gong, X.; Chen, Y.; Jiang, M.; Yang, T.; Zhao, G. Short-chain fatty acids regulate the immune responses via G protein-coupled receptor 41 in bovine rumen epithelial cells. *Front. Immunol.* **2019**, *10*, 2042. [CrossRef]
47. Ali, I.; Li, C.; Kuang, M.; Shah, A.U.; Shafiq, M.; Ahmad, M.A.; Abdalmegeed, D.; Li, L.; Wang, G. Nrf2 Activation and NF-κB & caspase/bax signaling inhibition by sodium butyrate alleviates LPS-induced cell injury in bovine mammary epithelial cells. *Mol. Immunol.* **2022**, *148*, 54–67.
48. Liu, J.; Chang, G.; Huang, J.; Ma, N.; Wang, Y.; Roy, A.C.; Shen, X. Sodium butyrate pretreatment mitigates lipopolysaccharide-induced inflammation through the TLR4/NF-κB signaling pathway in bovine embryo trachea cells. *Anim. Biotechnol.* **2022**, *33*, 1571–1581. [CrossRef]
49. Li, Y.; Liu, J.; Cui, Y.; Cao, Y.; Xu, P.; Kan, X.; Guo, W.; Fu, S. Sodium butyrate attenuates bovine mammary epithelial cell injury by inhibiting the formation of neutrophil extracellular traps. *Int. Immunopharmacol.* **2022**, *110*, 109009. [CrossRef]
50. Klobucher, K.N.; Badger, R.; Foxall, T.; Erickson, P.S. Effect of sodium butyrate, monensin, and butyric acid on the viability of *Eimeria bovis* sporozoites and their degree of damage to a bovine epithelial cell line. *J. Anim. Sci.* **2022**, *100*, skac360. [CrossRef]
51. Klobucher, K.N.; Stahl, T.C.; Islam, T.; Gray, A.S.; Curreri, S.I.; Erickson, P.S. Supplementing sodium butyrate to limit-fed heifers: Effects on growth, coccidiosis, urinary purine derivatives, and apparent total-tract nutrient digestibility. *J. Dairy Sci.* **2023**, *106*, 6894–6902. [CrossRef] [PubMed]
52. Liu, W.; La, A.L.; Evans, A.; Gao, S.; Yu, Z.; Bu, D.; Ma, L. Supplementation with sodium butyrate improves growth and anti-oxidant function in dairy calves before weaning. *J. Anim. Sci. Biotechnol.* **2021**, *12*, 1–9. [CrossRef] [PubMed]
53. Liu, S.; Wu, J.; Wu, Z.; Alugongo, G.M.; Khan, M.Z.; Li, J.; Xiao, J.; He, Z.; Ma, Y.; Li, S.; et al. Tributyrin administration improves intestinal development and health in pre-weaned dairy calves fed milk replacer. *Anim. Nutr.* **2022**, *10*, 399–411. [CrossRef]
54. Gao, J.; Dong, J.; Sun, Z.; Wang, T.; Guan, Y.; Sun, Y.; Qin, G.; Zhang, X.; Zhen, Y. Effects of antimicrobial peptide and tributyrin on fecal microflora and blood indices of female calves. *Food Sci. Nutr.* **2023**, *11*, 5248–5257. [CrossRef]
55. Murayama, K.; Fukui, T.; Isobe, N.; Oba, M.; Sugino, T. Effects of tributyrin supplementation in calf starter on growth, blood parameters, and development of rumen and small intestine in Holstein calves fed milk replacers with different fatty acid profile. *Anim. Feed Sci. Technol.* **2024**, *308*, 115883. [CrossRef]
56. Murayama, K.; Fukui, T.; Kushibiki, S.; Sakamoto, K.; Inouchi, K.; Sugino, T. Effects of medium-chain fatty acids and tributyrin supplementation in milk replacers on growth performance, blood metabolites, and hormone concentrations in Holstein dairy calves. *J. Dairy Sci.* **2023**, *106*, 4599–4607. [CrossRef]
57. Liu, S.; Ma, J.Y.; Zhou, J.; Wu, J.D.; Li, J.H.; Alugongo, G.M.; Xiao, J.X.; Wang, J.J.; Wang, Y.J.; Wang, W.; et al. Tributyrin supplementation in pasteurized waste milk: Effects on growth performance, health, and blood parameters of dairy calves. *J. Dairy Sci.* **2021**, *104*, 12496–12507. [CrossRef]

58. Sun, X.; Luo, S.; Jiang, C.; Tang, Y.; Cao, Z.; Jia, H.; Xu, Q.; Zhao, C.; Loor, J.J.; Xu, C. Sodium butyrate reduces bovine mammary epithelial cell inflammatory responses induced by exogenous lipopolysaccharide by inactivating NF-κB signaling. *J. Dairy Sci.* **2020**, *103*, 8388–8397. [CrossRef]
59. Jiang, L.; Wang, J.; Liu, Z.; Jiang, A.; Li, S.; Wu, D.; Zhang, Y.; Zhu, X.; Zhou, E.; Wei, Z.; et al. Sodium butyrate alleviates lipopolysaccharide-induced inflammatory responses by downregulation of NF-κB, NLRP3 signaling pathway, and activating histone acetylation in bovine macrophages. *Front. Vet. Sci.* **2020**, *7*, 579674. [CrossRef]
60. Guo, W.; Liu, J.; Sun, J.; Gong, Q.; Ma, H.; Kan, X.; Cao, Y.; Wang, J.; Fu, S. Butyrate alleviates oxidative stress by regulating NRF2 nuclear accumulation and H3K9/14 acetylation via GPR109A in bovine mammary epithelial cells and mammary glands. *Free Radic. Biol. Med.* **2020**, *152*, 728–742. [CrossRef]
61. Silva, L.G.; Ferguson, B.S.; Avila, A.S.; Faciola, A.P. Sodium propionate and sodium butyrate effects on histone deacetylase (HDAC) activity, histone acetylation, and inflammatory gene expression in bovine mammary epithelial cells. *J. Anim. Sci.* **2018**, *96*, 5244–5252. [CrossRef] [PubMed]
62. Xu, T.; Ma, N.; Wang, Y.; Shi, X.; Chang, G.; Loor, J.J.; Shen, X. Sodium butyrate supplementation alleviates the adaptive response to inflammation and modulates fatty acid metabolism in lipopolysaccharide-stimulated bovine hepatocytes. *J. Agric. Food Chem.* **2018**, *66*, 6281–6290. [CrossRef] [PubMed]
63. Frieten, D.; Gerbert, C.; Koch, C.; Dusel, G.; Eder, K.; Hoeflich, A.; Mielenz, B.; Hammon, H.M. Influence of ad libitum milk replacer feeding and butyrate supplementation on the systemic and hepatic insulin-like growth factor I and its binding proteins in Holstein calves. *J. Dairy Sci.* **2018**, *101*, 1661–1672. [CrossRef] [PubMed]
64. Gerbert, C.; Frieten, D.; Koch, C.; Dusel, G.; Eder, K.; Stefaniak, T.; Bajzert, J.; Jawor, P.; Tuchscherer, A.; Hammon, H.M. Effects of ad libitum milk replacer feeding and butyrate supplementation on behavior, immune status, and health of Holstein calves in the postnatal period. *J. Dairy Sci.* **2018**, *101*, 7348–7360. [CrossRef]
65. O'Hara, E.; Kelly, A.; McCabe, M.S.; Kenny, D.A.; Guan, L.L.; Waters, S.M. Effect of a butyrate-fortified milk replacer on gastrointestinal microbiota and products of fermentation in artificially reared dairy calves at weaning. *Sci. Rep.* **2018**, *8*, 14901. [CrossRef]
66. Fukumori, R.; Oba, M.; Izumi, K.; Otsuka, M.; Suzuki, K.; Gondaira, S.; Higuchi, H.; Oikawa, S. Effects of butyrate supplementation on blood glucagon-like peptide-2 concentration and gastrointestinal functions of lactating dairy cows fed diets differing in starch content. *J. Dairy Sci.* **2020**, *103*, 3656–3667. [CrossRef]
67. Dell'Anno, M.; Scaglia, E.; Reggi, S.; Grossi, S.; Rossi, C.A.; Frazzini, S.; Caprarulo, V.; Rossi, L. Evaluation of tributyrin supplementation in milk replacer on diarrhoea occurrence in preweaning Holstein calves. *Animal* **2023**, *17*, 100791. [CrossRef]
68. Wu, D.; Zhang, Z.; Song, Q.; Jia, Y.; Qi, J.; Xu, M. Modulating gastrointestinal microbiota in preweaning dairy calves: Dose-dependent effects of milk-based sodium butyrate supplementation. *Microorganisms* **2024**, *12*, 333. [CrossRef]
69. Wan, F.; Wen, X.; Zhao, H.; Tang, S.; Wang, M.; Yi, B.; Chen, L.; Lu, Y.; Zhong, R.; Zhang, H. Chemically protected sodium butyrate supplementation improves anti-inflammatory and antioxidant capacities potentially through modulating gut microbiota and short-chain fatty acids levels in piglets. *J. Funct. Foods* **2024**, *121*, 106434. [CrossRef]
70. Han, Y.; Tang, C.; Zhao, Q.; Fan, S.; Yang, P.; Zhang, J. Butyrate mitigates lipopolysaccharide-induced intestinal morphological changes in weanling piglets by regulating the microbiota and energy metabolism, and alleviating inflammation and apoptosis. *Microorganisms* **2022**, *10*, 2001. [CrossRef]
71. Li, W.; Lan, T.; Ding, Q.; Ren, Z.; Tang, Z.; Tang, Q.; Peng, X.; Xu, Y.; Sun, Z. Effect of low protein diets supplemented with sodium butyrate, medium-chain fatty acids, or n-3 polyunsaturated fatty acids on the growth performance, immune function, and microbiome of weaned piglets. *Int. J. Mol. Sci.* **2023**, *24*, 17592. [CrossRef] [PubMed]
72. Wang, C.C.; Wu, H.; Lin, F.H.; Gong, R.; Xie, F.; Peng, Y.; Feng, J.; Hu, C.H. Sodium butyrate enhances intestinal integrity, inhibits mast cell activation, inflammatory mediator production, and JNK signaling pathway in weaned pigs. *Innate Immun.* **2018**, *24*, 40–46. [CrossRef] [PubMed]
73. Barbosa, K.A.; Genova, J.L.; Pazdziora, M.L.; Hennig, J.F.; Azevedo, L.B.; Veiga, B.R.; Rodrigues, G.D.; Carvalho, S.T.; Paiano, D.; Saraiva, A.; et al. The role of dietary monoglycerides and tributyrin in enhancing performance and intestinal health function in nursery piglets. *Ital. J. Anim. Sci.* **2023**, *22*, 626–638. [CrossRef]
74. Zong, Q.; Li, K.; Qu, H.; Hu, P.; Xu, C.; Wang, H.; Wu, S.; Wang, S.; Liu, H.Y.; Cai, D.; et al. Sodium butyrate ameliorates deoxynivalenol-induced oxidative stress and inflammation in the porcine liver via NR4A2-mediated histone acetylation. *J. Agric. Food Chem.* **2023**, *71*, 10427–10437. [CrossRef] [PubMed]
75. Correia, A.M.; Genova, J.L.; Kim, S.W.; Abranches, F.F.; Rocha, G.C. Autolyzed yeast and sodium butyrate supplemented alone to diets promoted improvements in performance, intestinal health, and nutrient transporter in weaned piglets. *Sci. Rep.* **2024**, *14*, 11885. [CrossRef]
76. Kovanda, L.; Park, J.; Park, S.; Kim, K.; Li, X.; Liu, Y. Dietary butyrate and valerate glycerides impact diarrhea severity and immune response of weaned piglets under ETEC F4-ETEC F18 coinfection conditions. *J. Anim. Sci.* **2023**, *101*, skad401. [CrossRef]

77. López-Colom, P.; Castillejos, L.; Rodríguez-Sorrento, A.; Puyalto, M.; Mallo, J.J.; Martín-Orúe, S.M. Impact of in-feed sodium butyrate or sodium heptanoate protected with medium-chain fatty acids on gut health in weaned piglets challenged with *Escherichia coli* F4+. *Arch. Anim. Nutr.* **2020**, *74*, 271–295. [CrossRef]
78. Tian, M.; Li, L.; Tian, Z.; Zhao, H.; Chen, F.; Guan, W.; Zhang, S. Glyceryl butyrate attenuates enterotoxigenic *Escherichia coli*-induced intestinal inflammation in piglets by inhibiting the NF-κB/MAPK pathways and modulating the gut microbiota. *Food Funct.* **2022**, *13*, 6282–6292. [CrossRef]
79. You, C.; Xu, Q.; Chen, J.; Xu, Y.; Pang, J.; Peng, X.; Tang, Z.; Sun, W.; Sun, Z. Effects of different combinations of sodium butyrate, medium-chain fatty acids, and omega-3 polyunsaturated fatty acids on the reproductive performance of sows and biochemical parameters, oxidative status, and intestinal health of their offspring. *Animals* **2023**, *13*, 1093. [CrossRef]
80. Zhong, X.; Zhang, Z.; Wang, S.; Cao, L.; Zhou, L.; Sun, A.; Zhong, Z.; Nabben, M. Microbial-driven butyrate regulates jejunal homeostasis in piglets during the weaning stage. *Front. Microbiol.* **2019**, *9*, 3335. [CrossRef]
81. Bernad-Roche, M.; Bellés, A.; Grasa, L.; Casanova-Higes, A.; Mainar-Jaime, R.C. Effects of dietary supplementation with protected sodium butyrate on gut microbiota in growing-finishing pigs. *Animals* **2021**, *11*, 2137. [CrossRef] [PubMed]
82. Han, Y.; Zhao, Q.; Tang, C.; Li, Y.; Zhang, K.; Li, F.; Zhang, J. Butyrate mitigates weanling piglets from lipopolysaccharide-induced colitis by regulating microbiota and energy metabolism of the gut–liver axis. *Front. Microbiol.* **2020**, *11*, 588666. [CrossRef]
83. Sun, W.; Sun, J.; Li, M.; Xu, Q.; Zhang, X.; Tang, Z.; Chen, J.; Zhen, J.; Sun, Z. The effects of dietary sodium butyrate supplementation on the growth performance, carcass traits, and intestinal microbiota of growing-finishing pigs. *J. Appl. Microbiol.* **2020**, *128*, 1613–1623. [CrossRef]
84. Wang, S.; Zhang, C.; Yang, J.; Wang, X.; Wu, K.; Zhang, B.; Zhang, J.; Yang, A.; Rajput, S.A.; Qi, D. Sodium butyrate protects the intestinal barrier by modulating intestinal host defense peptide expression and gut microbiota after a challenge with deoxynivalenol in weaned piglets. *J. Agric. Food Chem.* **2020**, *68*, 4515–4527. [CrossRef]
85. Lin, F.; Li, X.; Wen, J.; Wang, C.; Peng, Y.; Feng, J.; Hu, C. Effects of coated sodium butyrate on performance, diarrhea, intestinal microflora, and barrier function of pigs during the first 2-week post-weaning. *Anim. Feed Sci. Technol.* **2020**, *263*, 114464. [CrossRef]
86. Upadhaya, S.D.; Jiao, Y.; Kim, Y.M.; Lee, K.Y.; Kim, I.H. Coated sodium butyrate supplementation to a reduced nutrient diet enhanced the performance and positively impacted villus height and fecal and digesta bacterial composition in weaner pigs. *Anim. Feed Sci. Technol.* **2020**, *265*, 114534. [CrossRef]
87. Chen, X.; Xu, J.; Su, Y.; Zhu, W. Effects of intravenous infusion with sodium butyrate on colonic microbiota, intestinal development, and mucosal immune-related gene expression in normal growing pigs. *Front. Microbiol.* **2018**, *9*, 1652. [CrossRef]
88. Feng, W.; Wu, Y.; Chen, G.; Fu, S.; Li, B.; Huang, B.; Wang, D.; Wang, W.; Liu, J. Sodium butyrate attenuates diarrhea in weaned piglets and promotes tight junction protein expression in colon in a GPR109A-dependent manner. *Cell. Physiol. Biochem.* **2018**, *47*, 1617–1629. [CrossRef]
89. Chen, J.; Xu, Q.; Li, Y.; Tang, Z.; Sun, W.; Zhang, X.; Sun, J.; Sun, Z. Comparative effects of dietary supplementations with sodium butyrate, medium-chain fatty acids, and n-3 polyunsaturated fatty acids in late pregnancy and lactation on the reproductive performance of sows and growth performance of suckling piglets. *J. Anim. Sci.* **2019**, *97*, 4256–4267. [CrossRef]
90. Soltani, M.; Kazemi-Bonchenari, M.; Khaltabadi-Farahani, A.H.; Afsarian, O. Interaction of forage provision (alfalfa hay) and sodium butyrate supplementation on performance, structural growth, blood metabolites, and rumen fermentation characteristics of lambs during pre-weaning period. *Anim. Feed Sci. Technol.* **2017**, *230*, 77–86. [CrossRef]
91. Górka, P.; Śliwiński, B.; Flaga, J.; Olszewski, J.; Wojciechowski, M.; Krupa, K.; Godlewski, M.M.; Zabielski, R.; Kowalski, Z.M. Effect of exogenous butyrate on the gastrointestinal tract of sheep. I. Structure and function of the rumen, omasum, and abomasum. *J. Anim. Sci.* **2018**, *96*, 5311–5324. [CrossRef] [PubMed]
92. Górka, P.; Śliwiński, B.; Flaga, J.; Olszewski, J.; Nawrocka, P.; Sobkowiak, K.; Miltko, R.; Godlewski, M.M.; Zabielski, R.; Kowalski, Z.M. Effect of exogenous butyrate on the gastrointestinal tract of sheep. II. Hydrolytic activity in the rumen and structure and function of the small intestine. *J. Anim. Sci.* **2018**, *96*, 5325–5335. [CrossRef] [PubMed]
93. Sun, M.; Ji, W.; Ye, H.; Cai, Y.; Yun, Y.; Wei, X.; Wang, C.; Mao, H. Sodium butyrate administration improves intestinal development of suckling lambs. *J. Anim. Sci.* **2024**, *102*, skae028. [CrossRef] [PubMed]
94. Świerk, S.; Przybyło, M.; Wajsman, I.; Molik, E.; Miltko, R.; Górka, P. Effect of increased intake of concentrates and sodium butyrate supplementation on reticulorumen macroanatomy and reticulorumen fermentation in growing rams. *Animals* **2023**, *17*, 100899. [CrossRef]
95. Świerk, S.; Przybyło, M.; Flaga, J.; Szczepanik, K.; Białek, W.; Flieger, P.; Górka, P. Effect of butyrate sources in a high-concentrate diet on rumen structure and function in growing rams. *Animals* **2024**, *18*, 101285. [CrossRef]
96. Świerk, S.; Przybyło, M.; Flaga, J.; Szczepanik, K.; Garus-Piętak, A.; Biernat, W.; Molik, E.; Wojtysiak, D.; Miltko, R.; Górka, P. Effect of increased intake of concentrates and sodium butyrate supplementation on ruminal epithelium structure and function in growing rams. *Animals* **2023**, *17*, 100898. [CrossRef]

97. Mohamed, M.Y.; Saba, F.E.; Gomaa, A.A.; Desoky, A.L.; Almwafy, A.A.; AbdEl-Salam, O.M.; Ibrahim, E.M. Reproductive tract development and histomorphometric analysis of testes in neonatal Zaraibi kids raised on milk replacer supplemented with sodium butyrate. *J. Anim. Physiol. Anim. Nutr.* **2020**, *104*, 812–822. [CrossRef]
98. Fan, D.; Fu, Y.; Zhang, J.; Bi, Y.; Ma, T.; Diao, Q.; Zhang, N. Sheep-derived butyrate-producing *Clostridium beijerinckii* R8 alleviates diarrhea by shaping the gut microbiota of goat kids. *Anim. Nutr.* **2024**, *19*, 13–24. [CrossRef]
99. Zhao, F.; He, W.; Wu, T.; Elmhadi, M.; Jiang, N.; Zhang, A.; Guan, P. Supplementation of coated sodium butyrate relieved weaning stress and reshaped microbial flora in weaned lambs. *Front. Vet. Sci.* **2024**, *11*, 1423920. [CrossRef]
100. Mammeri, M.; Obregón, D.A.; Chevillot, A.; Polack, B.; Julien, C.; Pollet, T.; Cabezas-Cruz, A.; Adjou, K.T. *Cryptosporidium parvum* infection depletes butyrate producer bacteria in goat kid microbiome. *Front. Microbiol.* **2020**, *11*, 548737. [CrossRef]
101. Wang, K.; Chen, X.; Zuo, L.; Pan, C.; Liu, G.; Zhang, X.; Du, J.; Zhang, C.; Zhang, B.; Wang, Z.; et al. Dietary sodium butyrate protects lipopolysaccharide-induced inflammatory response in lambs through inhibiting TLR4/NF-κB signaling pathway. *Ital. J. Anim. Sci.* **2021**, *20*, 1398–1409. [CrossRef]
102. Chang, G.; Ma, N.; Zhang, H.; Wang, Y.; Huang, J.; Liu, J.; Dai, H.; Shen, X. Sodium butyrate modulates mucosal inflammation injury mediated by GPR41/43 in the cecum of goats fed a high concentration diet. *Front. Physiol.* **2019**, *10*, 1130. [CrossRef] [PubMed]
103. Chishti, G.A.; Felix, T.L. Effects of supplemental calcium butyrate on lamb growth, plasma metabolites, and hepatic gluconeogenic and inflammatory gene expression changes. *Anim. Feed Sci. Technol.* **2020**, *270*, 114678. [CrossRef]
104. Tsugami, Y.; Suzuki, N.; Nii, T.; Isobe, N. Sodium acetate and sodium butyrate differentially upregulate antimicrobial component production in mammary glands of lactating goats. *J. Mammary Gland Biol. Neoplasia* **2022**, *27*, 133–144. [CrossRef]
105. Li, Z.; Wang, X.; Wang, W.; An, R.; Wang, Y.; Ren, Q.; Xuan, J. Benefits of tributyrin on growth performance, gastrointestinal tract development, ruminal bacteria, and volatile fatty acid formation of weaned Small-Tailed Han lambs. *Anim. Nutr.* **2023**, *15*, 187–196. [CrossRef]
106. Ma, N.; Abaker, J.A.; Bilal, M.S.; Dai, H.; Shen, X. Sodium butyrate improves antioxidant stability in sub-acute ruminal acidosis in dairy goats. *BMC Vet. Res.* **2018**, *14*, 275. [CrossRef]
107. Zhang, K.; Meng, M.; Gao, L.; Tu, Y.; Bai, Y. Sodium butyrate improves high-concentrate-diet-induced impairment of ruminal epithelium barrier function in goats. *J. Agric. Food Chem.* **2018**, *66*, 8729–8736. [CrossRef]
108. Liu, L.; Sun, D.; Mao, S.; Zhu, W.; Liu, J. Infusion of sodium butyrate promotes rumen papillae growth and enhances expression of genes related to rumen epithelial VFA uptake and metabolism in neonatal twin lambs. *J. Anim. Sci.* **2019**, *97*, 909–921. [CrossRef]
109. Soomro, J.; Lu, Z.; Gui, H.; Zhang, B.; Shen, Z. Synchronous and time-dependent expression of cyclins, cyclin-dependent kinases, and apoptotic genes in the rumen epithelia of butyrate-infused goats. *Front. Physiol.* **2018**, *9*, 496. [CrossRef]
110. Lee, C.; Kim, B.G.; Kim, J.H.; Chun, J.; Im, J.P.; Kim, J.S. Sodium butyrate inhibits the NF-κB signaling pathway and histone deacetylation, and attenuates experimental colitis in an IL-10 independent manner. *Int. Immunopharmacol.* **2017**, *51*, 47–56. [CrossRef]
111. Recharla, N.; Geesala, R.; Shi, X.-Z. Gut microbial metabolite butyrate and its therapeutic role in inflammatory bowel disease: A literature review. *Nutrients* **2023**, *15*, 2275. [CrossRef] [PubMed]
112. Siddiqui, M.T.; Cresci, G.A. The immunomodulatory functions of butyrate. *J. Inflamm. Res.* **2021**, *14*, 6025–6041. [CrossRef] [PubMed]
113. De Zoeten, E.F.; Wang, L.; Sai, H.; Dillmann, W.H.; Hancock, W.W. Inhibition of HDAC9 increases T regulatory cell function and prevents colitis in mice. *Gastroenterology* **2010**, *138*, 583–594. [CrossRef]
114. Zhang, S.Y.; Zhang, L.Y.; Wen, R.; Yang, N.; Zhang, T.N. Histone deacetylases and their inhibitors in inflammatory diseases. *Biomed. Pharmacother.* **2024**, *179*, 117295. [CrossRef]
115. Liu, L.; Li, L.; Min, J.; Wang, J.; Wu, H.; Zeng, Y.; Chen, S.; Chu, Z. Butyrate interferes with the differentiation and function of human monocyte-derived dendritic cells. *Cell. Immunol.* **2012**, *277*, 66–73. [CrossRef]
116. Zaiatz-Bittencourt, V.; Jones, F.; Tosetto, M.; Scaife, C.; Cagney, G.; Jones, E.; Doherty, G.A.; Ryan, E.J. Butyrate limits human natural killer cell effector function. *Sci. Rep.* **2023**, *13*, 2715. [CrossRef]
117. Shin, J.H.; Li, R.W.; Gao, Y.; Baldwin, R.; Li, C.J. Genome-wide ChIP-seq mapping and analysis reveal butyrate-induced acetylation of H3K9 and H3K27 correlated with transcription activity in bovine cells. *Funct. Integr. Genom.* **2012**, *12*, 119–130. [CrossRef]

biomolecules

MDPI

Review

Anti-Diabetic Therapies and Cancer: From Bench to Bedside

Dimitris Kounatidis [1], Natalia G. Vallianou [2], Irene Karampela [3], Eleni Rebelos [1], Marina Kouveletsou [1], Vasileios Dalopoulos [2], Petros Koufopoulos [2], Evanthia Diakoumopoulou [1], Nikolaos Tentolouris [1] and Maria Dalamaga [4,*]

1 Diabetes Center, First Department of Propaedeutic Internal Medicine, Medical School, Laiko General Hospital, National and Kapustina University of Athens, 11527 Athens, Greece; dimitriskounatidis82@outlook.com (D.K.); eleni.rebelos@utu.fi (E.R.); marinakouveletsou@gmail.com (M.K.); evitadiak@gmail.com (E.D.); ntentol@med.uoa.gr (N.T.)

2 First Department of Internal Medicine, Sismanogleio General Hospital, 15126 Athens, Greece; natalia.vallianou@hotmail.com (N.G.V.); billydalo@hotmail.gr (V.D.); peterkouf13@gmail.com (P.K.)

3 2nd Department of Critical Care, Medical School, Attikon General University Hospital, University of Athens, 1 Rimini str., 12461 Athens, Greece; eikaras1@gmail.com

4 Department of Biological Chemistry, National and Kapodistrian University of Athens, 75 Mikras Asias str., 11527 Athens, Greece

* Correspondence: madalamaga@med.uoa.gr

Abstract: Diabetes mellitus (DM) is a significant risk factor for various cancers, with the impact of anti-diabetic therapies on cancer progression differing across malignancies. Among these therapies, metformin has gained attention for its potential anti-cancer effects, primarily through modulation of the AMP-activated protein kinase/mammalian target of rapamycin (AMPK/mTOR) pathway and the induction of autophagy. Beyond metformin, other conventional anti-diabetic treatments, such as insulin, sulfonylureas (SUs), pioglitazone, and dipeptidyl peptidase-4 (DPP-4) inhibitors, have also been examined for their roles in cancer biology, though findings are often inconclusive. More recently, novel medications, like glucagon-like peptide-1 (GLP-1) receptor agonists, dual GLP-1/glucose-dependent insulinotropic polypeptide (GIP) agonists, and sodium-glucose co-transporter-2 (SGLT-2) inhibitors, have revolutionized DM management by not only improving glycemic control but also delivering substantial cardiovascular and renal benefits. Given their diverse metabolic effects, including anti-obesogenic properties, these novel agents are now under meticulous investigation for their potential influence on tumorigenesis and cancer advancement. This review aims to offer a comprehensive exploration of the evolving landscape of glucose-lowering treatments and their implications in cancer biology. It critically evaluates experimental evidence surrounding the molecular mechanisms by which these medications may modulate oncogenic signaling pathways and reshape the tumor microenvironment (TME). Furthermore, it assesses translational research and clinical trials to gauge the practical relevance of these findings in real-world settings. Finally, it explores the potential of anti-diabetic medications as adjuncts in cancer treatment, particularly in enhancing the efficacy of chemotherapy, minimizing toxicity, and addressing resistance within the framework of immunotherapy.

Keywords: cancer; chronic low-grade inflammation; diabetes mellitus; doxorubicin-induced cardiomyopathy; GLP-1 receptor agonists; immune check point inhibitors; metformin; SGLT-2 inhibitors; tirzepatide; tumor microenvironment

1. Introduction

Diabetes mellitus (DM), affecting approximately 537 million adults globally and projected to increase to 643 million by 2030, represents a significant global health crisis [1]. Notably, DM has long been recognized for its intricate bidirectional relationship with malignancy, influencing cancer prognosis and outcomes [2]. Current evidence shows cancer as the second leading cause of mortality in type 1 diabetes mellitus (T1DM), with nearly

20% of cancer patients concurrently diagnosed with DM [3,4]. This relationship is believed to stem from shared pathophysiological mechanisms, including hyperinsulinemia, chronic low-grade inflammation, and oxidative stress, promoting a cellular environment conducive to tumorigenesis [5]. Within this framework, anti-diabetic drugs have garnered attention, with studies suggesting both protective and carcinogenic potentials across various agents [6]. While certain drugs are contraindicated by the World Health Organization (WHO) for specific cancer types, conflicting studies complicate these guidelines [7].

Among these therapies, metformin has gained attention for its potential anti-cancer properties, primarily attributed to the activation of the AMP-activated protein kinase (AMPK) pathway, alongside its role in glucose regulation [8]. Other conventional anti-diabetic drugs have also been scrutinized, with evidence linking hyperinsulinemia to cancer risk, raising concerns about insulin therapy's carcinogenic potential, though the data remain ambiguous [9]. Additionally, pioglitazone, a peroxisome proliferator-activated receptor gamma (PPAR-γ) agonist, has been associated with an elevated bladder cancer risk, prompting caution in clinical guidelines, yet exhibits promising effects in cancers of the breast, lung, and colon [10].

The elucidation of the incretin effect has catalyzed the development of cutting-edge therapies that have markedly transformed the management of type 2 diabetes mellitus (T2DM). Mediated predominantly by the incretin hormones glucose-dependent insulinotropic polypeptide (GIP) and glucagon-like peptide-1 (GLP-1), this effect enhances insulin secretion in response to oral glucose intake, playing a vital role in maintaining glucose homeostasis. The enzyme dipeptidyl peptidase-4 (DPP-4) regulates the duration of incretin hormone activity, thereby influencing its metabolic effects [11]. Incretin-based therapies currently include DPP-4 inhibitors and GLP-1 receptor agonists (GLP-1 RAs), with the recent introduction of the dual GIP/GLP-1 agonist tirzepatide [12,13]. While GLP-1 RAs and tirzepatide have demonstrated essential cardiovascular (CV) benefits, concerns regarding their associations with thyroid and pancreatic cancers have emerged, leading to cautionary guidelines. Nonetheless, given their impact on weight reduction, these therapies may, under certain conditions, offer protective effects against malignancy [14,15].

Sodium-glucose co-transporter-2 (SGLT-2) inhibitors, a distinct class of anti-diabetic drugs, lower blood glucose levels by promoting glycosuria through the inhibition of the SGLT-2 co-transporter in the proximal renal tubules [16]. Noted for their versatility in T2DM management and benefits in heart failure and chronic kidney disease (CKD), SGLT-2 inhibitors were also explored for their anti-cancer qualities, with promising preclinical and clinical findings [17].

Through their diverse mechanisms of action, glucose-lowering therapies modulate key oncogenic signaling pathways and significantly alter the tumor microenvironment (TME). The TME plays a critical role in cancer biology, significantly influencing tumor initiation, progression, and metastasis. Comprised of a complex network of cellular and non-cellular elements, the TME fosters a niche that supports tumor growth and survival through mechanisms such as immune evasion, angiogenesis, and the release of pro-tumorigenic cytokines and growth factors. This network is shaped by interactions among immune cells, including T and B lymphocytes, natural killers (NK) cells, dendritic cells (DCs), and macrophages, as well as stromal components such as cancer-associated fibroblasts (CAFs), endothelial cells, and adipocytes. Additionally, extracellular matrix components and non-cellular factors like exosomes also contribute to the formation of a pro-tumorigenic TME [18]. However, despite the breadth of research, the overall impact of anti-diabetic medications on cancer risk and progression remains inconclusive, necessitating further investigation [7,19].

This narrative review explored the intersection between anti-diabetic treatments and cancer biology, focusing on how these treatments influence cancer-related signaling pathways and the TME. It further examined the translation of experimental findings into clinical practice, highlighting their potential to impact therapeutic outcomes, including improving chemotherapy efficacy, reducing toxicity, and overcoming resistance in immunotherapy.

2. Anti-Diabetic Therapies and Their Impact on Cancer Reprogramming: Insights from Experimental Studies

2.1. Metformin

Among the various classes of anti-DM medications, metformin stands out as the most extensively studied agent in relation to its effects on carcinogenesis. Over the years, substantial preclinical evidence has illuminated metformin's capacity to inhibit tumor growth across various cancers. This effect is primarily mediated by the activation of AMPK, a key regulator of cellular energy homeostasis. AMPK not only influences glucose and lipid metabolism but also directly impacts tumor bioenergetics by inhibiting the mammalian target of rapamycin (mTOR) pathway, which governs cell growth and proliferation [20]. This inhibition occurs through phosphorylation targeting either the raptor subunit of mTOR or tuberous sclerosis complex 2 (TSC2), while mTOR degradation is enhanced by unc-51-like kinase 1 (ULK1), an autophagy regulator. Additionally, AMPK promotes the phosphorylation and degradation of proteins like p70 ribosomal protein S6 kinase (p70S6K), critical for mRNA translation, further inhibiting the mTOR pathway [21,22]. Moreover, AMPK upregulation disrupts insulin signaling by reducing insulin receptor substrate-1 (IRS-1) phosphorylation, interrupting the phosphatidylinositol 3-kinase/protein kinase B (PI3K/Akt)/mTOR pathway, a known oncogenic driver [23]. While AMPK activation plays a central role in metformin's anticancer effects, research suggests that metformin can also provide AMPK-independent anti-tumor benefits.

In breast cancer, metformin reduces tumor cell proliferation, partly by activating AMPK and inhibiting key elements like phosphorylated 4E-BP1 in the mTOR pathway. It also downregulates cyclin D1, halting cell cycle progression, and suppresses cyclooxygenase-2 (COX-2) expression, lowering the risks of lymphatic and distant metastasis [24]. Metformin also showed promise in ER^+ tumors by increasing cell survival under estrogen-deprived conditions and enhancing mitochondrial respiration via fatty acid oxidation, especially when used alongside hormone therapies [25]. The drug exerts a profound impact on CAFs within the TME by disrupting pro-tumorigenic reprogramming processes. Specifically, metformin inhibits hypoxia-inducible factor-1 alpha (HIF-1a) expression through the activation of phosphorylated AMPK (p-AMPK). This suppression is associated with a reduction in the expression of pro-inflammatory cytokines, such as interleukin-8 (IL-8), which are governed by HIF-1α signaling and play critical roles in facilitating cancer progression. The attenuation of HIF-1α expression and its downstream effects further suggests a potential modulation of glucose metabolism in CAFs, as HIF-1α is a central regulator of glycolytic pathways under hypoxic conditions [26].

Metformin affects glucose metabolism, lowers lactate production, and improves tumor vascularization, contributing to a more immune-permissive TME by increasing $CD8^+$ T cell infiltration [27,28]. Notably, metformin reprograms tumor-associated macrophages (TAMs) from the tumor-promoting M2 phenotype to the anti-tumorigenic M1 phenotype, thereby reducing angiogenesis and enhancing immune-mediated tumor suppression. This reprogramming is mediated through the activation of AMPK, which modulates macrophage polarization by interacting with the nuclear factor kappa-light-chain-enhancer of activated B cells' (NF-κB) signaling pathway. Metformin's upregulation of AMPK and its direct inhibition of mitochondrial Complex I activity may influence metabolic processes, such as glycolysis and oxidative phosphorylation (OXPHOS), favoring the metabolic profile associated with M1 polarization [29]. Other immune effects include boosted NK cell cytotoxicity and decreased immunosuppressive activity by myeloid-derived suppressor cells (MDSCs) [30]. These TME-modulating effects hold potential therapeutic benefits across various cancers characterized by an aggressive TME, including triple-negative breast cancer (TNBC) and colorectal cancer (CRC) [28–31].

Metformin also shows potential in treating other gynecological malignancies beyond breast cancer. In endometrial cancer, AMPK activation upregulated Ten-Eleven Translocation 2 methylcytosine dioxygenase 2 (TET2) gene expression, which inhibited cancer cell growth [32]. In cervical cancer, metformin's upregulation of zinc finger protein 36 (ZFP36), an anti-tumor gene associated with mTOR complex1 (mTORC1) suppression, highlights its regulatory potential beyond AMPK [33]. In ovarian cancer, metformin influenced the AMPK/glycogen synthase kinase-3β (GSK3β) axis, leading to cyclin D1 degradation and G1 phase cell cycle arrest, with potential epigenetic effects under normoglycemic conditions [34,35]. Additionally, metformin disrupted transforming growth factor-β (TGF-β) signaling between ovarian cancer cells and mesothelial cells, reducing pro-inflammatory signals, including IL-8, and HIF1α-driven invasion [36].

In non-small cell lung cancer (NSCLC), metformin reduced proliferation by downregulating key oncogenic markers like hairy and enhancer of split-1 (HES1) and regulated the development and DNA damage response 1 (REDD1) by modulating p-mTOR and p53 levels [37]. Similar metabolic reprogramming effects were observed in pancreatic and prostate cancers, where metformin, through AMPK and peroxisome proliferator-activated receptor gamma coactivator-1alpha (PGC-1α) upregulation, enhanced mitochondrial function, while curtailing glycolysis and tumor growth [38,39]. Additional AMPK/mTOR-independent anti-cancer mechanisms included increased ROS production, disrupted OXPHOS, and induced redox imbalance, promoting apoptosis, as observed in prostate cancer models [40].

In hepatocellular carcinoma (HCC), metformin disrupted lipid metabolism, facilitated apoptosis, and bolstered oxidative stress through p38 mitogen-activated protein kinase (p38MAPK) activation [41], while its anti-metastatic potential, even at low doses, was underscored by the inhibition of the AMPK, c-Jun N-terminal kinase (JNK), IL-8, and the matrix metalloproteinase (MMP) 9 signaling axis [42]. A zebrafish model of metabolic dysfunction-associated fatty liver disease (MAFLD) and HCC demonstrated that a high-fat diet (HFD) promotes liver cancer, angiogenesis, and immune dysregulation, which metformin reversed by altering macrophage polarization and restoring T cell density [43]. Similar benefits were observed in cholangiocarcinoma (CCA), where metformin altered tumor metabolism by reducing lactate production and downregulating lactate dehydrogenase A (LDHA) expression, thereby disrupting the Warburg effect. This phenomenon describes the preference of cancer cells for glycolysis as their primary energy source, even under normal oxygen conditions, which supports tumor growth and survival [44].

In recent years, spatial transcriptomics has emerged as a cutting-edge technique for evaluating drug effects on cancer outcomes, allowing the detailed mapping of gene expression within tumor tissues. This innovative approach enables researchers to dissect the intricate interactions within the TME, including the cellular and molecular changes induced by therapeutic agents. Notably, spatial transcriptomics was employed to evaluate the impact of metformin in cancer [45,46]. In pancreatic cancer, for instance, metformin was linked to improved survival and immune modulation, reducing pro-tumoral macrophage activity and enhancing anti-tumor DC function. This translated into significantly better overall survival and five-year survival rates among metformin users in the PREOPANC trial [45]. Additionally, transcriptomic studies in CRC revealed that metformin modulates miRNA interactions that target critical growth pathways, particularly PI3K/Akt and MAPK/extracellular signal-regulated kinase (ERK), highlighting its potential for the post-transcriptional regulation of cancer progression [46].

Despite the diverse mechanisms through which metformin positively affects carcinogenesis, the activation of AMPK is the central mechanism of its action. Figure 1 presents significant AMPK-mediated pathways that contribute to metformin's broad anti-cancer potential.

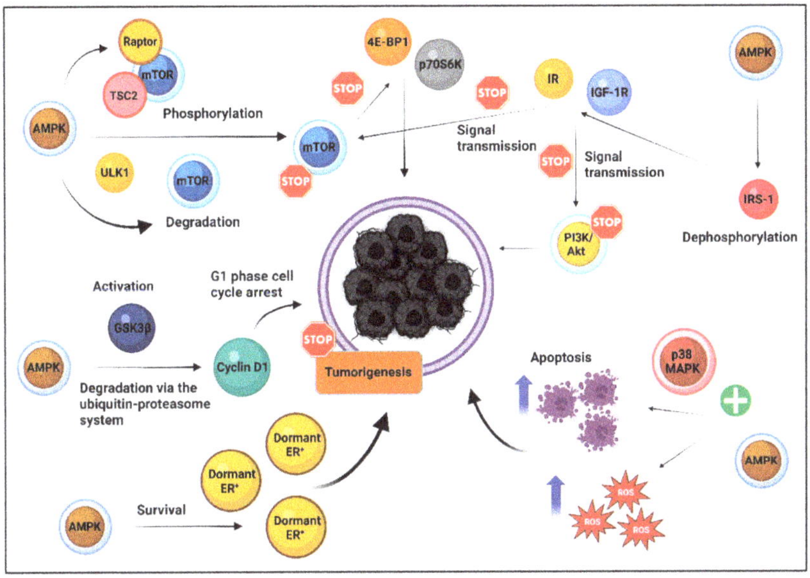

Figure 1. The impact of metformin on cancer-related signaling pathways via AMPK activation. AMPK activation by metformin disrupts key cancer-related signaling pathways, leading to inhibition of tumor growth and proliferation. AMPK deactivates the mTOR signaling pathway through phosphorylation and degradation mechanisms, inhibiting proteins such as p70S6K and 4E-BP1 involved in mRNA translation, thereby limiting cancer cell proliferation. Additionally, AMPK reduces phosphorylation of IRS-1, impairing signal transmission from IRs and IGF-1Rs, which disrupts the PI3K/Akt/mTOR signaling axis. In breast cancer, AMPK activation supports the survival of dormant ER^+ tumor cells under low-estrogen conditions. In ovarian cancer, AMPK influences the AMPK/GSK3β axis, triggering cyclin D1 degradation through the ubiquitin-proteasome system, resulting in cell cycle arrest at the G1 phase. AMPK activation also promotes the production of ROS and induces apoptosis, which collectively hinder tumor growth [20,22,23,25,34,41]. Abbreviations: 4E-BP1: Eukaryotic translation initiation factor 4E-binding protein 1; Akt: Protein kinase B; AMPK: AMP-activated protein kinase; ER^+: Estrogen receptor-positive; GSK3β: Glycogen synthase kinase 3 beta; IGF-1R: Insulin-like growth factor receptor; IR: Insulin receptor; IRS-1: Insulin receptor sub-strate-1; mTOR: Mammalian target of rapamycin; mRNA: Messenger RNA; p38MAPK: p38 mitogen-activated protein kinase; p70S6K: p70 ribosomal protein S6 kinase; PI3K: Phosphoinositide 3-kinase; Raptor: Regulatory-associated protein of mTOR; ROS: Reactive oxygen species; TSC2: Tuberous sclerosis complex 2; ULK1: Unc-51-like autophagy activating kinase 1. Created with www.BioRender.com.

2.2. *Insulin Therapy and Sulfonylureas*

2.2.1. Insulin Therapy

A robust body of experimental evidence notes insulin's role in fostering a pro-tumorigenic environment through its modulation of key cancer-associated signaling pathways. One prominent mechanism involves the mTOR pathway, where insulin enhances glucose uptake and glycogenolysis, processes that fuel cancer cell growth [47]. Additionally, insulin activates the IRS/PI3K/Akt and MAPK pathways, which govern cell cycle progression, anti-apoptotic responses, and angiogenesis [48]. Angiogenesis, the formation of new blood vessels, is particularly crucial for tumor progression, as it supplies essential nutrients and oxygen to the expanding tumor mass. Insulin's influence on these signaling pathways has been linked to a heightened potential for carcinogenesis across various cancer types. For instance, insulin promoted glioblastoma (GB) cell proliferation and survival through the activation of the Akt pathway [49]. Evidence indicates that both human insulin and insulin glargine can stimulate thyroid cell proliferation at elevated concentrations, increasing the phosphorylation of Akt and ERK1/2 in a dose-dependent manner, with insulin glargine showing prolonged effects. Notably, therapeutic doses of insulin did not appear to stimulate cell proliferation [50]. In lung cancer, insulin facilitated cell proliferation and migration by upregulating the PI3K/Akt pathway [51]. Similarly, in HCC, insulin enhanced cell growth and survival through the activation of both the PI3K/Akt and Rat sarcoma (Ras)/MAPK pathways [52]. Notably, in CRC, insulin drove tumor progression and metastasis by upregulating acetyl-CoA acetyltransferase 1 (ACAT1), an enzyme critical for cancer progression through its role in lipid metabolism and mitochondrial function, thereby supporting cancer cell growth, survival, and metastatic potential [53].

Existing evidence indicates that hyperinsulinemia plays a role in the progression of pancreatic cancer, as demonstrated by Zhang et al., who found that genetically reducing insulin production in mice led to an approximate 50% reduction in pancreatic intraepithelial neoplasia (PanIN) lesions [54]. Their research further suggests that managing insulin levels could inhibit pancreatic cancer by altering various cell types within the pancreatic microenvironment. Specifically, modifying *Ins2* gene dosage in a mouse model lacking the rodent-specific *Ins1* gene modestly reduced hyperinsulinemia across genders without substantially affecting glucose tolerance. This reduction in insulin was associated with fewer PanIN and acinar-to-ductal metaplasia lesions. Single-cell transcriptomics revealed that hyperinsulinemia influences several pancreatic cell types, particularly immune cells, by modifying protein synthesis pathways and key signaling cascades such as MAPK/ERK and PI3K/Akt [55].

In addition to these direct signaling effects, hyperinsulinemia can indirectly drive carcinogenesis, particularly in the context of obesity, a condition often accompanied by both hyperinsulinemia and insulin resistance. Obesity-related hyperinsulinemia leads to an imbalance in angiogenic factors, elevating pro-angiogenic agents, such as vascular endothelial growth factor (VEGF) and leptin, while reducing anti-angiogenic factors like adiponectin. This imbalance fostered a tumor-supportive microenvironment that promoted cancer progression and metastasis [56]. Moreover, chronic hyperinsulinemia arising from both endogenous overproduction and exogenous insulin administration was linked to persistent chronic low-grade inflammation. Elevated inflammatory cytokines, including IL-6 and tumor necrosis factor-alpha (TNF-α), played pivotal roles in creating an immunosuppressive, pro-angiogenic environment that further accelerated tumor growth and metastasis [57].

Insulin-like growth factor (IGF) signaling also plays a crucial role in cancer development. Due to the structural similarity between insulin and IGF-1, elevated insulin levels under hyperinsulinemic conditions can activate the IGF-1 receptor (IGF-1R), amplifying mitogenic and anti-apoptotic effects. Both insulin and IGF-1R signaling activated downstream pathways that regulate cell growth, proliferation, survival, and metabolism, processes that can be exploited by cancer cells to sustain tumor development [58]. Experimental models demonstrated that disrupting insulin receptor (IR) signaling, through either genetic manipulation or pharmacological inhibition, can significantly reduce tumor growth and metastasis. For example, in mouse models of mammary carcinoma, hyperinsulinemia increased tumor growth and metastasis, while treatment with insulin sensitizers or IR/IGF-1R inhibitors mitigated these effects. Likewise, silencing the IR in breast cancer cells reduced their metastatic potential [59].

Notably, concerns have been raised regarding the potential cancer-promoting effects of insulin analogs, particularly long-acting formulations such as insulin glargine. Due to structural modifications, these analogs may exhibit altered affinities for IR and IGF-1R, possibly enhancing mitogenic signaling. While in vitro studies suggested insulin glargine may have greater mitogenic potency compared to regular insulin, in vivo data remain inconclusive, and no definitive conclusions about its carcinogenic risk were established [60].

2.2.2. Sulfonylureas

Experimental studies on the anti-cancer efficacy of sulfonylureas (SUs) have shown encouraging results, with glibenclamide emerging as one of the most thoroughly studied SUs. In various preclinical models, glibenclamide has consistently demonstrated notable anti-cancer effects. For instance, in breast cancer cell lines, glibenclamide induced cytostatic effects through the modulation of adenosine triphosphate (ATP)-sensitive potassium channels, resulting in cell cycle arrest at the G0/G1 phase. This process is associated with the upregulation of the cyclin-dependent kinase inhibitor p27 and downregulation of cyclin E, thereby inhibiting tumor growth [61].

Glibenclamide may also influence carcinogenesis by altering the TME, particularly in NSCLC. A recent study by Chen et al. found that glibenclamide inhibited the conversion of normal fibroblasts into CAFs by blocking sulfonylurea receptor 1 (SUR1) and upregulating let-7a-5p, a microRNA that inhibits the fibroblast-to-CAF transition by targeting TGF-β signaling. The transformation of normal fibroblasts into CAFs is often associated with metabolic reprogramming, including a shift toward glycolysis that supports rapid cell proliferation and biomass accumulation. Since CAFs promote tumor growth, angiogenesis, and immune evasion, their prevention can foster a more immune-permissive TME and enhance T lymphocyte activity against tumor cells [62].

In HCC, glibenclamide showed inhibitory effects on voltage-gated potassium (Kv) channels, leading to dose- and calcium-dependent reductions in cell adhesion and proliferation [63]. Prostate cancer models similarly showed glibenclamide's capacity to induce dose-dependent growth inhibition and apoptosis within hours of administration [64]. Additionally, glibenclamide impacted angiogenesis and metastasis; in ovarian cancer, it reduced cellular invasion and migration by suppressing the secretion of platelet-derived growth factor-AA (PDGF-AA), an essential factor in tumor angiogenesis [65].

Other SUs have also exhibited anti-tumor effects. Chlorpropamide and gliclazide, for instance, inhibited TNF-α production, a cytokine crucial for cell survival and apoptosis regulation. Glipizide was found to reduce tumor growth, angiogenesis, and metastasis in breast cancer and melanoma models [66,67]. Additionally, acetohexamide demonstrated efficacy in preventing breast cancer cell entry into lymphatic ducts, further suggesting a role for SUs in reducing metastasis [68].

2.3. *Pioglitazone*

Pioglitazone has also garnered attention for its complex role in cancer biology, particularly concerning urogenital malignancies. Bladder cancer remains a major issue in the context of pioglitazone therapy, with experimental evidence indicating that pioglitazone may induce DNA damage and malignant transformation in bladder cells, affecting gene expression and epithelial–mesenchymal transition (EMT). EMT is a crucial biological process in which epithelial cells lose their adhesion and polarity while acquiring mesenchymal properties, such as enhanced migratory capacity and invasiveness, significantly contributing to the progression and aggressiveness of many cancers [69]. This is linked to the activation of PPAR-γ signaling that further promotes de novo lipogenesis, potentially supporting tumor growth [70]. However, contradictory findings complicate this narrative, as recent data indicate that pioglitazone may actually protect against bladder cancer progression by downregulating oncogenic markers such as p53 and cyclin D1, without promoting malignant transformations in normal bladder cells [71]. Importantly, these observations are further supported by proteomic analyses [72]. Beyond bladder cancer, pioglitazone's impact on other urogenital malignancies, including kidney and prostate cancers, remains contentious. In prostate cancer, pioglitazone may reduce tumorigenesis by alleviating obesity-related inflammation [73], while in renal cell carcinoma (RCC), the drug induced apoptosis through caspase activation and the downregulation of anti-apoptotic proteins [74].

Research also extended pioglitazone's potential to a variety of other cancers. In breast cancer, it inhibited cell proliferation and migration, likely through modulation of the Janus kinase/signal transducer and activator of transcription 3 (JAK/STAT3) signaling pathway, while also increasing adiponectin levels [75]. In lung cancer, combining pioglitazone with celecoxib was shown to reduce tumor weight and improve survival by inhibiting NF-κB-mediated proliferation [76]. In NSCLC, pioglitazone targeted cancer cell metabolism, reducing extracellular acidification and glucose metabolism markers, thus limiting cancer cells' ability to adapt to the nutrient-poor, hypoxic TME [77].

Despite these promising results, preliminary evidence suggests that PPAR-γ activation may have a dual effect on tumor progression. It can exert anti-tumorigenic effects on cancer cells while promoting pro-tumorigenic outcomes within the TME, particularly in myeloid cells. For example, Li et al. demonstrated that although PPAR-γ activation inhibited cancer cell growth, it paradoxically promoted metastasis in lung cancer models. In two mouse models, pioglitazone facilitated metastasis to the liver, brain, and lungs by increasing the accumulation of pro-tumor arginase I-positive macrophages. This metastasis-promoting effect was shown to be PPAR-γ dependent in myeloid cells, as mice with myeloid-specific PPAR-γ deletion exhibited reduced metastatic spread [78].

Pioglitazone also exhibited anti-fibrotic and hepatoprotective effects in HCC models, acting through pathways such as the MAPK and AMPK [79]. In pancreatic cancer, pioglitazone reduced metastasis by altering inflammation-related gene expression, including CEA and COX-2 [80]. Additionally, it enhanced ROS formation and suppressed B cell lymphoma 2 protein (Bcl-2) expression in hypoxic cancer cells, highlighting its potential role in controlling metastasis [81]. In CRC, pioglitazone reduced cancer stem cell viability and promoted a mesenchymal-to-epithelial transition (MET) [82]. Moreover, pioglitazone influenced the immune landscape of the TME in CRC by inhibiting the IL-6/STAT3 pathway, a vital regulator of immune evasion. Its downregulation, reflected in reduced IL-6 mRNA and STAT3 activity, enhanced mature DC functionality, boosting $CD8^+$ T cell priming and increasing tumor-infiltrating lymphocytes (TILs). Pioglitazone also amplified T cell cytotoxicity, significantly reducing tumor cell viability. By targeting metabolic and inflammatory barriers to DC activity, it strengthened anti-tumor responses, as evidenced by elevated IL-12 and interferon-gamma (IFN-γ) levels in combination-treated groups [83].

In hematological cancers, pioglitazone showed promise in enhancing anti-leukemic effects by inhibiting the hyperactivated PI3K/Akt pathway [84]. In chronic myeloid leukemia (CML), it disrupted CD44, a surface protein essential for cancer stem cell adhesion and migration within the TME. By inhibiting CD44, pioglitazone reduced the invasiveness of CML cells, potentially limiting metastasis and disease progression [85].

2.4. *Incretin Mimetics*

2.4.1. DDP-4 Inhibitors

Despite their benefits in managing glycemic levels, DPP-4 inhibitors demonstrated inconsistent results in cancer research, likely due to the significant heterogeneity in DPP-4 expression across different tumor types and environments [86]. In breast cancer, DPP-4 inhibition was linked to enhanced metastasis through the activation of the C-X-C motif chemokine ligand 12/C-X-C chemokine receptor type 4 (CXCL12/CXCR4)/mTOR signaling, a pathway known to promote EMT [87]. This metastatic propensity was further exacerbated by the activation of the ROS–(nuclear factor erythroid 2-related factor 2)Nrf2–heme oxygenase-1 (HO-1) axis, which led to NF-κB activation and upregulation of inflammatory cytokines, such as IL-6. This pro-inflammatory response is thought to create an environment conducive to tumor progression by supporting angiogenesis and suppressing immune activity [88]. Notably, saxagliptin was associated with increased migratory and invasive capabilities in thyroid cancer cells via Nrf2 upregulation, which may synergize with oncogenic pathways like mTOR and heat shock factor 1 (HSF1) to promote cancer cell survival and spread [89]. Additionally, animal studies raised concerns regarding DPP-4 inhibitors' role in pancreatic β-cell proliferation, a process that could potentially lead to neoplastic transformation within the pancreatic tissue [90].

Conversely, in other cancer contexts, DPP-4 inhibition may impart anti-tumor benefits, particularly by countering CAF-driven growth. In scirrhous gastric cancer (SGC), DPP-4 released from CAFs was identified as a key factor enhancing cancer cell proliferation, likely through CXCR4 signaling [91]. Similarly, in endometrial cancer, sitagliptin was observed to limit proliferation, invasion, and tumorigenesis [92], while in CRC models, DPP-4 inhibition reduced cell proliferation, inhibited angiogenesis, and promoted apoptosis [93]. Notably, in CRC, DPP-4 may exert both tumor-promoting and tumor-suppressive effects, depending on its expression levels and the specific context within the TME. Elevated CD26 levels, the active form of DPP-4, were associated with a poor prognosis and enhanced metastatic potential in CRC patients [94].

Additionally, DPP-4 inhibition may enhance anti-tumor immunity through the recruitment of eosinophils, which play a critical role in tumor cell death and the amplification of anti-tumor immune responses within the TME. Research indicates that, by preserving chemokines like C-C motif chemokine ligand (CCL)11, sitagliptin promoted eosinophil infiltration, an effect seen in models of HCC and breast cancer. This eosinophilic migration is IL-33 dependent and synergizes with immune checkpoint therapies, suggesting a novel mechanism whereby DPP-4 inhibition can bolster eosinophil-driven anti-tumor responses [95]. DPP-4 inhibition may also play a therapeutic role in obesity-driven HCC. Elevated DPP-4 levels associated with a HFD were found to drive HCC growth, metastasis, and angiogenesis via CCL2 signaling. In both cell and animal models, DPP-4 inhibition through vildagliptin or genetic deletion reduced HCC progression, effectively disrupting the pro-angiogenic DPP-4/CCL2 pathway. Clinical data further underscored these findings, showing that higher serum DPP-4 levels correlated with a worse prognosis in HCC patients [96]. Intriguingly, DPP-4 inhibition may enhance anti-tumor immune responses by improving the functionality of type 1 conventional dendritic cells (cDC1s). Sitagliptin, for example, enhanced antigen presentation by cDC1s, thereby facilitating T cell activation and subsequent tumor suppression. Mechanistically, sitagliptin appeared to stabilize key signaling molecules essential for cDC1 activity and may help restore glucose availability, thus addressing the metabolic requirements of these immune cells in the TME [97].

2.4.2. GLP-1 and Dual GIP/GLP-1 Receptor Agonists

GLP-1 RAs demonstrated varied effects in cancer, with concerns remaining about their impact on thyroid and pancreatic malignancies. Notably, the use of GLP-1 RAs is associated with an increased incidence of medullary thyroid cancer (MTC). Animal studies indicated that GLP-1 agonists such as exenatide and liraglutide elevated calcitonin levels and induced C-cell hyperplasia in rodent models [98,99]. However, the implications for human health remain uncertain, as GLP-1 receptor expression in human C-cells was minimal [100]. Interestingly, recent evidence demonstrated liraglutide's inhibitory effects on cell growth and migration in both papillary and medullary thyroid cancers by modulating the PI3K/Akt/mTOR pathway in a dose- and time-dependent manner [101].

Concerns also extend to the association between GLP-1 RAs and pancreatic cancer, as chronic pancreatitis, a potential side effect of GLP-1 RAs, is linked to an increased risk of malignancy [102]. However, liraglutide was found to inhibit pancreatic cancer growth and promote apoptosis by suppressing the PI3K/Akt and ERK1/2 signaling pathways [103]. Furthermore, GLP-1 agonism can modulate calcium signaling and cadherin expression, which are crucial for cell adhesion and mobility, thereby inhibiting the proliferation and migration of pancreatic cancer cells [104]. Additionally, GLP-1 RAs demonstrated protective effects on pancreatic β-cells in hyperglycemic states, potentially reducing apoptosis through the modulation of miR-139-5p and IRS-1 suppression [105].

In addition to thyroid and pancreatic cancers, the impact of GLP-1 RAs was explored in various other malignancies, with experimental studies revealing promising outcomes. For instance, exenatide inhibited glioma cell migration and invasion via the GLP-1R/sirtuin 3 (SIRT3) pathway [106], while in breast cancer, it reduced proliferation and promoted apoptosis by inhibiting NF-κB nuclear translocation [107]. Notably, liraglutide significantly debilitated breast cancer cell growth in the presence of a conditioned medium from obese adipose tissue, implicating adipokines in its mechanism [108]. However, at high concentrations, liraglutide may facilitate the progression of TNBC through the NADPH oxidase 4 (NOX4)/ROS/VEGF signaling pathway, highlighting the importance of dosage and cancer phenotype [109]. Similar beneficial effects in breast cancer were observed with dulaglutide, which demonstrated its anti-tumor properties by reactivating tumor suppressor genes through promoter demethylation [110].

The activation of AMPK by GLP-1 RAs further contributed to reduced cancer cell viability, particularly in glycolytic metabolism [111]. In endometrial carcinoma, both liraglutide and tirzepatide may suppress cell proliferation and exhibit anti-tumor effects, respectively, with tirzepatide exhibiting essential anti-obesity and anti-tumorigenic effects, potentially modulating metabolic and immune pathways relevant to tumor growth [112,113]. Research also suggests that exenatide may reduce the aggressiveness of ovarian cancer, as it appeared to modulate key cellular behaviors, including reducing cancer cell migration and promoting apoptosis through caspase activation, thereby directly influencing cancer cell dynamics within the TME. Furthermore, it modulated extracellular matrix remodeling, a vital factor in metastasis, by regulating the expression of MMP-2 and MMP-9 and their tissue inhibitors (TIMP-1 and TIMP-2). Additionally, exenatide decreased the production of adhesion molecules, such as intercellular adhesion molecule-1 (ICAM-1) and vascular cell adhesion molecule-1 (VCAM-1), limiting cancer cell adherence to the endothelial lining and thus inhibiting metastatic spread. In TNF-α-stimulated endothelial cells, the reduction in MMP-1 and MMP-9 production suggests that GLP-1 RAs may alter the TME's capacity to facilitate cancer cell invasion and spread [114].

Both liraglutide and semaglutide showed potential in preventing disease progression and reducing tumor burden in HCC [115,116]. GLP-1 RAs also offer potential benefits for immune functionality, particularly by reversing obesity-induced impairments in NK cell activity. Preclinical data indicated that liraglutide enhanced NK cell-mediated cytotoxicity against HCC cells, independently of $CD8^+$ T cells. Its anti-cancer effects were tied to the inhibition of the IL-6/STAT3 axis, often linked with unchecked cellular proliferation and immune evasion mechanisms. Beyond immune modulation, liraglutide likely impacted the TME's metabolic landscape, potentially enhancing anti-cancer immune responses by modulating metabolic pathways such as glycolysis, thereby strengthening NK cell-mediated tumor suppression [117]. Similar immunomodulatory effects were also observed with semaglutide. Clinical findings by De Barra et al. demonstrated that six months of semaglutide therapy in obese patients enhanced NK cell cytotoxicity, increasing the production of interferon-γ and granzyme B, both markers of NK cell activity. This improvement was linked to activation of the CD98-mTOR-glycolysis axis, a pivotal pathway for NK cell cytokine production, and notably, these enhancements in NK cell function were independent of weight reduction [118].

Furthermore, GLP-1 RAs inhibited cell cycle progression and promoted apoptosis in CRC via the PI3K/Akt/mTOR pathway [119]. Preliminary evidence also suggested GLP-1 receptor expression in metastatic castrate-resistant prostate cancer, indicating opportunities for further research in urinary tract malignancies [120]. Genetic evidence further outlined the role of GLP-1 signaling in cancer immunology. Zhu et al. examined GLP-1 signaling-related genes across 33 cancer types, identifying significant associations between GLP-1 signaling and immune cell infiltration levels, including $CD4^+$ T cells, NK cells, neutrophils, DCs, and macrophages. A lower GLP-1 signaling score, derived from gene enrichment analysis, was correlated with reduced immune cell infiltration, poorer survival rates, and lower responsiveness to immunotherapy across multiple cancer types. In CRC specifically, semaglutide reduced cellular migration, suggesting additional applications of GLP-1 RAs in impeding cancer spread [121].

2.5. SGLT-2 Inhibitors

A growing body of experimental research has highlighted the significant overexpression of SGLT-2 in malignant cells across various cancer types. Evidence suggests that targeting SGLT-2 can hinder tumor progression by disrupting the metabolic and signaling pathways critical for cancer cell survival. However, current data face challenges in clearly defining the efficacy of SGLT-2 inhibition across different malignancies, primarily due to notable differences in their metabolic pathways. The metabolism of each SGLT-2 inhibitor involves distinct enzymatic processes that affect drug inactivation and excretion, influencing stability, duration, and, ultimately, therapeutic outcomes. These metabolic discrepancies complicate efforts to establish universal conclusions regarding the anti-cancer potential of SGLT-2 inhibition. Nevertheless, SGLT-2 inhibition remains a promising therapeutic approach, demonstrating groundbreaking results in several malignancies, as supported by numerous experimental studies [122].

For instance, in glioblastoma, canagliflozin inhibited cell proliferation and glucose uptake, especially at 40 μM concentrations. Mechanistically, canagliflozin promoted AMPK phosphorylation while suppressing p70 S6 kinase and S6 ribosomal protein, both fundamental in cancer growth pathways [123]. In the context of thyroid cancer, canagliflozin showed promise in modulating the TME by downregulating pro-tumor chemokines such as CXCL8 and CCL2. By suppressing these chemokines, canagliflozin limited inflammatory cell recruitment into the TME, potentially reducing immune suppression and enhancing anti-tumor immunity. Canagliflozin also affected the survival and proliferation of both cancerous and endothelial cells, reducing IL-6 levels while increasing CXCL10 expression, indicating a shift toward an anti-tumor TME that may enhance immune response against cancer cells [124].

Contrary to concerns of breast cancer promotion by SGLT-2 inhibitors, experimental data suggest beneficial outcomes. Recent findings suggest that SGLT-2 inhibitors may offer a novel approach for targeting glucose metabolism in the TME of breast cancer. By restricting glucose availability, SGLT-2 silencing not only inhibited tumor cell proliferation but may also alleviate immunosuppressive conditions associated with high-glucose states, thus enhancing immune cell functionality. Mathematical models indicated that the optimal dosing of SGLT-2 inhibitors can reduce or eradicate tumor cells while sparing normal cells [125]. Novel insights revealed promising therapeutic effects of specific SGLT-2 inhibitors. Dapagliflozin, for example, demonstrated effectiveness in reducing breast cancer cell viability and triggering early apoptosis, primarily through the modulation of glucose metabolism. Likewise, empagliflozin exhibited anti-inflammatory properties that may influence tumor progression by altering the metabolic profile of the TME [126]. Moreover, empagliflozin acted as an miR-128-3p mimic, reducing CD44 expression under hypoxic conditions and fostering differentiation. In vivo, it inhibited tumor growth, limited lung metastasis, and elevated oxidative stress markers, potentially sensitizing cells to ferroptosis [127,128]. Data further indicated that SGLT-2 expression was significantly elevated in estrogen-sensitive breast cancer cells but remained undetectable in normal mammary tissue. The inhibition of SGLT-2 by ipragliflozin led to decreased cell proliferation and DNA synthesis, an effect negated by SGLT-2 knockdown. This mechanism appeared to involve disrupted glucose and sodium transport, resulting in reduced intracellular sodium influx, membrane hyperpolarization, mitochondrial destabilization, and the subsequent inhibition of cancer cell growth [129].

Moreover, SGLT-2 inhibition showed promise in HCC, with Shiba et al. demonstrating that canagliflozin mitigated metabolic dysfunction-associated steatohepatitis (MASH)-related HCC in a Western diet-fed mouse model by reducing hepatic steatosis, fibrosis, and tumor occurrence while enhancing adipose tissue health, promoting a "healthy expansion" profile marked by a lower oxidative stress index [130]. In HCC, tumor progression was closely linked to the WNT/β-catenin signaling pathway's enhancement of aerobic glycolysis in malignant cells. Hung et al. reported that canagliflozin inhibited glucose uptake by targeting multiple glucose transporters, including GLUT1, thereby reducing

HCC cell survival and colony formation. This effect was achieved through the disruption of β-catenin signaling, promoting its proteasomal degradation and inhibiting dephosphorylation, ultimately impairing tumor growth and extending survival [131]. HCC, characterized by metabolic reprogramming, hypoxia, and dysregulated signaling pathways, depends on hypoxia to drive tumor progression through angiogenesis, EMT, and glycolysis. Within the hypoxic TME of HCC, canagliflozin disrupted pro-tumor pathways by downregulating VEGF. Additionally, it suppressed glycolysis-associated proteins and EMT processes, essential for tumor progression and metastasis. By targeting the Akt/mTOR pathway, canagliflozin inhibited the accumulation of HIF-1α, reducing the metabolic and signaling adaptations critical for tumor survival and proliferation [132]. Similarly, in NSCLC, canagliflozin disrupted HIF-1α stabilization, interfering with mitochondrial function and survival pathways through mTOR inhibition and histone deacetylase 2 (HDAC2) suppression [133].

In gastric cancer, dapagliflozin reduced tumor proliferation by downregulating OTU deubiquitinase 5 (OTUD5), leading to yes-associated protein 1 (YAP1) degradation, a key protein in tumor progression [134]. Canagliflozin also demonstrated efficacy in CRC, disrupting cellular metabolism and inducing endoplasmic reticulum (ER) stress, which promotes autophagy and apoptosis through the upregulation of SIRT3 [135]. In pancreatic cancer models, canagliflozin's inhibition of the PI3K/Akt/mTOR pathway highlighted its anti-tumor potential by hindering glycolysis [136].

Investigations into urinary tract cancers showed that dapagliflozin exerted cytotoxic effects in RCC by reducing glucose uptake, regulating cell cycle progression, and promoting apoptosis in RCC cells [137]. Notably, canagliflozin also reduced cell growth in prostate cancer by curtailing mitochondrial respiration and ATP production, which parallels the metabolic effects seen with metformin. This mechanism involves AMPK activation, reduced lipid synthesis, and limited glucose uptake [138]. Despite clinical concerns about bladder cancer risks with SGLT-2 inhibitors, preclinical findings generally showed no increased bladder cancer incidence with dapagliflozin, even at high doses [139]. Comparable benefits were also documented in rarer malignancies, such as osteosarcoma, where SGLT-2 inhibition appeared to activate the stimulator of interferon genes (STING) pathway, subsequently upregulating interferon regulatory factor 3/interferon-beta (IRF3/IFN-β) signaling. The activation of this pathway within the TME enhanced immune surveillance and tumor suppression, promoting interferon production and other immune-activating signals that facilitate immune cell recruitment to target cancerous tissues [140]. Figure 2 depicts the anti-tumor effects of SGLT-2 inhibitors across various cancer models, as demonstrated in animal studies.

Across various therapeutic classes, anti-diabetic medications exhibit the capacity to modulate cancer-related signaling pathways and impact the TME, with promising immunomodulatory effects. While metformin stands out with the most robust data supporting its anti-cancer potential, emerging research on newer agents, such as GLP-1 RAs and SGLT-2 inhibitors, is also encouraging, though still in its nascent stages. Conversely, evidence for DDP-4 inhibitors and pioglitazone remains mixed, showing potential benefits only in specific cancer contexts, and findings on exogenous insulin administration lack clarity. The divergent results around hyperinsulinemia's malignancy-promoting effects and the complex role of insulin secretagogues, like SUs, further complicate the landscape. This complexity highlights the need for further research to fully understand how anti-diabetic treatments affect cancer biology and their potential therapeutic applications. Table 1 summarizes key molecular mechanisms by which anti-diabetic drugs influence cancer development and progression across various cancer types. Additionally, Figure 3 illustrates diverse mechanisms by which these agents may influence cancer development and progression through their effects on the TME.

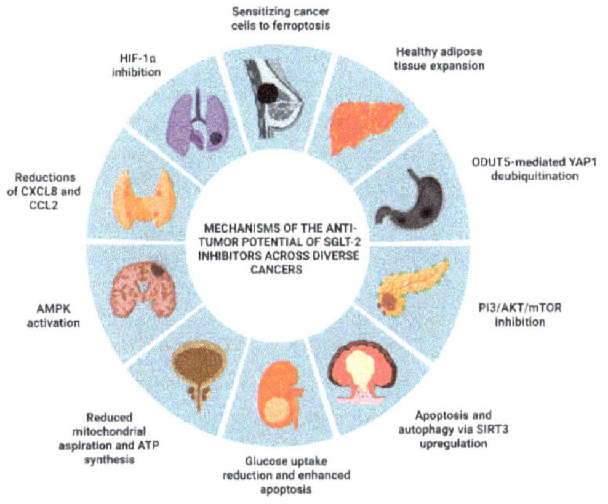

Figure 2. Anti-tumor activity of SGLT-2 inhibitors in cancer models: insights from animal studies [123–125,127,130,133,136–138]. Abbreviations: AMPK: AMP-activated protein kinase; ATP: Adenosine triphosphate; CCL2: C-C motif chemokine ligand 2; CXCL8: C-X-C motif chemokine ligand 8; HIF-1α: Hypoxia-inducible factor 1-alpha; mTOR: Mammalian target of rapamycin; ODUT5: OTU deubiquitinase 5; PI3K: Phosphoinositide 3-kinase; SGLT-2: Sodium-glucose co-transporter 2; SIRT3: Sirtuin 3; YAP1: Yes-associated protein 1. Created with www.BioRender.com.

Table 1. The impact of anti-diabetic drugs on cancer development and progression across different cancer types.

Cancer Type	Anti-Diabetic Drugs and Their Impact on Cancer-Related Signaling
Brain cancer	- Metformin targets GB-initiating cells through the AMPK-FOXO3 axis [141] - Insulin facilitates GB proliferation and survival by activating the Akt pathway [49] - Glibenclamide inhibits GB growth by promoting intracellular acidification through downregulation of Kir4.1 and MCT1 expression at specific doses [142] - Pioglitazone's impact on GB is heterogeneous; significantly dampens cell viability and proliferation in only a subset, without increasing differentiation or affecting the Wnt/β-catenin pathway to a notable extent [143] - Linagliptin exerts anti-tumor effects on GB cells potentially through modulation of proteins involved in the cell cycle and adhesion via phosphorylated NF-kB regulation [144] - Exenatide attenuates GB cell migration and invasion through the GLP-1R/SIRT3 signaling pathway [106] - Canagliflozin reduces GB cell proliferation and glucose uptake by favoring AMPK phosphorylation and suppressing p70 S6 kinase and S6 ribosomal protein activity [123]
Thyroid cancer	- At concentrations of 0.1 mM and above, metformin increases the percentage of apoptotic cells and promotes G0/G1 phase cell cycle arrest, with no effect on the DNA repair response at concentrations as low as 0.3 mM [145] - Both human insulin and insulin glargine facilitate thyroid cell proliferation at high doses, enhancing the phosphorylation of IR, Akt, and ERK1/2 in a dose-dependent manner, with glargine showing longer-lasting effects; however, therapeutic doses do not stimulate cell proliferation [50] - PPAR-γ agonistic effect of pioglitazone prevents metastatic thyroid cancer by promoting adipocyte-like trans-differentiation of thyroid carcinoma cells [146] - Saxagliptin boosts migration and invasion of PTC through the Nrf2/HO-1 signaling pathway [89] - Liraglutide may mitigate cell growth and migration in both PTC and MTC by modulating the PI3K/Akt/mTOR pathway in a dose- and time-dependent manner [101] - Canagliflozin reduces cell viability and colony formation while lowering pro-tumor chemokines, such as CXCL8 and CCL2, which are involved in enhanced cell migration and endothelial proliferation [124]
Breast cancer	- Metformin promotes survival of dormant ER^+ by upregulating AMPK [25] - Insulin can regulate breast cancer growth by activating both insulin and IGF receptors [147] - Glipizide blocks angiogenesis by modulating VEGF/VEGFR2 signaling [67] - Pioglitazone mitigates cancer cell proliferation and migration by modulating the JAK/STAT3 signaling pathway [75] - DPP-4i favors EMT and metastatic potential through the CXCL12/CXCR4 signaling pathway [87] - At high concentrations, liraglutide may promote cancer progression via the NOX4/ROS/VEGF signaling pathway [109] - Ipragliflozin attenuates cell proliferation by modifying membrane dynamics and simulating glucose deprivation effects [129]

Table 1. *Cont.*

Cancer Type	Anti-Diabetic Drugs and Their Impact on Cancer-Related Signaling
Lung cancer	- Metformin reduces oncogenic markers like HES1 and REDD1, while modulating p-mTOR and p53 levels, resulting in diminished tumor cell proliferation [37] - Insulin favors the proliferation and migration of cancer cells by upregulating the PI3K/Akt pathway [51] - Glibenclamide targets SUR1 to inhibit cell growth, migration, and EMT by reducing p70S6K activity and increasing the tumor suppressor KLF4 [148] - In combination with celecoxib, pioglitazone decreases tumor weight and increases survival by mitigating NF-κB-mediated proliferation [76] - Vildagliptin inhibits lung metastases by hindering autophagy, promoting apoptosis, and modulating the cell cycle [149] - Liraglutide hinders lung cancer cell proliferation, migration, and EMT, while also demonstrating anti-aging effects by debilitating cellular senescence and ER stress] [150] - Canagliflozin disrupts HIF-1α stabilization, impairing mitochondrial function and survival pathways by inhibiting mTOR and HDAC2 [133]
Hepatocellular carcinoma	- Metformin activates AMPK to disrupt lipid metabolism, enhances apoptosis and oxidative stress via p38MAPK activation [41] - Insulin promotes cell proliferation and survival by activating PI3K/Akt and Ras/MAPK pathways [52] - Glibenclamide inhibits Kv channels, resulting in a dose- and calcium-dependent decrease in the adhesion and proliferation of tumor cells [63] - Pioglitazone shows anti-fibrotic and hepatoprotective effects, potentially by modulating the MAPK and AMPK signaling pathways [79] - Linagliptin inhibits tumor cell growth by modulating ADORA3, inducing apoptosis and increasing cAMP levels [151] - Liraglutide may enhance anti-tumor immune responses through the IL-6/STAT3 signaling pathway [117] - Canagliflozin may mitigate liver steatosis, fibrosis, and tumor development, while promoting healthier adipose tissue with lower oxidative stress [130]
Pancreatic cancer	- Metformin facilitates apoptosis in cancer cells by modulating histone acetyltransferases (PCAF, p300, CBP) and SIRT1 expression [152] - Insulin induces the growth and fibrosing responses of PaSC through activation of the IR/IGF-1R, which in turn enhance Akt/mTOR/p70S6K signaling and reduce FOXO1, contributing to cell proliferation and extracellular matrix production [153] - Gliclazide favors DNA repair in cancer cells by stimulating NER and non-NHEJ pathways, without affecting these processes in normal human cells [154] - Pioglitazone blocks metastasis by altering inflammation-related gene expression, including CEA and COX-2 [80] - Saxagliptin promotes β-cell proliferation by elevating stromal cell-derived factor-1α [90] - Liraglutide inhibits cancer growth and promotes apoptosis by downregulating the PI3K/Akt and ERK1/2 signaling pathways [103] - Canagliflozin exhibits anti-tumor activity in cancer cells by downregulating the PI3K/Akt/mTOR pathway and impairing glycolysis [136]
Colorectal cancer	- Metformin mitigates cancer cell proliferation by targeting INHBA, inhibiting TGF-β/PI3K/Akt signaling and causing G1/S cell cycle arrest [155] - Insulin promotes cancer progression and metastasis by upregulating ACAT1 [53] - Glibenclamide and miR-223, by inhibiting NLRP3 in cancer cells, mitigate cell growth and migration, but miR-223 has a more pronounced effect on apoptosis and cytokine secretion, though neither fully prevents metastasis [156] - Pioglitazone mitigates cancer stem cell viability and increases MET, aiding in cancer suppression [82] - Sitagliptin attenuates cancer metastasis by reducing cell invasion, motility, and EMT through DPP4-dependent mechanisms [94] - GLP-1 RAs diminish cell cycle progression and promote apoptosis through the PI3K/Akt/mTOR pathway [119] - Canagliflozin shows potential efficacy by disrupting cellular metabolism and inducing ER stress, which in turn promotes autophagy and apoptosis via SIRT3 upregulation [135]
Bladder cancer	- Metformin suppresses the migration and proliferation of cancer cells and promotes apoptosis, likely by inhibiting the PI3K/Akt/mTOR pathway [157] - Both high-dose human insulin and insulin glargine promote tumor cell proliferation by activating Akt phosphorylation through a PI3K-independent pathway [158] - Pioglitazone causes DNA damage and promotes malignant transformation by altering gene expression and inducing EMT [69] - DDP-4i may reduce tumor aggressiveness by impairing cancer cell viability, proliferation, migration, and invasion [159] - Empagliflozin induces dysplastic changes in urothelium, with decreased expression of CK-7 and CK-8 and increased expression of CK-20 and Ki-67, suggesting heightened proliferative activity [160]
Prostate cancer	- Metformin induces oxidative stress in cancer cells by promoting ROS and disrupting oxidative phosphorylation, leading to redox imbalance and enhanced apoptosis [40] - Glipizide inhibits angiogenesis via the HMGIY/Angiopoietin-1 signaling pathway, without affecting cancer cell proliferation [161] - Pioglitazone alleviates inflammation in periprostatic WAT, potentially impacting cancer progression by modulating adipose tissue-related inflammatory responses [73] - Exenatide may lower tumor cancer progression through the inhibition of ERK-MAPK signaling [162] - Canagliflozin inhibits tumor cell growth by mitigating mitochondrial respiration and ATP production. This mechanism involves activation of AMPK, reduced lipid synthesis, and decreased glucose uptake [138].

Table 1. *Cont.*

Cancer Type	Anti-Diabetic Drugs and Their Impact on Cancer-Related Signaling
Ovarian cancer	- Metformin alleviates cell proliferation by activating the AMPK/GSK3β pathway, resulting in cyclin D1 degradation and G1 phase cell cycle arrest [34] - Through IGF-I, insulin induces COX-2 expression, which enhances VEGF production and PGE2 biosynthesis. This process activates PI3K, MAPK, and PKC pathways and promotes SKOV-3 cell migration by favoring uPA over PAI-1 via the PI3K/AKT pathway [163]. - Glibenclamide dampens angiogenesis and metastasis in ovarian cancer by reducing cellular invasion and migration through suppression of PDGF-AA secretion [65] - Ciglitazone and troglitazone reduce ovarian cancer cell proliferation, while pioglitazone has no effect, suggesting that the observed impact is PPAR-γ independent [164] - Sitagliptin promotes apoptosis via caspase 3/7 activation and suppresses migration and invasiveness in SKOV-3 cells, while also reducing the production of MMPs and TIMPs [165] - Exenatide may lower ovarian cancer aggressiveness by debilitating cell migration, inducing apoptosis, and modulating metalloproteinase expression [114]
Endometrial cancer	- Metformin activates AMPK, leading to enhanced TET2 gene expression and suppression of cancer cell growth [32] - Insulin stimulates aromatase activity in both endometrial glands and stroma, suggesting that hyperinsulinemia may increase the risk of estrogen-dependent endometrial neoplasia by enhancing local estrogen production [166] - Pioglitazone demonstrates significant dose-dependent anti-cancer activity by improving body weight, survival time, and uterine tissue weight [167] - DPP-4 overexpression accelerates carcinoma progression by enhancing cell proliferation, invasion, and HIF-1a-VEGFA signaling, while pharmacological inhibition with sitagliptin shows potential as an effective therapeutic strategy [92] - Tirzepatide mitigates tumor growth by altering glycolysis and ErbB signaling in obese mice, while enhancing glycosylation and phospholipase D signaling in lean mice [113]
Other malignancies	- Metformin inhibits the invasion and proliferation of cervical cancer cells by regulating the insulin signaling pathway and upregulating the expression of the tumor suppressor IGFBP7 [168] - Metformin mitigates MPM cell proliferation and promotes apoptosis by decreasing Notch1 activation [169] - Sitagliptin affects gastric cancer cell proliferation by reducing MAGE-A3 expression through the inactivation of YAP [170] - Dapagliflozin exerts cytotoxic effects in RCC by decreasing glucose uptake, disrupting cell cycle progression, and promoting apoptosis [137]

Abbreviations: ACAT1: acetyl-CoA acetyltransferase 1; ADORA3: adenosine A3 receptor; Akt: protein kinase B; AMPK: AMP-activated protein kinase; ATP: adenosine triphosphate; cAMP: cyclic adenosine monophosphate; CBP: CREB-binding protein; CCL2: C-C motif chemokine ligand 2; COX-2: cyclooxygenase-2; CXCR4: C-X-C chemokine receptor type 4; CXCL8: C-X-C motif chemokine ligand 8; CXCL12: C-X-C motif chemokine ligand 12; DPP-4: dipeptidyl peptidase 4; DDP-4i: dipeptidyl peptidase-4 inhibitor; EMT: epithelial-to-mesenchymal transition; ER: endoplasmic reticulum; ERK1/2: extracellular signal-regulated kinases 1 and 2; FOXO1: Forkhead box O1; FOXO3: Forkhead box O3; GB: glioblastoma; GLP-1 RAs: glucagon-like peptide-1 receptor agonists; GLP-1R: glucagon-like peptide-1 receptor; GSK3β: glycogen synthase kinase 3 beta; HES1: hairy and enhancer of split-1; HGF: hepatocyte growth factor; HIF-1α: hypoxia-inducible factor 1-alpha; HMGIY: high-mobility group protein I(Y); IGF: insulin-like growth factor; IGF-I: insulin-like growth factor I; IGF-1R: insulin-like growth factor 1 receptor; IGFBP7: insulin-like growth factor-binding protein 7; IL-6: interleukin-6; INHBA: inhibin subunit beta A; IR: insulin receptor; JAK: Janus kinase; Ki-67: antigen Kiel 67; KLF4: Kruppel-like factor 4; MAGE-A3: melanoma-associated antigen 3; MAPK: mitogen-activated protein kinase; MCT1: monocarboxylate transporter 1; MET: mesenchymal-to-epithelial transition; MMPs: matrix metalloproteinases; MMP-2: matrix metalloproteinase-2; mM: millimolar; mTOR: mechanistic target of rapamycin; NER: nucleotide excision repair; NHEJ: non-homologous end joining; NF-κB: nuclear factor kappa-light-chain-enhancer of activated B cells; NLRP3: NLR family pyrin domain containing 3; Notch1: neurogenic locus notch homolog protein 1; NOX4: NADPH oxidase 4; p38MAPK: p38 mitogen-activated protein kinase; p70 S6 kinase: 70 kDa ribosomal protein S6 kinase; PAI-1: plasminogen activator inhibitor-1; PaSC: pancreatic stellate cell; PGE2: prostaglandin E2; PI3K: phosphoinositide 3-kinase; PKC: protein kinase C; p-mTOR: phosphorylated mechanistic target of rapamycin; PPAR-γ: peroxisome proliferator-activated receptor gamma; PTC: papillary thyroid carcinoma; Ras: rat sarcoma; RCC: renal cell carcinoma; REDD1: regulated in development and DNA damage responses 1; ROS: reactive oxygen species; SIRT: sirtuin; SKOV-3: a human ovarian cancer cell line; STAT3: signal transducer and activator of transcription 3; TET2: Ten-Eleven Translocation methylcytosine dioxygenase 2; TGF-β: transforming growth factor beta; TIMP: tissue inhibitor of metalloproteinases; VEGFA: vascular endothelial growth factor A; VEGF: vascular endothelial growth factor; VEGFR2: vascular endothelial growth factor receptor 2; WAT: white adipose tissue; Wnt: wingless-type MMTV integration site family; YAP: yes-associated protein.

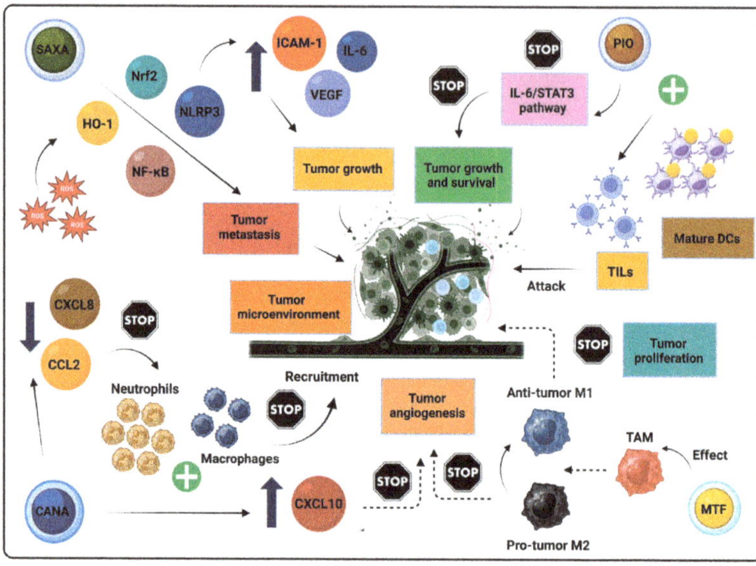

Figure 3. Schematic illustration of the diverse mechanisms through which anti-diabetic pharmacotherapy may affect tumor microenvironment dynamics. Metformin promotes the conversion of TAMs into an M1 phenotype, thereby enhancing anti-proliferative activity and inhibiting angiogenic processes. Pioglitazone modifies immune responses through the targeting of the IL-6/STAT3 signaling pathway, leading to increased activation of TILs, which are essential for recognizing and attacking tumor cells in the presence of mature dendritic cells. Additionally, canagliflozin reduces the release of the inflammatory chemokines CXCL8 and CCL2, which may limit the recruitment of immune cells while simultaneously increasing the chemokine CXCL10, ultimately inhibiting tumor angiogenesis. Conversely, DPP-4 inhibitors may promote metastatic processes by activating the ROS-mediated Nrf2/HO-1/NF-κB/NLRP3 signaling axis, resulting in elevated production of inflammatory cytokines, adhesion molecules, and angiogenic factors, including IL-6, ICAM-1, and VEGF [29,83,88,124]. Abbreviations: CANA: Canagliflozin; CCL2: C-C motif chemokine ligand 2; CXCL8: C-X-C motif chemokine ligand 8; CXCL10: C-X-C motif chemokine ligand 10; DCs: Dendritic cells; DPP-4: Dipeptidyl peptidase-4; HO-1: Heme oxygenase-1; ICAM-1: Intercellular adhesion molecule 1; IL-6: Interleukin 6; M1 phenotype: Macrophage 1 phenotype; M2 phenotype: Macrophage 2 phenotype; MTF: Metformin; NLRP3: NOD-like receptor family pyrin domain containing 3; NF-κB: Nuclear factor kappa-light-chain-enhancer of activated B cells; Nrf2: Nuclear factor erythroid 2-related factor 2; PIO: Pioglitazone; ROS: Reactive oxygen species; SAXA: Saxagliptin; STAT3: Signal transducer and activator of transcription 3; TILs: Tumor-infiltrating lymphocytes; TAMs: Tumor-associated mac-rophages; VEGF: Vascular endothelial growth factor. Created with www.BioRender.com.

3. Enhancing Chemotherapy: The Integration of Anti-Diabetic Treatments in Improving Therapeutic Outcomes

3.1. Exploring Anti-Diabetic Therapies as a Strategy to Diminish Chemotherapy-Induced Cardiotoxicity

Anthracycline-related cardiotoxicity, particularly linked to the widespread use of doxorubicin (DOX) in cancer treatment, remains a significant clinical challenge. Emerging evidence indicates that many anti-diabetic agents, known for their pleiotropic effects beyond glycemic control, may confer cardioprotective benefits. This has sparked growing interest in their potential role in mitigating DOX-induced cardiac damage, a protective effect demonstrated primarily in experimental studies and, to a lesser extent, in clinical settings.

3.1.1. Metformin

Metformin demonstrates significant therapeutic potential in mitigating DOX-induced cardiotoxicity through its versatile effects on inflammation, oxidative stress, apoptosis, and mitochondrial dysfunction. Its anti-inflammatory properties are characterized by the downregulation of the high-mobility group box 1 (HMGB1)/NF-κB/NLRP3 signaling axis, alongside the modulation of AMPK and MAPK pathways, both of which play crucial roles in cellular stress responses [171,172]. Notably, AMPK upregulation can be achieved even at low doses [173], and metformin may also enhance AMPK activation indirectly by promoting the expression of adiponectin and its receptors [174]. In addition to its anti-inflammatory effects, metformin pre-treatment appears to confer protective benefits by restoring critical anti-oxidant enzyme activity, inhibiting caspase-mediated apoptosis, and reducing DNA fragmentation [175]. Furthermore, recent investigations underscored metformin's ability to modulate autophagy and mitophagy pathways, normalizing the expression of essential autophagy-related proteins such as beclin-1, microtubule-associated protein 1A/1B light chain 3B (LC3B-II), and p62, which are pivotal for maintaining cellular homeostasis [176]. Comparable cardioprotective effects were documented in cases of trastuzumab-induced cardiotoxicity, which mirrors the cardiotoxic mechanisms associated with DOX [177]. Conversely, recent animal data failed to replicate the cardioprotective benefits of metformin in the context of cardiotoxicity stemming from combination chemotherapy regimens, including cyclophosphamide, methotrexate, and 5-fluorouracil (5-FU) [178].

A recent meta-analysis of animal studies underpinned the centrality of metformin's anti-oxidative mechanisms in its cardioprotective role, providing further support for its potential in mitigating DOX-induced cardiotoxicity [179]. However, clinical evidence remains inconclusive. In a randomized controlled trial involving breast cancer patients, metformin did not prevent myocardial injury, as evidenced by elevated high-sensitivity troponin-I levels and reductions in left ventricular ejection fraction (LVEF). Nonetheless, metformin was observed to play a significant role in preserving mitochondrial function during DOX treatment [180].

3.1.2. Other Conventional Anti-Diabetic Treatments

Research further underscored pioglitazone's potential in mitigating DOX-induced cardiomyopathy, though its cardioprotective effects remain incomplete. Animal studies suggested that pioglitazone administration attenuates DOX-induced left ventricular dysfunction, particularly during the acute and chronic phases, while offering limited protection in the hyperacute phase [181]. The cardiotoxicity associated with DOX was linked to elevated miR-130a levels and reduced PPAR-γ activity. Targeting miR-130a showed promise in mitigating cardiac damage by restoring PPAR-γ function in cardiomyocytes, thereby exerting anti-inflammatory and anti-apoptotic effects. Pioglitazone contributed to this protective mechanism by increasing PPAR-γ expression and decreasing miR-130a levels, although it failed to fully prevent DOX-induced cardiotoxicity [182,183].

Moreover, recent data underlined pioglitazone's ability to reverse key biochemical markers associated with DOX-induced cardiomyopathy, such as altered thyroid hormone fractions and elevated troponin levels [184]. However, concerns persist regarding the potential exacerbation of cardiac injury via PPAR-γ agonism. Some reports suggested that PPAR-γ agonism may contribute to structural damage in heart tissue and alter critical biochemical markers, raising doubts about pioglitazone's cardioprotective capacity [185]. While pioglitazone itself is not considered directly cardiotoxic, its activation of PPAR-γ could, under certain conditions, aggravate cardiac dysfunction. These findings indicate that the therapeutic use of pioglitazone in managing DOX-induced cardiomyopathy requires careful consideration of both its potential benefits and associated risks. Lastly, data on the role of SUs in chemotherapy-induced cardiotoxicity are lacking. However, one experimental study suggested that glibenclamide may exacerbate Adriamycin-promoted cardiotoxicity by activating oxidative stress-induced ER stress [186].

3.1.3. Incretin Mimetics

Limited evidence also suggests that DPP-4 inhibitors may offer cardioprotective effects against DOX-induced cardiotoxicity. Similar to metformin, sitagliptin was shown to suppress pro-inflammatory cytokines, such as TNF-α, and inhibit NF-κB activation. Additionally, it mitigated oxidative damage by reducing lipid peroxidation and preserving anti-oxidant defenses in cardiac tissue. Sitagliptin also inhibited the upregulation of pro-apoptotic proteins [187]. Linagliptin also demonstrated comparable benefits, particularly in preserving myocardial fiber structure and reducing oxidative stress, evidenced by lower levels of malondialdehyde (MDA), a marker of lipid peroxidation, and preserved glutathione peroxidase (GPx) activity [188].

Preliminary research indicated that GLP-1 RAs may also have significant potential in mitigating DOX-associated cardiomyopathy. This potential was attributed to liraglutide's capacity to reduce inflammation and apoptosis, mediated by the upregulation of AMPK activity, which may attenuate markers of cardiac injury and support cellular anti-oxidant systems [189]. However, conflicting data suggested that liraglutide may not effectively inhibit critical inflammatory mediators such as NF-κB. Furthermore, it appeared ineffective in restoring mitochondrial dynamics, as evidenced by unaltered levels of mitochondrial proteins, including protein optic atrophy-1 (OPA-1), mitofusin-2 (MFN-2), dynamin-related protein 1 (DRP-1), and topoisomerase 2β [190]. In contrast, recent evidence for semaglutide suggested cardioprotective qualities, with moderate to high dosages correlating with reductions in troponin and lactate dehydrogenase (LDH) levels, alongside improved histopathological outcomes [191]. Lastly, tirzepatide was shown to exhibit anti-inflammatory and anti-oxidant activities in the context of DOX-induced cardiomyopathy, likely through the activation of the PI3K/Akt signaling pathway [192].

3.1.4. SGLT-2 Inhibitors

Among the various classes of anti-diabetic agents, SGLT-2 inhibitors have emerged as the most cardioprotective, making their potential to mitigate chemotherapy-related cardiotoxicity especially promising. A growing body of preclinical evidence notes this cardioprotective capacity, with distinct mechanisms of action identified across the major SGLT-2 inhibitors [193,194]. Notably, dapagliflozin was shown to exert pleiotropic benefits, largely through the activation of the PI3K/Akt signaling pathway. This activation led to a reduction in oxidative stress, suppression of inflammation, attenuation of cardiac hypertrophy, and mitigation of myocardial fibrosis. Furthermore, dapagliflozin showed efficacy in ameliorating mitochondrial dysfunction, alongside notable anti-arrhythmic properties, the latter attributed to its ability to mitigate DOX-induced sarcolemma injury and myocardial necrosis [195,196]. Empagliflozin likewise exhibited robust cardioprotective features, particularly in preventing adverse cardiac remodeling, even in the absence of DM [197]. Its additional anti-oxidant properties, potentially mediated via the inhibition of the JNK/STAT3 signaling cascade, further underscored its therapeutic promise in mitigating chemotherapy-induced cardiotoxicity [198]. Similar protective mechanisms also were observed with canagliflozin, which notably reduced cisplatin-associated cardiotoxic impacts [199]. Uniquely, canagliflozin extended its protective repertoire by mitigating carfilzomib-induced endothelial apoptosis through an AMPK-dependent pathway [200].

Beyond their anti-inflammatory and anti-oxidant properties, SGLT-2 inhibitors appeared to counteract chemotherapy-induced cardiotoxicity through a profound modulation of cellular energy metabolism. These agents activated key nutrient deprivation pathways, such as AMPK and sirtuins, while concurrently inhibiting nutrient excess signaling, including Akt/mTOR. This dual regulatory mechanism fostered enhanced autophagy, stimulated mitochondrial biogenesis via PGC-1α activation, and promoted ketogenesis, collectively supporting myocardial resilience under stress [201–203]. Supporting this mechanistic foundation, observational data corroborated the cardioprotective effects of SGLT-2 inhibitors in clinical settings. For example, a study by Gongora et al. involving 3033 diabetic cancer patients receiving anthracycline chemotherapy revealed a significant reduction in cardiac events (3% vs. 20%) and overall mortality (9% vs. 43%) in those treated with SGLT-2 inhibitors compared to controls [204]. Notably, recent research by Liu et al. suggested that empagliflozin may offer cardioprotective benefits in humans exposed to sorafenib by influencing pathways related to inflammation, fibrosis, DNA damage,and ferroptosis [205].

In light of these findings, it is evident that anti-diabetic therapies, particularly metformin and SGLT-2 inhibitors, may offer substantial cardioprotective gains for oncology patients undergoing DOX treatment. This is unsurprising, as inflammation, oxidative stress, apoptosis, and mitochondrial dysfunction, key drivers of DOX-induced cardiomyopathy [206], are pathways that anti-diabetic agents were shown to modulate in experimental models. Moving forward, well-designed randomized clinical trials are essential to confirm these cardioprotective impacts and determine whether glucose-lowering treatments can be effectively repurposed to prevent chemotherapy-induced cardiotoxicity. Figure 4 illustrates the key molecular mechanisms by which various anti-diabetic agents mitigate DOX-induced cardiotoxicity through the modulation of distinct signaling pathways.

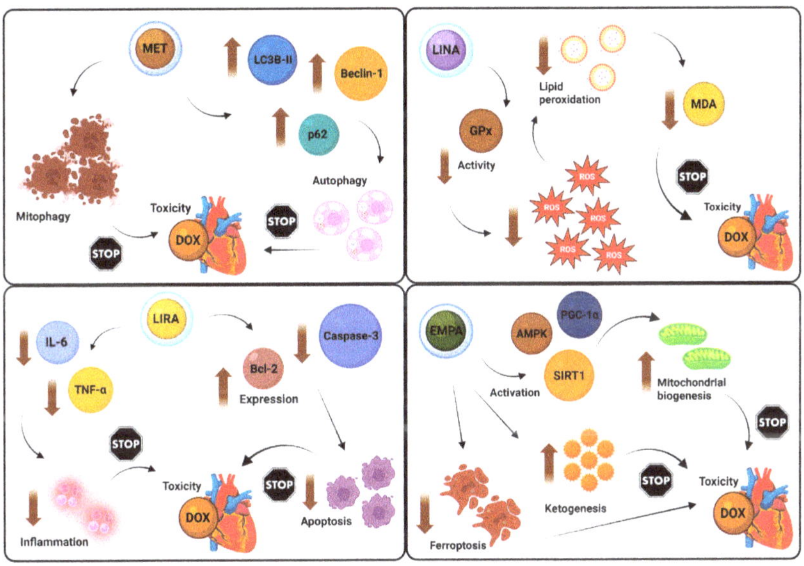

Figure 4. Schematic presentation of signaling pathways influenced by anti-diabetic drugs to counteract doxorubicin cardiotoxicity. Metformin may enhance cardiac function by modulating autophagy and mitophagy pathways, which is evidenced by the normalization of autophagy markers such as beclin-1, LC3B-II, and p62. Liraglutide demonstrates the ability to reduce inflammation and apoptosis by decreasing the levels of pro-inflammatory cytokines (IL-6 and TNF-α), downregulating the pro-apoptotic caspase-3, and upregulating the anti-apoptotic protein Bcl-2. On the other hand, linagliptin exhibits antioxidant properties, potentially reducing oxidative stress through the reduction of GPx activity, thereby limiting ROS generation, decreasing lipid peroxidation, and mitigating MDA formation. This process preserves cellular integrity and improves myocardial fiber structure, which is crucial in alleviating DOX-induced cardiomyopathy. Lastly, empagliflozin is illustrated to potentially mitigate DOX-related cardiac injury by enhancing mitochondrial biogenesis through the activation of the AMPK/SIRT1/PGC-1α pathway. Additionally, it may reduce ferroptosis and enhance ketogenesis, contributing further to its cardioprotective effects [176,188,189,199–201]. Abbreviations: AMPK: AMP-activated protein kinase; Bcl-2: B cell lymphoma 2; DOX: Doxorubicin; EMPA: Empagliflozin; GPx: Glutathione peroxidase; IL-6: Interleukin 6; LC3B-II: Microtubule-associated protein 1A/1B light chain 3B, form II; LINA: Linagliptin; LIRA: Liraglutide: MDA: Malondialdehyde; MTF: Metformin; PGC-1α: Peroxisome proliferator-activated receptor gamma coactivator 1-alpha; ROS: Reactive oxygen species; SIRT1: Sirtuin 1; TNF-α: Tumor necrosis factor-alpha. Created with www.BioRender.com.

3.2. Anti-Diabetic Agents as Modulators of Chemotherapy-Induced Toxicities: Beyond Cardiotoxicity

Beyond their cardioprotective benefits, anti-diabetic therapies also play a vital role in mitigating other chemotherapy-related toxicities, particularly those affecting the kidneys and liver. These protective effects are driven by shared mechanisms, including the modulation of inflammation, oxidative stress, apoptosis, and mitochondrial dysfunction, processes that contribute to both cardiac and non-cardiac toxicities. Anti-diabetic agents have shown significant promise in reducing chemotherapy-induced acute kidney injury (AKI), especially in cases involving cisplatin. Among these, metformin emerged as a key protective agent, known for its ability to activate AMPK signaling and enhance autophagic flux in renal tubular cells, mechanisms that reduce apoptosis and preserve renal function [207]. Similarly, gliclazide and pioglitazone, which act through anti-oxidant and anti-inflammatory pathways, also demonstrated potent nephroprotective properties [208,209]. Additionally, DPP-4 inhibitors showed promise in preclinical models, mitigating renal damage by modulating key inflammatory and fibrotic pathways, such as the MAPK/JNK signaling pathway and the NLRP3 inflammasome [210,211]. Furthermore, GLP-1 RAs, such as liraglutide, particularly in combination with curcumin, exhibited nephroprotective effects by enhancing the Nrf2/HO-1 pathway and regulating GSK-3β activity [212]. Moreover, SGLT-2 inhibitors, including dapagliflozin and empagliflozin, offer renal protection by reducing oxidative stress, renal fibrosis, and apoptosis, particularly in cases of doxorubicin-induced nephrotoxicity [213,214].

Anti-diabetic treatments also showed promise in managing chemotherapy-related liver injury. Metformin lowered doxorubicin accumulation in the liver via the inhibition of organic cation transporters [215], while gliclazide and pioglitazone provided hepatoprotection against cisplatin and tamoxifen-induced liver injury by modulating oxidative stress and inflammation [216,217]. SGLT-2 inhibitors further shielded liver tissue by preserving structural integrity and reducing oxidative stress, with dapagliflozin showing enhanced efficacy in combination with silymarin [218]. Glucose-lowering treatments further showed protective effects in various organ systems beyond the kidneys and liver. Metformin reduced cisplatin-induced ototoxicity by promoting SIRT3 expression and restoring glucose uptake in auditory cells [219]. In chemotherapy-induced peripheral neuropathy, DPP-4 inhibitors like alogliptin, along with pioglitazone, alleviated neurotoxicity through anti-inflammatory and anti-oxidant actions [220,221]. Pioglitazone also reduced cisplatin-induced testicular damage by downregulating the toll-like receptor 4/myeloid differentiation factor 88 (TLR4/MyD88)/NF-κB signaling pathway [222], while liraglutide improved reproductive function following doxorubicin exposure [223]. Additionally, metformin was shown to reduce cisplatin-induced genotoxicity in bone marrow cells by decreasing micronucleated erythrocytes and oxidative stress, while also improving hematological profiles [224].

While preclinical data strongly support the protective outcomes of anti-diabetic agents against chemotherapy-induced toxicities, clinical translation remains in its early stages, though initial findings are promising. A randomized controlled trial investigating metformin in non-DM breast cancer patients receiving Adriamycin-cyclophosphamide plus paclitaxel demonstrated a significant reduction in peripheral neuropathy, oral mucositis, and fatigue. Additionally, metformin preserved cardiac function and reduced the risk of developing fatty liver, suggesting its broad potential in mitigating multiple chemotherapy-related toxicities [225]. Further clinical evidence highlighted metformin's efficacy in reducing paclitaxel-induced peripheral neuropathy, improving treatment tolerance, and potentially enhancing long-term outcomes [226].

In DM cancer patients, the nephroprotective results of DPP-4 inhibitors also were demonstrated. A study by Iwakura et al. found that DPP-4 inhibitor users experienced a significantly lower decline in estimated glomerular filtration rate (eGFR) two weeks after cisplatin treatment and a reduced incidence of AKI compared to non-users (25% vs. 64%) [227]. These early clinical findings suggest that anti-diabetic drugs may offer significant adjunctive benefits in reducing chemotherapy-induced toxicities. However, further clinical trials are necessary to validate their efficacy and expand their use in the oncological settings.

3.3. Decoding the Impact of Anti-Diabetic Therapies on Chemotherapy Outcomes Beyond Toxicity Mitigation

3.3.1. Traditional Anti-Diabetic Treatments

Metformin showed promising results in both preclinical and clinical studies across various malignancies, notably in gynecological cancers [228,229]. Of special interest is the synergistic effect of metformin with doxorubicin in breast cancer, where it appears to facilitate the selective eradication of cancer stem cells, significantly enhancing the therapeutic impact of chemotherapy [230]. In ovarian cancer, metformin, when combined with carboplatin, demonstrated adjunctive effects by inhibiting the Akt/mTOR signaling pathway [231]. Notably, metformin may also overcome cisplatin resistance in ovarian cancer cells by inducing autophagy [232]. Data from non-DM breast cancer patients (stages II and III) showed that the combination of metformin with neoadjuvant chemotherapy resulted in a lower residual cancer burden score (40.7% vs. 68.8% in class 3 patients) [233].

Metformin's potential was further explored in CRC, where combination therapy with 5-FU showed tumor size reduction through the modulation of NF-κB signaling [234]. There is also compelling evidence suggesting that metformin may enhance the chemosensitivity of CRC cells to the oxaliplatin-based regimen (FuOx). In vitro studies demonstrated that metformin significantly improved FuOx's efficacy by inhibiting cell proliferation, colony formation, and migration, while also inducing cell cycle arrest and promoting late apoptosis through the regulation of mitochondrial proteins. These findings were further corroborated by in vivo experiments, where the combination of metformin and FuOx resulted in a significantly greater reduction in tumor volume compared to either treatment alone [235]. However, metformin's efficacy appeared limited in advanced pancreatic cancer, where its addition to gemcitabine and erlotinib regimens did not improve patient outcomes [236].

Preliminary and clinical data also suggested a role for metformin in hematological malignancies. For instance, in acute myeloid leukemia, metformin was shown to enhance the anti-tumor effect of a co-administration with cytarabine by inhibiting the mTOR complex1 (mTORC1/p70S6) kinase signaling pathway [237]. Similarly, in lymphoma, the addition of metformin to a regimen of rituximab and anthracyclines significantly improved both progression-free and overall survivals, compared to other anti-hyperglycemic agents. This study further demonstrated that metformin could potentiate the anti-diffuse large B cell lymphoma (DLBCL) effects of DOX and rituximab [238].

Synergistic results of pioglitazone were also reported, particularly in the context of breast cancer treatment. In combination with cisplatin, pioglitazone showed anti-apoptotic features, offering a promising adjunct therapy for women with TNBP. Compared to cisplatin monotherapy, co-administration with pioglitazone significantly enhanced tumor cell apoptosis in a dose-dependent manner, as indicated by reduced expression of the anti-apoptotic protein Bcl-2 and increased expression of pro-apoptotic proteins such as cleaved-poly-ADP ribose polymerase (PARP) and caspase-9 [239].

A significant body of research further explored the role of DPP-4 inhibition in chemotherapy outcomes, yielding mixed findings [86,240]. Several endogenous DPP-4 substrates, such as CXCL12, human granulocyte-macrophage colony-stimulating factor (GM-CSF), and IL-3, may have diverse consequences on chemotherapy-induced damage to healthy tissues [241]. For instance, certain studies suggested that DPP-4 inhibition may contribute to breast cancer chemoresistance by increasing levels of CXCL12, leading to the induction of EMT in tumors [242]. However, other findings highlight the potential of targeting DPP-4 to inhibit angiogenesis in 5-FU-resistant colon cancer. In this context, exosomal DPP-4 was proposed as a promising prognostic marker [243].

3.3.2. Novel Anti-Diabetic Therapies

Similar favorable outcomes in chemotherapy efficacy were observed with exenatide, which demonstrated the potential to alleviate resistance to enzalutamide in advanced prostate cancer. In a murine experimental model, the combination of exenatide and enzalutamide significantly suppressed tumor growth compared to enzalutamide monotherapy. This combinatorial treatment not only mitigated the enzalutamide-induced invasion and migration of prostate cancer cells but also reduced the levels of Akt and mTOR, which were activated in response to enzalutamide. Importantly, the dual therapy resulted in a further decrease in nuclear androgen receptor localization, despite exenatide alone exerting no effect on nuclear androgen receptor levels [244].

Recent research also underlined the potential of combining SGLT-2 inhibitors with cisplatin therapy as a promising strategy to overcome cisplatin resistance in hepatoblastoma. Specifically, dapagliflozin, through SGLT-2 inhibition, reduced glucose uptake, a critical factor in the development of cisplatin resistance [245]. Similarly, canagliflozin showed therapeutic promise in HCC by inhibiting cell proliferation and migration through mechanisms that extended beyond its primary SGLT-2 target. Notably, canagliflozin induced the downregulation of pyruvate kinase M2 (PKM2), leading to c-Myc degradation, which disrupted glutamine metabolism and induced glutamine starvation. This metabolic alteration promoted ferroptosis, sensitizing cancer cells to cisplatin [246].

Furthermore, the combination of dapagliflozin and canagliflozin was reported to enhance the anti-cancer efficacy of paclitaxel in ovarian cancer and oral squamous cell carcinoma. Mechanistic insights from studies on ovarian cancer cells indicated that canagliflozin augmented paclitaxel-induced apoptosis and DNA damage while impairing the spindle assembly checkpoint (SAC) through the downregulation of cyclin B1 and phosphorylated BUB1 mitotic checkpoint kinase 1 (BUBR1). This ultimately accelerated premature mitotic exit and contributed to the accumulation of aneuploid cells [247]. Recently, Karim et al. investigated the synergistic effects of prominent SGLT-2 inhibitors, canagliflozin, dapagliflozin, and empagliflozin, when used in conjunction with doxorubicin to address chemotherapy resistance. Their findings demonstrated that the combination of canagliflozin and doxorubicin significantly enhanced cytotoxicity in breast cancer cells, attributed to a reduction in glucose consumption and lower levels of intracellular ATP and lactate, both of which are essential for cancer cell proliferation [248]. While emerging preclinical studies support the potential of SGLT-2 inhibitors to enhance chemotherapy efficacy across various cancers, current clinical data remain limited. Notably, a case report showed that the combination of SGLT-2 inhibitors with cetuximab can reduce tumor size and carcinoembryonic antigen levels in metastatic CRC, warranting further investigation into these combinations in clinical settings [249].

Overall, the aforementioned findings highlight the promising role of anti-diabetic drugs in overcoming drug resistance and improving therapeutic outcomes. However, to substantiate these observations and explore their broader applicability, comprehensive future clinical studies are essential.

4. Unleashing the Benefits of Immunotherapy with Anti-Diabetic Treatment: Emerging Preclinical Insights with Potential Future Clinical Applications

Beyond their role in improving chemotherapeutic outcomes, anti-diabetic drugs are increasingly recognized for their potential to enhance the efficacy of immunotherapies. The beneficial impact of anti-diabetic medications in this context is likely due to their ability to reprogram the TME, promoting a cellular milieu less conducive to tumor progression. Current research is exploring how these agents may emulate or augment the effects of immune checkpoint inhibitors (ICIs), such as programmed cell death protein-1 (PD-1) and programmed death-ligand-1 (PD-L1) inhibitors, with emerging evidence suggesting that they may also interact synergistically with anti-cytotoxic T lymphocyte-associated protein-4 (CTLA-4) therapies.

4.1. Metformin

Metformin has emerged as a powerful adjunct in enhancing the efficacy of ICI therapies, particularly those targeting the PD-1 pathway. This enhancement is primarily achieved through metformin's modulation of the TME, by promoting the infiltration of $CD4^+$ T cells essential for anti-tumor responses, while simultaneously diminishing the population of regulatory T cells (Tregs) that typically exert immunosuppressive effects. Through the upregulation of IL-17A, metformin reinforces immune activation, and it also induces epigenetic modifications in Treg DNA to curtail their suppressive capabilities. Additionally, metformin activates AMPK, initiating a signaling cascade involving SIRT2 that leads to the downregulation of CCR8 on Tregs. This attenuation of CCR8 expression diminishes immune evasion, thereby enhancing anti-tumor immunity and augmenting the therapeutic efficacy of anti-PD-1 therapies [250]. In addition to its effects on the TME, metformin influences immune checkpoint gene expression across several malignancies, including breast cancer and CRC [251]. Specifically, it was shown to reduce PD-L1 expression through various mechanisms, including PD-L1 degradation in the ER and activation of the Hippo signaling pathway [252,253]. By downregulating PD-L1, metformin alleviates the inhibitory signaling that typically constrains cytotoxic T lymphocyte activity, thereby enhancing the potential for synergistic effects with CTLA-4 inhibitors [252].

In recent years, studies emphasized the important relationship between metformin and the gut microbiome, suggesting that these interactions may enhance the effectiveness of ICI therapies [253]. While the precise mechanisms are still under investigation, metformin's high concentrations in the intestines relative to its systemic levels indicate that it plays a crucial role in shaping gut microbiota [254]. The drug promotes a balanced composition of gut microbiota, facilitating immune regulation through various mechanisms, including the influence of bile acid dehydroxylation and the activation of intestinal Farnesoid X receptor (FXR) signaling, which collectively help mitigate metabolic dysregulation [255]. Additionally, metabolites derived from the gut, such as short-chain fatty acids (SCFAs) and bile acids, contribute significantly to immune regulation. This suggests that metformin's ability to modulate the microbiome may create a supportive environment for improved anti-tumor immunity when used alongside ICIs [256]. However, inter-individual variability in the concentrations of metformin within the gut and the composition of the microbiome complicate the reliable translation of these microbiota-mediated benefits into clinical practice [251].

Emerging clinical evidence notes metformin's potential to enhance survival outcomes in patients undergoing ICI therapy. A retrospective study in Taiwan involving 878 diabetic patients demonstrated significantly prolonged overall survival (OS) and progression-free survival (PFS) among metformin users compared to non-users (OS: 15.4 vs. 6.1 months; PFS: 5.1 vs. 1.9 months), even after adjusting for confounding factors such as age and cancer stage [257]. Similarly, in a multicenter cohort of 516 patients with solid tumors treated with ICIs, metformin showed specific survival benefits in the lung cancer subgroup, correlating with improved OS and PFS [258]. Furthermore, bioinformatic analyses identified five metformin-targeted genes in NSCLC, all downregulated in cancerous tissue and associated with an increased infiltration of immune cells, including $CD4^+$ and $CD8^+$ T cells [251]. While metformin's role as an adjunct in immunotherapy is increasingly recognized, further investigation is required to fully elucidate its mechanisms and optimize its clinical application.

4.2. Other Conventional Anti-Diabetic Therapies

In a similar vein, pioglitazone showed significant potential in enhancing immunotherapy efficacy, particularly in CRC. Pioglitazone promoted PD-L1 degradation, enhancing immune-mediated tumor recognition. Jia et al. demonstrated that pioglitazone reduced PD-L1 protein levels in cancer cells through autophagy, independent of changes in gene expression. This was achieved via the activation of PPAR-γ, which facilitated PD-L1 degradation by promoting its translocation to the lysosome, thereby increasing the interaction between PPAR-γ and PD-L1. When combined with PD-1 inhibitors, pioglitazone enhanced treatment efficacy by reducing PD-L1 levels and promoting the infiltration of cytotoxic $CD8^+$ T cells, thereby amplifying the anti-tumor immune response in CRC [259]. On the other hand, the impact of SUs on immunotherapy outcomes is less well documented, although recent research suggests that glibenclamide may enhance the therapeutic efficacy of PD-1 blockade. Specifically, glibenclamide acted as an agonist of the transketolase 1 gene (TKTL1), which plays a crucial role in mediating the immune response to PD-1 therapy in kidney clear cell carcinoma. By increasing TKTL1 expression, glibenclamide promoted immune cell recruitment within the TME, potentially augmenting the effectiveness of anti-PD-1 treatment [260].

DDP-4 inhibitors further showed promising potential in modulating immune responses and improving therapeutic outcomes for cancer patients who exhibited resistance to conventional PD-L1 treatments. Recent evidence suggests that combining anagliptin with PD-L1 antibodies led to significantly improved outcomes in NSCLC compared to PD-L1 blockade alone. This enhancement was primarily due to anagliptin's ability to inhibit macrophage formation and prevent M2 polarization within the TME, achieved through the reduction in ROS production and the suppression of signaling pathways that promote monocyte-to-macrophage differentiation [261]. A recent investigation evaluated the effects of six commonly prescribed anti-diabetic medications on the efficacy of anti-PD-1 immunotherapy in syngeneic mouse models of CRC and melanoma, revealing differential impacts on tumor suppression. Notably, sitagliptin was found to enhance the efficacy of anti-PD-1 therapy, whereas metformin exhibited a neutral effect. In contrast, glimepiride, pioglitazone, and insulin were associated with decreased tumor inhibition [262].

4.3. *Novel Anti-Diabetic Treatments*

4.3.1. GLP-1 and Dual GIP/GLP-1 Receptor Agonists

Although direct data on the impact of GLP-1 RAs on ICI therapy are lacking, emerging evidence suggests potential benefits. Preliminary findings indicated that liraglutide may enhance the anti-tumor efficacy of PD-1 inhibitors by targeting neutrophil extracellular traps (NETs) [263]. NETs, composed of DNA, histones, and proteins, are released by neutrophils in response to tissue damage and contribute to cancer progression by creating a pro-inflammatory environment that fosters tumor growth, metastasis, and immune evasion [264]. Chen et al. demonstrated that liraglutide significantly reduced NET markers, including myeloperoxidase (MPO), double-stranded DNA (dsDNA), and elastase, in lung and liver cancer models, while inhibiting NET formation by reducing ROS. When combined with PD-1 inhibitors, liraglutide facilitated sustained $CD8^+$ T cell responses, thereby offering protection against tumor recurrence [263].

The paradoxical relationship between obesity and improved outcomes in certain cancers treated with ICIs, such as melanoma and RCC, presents a complex and evolving landscape. Although the underlying mechanisms are not fully understood, obesity appears to influence several key factors, including inflammation, cancer cell metabolism, and angiogenesis [265]. Human tissue research indicated that obesity is associated with reductions in OXPHOS, creating a state of metabolic rest that may enhance the effectiveness of cancer therapies [266]. In preclinical models of RCC, obesity was linked to the decreased expression of fatty acid synthase, a critical enzyme in lipid metabolism that supports cancer cell survival and proliferation. Additionally, obesity seemed to increase immune cell infiltration and hypoxia in tumor-adjacent adipose tissue, potentially boosting the immune response [267]. Interestingly, clinical data from patients with stage 4 cancer treated with anti-PD-1/PD-L1 inhibitors showed that overweight or obese patients often experienced more favorable outcomes, such as prolonged time to treatment failure, compared to those with a normal body mass index (BMI) [268].

This phenomenon, often referred to as the "obesity paradox", highlights the potential benefits of excess weight in enhancing the immune response during ICI therapy [265]. Thus, this paradox raises important questions about the impact of weight loss induced by GLP-1 analogs and GIP/GLP-1 agonists, which are commonly prescribed for their weight-reducing effects. It remains crucial to investigate whether weight loss from these therapies could inadvertently reduce the therapeutic benefits observed in obese patients undergoing immunotherapy for certain cancer types.

4.3.2. SGLT-2 Inhibitors

Recent experimental studies revealed the critical role of SGLT-2 in regulating immune checkpoint activities within cancer cells, suggesting that SGLT-2 inhibitors may serve as effective modulators of anti-tumor immunity. The immunomodulatory properties of these drugs were demonstrated across various agents, with canagliflozin emerging as a prominent example. Canagliflozin exhibited a dual role in both glycemic control and immunoregulation by impeding T cell activation, proliferation, and effector functions through the inhibition of key signaling pathways, such as ERK and mTORC1. This inhibition reduced c-Myc levels, a transcription factor essential for metabolic processes critical to T cell function [269]. In cancer therapy, canagliflozin's favorable immunomodulatory effects were noted through its interaction with PD-L1. By disrupting the SGLT-2/PD-L1 axis, canagliflozin promoted the degradation of PD-L1 via the ubiquitin-proteasome pathway, thereby enhancing T cell-mediated cytotoxicity. This action limited tumor progression and improved the efficacy of immune checkpoint blockade therapies, akin to the effects observed with PD-1 monoclonal antibodies [270].

Nevertheless, SGLT-2 silencing extended to other immune checkpoints, such as CTLA-4. Dapagliflozin and empagliflozin were shown to mitigate the adverse effects associated with ipilimumab, primarily by modulating key inflammatory mediators [271,272]. In hyperglycemic states, empagliflozin not only reduced ROS formation but also decreased the production of cytokines and growth factors, including IL-6, TGF-β, VEGF, and leukotrienes. This reduction led to a downregulation of NF-κB and NLRP3 inflammasome activity [272]. Importantly, clinical data suggested that patients receiving ICIs who concurrently were using SGLT-2 inhibitors experienced a significant reduction in all-cause mortality compared to non-users (21% vs. 59%) over a long-term follow-up of nearly two years. This survival benefit persisted despite no significant impact on major adverse cardiovascular events (MACE) [273].

These findings suggest that incorporating SGLT-2 inhibitors into cancer immunotherapy regimens holds great potential for enhancing patient outcomes by leveraging their immunomodulatory properties. However, to fully harness this potential, future clinical studies are necessary to elucidate the precise mechanisms driving these effects and to optimize treatment strategies for patients undergoing ICI therapy. Figure 5 illustrates the mechanisms by which anti-diabetic drugs may modulate immune responses, contributing to enhanced anti-PD-1 therapeutic outcomes.

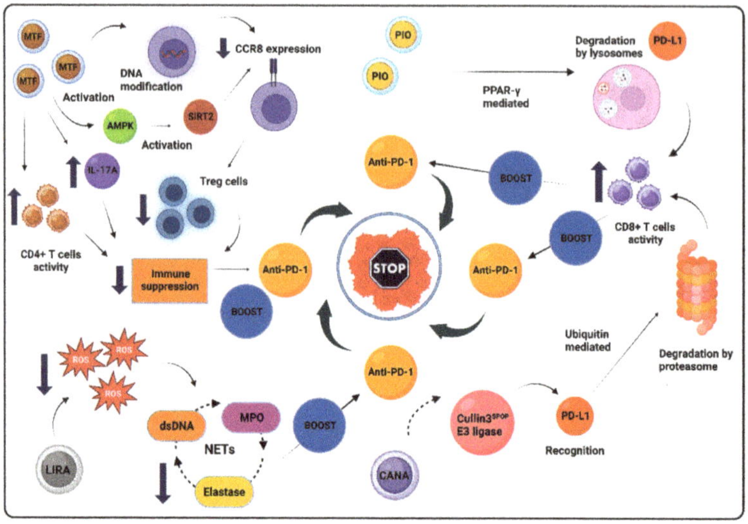

Figure 5. Modulation of immune responses by anti-diabetic drugs in bolstering anti-PD-1 immunotherapy outcomes. Metformin enhances the infiltration of $CD4^+$ T cells, crucial for initiating and sustaining anti-tumor immunity, while concurrently reducing the population of Tregs, which suppress immune responses within the TME. Moreover, metformin increases IL-17A levels and induces epigenetic modifications in Tregs, resulting in reductions in their immunosuppressive capabilities. By activating AMPK and triggering a downstream signaling cascade involving SIRT2, metformin promotes the downregulation of CCR8 on tumor-infiltrating Tregs. This downregulation decreases immune evasion and boosts anti-tumor responses, particularly when combined with PD-1 inhibitors. Pioglitazone and canagliflozin also contribute to improved ICI efficacy by promoting the degradation of PD-L1, a checkpoint protein that inhibits T cell activity. Pioglitazone acts as a PPAR-γ agonist, facilitating PD-L1 localization to lysosomes, while canagliflozin enhances PD-L1 recognition by the Cullin3SPOP E3 ligase for ubiquitination and proteasomal degradation. This degradation leads to increased infiltration of cytotoxic $CD8^+$ T cells, thereby reinforcing the anti-tumor immune response. Additionally, liraglutide, through its GLP-1 receptor agonism, mitigates oxidative stress, resulting in reduced production of NETs, which can promote inflammation and tumor progression. By decreasing NET levels, liraglutide aids in alleviating tumor-promoting inflammation, thereby further enhancing the efficacy of ICIs [250,259,263,270]. Abbreviations: AMPK: AMP-activated protein kinase; Anti-PD-1: Anti-programmed cell death protein 1; CANA: Canagliflozin; CCR8: C-C chemokine receptor 8; $CD4^+$ T cells: Cluster of differentiation 4 positive T cells; $CD8^+$ T cells: Cluster of differentiation 8 positive T cells; dsDNA: Double-stranded deoxyribonucleic acid; IL-17A: Interleukin-17A; ICI: Immune checkpoint inhibitor PD-L1: Programmed death-ligand 1; LIRA: Liraglutide; MPO: Myeloperoxidase; MTF: Metformin; NETs: Neutrophil extracellular traps; PIO: Pioglitazone; PPAR-γ: Peroxisome proliferator-activated receptor gamma; ROS: Reactive oxygen species; SIRT2: Sirtuin 2; TME: Tumor microenvironment; Tregs: Regulatory T cells. Created with www.BioRender.com.

5. Translating Preclinical Insights into Clinical Cancer Outcomes: Can We Draw Definitive Conclusions?

5.1. Metformin

In recent years, extensive clinical research explored metformin's potential role in influencing cancer risk and progression. These investigations addressed both the overall incidence of malignancies and specific cancer types, yielding heterogeneous findings. A particular focus was placed on metformin's association with breast cancer outcomes. While certain studies suggested a potential reduction in cancer incidence and mortality relative to other anti-diabetic therapies, others found no statistically significant association between metformin use and cancer prognosis [274]. Importantly, studies indicating a favorable effect of metformin often emphasized its time- and dose-dependent influence. Moreover, the post-diagnosis administration of metformin was correlated with lower mortality rates across various cancers, including breast, lung, and endometrial malignancies. Interestingly, long-term metformin administration, specifically over a period exceeding five years, was linked to a diminished risk of brain tumors [275,276].

Metformin's potential therapeutic benefits extended to gastrointestinal malignancies, with evidence suggesting a positive effect on HCC, pancreatic cancer, and CRC. For example, a recent study by Tarhini et al. reported enhanced OS and disease-free survival in patients with T2DM and CRC who received metformin therapy. However, it is crucial to consider that patients in the metformin cohort were generally younger and exhibited higher body mass indices compared to non-users, which may have confounded the outcomes [277]. Conversely, data from the Korean National Health Insurance Service-National Health Screening Cohort indicated an increased risk of pancreatic cancer among diabetic women treated with metformin, while no significant risk difference was observed between metformin users and non-users among diabetic men [278].

The impact of metformin on urogenital cancers remains ambiguous. Some evidence suggests a potential reduction in the incidence of kidney cancer with metformin use, though its effects on prostate cancer are more nuanced [279]. While metformin did not appear to elevate the overall risk of prostate cancer, it was associated with a 14% increase in the incidence of low-grade prostate cancer and a 25% reduction in the risk of high-grade disease [280]. Similarly, research on bladder, ovarian, and endometrial cancers yielded divergent results, with studies reporting positive, neutral, or even contradictory outcomes [281–285].

Meta-analyses provided further insights into the relationship between metformin and cancer, though inconsistencies persist. A meta-analysis conducted by O'Connor et al., which synthesized data from 166 studies on cancer incidence, suggested that metformin use was associated with a reduced overall cancer risk, particularly in gastrointestinal, urinary tract, and hematologic malignancies. Additionally, some studies supported a positive association between metformin use and a reduced risk of multiple myeloma in DM individuals [286,287]. However, a more recent meta-analysis by Mesquita et al. contradicted these findings, reporting no significant anti-cancer effects of metformin in T2DM patients, regardless of body weight or prediabetic status [288]. Although certain meta-analyses suggested metformin may not be effective in preventing certain cancers, such as endometrial cancer, the drug may confer secondary benefits, including a reduced mortality risk and prolonged progression-free survival [289].

5.2. Insulin Therapy and Sulfonylureas

The association among insulin, oral insulin secretagogues, and cancer risk was as a significant focus of research, particularly in the preceding years. As previously discussed, hyperinsulinemia may act as a facilitator in cancer development, prompting numerous clinical studies to investigate whether therapies that elevate insulin levels contribute to tumorigenesis in DM patients. A pivotal study by Vicentini et al. identified a 20% increase in cancer incidence, particularly among T2DM subjects undergoing insulin therapy, when compared to non-DM individuals. This increased incidence was especially pronounced for liver, pancreatic, and bladder cancers. Notably, T1DM patients also exhibited a heightened risk of bladder cancer, suggesting that the oncogenic potential of insulin therapy may extend beyond T2DM populations [290].

Further research examined the cancer risks associated with different insulin secretagogues. Data from Taiwan's National Health Insurance database, which encompassed over 8000 cancer cases in T2DM patients, demonstrated that insulin use was linked to the highest overall cancer risk, followed by glinides and SUs. Notably, while first- and second-generation SUs were associated with a moderate increase in cancer risk, third-generation agents such as glimepiride did not show a significant association with cancer. The observed variability in risk profiles underscores the differential oncogenic potential of specific agents, with liver cancer being particularly associated with SU use [291].

Comparative analyses between insulin secretagogues and metformin consistently revealed a more favorable cancer risk profile for the latter. A population-based cohort study tracking over 10,000 patients reported higher cancer-related mortality among SU and insulin users compared to those receiving metformin. During a 5.4-year follow-up period, cancer-related mortality was 4.9% among SU users, 5.8% among insulin users, and 3.5% among metformin users, suggesting that metformin may offer protective effects against cancer, while insulin-promoting therapies may elevate oncogenic risk. However, the precise mechanisms underlying these observations remain unclear, leaving it uncertain whether the adverse outcomes were primarily attributable to insulin-promoting therapies or whether metformin's protective properties were responsible [292].

Despite growing evidence, the relationship between SUs and cancer risk remains inconclusive. For instance, a systematic review by Chen et al., encompassing data from 77 clinical studies, did not confirm a definitive association between SU use and increased cancer risk. Moreover, randomized controlled trials (RCTs) did not consistently demonstrate significant oncogenic risks linked to SUs, and in some instances, these agents were even associated with protective effects against certain cancers, such as prostate cancer, particularly in low-risk populations. The inconsistency in findings may be due to variations in study design, patient demographics, and the pharmacological characteristics of different SUs [293]. For example, gliclazide, which stimulates insulin release in response to meals, may result in lower overall insulin exposure compared to glibenclamide, which maintains prolonged systemic insulin levels. Since hyperinsulinemia is a key driver of tumor growth, agents like gliclazide that limit sustained insulin exposure may carry a lower oncogenic risk than those that promote persistent hyperinsulinemia [56,294].

5.3. *Pioglitazone*

Numerous clinical studies and meta-analyses, including real-world data, implicated pioglitazone in a possible increased risk of bladder cancer, with this risk appearing to be dose- and time-dependent. Variability across geographical regions and potential influence from study funding sources further complicated these findings [295–297]. Early concerns were raised by the PROactive trial, a pivotal study on pioglitazone, which reported a higher incidence of bladder cancer in the pioglitazone group compared to the placebo group, although no significant difference was observed in the overall incidence of malignancies [298]. Importantly, this increased risk seemed to be drug specific, as rosiglitazone, another thiazolidinedione, was not significantly associated with bladder cancer [299].

However, emerging studies cast doubt on the validity of this association, with some reporting no significant link between pioglitazone use and bladder carcinogenesis, even in ethnically diverse populations [300,301]. For instance, a recent case-control study involving 6440 Asian-Indian patients with T2DM found no increased incidence of bladder cancer among pioglitazone users. Despite these findings, the study's authors recommended caution, particularly in patients with a history of hematuria, and identified age over 58 as a significant risk factor for bladder cancer [300]. In light of the ongoing debate, the U.S. Food and Drug Administration (FDA) continues to advise against pioglitazone use in individuals with a history of hematuria or bladder cancer. Conflicting evidence also exists regarding pioglitazone's association with other urinary tract cancers, such as prostate cancer, where some studies indicated a possible increased risk, while renal cancer showed neutral findings [302–304].

Pioglitazone's role in other malignancies was also investigated, though the results remain inconclusive. Research on breast and lung cancers produced mixed findings, with some studies suggesting potential therapeutic benefits, while others report no significant association [302,305]. Pioglitazone was evaluated for its potential role in lung cancer chemoprevention, especially among high-risk current and former smokers. While a trend towards decreased Ki-67 labeling in former smokers with baseline dysplasia treated with pioglitazone was observed, no overall improvement in endobronchial histology was detected, though histological improvements were noted in certain lesions [306].

Moreover, pioglitazone may have a protective effect against certain gastrointestinal cancers, particularly HCC, likely due to its positive impact on liver steatosis and fibrosis [307]. Some evidence suggests that prolonged pioglitazone use may be inversely correlated with HCC risk, especially in T2DM subjects with predisposing risk factors [308]. Meta-analyses similarly point to a potential protective role for pioglitazone in CRC [309]. While PPAR-γ agonism may reduce the risk of some cancers, such as breast and prostate cancers, its relationship with pancreatic cancer remains unclear. Although some studies suggested a heightened risk, a recent meta-analysis found no significant association. By contrast, insulin and SUs demonstrated a stronger association with increased pancreatic cancer risk in DM patients, with variations noted across ethnic populations [310,311].

5.4. *Incretin Mimetics*

5.4.1. DDP-4 Inhibitors

DPP-4 inhibitors were a subject of considerable debate, particularly regarding their potential association with thyroid and pancreatic cancers. While preclinical studies raised concerns about a possible link between DPP-4 inhibition and increased cancer risk, clinical meta-analyses did not substantiate these findings. Specifically, no statistically significant increase in thyroid cancer incidence was observed among patients treated with DPP-4 inhibitors, despite experimental data suggesting a potential risk [312]. Similarly, the association between DPP-4 inhibitors and pancreatic cancer remains inconclusive. Although concerns were raised, clinical studies largely failed to demonstrate a clear, consistent elevation in pancreatic cancer incidence among patients using these drugs [313,314].

On a more positive note, some clinical data indicated the potential benefits of DPP-4 silencing for certain cancer types, particularly CRC cancer. Diabetic patients treated with DPP-4 inhibitors, such as sitagliptin, showed improved disease-free survival rates and lower recurrence rates for CRC compared to those treated with other anti-diabetic agents, including metformin [315]. In addition to CRC, evidence points to a potential protective effect of DPP-4 inhibitors against melanoma. Compared to SUs, DPP-4 inhibitors were associated with a 23% reduction in melanoma risk, although no significant difference was observed regarding non-melanoma skin cancers [316].

However, not all studies support favorable outcomes across cancer types. For instance, investigations into breast cancer did not demonstrate significant survival advantages in patients treated with DPP-4 inhibitors [317]. Similarly, a retrospective analysis of metastatic RCC found no substantial clinical benefit from DPP-4 inhibition, further highlighting the variability in cancer-related outcomes associated with these drugs [318].

5.4.2. GLP-1 RAs

Through the last years, concerns emerged regarding the potential link between GLP-1RAs and the development of thyroid cancer, particularly MTC [319]. Despite these worries, several RCTs failed to demonstrate significant elevations in calcitonin levels or an increased risk of MTC among patients treated with GLP-1RAs [320,321]. However, retrospective observational studies and pharmacovigilance reports suggested a possible association between GLP-1RA use and thyroid malignancies, which complicated the interpretation of clinical safety data [322,323]. As previously mentioned, preclinical studies showed that GLP-1 receptors were expressed in thyroid C-cells and that GLP-1 receptor agonism can promote C-cell hyperplasia and tumorigenesis in rodent models. Importantly, these effects appeared to be less pronounced in humans [98–100]. A recent meta-analysis encompassing 64 clinical trials, 26 of which included thyroid cancer cases, identified a moderate relative risk increase for thyroid cancer associated with GLP-1RA treatment. Nevertheless, the absolute risk increment was small, and no statistically significant correlation was found for specific thyroid cancers, including PTC and MTC [324]. Other meta-analyses reported null associations between GLP-1RA therapy and overall thyroid cancer risk, further complicating the interpretation of these findings. This inconsistency suggests that while the potential risk may exist, it is not yet definitively established [325].

Fears regarding the potential link between GLP-1RAs and pancreatic cancer have also garnered attention, primarily due to the known association between GLP-1RAs and pancreatitis, which is a risk factor for pancreatic malignancy [326]. However, accumulating clinical evidence has largely refuted a causal link between GLP-1RAs and pancreatic cancer [327,328], with some studies even suggesting a potential protective effect [329]. A notable large-scale population-based cohort study with a seven-year follow-up demonstrated no increased incidence of pancreatic cancer among GLP-1RA users. However, the study authors recommended continued surveillance beyond the follow-up period due to the long latency period of pancreatic cancer [330].

Beyond thyroid and pancreatic cancers, research examined the broader impact of GLP-1RAs on various other malignancies. Emerging evidence suggests that GLP-1RAs may confer protective effects against certain cancers when compared to other anti-diabetic therapies. For instance, GLP-1RAs were associated with a reduced risk of HCC and hepatic decompensation in adults with T2DM [331]. Similarly, a nationwide Danish cohort study reported a lower incidence of prostate cancer in individuals using GLP-1RAs compared to those on basal insulin, particularly in older adults and those with preexisting CV disease [332]. Further supporting these potential anti-cancer benefits, a large, real-world database analysis of 112,000 patients treated with GLP-1RAs found reductions in the risks of lung, colon, bladder, and prostate cancers [333]. A comprehensive retrospective cohort study utilizing a nationwide electronic health records database of over 1.6 million diabetic patients revealed that GLP-1RAs were associated with significantly reduced risks for 10 out of 13 organ-associated cancers, when compared to insulin. Notable reductions were observed in gallbladder cancer, meningioma, and pancreatic cancer. However, the study did not find significant associations between GLP-1RAs and breast or thyroid cancers, while an increased risk of kidney cancer was observed when compared to metformin [334].

5.4.3. Tirzepatide

Current evidence on the role of tirzepatide in cancer development remains limited and inconclusive. A recent meta-analysis of phases 2 and 3 randomized RCTs investigating tirzepatide in T2DM patients found no significant increase in cancer risk associated with its use. This meta-analysis, which included nine RCTs, did not identify a statistically significant elevation in the incidence of any type of cancer, either as a primary or secondary outcome. However, it is important to consider the study's limitations, including small sample sizes, relatively short durations (ranging from 36 to 72 weeks), and the limited number of reported cancer events. These constraints necessitate cautious interpretation of the findings, as the relatively short follow-up periods may not capture the full latency period for many cancers [335]. Moreover, data from the U.S. Food and Drug Administration Adverse Event Reporting System (FAERS) sparked issues about a potential association between tirzepatide and an increased reporting of MTC. Specifically, there was a 13.67-fold higher likelihood of reported MTC cases in patients treated with tirzepatide compared to other medications. While this signal warrants attention, it is important to note that adverse event reporting databases are subject to various biases and confounders and do not establish causality. The overall safety profile of tirzepatide appears to be similar to that of other GLP-1RAs, indicating that, apart from the potential increased risk of MTC, tirzepatide's cancer risk seems aligned with established therapies in its class [336].

5.5. SGLT-2 Inhibitors

In contrast to the well-documented cardioprotective and nephroprotective effects of the SGLT-2 inhibitors, worries emerged about potential cancer risks, specifically regarding breast and bladder cancers. Early preclinical studies involving dapagliflozin raised fears due to observed, though statistically non-significant, increases in breast cancer incidence in females and bladder cancer incidence in males. These considerations were significant enough to delay dapagliflozin's approval by the U.S. FDA in 2012, underscoring the need for further investigation into whether these risks translate to clinical outcomes [337]. Recent evidence generally alleviated questions regarding breast cancer risk. A large-scale study following 2154 breast cancer cases for an average of 2.2 years found no significant association between SGLT-2 inhibitors and breast cancer risk, irrespective of patient age or medication duration. The study included comparisons to DPP-4 inhibitors, with similar neutral findings regarding breast cancer [338]. Furthermore, a population-based cohort analysis of 60,112 T2DM patients revealed a significant reduction in breast cancer risk associated with dapagliflozin, suggesting that early issues may have been due to detection bias rather than an actual increase in incidence [339]. Despite this growing body of evidence suggesting a neutral or even protective effect, long-term data remain crucial to fully confirm the absence of risk [337,340].

Bladder cancer, particularly in relation to empagliflozin, is also a subject of ongoing debate. The EMPA-REG OUTCOME (Empagliflozin Cardiovascular Outcome Event Trial in Type 2 Diabetes Mellitus Patients–Removing Excess Glucose) clinical trial reported a small number of bladder cancer cases in empagliflozin users, with none reported in the placebo or comparator groups [341,342]. The low absolute number of events complicates interpretation, as bladder cancer typically progresses subtly and over an extended period, often spanning many years. Moreover, the increased glucose excretion caused by SGLT-2 inhibitors could potentially result in more frequent urinary tract infections and genital issues, leading to earlier cancer detection [340]. A meta-analysis encompassing 76 trials with over 116,000 participants similarly concluded that SGLT-2 inhibition is unlikely to significantly impact the risk of breast or bladder cancers, adding to the growing consensus that initial safety concerns may have been overstated [343].

Beyond breast and bladder cancers, emerging evidence points to potential protective effects of SGLT-2 inhibitors against other malignancies. A Mendelian randomization study suggested a lower risk of prostate cancer associated with SGLT-2 inhibition, potentially related to reductions in prostate-specific antigen (PSA) levels and uridine metabolism [344]. Similarly, canagliflozin demonstrated a protective effect against gastrointestinal cancers, although these findings require further validation in larger clinical trials [345]. Interestingly, a recent meta-analysis of 58 studies involving over 113,000 participants found no significant increase in overall cancer risk with these agents, though specific drug differences were noted. Ertugliflozin was associated with a higher overall incidence of cancer, while dapagliflozin demonstrated a 47% reduction in bladder cancer risk and a 26% reduction in respiratory cancer risk. However, dapagliflozin and ertugliflozin were linked to a potentially elevated risk of kidney cancer compared to placebo [346].

In conclusion, the impact of anti-diabetic drugs on cancer risk and outcomes presents a complex landscape requiring careful consideration. Metformin showed promise in lowering cancer risk and enhancing survival across various cancers. Despite these encouraging data, it has yet to receive formal approval for this purpose. This lack of endorsement largely stems from inconsistent study findings, which may be influenced by factors such as dosage, treatment duration, and individual patient characteristics. Interpreting these results warrants thoughtful analysis, as observational studies and meta-analyses often display high variability and are susceptible to biases, including immortal time bias, which can distort outcomes. Notably, a large phase 3 randomized trial involving 3649 women treated over five years found no significant benefit from metformin as an adjunct therapy for breast cancer in terms of disease-free or overall survival [347].

Evidence suggests hyperinsulinemia as a potential risk factor for cancer development, yet findings remain conflicting. This uncertainty may stem from the distinction between the controlled therapeutic use of exogenous insulin and the endogenous overproduction observed in insulin resistance. Exogenous insulin therapy aims to maintain euglycemia, thereby potentially mitigating the harmful effects of hyperglycemia, such as inflammation, immune dysfunction, and tumor promotion. Additionally, treatments like oral insulin secretagogues (e.g., SUs) showed favorable impacts on certain cancers, highlighting the complex and context-dependent role of insulin in cancer dynamics. Furthermore, the association between pioglitazone and cancer remains contentious, with bladder cancer being the most reported concern. Recent studies raised questions about this association, yet caution is still advised, especially for patients with a history of hematuria or bladder cancer.

Incretin-based therapies present a dynamic area of research, showing both therapeutic benefits and ongoing safety concerns. Although large-scale studies like LEADER (Liraglutide Effect and Action in Diabetes) have alleviated initial worries regarding the risk of thyroid malignancies, concerns remain, particularly for individuals with a family history of MEN2 [348]. Newer agents, such as tirzepatide, appear to pose no overall increased cancer risk, yet continued monitoring is vital, particularly regarding associations with MTC. Finally, early concerns regarding SGLT-2 inhibitors and increased risks of breast and bladder cancers have largely been dispelled. On the other hand, drug-specific effects, notably dapagliflozin's protective benefits against bladder cancer and ertugliflozin's potential link to renal cancer, highlight the importance of individualized risk assessments.

6. Conclusions

In recent years, anti-diabetic medications have garnered significant attention for their potential impact on cancer biology and therapeutic outcomes. Among these treatments, metformin stands out as the most extensively studied, with substantial evidence supporting its role in inhibiting tumor cell viability across various cancer types. In contrast, the efficacy of other glucose-lowering conventional therapies, such as insulin, oral insulin secretagogues, pioglitazone, and DPP-4 inhibitors, remains controversial or, in some cases, unfavorable. Furthermore, early findings suggest that novel therapies, particularly SGLT-2 inhibitors, may offer notable anti-cancer benefits, likely due to their broad pleiotropic effects. Despite these promising developments, conclusive evidence regarding the role of anti-diabetic therapies in cancer prevention and treatment remains elusive. To date, no anti-diabetic drug has been formally approved for anti-cancer use. Furthermore, even in cases like pioglitazone and GLP-1 analogues, where their use is discouraged in specific malignancies, some studies present conflicting results. The divergent findings observed were likely influenced by discrepancies between animal models and human trials, as well as variations in clinical trial designs, dosing regimens, and patient demographics. These limitations highlight the need for caution but do not diminish the potential significance of anti-diabetic treatments in tumorigenesis. To move this field forward, future studies must address these challenges through robust methodologies, large and diverse participant cohorts, and extended follow-up periods. Such efforts are essential to fully understand the oncological implications of anti-diabetic therapies, ensuring that their integration into cancer care is both evidence based and tailored to patient safety.

Author Contributions: D.K. and N.G.V. conceived the idea of the review, organized its plan, and reviewed the study; D.K. wrote major parts of the manuscript; I.K., E.R. and M.K. wrote minor parts of the manuscript; V.D. and P.K. performed the literature search and prepared all references; D.K. made the figures and prepared the table; E.D., N.T. and M.D. reviewed, edited, and supervised the manuscript. All authors have read and agreed to the published version of the manuscript.

Funding: This research received no external funding.

Institutional Review Board Statement: Not applicable.

Informed Consent Statement: Not applicable.

Data Availability Statement: Not applicable.

Conflicts of Interest: The authors declare no conflicts of interest.

Abbreviations

4E-BP1: Eukaryotic translation initiation factor 4E-binding protein 1; 5-FU: 5-fluorouracil; ACAT1: Acetyl-CoA acetyltransferase 1; ADORA3: Adenosine A3 receptor; AKI: Acute kidney injury; Akt: Protein kinase B; AMPK: AMP-activated protein kinase; ATP: Adenosine triphosphate; Bcl-2: B cell lymphoma 2; BMI: Body mass index; BUBR1: BUB1 mitotic checkpoint kinase 1; CAFs: Cancer-associated fibroblasts; cAMP: Cyclic adenosine monophosphate; CANA: Canagliflozin; CBP: CREB-binding protein; CCL2: C-C motif chemokine ligand 2; $CD4^+$ T cells: Cluster of differentiation 4 positive T cells; $CD8^+$ T cells: Cluster of differentiation 8 positive T cells; CKD: Chronic kidney disease; CML: Chronic myeloid leukemia; COX-2: Cyclooxygenase-2; CRC: Colorectal carcinoma; CV: Cardiovascular; CXCL: C-X-C motif chemokine ligand 12; CXCL8: C-X-C motif chemokine ligand 8; CXCL12: C-X-C motif chemokine ligand 12; CXCR: C-X-C chemokine receptor; CXCR4: C-X-C chemokine receptor type 4; DAPA: Dapagliflozin; DC: Dendritic cell; DLBCL: Diffuse large B cell lymphoma; DM: Diabetes mellitus; EMT: Epithelial-to-mesenchymal transition; ER: Endoplasmic reticulum; ER^+: Estrogen receptor-positive; ERK: Extracellular signal-regulated kinase; DOX: Doxorubicin; DPP-4: Dipeptidyl peptidase-4; DPP-4i: Dipeptidyl peptidase-4 inhibitor; dsDNA: Double-stranded DNA; eGFR: Estimated glomerular filtration rate; EMPA: Empagliflozin; EMPA-REG OUTCOME: The Empagliflozin Cardiovascular Outcome Event Trial in Type 2 Diabetes Mellitus Patients–Removing Excess Glucose; FAERS: U.S. Food and Drug Administration Adverse Event Reporting System; FDA: Food and Drug Administration; FOXO1: Forkhead box O1; FOXO3: Forkhead box O3; FXR: Farnesoid X receptor; GB: Glioblastoma; GIP: Glucose-dependent insulinotropic polypeptide; GLP-1 RAs: Glucagon-like peptide-1 receptor agonists; GLP-1R: Glucagon-like peptide-1 receptor; GM-CSF: Granulocyte-macrophage colony-stimulating factor; GPx: Glutathione peroxidase; GSK3β: Glycogen synthase kinase 3 beta; HCC: Hepatocellular carcinoma; HDAC2: Histone deacetylase 2; HES1: Hairy and enhancer of split-1; HGF: Hepatocyte growth factor; HIF-1α: Hypoxia-inducible factor 1-alpha; HMGB1: High-mobility group box 1; HMGIY: High-mobility group protein I(Y); HO-1: Heme oxygenase-1; HSF1: Heat shock factor 1; ICAM-1: Intercellular adhesion molecule-1; ICI: Immune checkpoint inhibitor; IFN-β: Interferon-beta; IFN-γ: Interferon-gamma; IGF: Insulin-like growth factor; IGF-1: Insulin-like growth factor 1; IGF-1R: Insulin-like growth factor receptor; IGFBP7: Insulin-like growth factor-binding protein 7; IL: Interleukin; INHBA: inhibin subunit beta A; IR: Insulin receptor; IRF3: Interferon regulatory factor 3; IRS-1: Insulin receptor substrate-1; JAK: Janus kinase; JNK: c-Jun N-terminal kinase; Ki-67: Antigen Kiel 67; KLF4: Kruppel-like factor 4; Kv: Voltage-gated potassium; LDH: Lactate dehydrogenase; LEADER: Liraglutide Effect and Action in Diabetes trial; LINA: Linagliptin; LIRA: Liraglutide; LVEF: Left ventricular ejection fraction; MACE: Major adverse cardiovascular events; MAFLD: Metabolic dysfunction-associated fatty liver disease; MAGE-A3: Melanoma-associated antigen 3; MAPK: Mitogen-activated protein kinase; MCT1: Monocarboxylate transporter 1; MDA: Malondialdehyde; MDSCs: Myeloid-derived suppressor cells; MET: Mesenchymal-to-epithelial transition; miR: MicroRNA; mM: millimolar; MMP: Matrix met-alloproteinase; MTC: Medullary thyroid cancer; mRNA: Messenger RNA; MTF: Metformin; mTOR: Mammalian target of rapamycin; mTORC1: Mammalian target of rapamycin complex 1; MyD88: Myeloid differentiation factor 88; NER: Nucleotide excision repair; NETs: Neutrophil ex-tracellular traps; NF-κB: Nuclear factor kappa-light-chain-enhancer of activated B cells; NHEJ: Non-homologous end joining; NK: Natural killer; NLRP3: NLR family pyrin domain containing 3; Notch1: Neurogenic locus notch homolog protein 1; NOX4: NADPH oxidase 4; Nrf2: Nuclear factor erythroid 2-related factor 2; NSCLC: Non-small cell lung cancer; OPA-1: Protein optic atrophy-1; OS: Overall survival; OTUD5: OTU deubiquitinase 5; OXPHOS: Oxidative phosphorylation; p-AMPK: Phosphorylated AMPK;p38MAPK: p38 mitogen-activated protein kinase; p70 S6 kinase: 70 kDa ribosomal protein S6 kinase; PAI-1: Plasminogen activator inhibitor-1; PanIN: Pancreatic intraepithelial neoplasia; PARP: Cleaved-poly-ADP ribose polymerase; PaSC: Pancreatic stellate cell; PD-1: Programmed cell death protein-1; PDGF-AA: Platelet-derived growth factor-AA; PD-L1: Programmed death-ligand-1; PGC-1α: Peroxisome proliferator-activated receptor gamma coactivator-1 alpha; PGE2: Prostaglandin E2; PFS: Progression-free survival; PI3K: Phosphoinositide 3-kinase; PIO: Pioglitazone; PKC: Protein kinase C; PKM2: Pyruvate kinase M2; p-mTOR: Phosphorylated mechanistic target of rapamycin; PPAR-γ: Peroxisome proliferator-activated receptor gamma; PSA: Prostate-specific antigen; PTC: Papillary thyroid carcinoma; Raptor: Regulatory-associated protein of mTOR; Ras: Rat sarcoma; RCC: Renal cell carcinoma; REDD1: Regulated in development and DNA damage response 1; ROS: Reactive oxygen species; SAC: Spindle assembly checkpoint; SAXA: Saxagliptin; SCFAs: Short-chain fatty acids; SGLT-2: Sodium-glucose co-transporter-2; SIRT: sirtuin; SKOV-3: a human ovarian cancer cell line; STAT3: Signal transducer and activator of transcription 3; STING: Stimulator of interferon genes; SUs: Sulfonylureas; SUR1: Sulfonylurea receptor 1; T1DM: Type 1 diabetes mellitus; T2DM: Type 2 diabetes mellitus; TAMs: Tumor-associated macrophages; TET2: Ten-Eleven Translocation methyl-cytosine dioxygenase 2; TGF-β: Transforming growth factor-beta; TIL: Tumor-infiltrating lympho-cyte; TIMP: Tissue inhibitor of metalloproteinases; TLR4: Toll-like receptor 4; TME: Tumor microen-vironment; TNBC: Triple-negative breast cancer; TNF: Tumor necrosis factor; Tregs: Regulatory T cells; TSC2: Tuberous sclerosis complex 2; ULK1: Unc-51-like autophagy activating kinase 1; VCAM-1: Vascular cell adhesion molecule-1; VEGF: Vascular endothelial growth factor; VEGFA: Vascular endothelial growth factor A; VEGFR2: Vascular endothelial growth factor receptor 2; WAT: White adipose tissue; WHO: World Health Organization; Wnt: Wingless-type MMTV integration site family; YAP1: Yes-associated protein 1; ZFP36: Zinc finger protein 36.

References

1. Hossain, M.J.; Al-Mamun, M.; Islam, M.R. Diabetes mellitus, the fastest growing global public health concern: Early detection should be focused. *Health Sci. Rep.* **2024**, *7*, e2004. [CrossRef] [PubMed]
2. Zhu, B.; Qu, S. The Relationship Between Diabetes Mellitus and Cancers and Its Underlying Mechanisms. *Front. Endocrinol.* **2022**, *13*, 800995. [CrossRef] [PubMed]
3. Cignarelli, A.; Genchi, V.A.; Caruso, I.; Natalicchio, A.; Perrini, S.; Laviola, L.; Giorgino, F. Diabetes and cancer: Pathophysiological fundamentals of a 'dangerous affair'. *Diabetes Res. Clin. Pract.* **2018**, *143*, 378–388. [CrossRef] [PubMed]
4. Pliszka, M.; Szablewski, L. Associations between Diabetes Mellitus and Selected Cancers. *Int. J. Mol. Sci.* **2024**, *25*, 7476. [CrossRef]
5. González, P.; Lozano, P.; Ros, G.; Solano, F. Hyperglycemia and Oxidative Stress: An Integral, Updated and Critical Overview of Their Metabolic Interconnections. *Int. J. Mol. Sci.* **2023**, *24*, 9352. [CrossRef] [PubMed]
6. Dhas, Y.; Biswas, N.M.R.D.; Jones, L.D.; Ashili, S. Repurposing metabolic regulators: Antidiabetic drugs as anticancer agents. *Mol. Biomed.* **2024**, *5*, 40. [CrossRef]
7. Liu, Y.C.; Nguyen, P.A.; Humayun, A.; Chien, S.C.; Yang, H.C.; Asdary, R.N.; Syed-Abdul, S.; Hsu, M.H.; Moldovan, M.; Yen, Y.; et al. Does long-term use of antidiabetic drugs changes cancer risk? *Medicine* **2019**, *98*, e17461. [CrossRef]
8. Choi, Y.K.; Park, K.G. Metabolic roles of AMPK and metformin in cancer cells. *Mol. Cells* **2013**, *36*, 279–287. [CrossRef]
9. Home, P. Insulin therapy and cancer. *Diabetes Care* **2013**, *36* (Suppl. S2), S240–S244. [CrossRef]
10. Yousefnia, S.; Momenzadeh, S.; Forootan, F.S.; Ghaedi, K.; Esfahani, M.H.N. The influence of peroxisome proliferator-activated receptor γ (PPARγ) ligands on cancer cell tumorigenicity. *Gene* **2018**, *649*, 14–22. [CrossRef]
11. Nauck, M.A.; Meier, J.J. The incretin effect in healthy individuals and those with type 2 diabetes: Physiology, pathophysiology, and response to therapeutic interventions. *Lancet Diabetes Endocrinol.* **2016**, *4*, 525–536. [CrossRef] [PubMed]
12. Saini, K.; Sharma, S.; Khan, Y. DPP-4 inhibitors for treating T2DM—Hype or hope? an analysis based on the current literature. *Front. Mol. Biosci.* **2023**, *10*, 1130625. [CrossRef] [PubMed]
13. Andraos, J.; Muhar, H.; Smith, S.R. Beyond glycemia: Comparing tirzepatide to GLP-1 analogues. *Rev. Endocr. Metab. Disord.* **2023**, *24*, 1089–1101. [CrossRef] [PubMed]
14. Fareed, A.; Hussain, A. The Expanding Role of GLP-1: From Diabetes Management to Cancer Treatment. *Clin. Med. Insights Endocrinol. Diabetes* **2023**, *16*, 11795514231213566. [CrossRef]
15. Popoviciu, M.S.; Păduraru, L.; Yahya, G.; Metwally, K.; Cavalu, S. Emerging Role of GLP-1 Agonists in Obesity: A Comprehensive Review of Randomised Controlled Trials. *Int. J. Mol. Sci.* **2023**, *24*, 10449. [CrossRef]
16. Vallianou, N.G.; Tsilingiris, D.; Kounatidis, D.; Lempesis, I.G.; Karampela, I.; Dalamaga, M. Sodium-glucose cotransporter-2 inhibitors in obesity and associated cardiometabolic disorders: Where do we stand? *Pol. Arch. Intern. Med.* **2022**, *132*, 16342. [CrossRef]
17. Basak, D.; Gamez, D.; Deb, S. SGLT2 Inhibitors as Potential Anticancer Agents. *Biomedicines* **2023**, *11*, 1867. [CrossRef]
18. De Visser, K.E.; Joyce, J.A. The evolving tumor microenvironment: From cancer initiation to metastatic outgrowth. *Cancer Cell* **2023**, *41*, 374–403. [CrossRef]
19. Søndergaard, C.S.; Esquivel, P.N.; Dalamaga, M.; Magkos, F. Use of Antihyperglycemic Drugs and Risk of Cancer in Patients with Diabetes. *Curr. Oncol. Rep.* **2023**, *25*, 29–40. [CrossRef]
20. Zamanian, M.Y.; Golmohammadi, M.; Yumashev, A.; Hjazi, A.; Toama, M.A.; AbdRabou, M.A.; Gehlot, A.; Alwaily, E.R.; Shirsalimi, N.; Yadav, P.K.; et al. Effects of metformin on cancers in experimental and clinical studies: Focusing on autophagy and AMPK/mTOR signaling pathways. *Cell Biochem. Funct.* **2024**, *42*, e4071. [CrossRef]
21. Jung, C.H.; Seo, M.; Otto, N.M.; Kim, D.H. ULK1 inhibits the kinase activity of mTORC1 and cell proliferation. *Autophagy* **2011**, *7*, 1212–1221. [CrossRef] [PubMed]

22. Holczer, M.; Hajdú, B.; Lőrincz, T.; Szarka, A.; Bánhegyi, G.; Kapuy, O. A Double Negative Feedback Loop between mTORC1 and AMPK Kinases Guarantees Precise Autophagy Induction upon Cellular Stress. *Int. J. Mol. Sci.* **2019**, *20*, 5543. [CrossRef] [PubMed]
23. Chopra, I.; Li, H.F.; Wang, H.; Webster, K.A. Phosphorylation of the insulin receptor by AMP-activated protein kinase (AMPK) promotes ligand-independent activation of the insulin signalling pathway in rodent muscle. *Diabetologia* **2012**, *55*, 783–794. [CrossRef] [PubMed]
24. Shi, B.; Hu, X.; He, H.; Fang, W. Metformin suppresses breast cancer growth via inhibition of cyclooxygenase-2. *Oncol. Lett.* **2021**, *22*, 615. [CrossRef] [PubMed]
25. Hampsch, R.A.; Wells, J.D.; Traphagen, N.A.; McCleery, C.F.; Fields, J.L.; Shee, K.; Dillon, L.M.; Pooler, D.B.; Lewis, L.D.; Demidenko, E.; et al. AMPK Activation by Metformin Promotes Survival of Dormant ER+ Breast Cancer Cells. *Clin. Cancer Res.* **2020**, *26*, 3707–3719. [CrossRef]
26. Shao, S.; Zhao, L.; An, G.; Zhang, L.; Jing, X.; Luo, M.; Li, W.; Meng, D.; Ning, Q.; Zhao, X.; et al. Metformin suppresses HIF-1α expression in cancer-associated fibroblasts to prevent tumor-stromal cross talk in breast cancer. *FASEB J.* **2020**, *34*, 10860–10870. [CrossRef]
27. Repas, J.; Peternel, L.; Sourij, H.; Pavlin, M. Low glucose availability potentiates the effects of metformin on model T cell activation and exhaustion markers in vitro. *Front. Endocrinol.* **2023**, *14*, 1216193. [CrossRef]
28. Li, G.Y.; Feng, Y.Q.; Jia, Y.F.; Wang, K.F.; Li, Y.; Zhang, S.J.; Han, S.X.; Wang, J.C. Metformin enhances T lymphocyte anti-tumor immunity by increasing the infiltration via vessel normalization. *Eur. J. Pharmacol.* **2023**, *944*, 175592. [CrossRef]
29. Chiang, C.F.; Chao, T.T.; Su, Y.F.; Hsu, C.C.; Chien, C.Y.; Chiu, K.C.; Shiah, S.G.; Lee, C.H.; Liu, S.Y.; Shieh, Y.S. Metformin-treated cancer cells modulate macrophage polarization through AMPK-NF-κB signaling. *Oncotarget* **2017**, *8*, 20706–20718. [CrossRef]
30. Wu, Z.; Zhang, C.; Najafi, M. Targeting of the tumor immune microenvironment by metformin. *J. Cell Commun. Signal.* **2022**, *16*, 333–348. [CrossRef]
31. Cao, Y.; Wo, M.; Xu, C.; Fei, X.; Jin, J.; Shan, Z. An AMPK agonist suppresses the progress of colorectal cancer by regulating the polarization of TAM to M1 through inhibition of HIF-1α and mTOR signal pathway. *J. Cancer Res. Ther.* **2023**, *19*, 1560–1567. [CrossRef] [PubMed]
32. Zhang, J.; Kuang, L.; Li, Y.; Wang, Q.; Xu, H.; Liu, J.; Zhou, X.; Li, Y.; Zhang, B. Metformin Regulates TET2 Expression to Inhibit Endometrial Carcinoma Proliferation: A New Mechanism. *Front. Oncol.* **2022**, *12*, 856707. [CrossRef] [PubMed]
33. De la Cruz-López, K.G.; Alvarado-Ortiz, E.; Valencia-González, H.A.; Beltrán-Anaya, F.O.; Zamora-Fuentes, J.M.; Hidalgo-Miranda, A.; Ortiz-Sánchez, E.; Espinal-Enríquez, J.; García-Carrancá, A. Metformin induces ZFP36 by mTORC1 inhibition in cervical cancer-derived cell lines. *BMC Cancer* **2024**, *24*, 853. [CrossRef] [PubMed]
34. Gwak, H.; Kim, Y.; An, H.; Dhanasekaran, D.N.; Song, Y.S. Metformin induces degradation of cyclin D1 via AMPK/GSK3β axis in ovarian cancer. *Mol. Carcinog.* **2017**, *56*, 349–358. [CrossRef] [PubMed]
35. Tang, G.; Guo, J.; Zhu, Y.; Huang, Z.; Liu, T.; Cai, J.; Yu, L.; Wang, Z. Metformin inhibits ovarian cancer via decreasing H3K27 trimethylation. *Int. J. Oncol.* **2018**, *52*, 1899–1911. [CrossRef]
36. Hart, P.C.; Kenny, H.A.; Grassl, N.; Watters, K.M.; Litchfield, L.M.; Coscia, F.; Blaženović, I.; Ploetzky, L.; Fiehn, O.; Mann, M.; et al. Mesothelial Cell HIF1α Expression Is Metabolically Downregulated by Metformin to Prevent Oncogenic Tumor-Stromal Crosstalk. *Cell Rep.* **2019**, *29*, 4086–4098.e6. [CrossRef]
37. Bond, N.L.S.; Dréau, D.; Marriott, I.; Bennett, J.M.; Turner, M.J.; Arthur, S.T.; Marino, J.S. Low-Dose Metformin Treatment Reduces In Vitro Growth of the LL/2 Non-small Cell Lung Cancer Cell Line. *Biomedicines* **2022**, *11*, 65. [CrossRef]
38. Liu, C.; Zhou, X.; Ju, H.; Zhang, Y. Inhibition of pyruvate carboxylase reverses metformin resistance by activating AMPK in pancreatic cancer. *Life Sci.* **2023**, *327*, 121817. [CrossRef]
39. Penfold, L.; Woods, A.; Pollard, A.E.; Arizanova, J.; Pascual-Navarro, E.; Muckett, P.J.; Dore, M.H.; Montoya, A.; Whilding, C.; Fets, L.; et al. AMPK activation protects against prostate cancer by inducing a catabolic cellular state. *Cell Rep.* **2023**, *42*, 112396. [CrossRef]
40. Dixon, S.; Tran, A.; Schrier, M.S.; Dong, J.; Deth, R.C.; Castejon, A.; Trivedi, M.S. Metformin-induced oxidative stress inhibits LNCaP prostate cancer cell survival. *Mol. Biol. Rep.* **2024**, *51*, 729. [CrossRef]
41. Park, D.; Lee, S.; Boo, H. Metformin Induces Lipogenesis and Apoptosis in H4IIE Hepatocellular Carcinoma Cells. *Dev. Reprod.* **2023**, *27*, 77–89. [CrossRef] [PubMed]
42. Zhao, C.; Zheng, L.; Ma, Y.; Zhang, Y.; Yue, C.; Gu, F.; Niu, G.; Chen, Y. Low-dose metformin suppresses hepatocellular carcinoma metastasis via the AMPK/JNK/IL-8 pathway. *Int. J. Immunopathol. Pharmacol.* **2024**, *38*, 3946320241249445. [CrossRef] [PubMed]
43. de Oliveira, S.; Houseright, R.A.; Graves, A.L.; Golenberg, N.; Korte, B.G.; Miskolci, V.; Huttenlocher, A. Metformin modulates innate immune-mediated inflammation and early progression of NAFLD-associated hepatocellular carcinoma in zebrafish. *J. Hepatol.* **2019**, *70*, 710–721. [CrossRef] [PubMed]
44. Tang, D.; Xu, L.; Zhang, M.; Dorfman, R.G.; Pan, Y.; Zhou, Q.; Zhou, L.; Wang, Y.; Li, Y.; Yin, Y.; et al. Metformin facilitates BG45-induced apoptosis via an anti-Warburg effect in cholangiocarcinoma cells. *Oncol. Rep.* **2018**, *39*, 1957–1965. [CrossRef]
45. Van Eijck, C.W.F.; Vadgama, D.; van Eijck, C.H.J.; Wilmink, J.W.; Dutch Pancreatic Cancer Group (DPCG). Metformin boosts antitumor immunity and improves prognosis in upfront resected pancreatic cancer: An observational study. *J. Natl. Cancer Inst.* **2024**, *116*, 1374–1383. [CrossRef]

46. Orang, A.; Marri, S.; McKinnon, R.A.; Petersen, J.; Michael, M.Z. Restricting Colorectal Cancer Cell Metabolism with Metformin: An Integrated Transcriptomics Study. *Cancers* **2024**, *16*, 2055. [CrossRef]
47. Leitner, B.P.; Siebel, S.; Akingbesote, N.D.; Zhang, X.; Perry, R.J. Insulin and cancer: A tangled web. *Biochem. J.* **2022**, *479*, 583–607. [CrossRef]
48. Hua, H.; Kong, Q.; Yin, J.; Zhang, J.; Jiang, Y. Insulin-like growth factor receptor signaling in tumorigenesis and drug resistance a challenge for cancer therapy. *J. Hematol. Oncol.* **2020**, *13*, 64. [CrossRef]
49. Gong, Y.; Ma, Y.; Sinyuk, M.; Loganathan, S.; Thompson, R.C.; Sarkaria, J.N.; Chen, W.; Lathia, J.D.; Mobley, B.C.; Clark, S.W.; et al. Insulin-mediated signaling promotes proliferation and survival of glioblastoma through Akt activation. *Neuro. Oncol.* **2016**, *18*, 48–57. [CrossRef]
50. Sheng, X.; Yao, K.; Shao, A.; Tu, S.; Zhang, X.; Chen, T.; Yao, D. The Role of Insulin Glargine and Human Insulin in the Regulation of Thyroid Proliferation Through Mitogenic Signaling. *Front. Endocrinol.* **2019**, *10*, 594. [CrossRef]
51. Jiang, J.; Ren, H.Y.; Geng, G.J.; Mi, Y.J.; Liu, Y.; Li, N.; Yang, S.Y.; Shen, D.Y. Oncogenic activity of insulin in the development of non-small cell lung carcinoma. *Oncol. Lett.* **2018**, *15*, 447–452. [CrossRef] [PubMed]
52. Kurniawan, J.; Teressa, M. Insulin Use and The Risk of Hepatocellular Carcinoma: Insights and Implications. *Acta Med. Indones.* **2024**, *56*, 107–113.
53. Chen, X.; Liang, H.; Song, Q.; Xu, X.; Cao, D. Insulin promotes progression of colon cancer by upregulation of ACAT1. *Lipids Health Dis.* **2018**, *17*, 122. [CrossRef] [PubMed]
54. Zhang, A.M.Y.; Magrill, J.; de Winter, T.J.J.; Hu, X.; Skovsø, S.; Schaeffer, D.F.; Kopp, J.L.; Johnson, J.D. Endogenous Hyperinsulinemia Contributes to Pancreatic Cancer Development. *Cell Metab.* **2019**, *30*, 403–404. [CrossRef]
55. Zhang, A.M.Y.; Chu, K.H.; Daly, B.F.; Ruiter, T.; Dou, Y.; Yang, J.C.C.; de Winter, T.J.J.; Chhuor, J.; Wang, S.; Flibotte, S.; et al. Effects of hyperinsulinemia on pancreatic cancer development and the immune microenvironment revealed through single-cell transcriptomics. *Cancer Metab.* **2022**, *10*, 5. [CrossRef]
56. Vallianou, N.; Kounatidis, D.; Christodoulatos, G.S.; Panagopoulos, F.; Karampela, I.; Dalamaga, M. Mycobiome and Cancer: What Is the Evidence? *Cancers* **2021**, *13*, 3149. [CrossRef]
57. Zhang, A.M.Y.; Wellberg, E.A.; Kopp, J.L.; Johnson, J.D. Hyperinsulinemia in Obesity, Inflammation, and Cancer. *Diabetes Metab. J.* **2021**, *45*, 285–311. [CrossRef]
58. Wang, P.; Mak, V.C.; Cheung, L.W. Drugging IGF-1R in cancer: New insights and emerging opportunities. *Genes Dis.* **2022**, *10*, 199–211. [CrossRef] [PubMed]
59. Ferguson, R.D.; Novosyadlyy, R.; Fierz, Y.; Alikhani, N.; Sun, H.; Yakar, S.; Leroith, D. Hyperinsulinemia enhances c-Myc mediated mammary tumor development and advances metastatic progression to the lung in a mouse model of type 2 diabetes. *Breast Cancer Res.* **2012**, *14*, R8. [CrossRef]
60. Sciacca, L.; Vella, V.; Frittitta, L.; Tumminia, A.; Manzella, L.; Squatrito, S.; Belfiore, A.; Vigneri, R. Long-acting insulin analogs and cancer. *Nutr. Metab. Cardiovasc. Dis.* **2018**, *28*, 436–443. [CrossRef]
61. Núñez, M.; Medina, V.; Cricco, G.; Croci, M.; Cocca, C.; Rivera, E.; Bergoc, R.; Martín, G. Glibenclamide inhibits cell growth by inducing G0/G1 arrest in the human breast cancer cell line MDA-MB-231. *BMC Pharmacol. Toxicol.* **2013**, *14*, 6. [CrossRef] [PubMed]
62. Chen, H.; Zhao, L.; Meng, Y.; Qian, X.; Fan, Y.; Zhang, Q.; Wang, C.; Lin, F.; Chen, B.; Xu, L.; et al. Sulfonylurea receptor 1-expressing cancer cells induce cancer-associated fibroblasts to promote non-small cell lung cancer progression. *Cancer Lett.* **2022**, *536*, 215611. [CrossRef]
63. Zhou, Q.; Kwan, H.Y.; Chan, H.C.; Jiang, J.L.; Tam, S.C.; Yao, X. Blockage of voltage-gated K+ channels inhibits adhesion and proliferation of hepatocarcinoma cells. *Int. J. Mol. Med.* **2003**, *11*, 261–266. [CrossRef]
64. Hendriks, A.M.; Schrijnders, D.; Kleefstra, N.; de Vries, E.G.E.; Bilo, H.J.G.; Jalving, M.; Landman, G.W.D. Sulfonylurea derivatives and cancer, friend or foe? *Eur. J. Pharmacol.* **2019**, *861*, 172598. [CrossRef]
65. Yasukagawa, T.; Niwa, Y.; Simizu, S.; Umezawa, K. Suppression of cellular invasion by glybenclamide through inhibited secretion of platelet-derived growth factor in ovarian clear cell carcinoma ES-2 cells. *FEBS Lett.* **2012**, *586*, 1504–1509. [CrossRef] [PubMed]
66. Nunes, F.B.; Alves-Filho, J.C.; Alves Bastos, C.M.; Tessele, P.M.; Caberlon, E.; Moreira, K.B.; Ferreira, T.M.; de Oliveira, J.R. Effect of the chlorpropamide and fructose-1,6-bisphosphate of soluble TNF receptor II levels. *Pharmacol. Res.* **2004**, *49*, 449–453. [CrossRef]
67. Mao, G.; Zheng, S.; Li, J.; Liu, X.; Zhou, Q.; Cao, J.; Zhang, Q.; Zheng, L.; Wang, L.; Qi, C. Glipizide Combined with ANP Suppresses Breast Cancer Growth and Metastasis by Inhibiting Angiogenesis through VEGF/VEGFR2 Signaling. *Anticancer Agents Med. Chem.* **2022**, *22*, 1735–1741. [CrossRef] [PubMed]
68. Kretschy, N.; Teichmann, M.; Kopf, S.; Atanasov, A.G.; Saiko, P.; Vonach, C.; Viola, K.; Giessrigl, B.; Huttary, N.; Raab, I.; et al. In vitro inhibition of breast cancer spheroid-induced lymphendothelial defects resembling intravasation into the lymphatic vasculature by acetohexamide, isoxsuprine, nifedipin and proadifen. *Br. J. Cancer* **2013**, *108*, 570–578. [CrossRef]
69. Atlı Şekeroğlu, Z.; Şekeroğlu, V.; Kontaş Yedier, S.; İlkun, E.; Liou, L.S. Increased DNA strand breaks and neoplastic transformation in human bladder cells treated with pioglitazone. *Environ. Mol. Mutagen.* **2021**, *62*, 143–154. [CrossRef]
70. Peng, T.; Wang, G.; Cheng, S.; Xiong, Y.; Cao, R.; Qian, K.; Ju, L.; Wang, X.; Xiao, Y. The role and function of PPARγ in bladder cancer. *J. Cancer* **2020**, *11*, 3965–3975. [CrossRef]

71. Yang, S.L.; Wang, J.J.; Chen, M.; Xu, L.; Li, N.; Luo, Y.L.; Bu, L.; Zhang, M.N.; Li, H.; Su, B.L. Pioglitazone Use and Risk of Bladder Cancer: An In Vitro Study. *Int. J. Med. Sci.* **2018**, *15*, 228–237. [CrossRef] [PubMed]
72. Shahid, M.; Kim, M.; Yeon, A.; Jin, P.; Kim, W.K.; You, S.; Kim, J. Pioglitazone Alters the Proteomes of Normal Bladder Epithelial Cells but Shows No Tumorigenic Effects. *Int. Neurourol. J.* **2020**, *24*, 29–40. [CrossRef] [PubMed]
73. Miyazawa, M.; Subbaramaiah, K.; Bhardwaj, P.; Zhou, X.K.; Wang, H.; Falcone, D.J.; Giri, D.D.; Dannenberg, A.J. Pioglitazone Inhibits Periprostatic White Adipose Tissue Inflammation in Obese Mice. *Cancer Prev. Res.* **2018**, *11*, 215–226. [CrossRef] [PubMed]
74. Jang, J.H.; Lee, T.J.; Sung, E.G.; Song, I.H.; Kim, J.Y. Pioglitazone mediates apoptosis in Caki cells via downregulating c-FLIP(L) expression and reducing Bcl-2 protein stability. *Oncol. Lett.* **2021**, *22*, 743. [CrossRef]
75. Jiao, X.X.; Lin, S.Y.; Lian, S.X.; Qiu, Y.R.; Li, Z.H.; Chen, Z.H.; Lu, W.Q.; Zhang, Y.; Deng, L.; Jiang, Y.; et al. The inhibition of the breast cancer by PPARγ agonist pioglitazone through JAK2/STAT3 pathway. *Neoplasma* **2020**, *67*, 834–842. [CrossRef]
76. Kiran, A.V.V.V.R.; Kumari, G.K.; Krishnamurthy, P.T. Preliminary evaluation of anticancer efficacy of pioglitazone combined with celecoxib for the treatment of non-small cell lung cancer. Invest. *New Drugs* **2022**, *40*, 1–9. [CrossRef]
77. Ciaramella, V.; Sasso, F.C.; Di Liello, R.; Corte, C.M.D.; Barra, G.; Viscardi, G.; Esposito, G.; Sparano, F.; Troiani, T.; Martinelli, E.; et al. Activity and molecular targets of pioglitazone via blockade of proliferation, invasiveness and bioenergetics in human NSCLC. *J. Exp. Clin. Cancer Res.* **2019**, *38*, 178. [CrossRef]
78. Li, H.; Sorenson, A.L.; Poczobutt, J.; Amin, J.; Joyal, T.; Sullivan, T.; Crossno, J.T., Jr.; Weiser-Evans, M.C.; Nemenoff, R.A. Activation of PPARγ in myeloid cells promotes lung cancer progression and metastasis. *PLoS ONE* **2011**, *6*, e28133. [CrossRef]
79. Li, S.; Ghoshal, S.; Sojoodi, M.; Arora, G.; Masia, R.; Erstad, D.J.; Lanuti, M.; Hoshida, Y.; Baumert, T.F.; Tanabe, K.K.; et al. Pioglitazone Reduces Hepatocellular Carcinoma Development in Two Rodent Models of Cirrhosis. *J. Gastrointest. Surg.* **2019**, *23*, 101–111. [CrossRef]
80. Ninomiya, I.; Yamazaki, K.; Oyama, K.; Hayashi, H.; Tajima, H.; Kitagawa, H.; Fushida, S.; Fujimura, T.; Ohta, T. Pioglitazone inhibits the proliferation and metastasis of human pancreatic cancer cells. *Oncol. Lett.* **2014**, *8*, 2709–2714. [CrossRef]
81. Huang, G.; Zhang, M.; Wang, M.; Xu, W.; Duan, X.; Han, X.; Ren, J. Pioglitazone, a peroxisome proliferator-activated receptor γ agonist, induces cell death and inhibits the proliferation of hypoxic HepG2 cells by promoting excessive production of reactive oxygen species. *Oncol. Lett.* **2024**, *27*, 160. [CrossRef] [PubMed]
82. Alamdar, N.; Farivar, S.; Baghaei, K.; Hamidieh, A.A.; Soltaninejad, H.; Aghdaei, H.A.; Zali, M.; Saltanatpour, Z. Effect of Pioglitazone and Cetuximab on Colon Cancer Stem-like Cell (CCSLCs) Properties. *Curr. Stem Cell Res. Ther.* **2024**, *19*, 1–8. [CrossRef] [PubMed]
83. Tokhanbigli, S.; Alavifard, H.; Asadzadeh Aghdaei, H.; Zali, M.R.; Baghaei, K. Combination of pioglitazone and dendritic cell to optimize efficacy of immune cell therapy in CT26 tumor models. *Bioimpacts* **2023**, *13*, 333–346. [CrossRef] [PubMed]
84. Esmaeili, S.; Yousefi, A.M.; Delshad, M.; Bashash, D. Synergistic effects of PI3K inhibition and pioglitazone against acute promyelocytic leukemia cells. *Mol. Genet. Genomic. Med.* **2023**, *11*, e2106. [CrossRef]
85. Prost, S.; Relouzat, F.; Spentchian, M.; Ouzegdouh, Y.; Saliba, J.; Massonnet, G.; Beressi, J.P.; Verhoeyen, E.; Raggueneau, V.; Maneglier, B.; et al. Erosion of the chronic myeloid leukaemia stem cell pool by PPARγ agonists. *Nature* **2015**, *525*, 380–383. [CrossRef]
86. Busek, P.; Duke-Cohan, J.S.; Sedo, A. Does DPP-IV Inhibition Offer New Avenues for Therapeutic Intervention in Malignant Disease? *Cancers* **2022**, *14*, 2072. [CrossRef]
87. Yang, F.; Takagaki, Y.; Yoshitomi, Y.; Ikeda, T.; Li, J.; Kitada, M.; Kumagai, A.; Kawakita, E.; Shi, S.; Kanasaki, K.; et al. Inhibition of Dipeptidyl Peptidase-4 Accelerates Epithelial-Mesenchymal Transition and Breast Cancer Metastasis via the CXCL12/CXCR4/mTOR Axis. *Cancer Res.* **2019**, *79*, 735–746. [CrossRef]
88. Li, R.; Zeng, X.; Yang, M.; Feng, J.; Xu, X.; Bao, L.; Ye, T.; Wang, X.; Xue, B.; Huang, Y. Antidiabetic DPP-4 Inhibitors Reprogram Tumor Microenvironment That Facilitates Murine Breast Cancer Metastasis Through Interaction With Cancer Cells via a ROS-NF-κB-NLRP3 Axis. *Front. Oncol.* **2021**, *11*, 728047. [CrossRef] [PubMed]
89. He, L.; Zhang, T.; Sun, W.; Qin, Y.; Wang, Z.; Dong, W.; Zhang, H. The DPP-IV inhibitor saxagliptin promotes the migration and invasion of papillary thyroid carcinoma cells via the NRF2/HO1 pathway. *Med. Oncol.* **2020**, *37*, 97. [CrossRef]
90. Li, C.J.; Sun, B.; Fang, Q.H.; Ding, M.; Xing, Y.Z.; Chen, L.M.; Yu, D.M. Saxagliptin Induces β-Cell Proliferation through Increasing Stromal Cell-Derived Factor-1α In Vivo and In Vitro. *Front. Endocrinol.* **2017**, *8*, 326. [CrossRef]
91. Kushiyama, S.; Yashiro, M.; Yamamoto, Y.; Sera, T.; Sugimoto, A.; Nishimura, S.; Togano, S.; Kuroda, K.; Okuno, T.; Miki, Y.; et al. Dipeptidyl Peptidase-4 from Cancer-associated Fibroblasts Stimulates the Proliferation of Scirrhous-type Gastric Cancer Cells. *Anticancer Res.* **2022**, *42*, 501–509. [CrossRef] [PubMed]
92. Yang, X.; Zhang, X.; Wu, R.; Huang, Q.; Jiang, Y.; Qin, J.; Yao, F.; Jin, G.; Zhang, Y. DPPIV promotes endometrial carcinoma cell proliferation, invasion and tumorigenesis. *Oncotarget* **2017**, *8*, 8679–8692. [CrossRef] [PubMed]
93. Yang, Y.; Li, J.; Lei, W.; Wang, H.; Ni, Y.; Liu, Y.; Yan, H.; Tian, Y.; Wang, Z.; Yang, Z.; et al. CXCL12-CXCR4/CXCR7 Axis in Cancer: From Mechanisms to Clinical Applications. *Int. J. Biol. Sci.* **2023**, *19*, 3341–3359. [CrossRef]
94. Varela-Calviño, R.; Rodríguez-Quiroga, M.; Dias Carvalho, P.; Martins, F.; Serra-Roma, A.; Vázquez-Iglesias, L.; Páez de la Cadena, M.; Velho, S.; Cordero, O.J. The mechanism of sitagliptin inhibition of colorectal cancer cell lines' metastatic functionalities. *IUBMB Life* **2021**, *73*, 761–773. [CrossRef]

95. Hollande, C.; Boussier, J.; Ziai, J.; Nozawa, T.; Bondet, V.; Phung, W.; Lu, B.; Duffy, D.; Paradis, V.; Mallet, V.; et al. Inhibition of the dipeptidyl peptidase DPP4 (CD26) reveals IL-33-dependent eosinophil-mediated control of tumor growth. *Nat. Immunol.* **2019**, *20*, 257–264. [CrossRef] [PubMed]
96. Qin, C.J.; Zhao, L.H.; Zhou, X.; Zhang, H.L.; Wen, W.; Tang, L.; Zeng, M.; Wang, M.D.; Fu, G.B.; Huang, S.; et al. Inhibition of dipeptidyl peptidase IV prevents high fat diet-induced liver cancer angiogenesis by downregulating chemokine ligand 2. *Cancer Lett.* **2018**, *420*, 26–37. [CrossRef]
97. Ng, I.I.; Zhang, J.; Tian, T.; Peng, Q.; Huang, Z.; Xiao, K.; Yao, X.; Ng, L.; Zeng, J.; Tang, H. Network-based screening identifies sitagliptin as an antitumor drug targeting dendritic cells. *J. Immunother. Cancer* **2024**, *12*, e008254. [CrossRef]
98. Bjerre Knudsen, L.; Madsen, L.W.; Andersen, S.; Almholt, K.; de Boer, A.S.; Drucker, D.J.; Gotfredsen, C.; Egerod, F.L.; Hegelund, A.C.; Jacobsen, H.; et al. Glucagon-like Peptide-1 receptor agonists activate rodent thyroid C-cells causing calcitonin release and C-cell proliferation. *Endocrinology* **2010**, *151*, 1473–1486. [CrossRef]
99. Van den Brink, W.; Emerenciana, A.; Bellanti, F.; Della Pasqua, O.; van der Laan, J.W. Prediction of thyroid C-cell carcinogenicity after chronic administration of GLP1-R agonists in rodents. *Toxicol. Appl. Pharmacol.* **2017**, *320*, 51–59. [CrossRef]
100. Rosol, T.J. On-target effects of GLP-1 receptor agonists on thyroid C-cells in rats and mice. *Toxicol. Pathol.* **2013**, *41*, 303–309. [CrossRef]
101. Zhang, X.; Zhang, L.; Wang, B.; Zhang, X.; Gu, L.; Guo, K.; Zhang, X.; Zhou, Z. GLP-1 receptor agonist liraglutide inhibits the proliferation and migration of thyroid cancer cells. *Cell. Mol. Biol.* **2023**, *69*, 221–225. [CrossRef] [PubMed]
102. Nauck, M.A.; Friedrich, N. Do GLP-1-based therapies increase cancer risk? *Diabetes Care* **2013**, *36* (Suppl. S2), S245–S252. [CrossRef] [PubMed]
103. Zhao, H.; Wei, R.; Wang, L.; Tian, Q.; Tao, M.; Ke, J.; Liu, Y.; Hou, W.; Zhang, L.; Yang, J.; et al. Activation of glucagon-like peptide-1 receptor inhibits growth and promotes apoptosis of human pancreatic cancer cells in a cAMP-dependent manner. *Am. J. Physiol. Endocrinol. Metab.* **2014**, *306*, E1431–E1441. [CrossRef] [PubMed]
104. Zhao, H.; Jiang, X.; Duan, L.; Yang, L.; Wang, W.; Ren, Z. Liraglutide suppresses the metastasis of PANC-1 co-cultured with pancreatic stellate cells through modulating intracellular calcium content. *Endocr. J.* **2019**, *66*, 1053–1062. [CrossRef] [PubMed]
105. Li, J.; Su, L.; Gong, Y.Y.; Ding, M.L.; Hong, S.B.; Yu, S.; Xiao, H.P. Downregulation of miR-139-5p contributes to the antiapoptotic effect of liraglutide on the diabetic rat pancreas and INS-1 cells by targeting IRS1. *PLoS ONE* **2017**, *12*, e0173576. [CrossRef]
106. Nie, Z.J.; Zhang, Y.G.; Chang, Y.H.; Li, Q.Y.; Zhang, Y.L. Exendin-4 inhibits glioma cell migration, invasion and epithelial-to-mesenchymal transition through GLP-1R/sirt3 pathway. *Biomed. Pharmacother.* **2018**, *106*, 1364–1369. [CrossRef]
107. Iwaya, C.; Nomiyama, T.; Komatsu, S.; Kawanami, T.; Tsutsumi, Y.; Hamaguchi, Y.; Horikawa, T.; Yoshinaga, Y.; Yamashita, S.; Tanaka, T.; et al. Exendin-4, a Glucagonlike Peptide-1 Receptor Agonist, Attenuates Breast Cancer Growth by Inhibiting NF-κB Activation. *Endocrinology* **2017**, *158*, 4218–4232. [CrossRef]
108. Alanteet, A.A.; Attia, H.A.; Shaheen, S.; Alfayez, M.; Alshanawani, B. Anti-Proliferative Activity of Glucagon-Like Peptide-1 Receptor Agonist on Obesity-Associated Breast Cancer: The Impact on Modulating Adipokines' Expression in Adipocytes and Cancer Cells. *Dose Response* **2021**, *19*, 1559325821995651. [CrossRef]
109. Liu, Z.Z.; Duan, X.X.; Yuan, M.C.; Yu, J.; Hu, X.; Han, X.; Lan, L.; Liu, B.W.; Wang, Y.; Qin, J.F. Glucagon-like peptide-1 receptor activation by liraglutide promotes breast cancer through NOX4/ROS/VEGF pathway. *Life Sci.* **2022**, *294*, 120370. [CrossRef]
110. Tatsch, J.M.; Furman, D.P.; Nobre, R.M.; Wurzer, K.M.; da Silva, L.C.; Picheth, G.F.; Ramos, E.A.; Acco, A.; Klassen, G. Dulaglutide as a demethylating agent to improve the outcome of breast cancer. *Epigenomics* **2023**, *15*, 1309–1322. [CrossRef]
111. Ligumsky, H.; Amir, S.; Arbel Rubinstein, T.; Guion, K.; Scherf, T.; Karasik, A.; Wolf, I.; Rubinek, T. Glucagon-like peptide-1 analogs activate AMP kinase leading to reversal of the Warburg metabolic switch in breast cancer cells. *Med. Oncol.* **2024**, *41*, 138. [CrossRef] [PubMed]
112. Zhu, X.X.; Feng, Z.H.; Liu, L.Z.; Zhang, Y. Liraglutide suppresses the proliferation of endometrial cancer cells through the adenosine 5′-monophosphate (AMP)-activated protein kinase signaling pathway. *Chin. Med. J.* **2021**, *134*, 576–578. [CrossRef] [PubMed]
113. Kong, W.; Deng, B.; Shen, X.; John, C.; Haag, J.; Sinha, N.; Lee, D.; Sun, W.; Chen, S.; Zhang, H.; et al. Tirzepatide as an innovative treatment strategy in a pre-clinical model of obesity-driven endometrial cancer. *Gynecol. Oncol.* **2024**, *191*, 116–123. [CrossRef]
114. Kosowska, A.; Gallego-Colon, E.; Garczorz, W.; Kłych-Ratuszny, A.; Aghdam, M.R.F.; Woz, M.; Witek, A.; Wróblewska-Czech, A.; Cygal, A.; Wojnar, J.; et al. Exenatide modulates tumor-endothelial cell interactions in human ovarian cancer cells. *Endocr. Connect.* **2017**, *6*, 856–865. [CrossRef]
115. Kojima, M.; Takahashi, H.; Kuwashiro, T.; Tanaka, K.; Mori, H.; Ozaki, I.; Kitajima, Y.; Matsuda, Y.; Ashida, K.; Eguchi, Y.; et al. Glucagon-Like Peptide-1 Receptor Agonist Prevented the Progression of Hepatocellular Carcinoma in a Mouse Model of Nonalcoholic Steatohepatitis. *Int. J. Mol. Sci.* **2020**, *21*, 5722. [CrossRef]
116. Hansen, H.H.; Pors, S.; Andersen, M.W.; Vyberg, M.; Nøhr-Meldgaard, J.; Nielsen, M.H.; Oró, D.; Madsen, M.R.; Lewinska, M.; Møllerhøj, M.B.; et al. Semaglutide reduces tumor burden in the GAN diet-induced obese and biopsy-confirmed mouse model of NASH-HCC with advanced fibrosis. *Sci. Rep.* **2023**, *13*, 23056. [CrossRef]
117. Lu, X.; Xu, C.; Dong, J.; Zuo, S.; Zhang, H.; Jiang, C.; Wu, J.; Wei, J. Liraglutide activates nature killer cell-mediated antitumor responses by inhibiting IL-6/STAT3 signaling in hepatocellular carcinoma. *Transl. Oncol.* **2021**, *14*, 100872. [CrossRef] [PubMed]
118. De Barra, C.; Khalil, M.; Mat, A.; O'Donnell, C.; Shaamile, F.; Brennan, K.; O'Shea, D.; Hogan, A.E. Glucagon-like peptide-1 therapy in people with obesity restores natural killer cell metabolism and effector function. *Obesity* **2023**, *31*, 1787–1797. [CrossRef]

119. Tong, G.; Peng, T.; Chen, Y.; Sha, L.; Dai, H.; Xiang, Y.; Zou, Z.; He, H.; Wang, S. Effects of GLP-1 Receptor Agonists on Biological Behavior of Colorectal Cancer Cells by Regulating PI3K/AKT/mTOR Signaling Pathway. *Front. Pharmacol.* **2022**, *13*, 901559. [CrossRef]
120. Stein, M.S.; Kalff, V.; Williams, S.G.; Murphy, D.G.; Colman, P.G.; Hofman, M.S. The GLP-1 receptor is expressed in vivo by human metastatic prostate cancer. *Endocr. Oncol.* **2024**, *4*, e230015. [CrossRef]
121. Zhu, C.; Lai, Y.; Liu, C.; Teng, L.; Zhu, Y.; Lin, X.; Fu, X.; Lai, Q.; Liu, S.; Zhou, X.; et al. Comprehensively prognostic and immunological analyses of GLP-1 signaling-related genes in pan-cancer and validation in colorectal cancer. *Front. Pharmacol.* **2024**, *15*, 1387243. [CrossRef] [PubMed]
122. Sun, M.; Sun, J.; Sun, W.; Li, X.; Wang, Z.; Sun, L.; Wang, Y. Unveiling the anticancer effects of SGLT-2i: Mechanisms and therapeutic potential. *Front. Pharmacol.* **2024**, *15*, 1369352. [CrossRef] [PubMed]
123. Shoda, K.; Tsuji, S.; Nakamura, S.; Egashira, Y.; Enomoto, Y.; Nakayama, N.; Shimazawa, M.; Iwama, T.; Hara, H. Canagliflozin Inhibits Glioblastoma Growth and Proliferation by Activating AMPK. *Cell. Mol. Neurobiol.* **2023**, *43*, 879–892. [CrossRef]
124. Coperchini, F.; Greco, A.; Croce, L.; Pignatti, P.; Muzza, M.; Petrosino, E.; Teliti, M.; Magri, F.; Rotondi, M. Canagliflozin reduces thyroid cancer cells migration in vitro by inhibiting CXCL8 and CCL2: An additional anti-tumor effect of the drug. *Biomed. Pharmacother.* **2024**, *170*, 115974. [CrossRef]
125. Alblowy, A.H.; Maan, N.; Ibrahim, A.A. Optimal control strategies for SGLT2 inhibitors as a novel anti-tumor agent and their effect on human breast cancer cells with the effect of time delay and hyperglycemia. *Comput. Biol. Med.* **2023**, *166*, 107552. [CrossRef]
126. Youssef, M.E.; Yahya, G.; Popoviciu, M.S.; Cavalu, S.; Abd-Eldayem, M.A.; Saber, S. Unlocking the Full Potential of SGLT2 Inhibitors: Expanding Applications beyond Glycemic Control. *Int. J. Mol. Sci.* **2023**, *24*, 6039. [CrossRef]
127. Nalla, L.V.; Khairnar, A. Empagliflozin mediated miR-128-3p upregulation promotes differentiation of hypoxic cancer stem-like cells in breast cancer. *Eur. J. Pharmacol.* **2023**, *943*, 175565. [CrossRef] [PubMed]
128. Nalla, L.V.; Khairnar, A. Empagliflozin drives ferroptosis in anoikis-resistant cells by activating miR-128-3p dependent pathway and inhibiting CD98hc in breast cancer. *Free Radic. Biol. Med.* **2024**, *220*, 288–300. [CrossRef] [PubMed]
129. Komatsu, S.; Nomiyama, T.; Numata, T.; Kawanami, T.; Hamaguchi, Y.; Iwaya, C.; Horikawa, T.; Fujimura-Tanaka, Y.; Hamanoue, N.; Motonaga, R.; et al. SGLT2 inhibitor ipragliflozin attenuates breast cancer cell proliferation. *Endocr. J.* **2020**, *67*, 99–106. [CrossRef]
130. Shiba, K.; Tsuchiya, K.; Komiya, C.; Miyachi, Y.; Mori, K.; Shimazu, N.; Yamaguchi, S.; Ogasawara, N.; Katoh, M.; Itoh, M.; et al. Canagliflozin, an SGLT2 inhibitor, attenuates the development of hepatocellular carcinoma in a mouse model of human NASH. *Sci. Rep.* **2018**, *8*, 2362. [CrossRef]
131. Hung, M.H.; Chen, Y.L.; Chen, L.J.; Chu, P.Y.; Hsieh, F.S.; Tsai, M.H.; Shih, C.T.; Chao, T.I.; Huang, C.Y.; Chen, K.F. Canagliflozin inhibits growth of hepatocellular carcinoma via blocking glucose-influx-induced β-catenin activation. *Cell Death Dis.* **2019**, *10*, 420. [CrossRef] [PubMed]
132. Luo, J.; Sun, P.; Zhang, X.; Lin, G.; Xin, Q.; Niu, Y.; Chen, Y.; Xu, N.; Zhang, Y.; Xie, W. Canagliflozin Modulates Hypoxia-Induced Metastasis, Angiogenesis and Glycolysis by Decreasing HIF-1α Protein Synthesis via AKT/mTOR Pathway. *Int. J. Mol. Sci.* **2021**, *22*, 13336. [CrossRef] [PubMed]
133. Biziotis, O.D.; Tsakiridis, E.E.; Ali, A.; Ahmadi, E.; Wu, J.; Wang, S.; Mekhaeil, B.; Singh, K.; Menjolian, G.; Farrell, T.; et al. Canagliflozin mediates tumor suppression alone and in combination with radiotherapy in non-small cell lung cancer (NSCLC) through inhibition of HIF-1α. *Mol. Oncol.* **2023**, *17*, 2235–2256. [CrossRef] [PubMed]
134. Ren, K.; Wang, X.; Ma, R.; Chen, H.; Min, T.; Ma, Y.; Xie, X.; Wang, W.; Deng, X.; Zhou, Z.; et al. Dapagliflozin suppressed gastric cancer growth via regulating OTUD5 mediated YAP1 deubiquitination. *Eur. J. Pharmacol.* **2024**, *983*, 177002. [CrossRef]
135. Anastasio, C.; Donisi, I.; Del Vecchio, V.; Colloca, A.; Mele, L.; Sardu, C.; Marfella, R.; Balestrieri, M.L.; D'Onofrio, N. SGLT2 inhibitor promotes mitochondrial dysfunction and ER-phagy in colorectal cancer cells. *Cell Mol. Biol. Lett.* **2024**, *29*, 80. [CrossRef]
136. Xu, D.; Zhou, Y.; Xie, X.; He, L.; Ding, J.; Pang, S.; Shen, B.; Zhou, C. Inhibitory effects of canagliflozin on pancreatic cancer are mediated via the downregulation of glucose transporter-1 and lactate dehydrogenase A. *Int. J. Oncol.* **2020**, *57*, 1223–1233. [CrossRef]
137. Kuang, H.; Liao, L.; Chen, H.; Kang, Q.; Shu, X.; Wang, Y. Therapeutic Effect of Sodium Glucose Co-Transporter 2 Inhibitor Dapagliflozin on Renal Cell Carcinoma. *Med. Sci. Monit.* **2017**, *23*, 3737–3745. [CrossRef]
138. Villani, L.A.; Smith, B.K.; Marcinko, K.; Ford, R.J.; Broadfield, L.A.; Green, A.E.; Houde, V.P.; Muti, P.; Tsakiridis, T.; Steinberg, G.R. The diabetes medication Canagliflozin reduces cancer cell proliferation by inhibiting mitochondrial complex-I supported respiration. *Mol. Metab.* **2016**, *5*, 1048–1056. [CrossRef]
139. Billger, M.; Kirk, J.; Chang, J.; Bédard, A.; Attalla, B.; Haile, S.; Söderberg, M. A study in a rat initiation-promotion bladder tumour model demonstrated no promoter/progressor potential of dapagliflozin. *Regul. Toxicol. Pharmacol.* **2019**, *103*, 166–173. [CrossRef]
140. Wu, W.; Zhang, Z.; Jing, D.; Huang, X.; Ren, D.; Shao, Z.; Zhang, Z. SGLT2 inhibitor activates the STING/IRF3/IFN-β pathway and induces immune infiltration in osteosarcoma. *Cell Death Dis.* **2022**, *13*, 523. [CrossRef]
141. Sato, A.; Sunayama, J.; Okada, M.; Watanabe, E.; Seino, S.; Shibuya, K.; Suzuki, K.; Narita, Y.; Shibui, S.; Kayama, T.; et al. Glioma-initiating cell elimination by metformin activation of FOXO3 via AMPK. *Stem Cells Transl. Med.* **2012**, *1*, 811–824. [CrossRef] [PubMed]

142. Guo, L.; Sheng, H.; Liu, Q.; Yang, Q.; Zhu, S. Effect of glibenclamide on viability and acid-base equilibrium of glioblastoma cells. *Chin. J. Pathophysiol.* **2017**, *12*, 1405–1410.
143. Cilibrasi, C.; Butta, V.; Riva, G.; Bentivegna, A. Pioglitazone Effect on Glioma Stem Cell Lines: Really a Promising Drug Therapy for Glioblastoma? *PPAR Res.* **2016**, *2016*, 7175067. [CrossRef] [PubMed]
144. Tsuji, S.; Kudo, U.; Hatakeyama, R.; Shoda, K.; Nakamura, S.; Shimazawa, M. Linagliptin Decreased Tumor Progression in a Glioblastoma Model. *Biochem. Biophys. Res. Commun.* **2024**, *711*, 149897. [CrossRef] [PubMed]
145. Kheder, S.; Sisley, K.; Hadad, S.; Balasubramanian, S.P. Effects of prolonged exposure to low dose metformin in thyroid cancer cell lines. *J. Cancer* **2017**, *8*, 1053–1061. [CrossRef]
146. Dobson, M.E.; Diallo-Krou, E.; Grachtchouk, V.; Yu, J.; Colby, L.A.; Wilkinson, J.E.; Giordano, T.J.; Koenig, R.J. Pioglitazone induces a proadipogenic antitumor response in mice with PAX8-PPARgamma fusion protein thyroid carcinoma. *Endocrinology* **2011**, *152*, 4455–4465. [CrossRef]
147. Monteiro, M.; Zhang, X.; Yee, D. Insulin promotes growth in breast cancer cells through the type I IGF receptor in insulin receptor deficient cells. *Exp. Cell Res.* **2024**, *434*, 113862. [CrossRef]
148. Xu, K.; Sun, G.; Li, M.; Chen, H.; Zhang, Z.; Qian, X.; Li, P.; Xu, L.; Huang, W.; Wang, X. Glibenclamide Targets Sulfonylurea Receptor 1 to Inhibit p70S6K Activity and Upregulate KLF4 Expression to Suppress Non-Small Cell Lung Carcinoma. *Mol. Cancer Ther.* **2019**, *18*, 2085–2096. [CrossRef]
149. Jang, J.H.; Baerts, L.; Waumans, Y.; De Meester, I.; Yamada, Y.; Limani, P.; Gil-Bazo, I.; Weder, W.; Jungraithmayr, W. Suppression of lung metastases by the CD26/DPP4 inhibitor Vildagliptin in mice. *Clin. Exp. Metastasis* **2015**, *32*, 677–687. [CrossRef]
150. Pu, Z.; Yang, Y.; Qin, S.; Li, X.; Cui, C.; Chen, W. The Effect of Liraglutide on Lung Cancer and Its Potential Protective Effect on High Glucose-Induced Lung Senescence and Oxidative Damage. *Front. Biosci.* **2023**, *28*, 259. [CrossRef]
151. Ayoub, B.M.; Attia, Y.M.; Ahmed, M.S. Structural re-positioning, in silico molecular modelling, oxidative degradation, and biological screening of linagliptin as adenosine 3 receptor (ADORA3) modulators targeting hepatocellular carcinoma. *J. Enzyme Inhib. Med. Chem.* **2018**, *33*, 858–866. [CrossRef] [PubMed]
152. Szymczak-Pajor, I.; Drzewoski, J.; Świderska, E.; Strycharz, J.; Gabryanczyk, A.; Kasznicki, J.; Bogdańska, M.; Śliwińska, A. Metformin Induces Apoptosis in Human Pancreatic Cancer (PC) Cells Accompanied by Changes in the Levels of Histone Acetyltransferases (Particularly, p300/CBP-Associated Factor (PCAF) Protein Levels). *Pharmaceuticals* **2023**, *16*, 115. [CrossRef] [PubMed]
153. Yang, J.; Waldron, R.T.; Su, H.Y.; Moro, A.; Chang, H.H.; Eibl, G.; Ferreri, K.; Kandeel, F.R.; Lugea, A.; Li, L.; et al. Insulin promotes proliferation and fibrosing responses in activated pancreatic stellate cells. *Am. J. Physiol. Gastrointest. Liver Physiol.* **2016**, *311*, G675–G687. [CrossRef]
154. Sliwinska, A.; Sliwinski, T.; Kasznicki, J.; Drzewoski, J. Effect of gliclazide on nucleotide excision repair (NER) and non-homologous DNA end joining (NHEJ) in normal and cancer cells. *J. Physiol. Pharmacol.* **2010**, *61*, 347–353.
155. Xiao, Q.; Xiao, J.; Liu, J.; Liu, J.; Shu, G.; Yin, G. Metformin suppresses the growth of colorectal cancer by targeting INHBA to inhibit TGF-β/PI3K/AKT signaling transduction. *Cell Death Dis.* **2022**, *13*, 202. [CrossRef] [PubMed]
156. Hamza, S.; Garanina, E.E.; Shkair, L.; Alsaadi, M.; Khaiboullina, S.F.; Tezcan, G. Implications of NLRP3 Suppression Using Glibenclamide and miR-223 against Colorectal Cancer. *Pharmaceuticals* **2024**, *17*, 299. [CrossRef]
157. Shen, Z.; Xue, D.; Wang, K.; Zhang, F.; Shi, J.; Jia, B.; Yang, D.; Zhang, Q.; Zhang, S.; Jiang, H.; et al. Metformin exerts an antitumor effect by inhibiting bladder cancer cell migration and growth, and promoting apoptosis through the PI3K/AKT/mTOR pathway. *BMC Urol.* **2022**, *22*, 79. [CrossRef]
158. Liu, S.; Li, Y.; Lin, T.; Fan, X.; Liang, Y.; Heemann, U. High dose human insulin and insulin glargine promote T24 bladder cancer cell proliferation via PI3K-independent activation of Akt. *Diabetes Res. Clin. Pract.* **2011**, *91*, 177–182. [CrossRef]
159. Liang, P.I.; Yeh, B.W.; Li, W.M.; Chan, T.C.; Chang, I.W.; Huang, C.N.; Li, C.C.; Ke, H.L.; Yeh, H.C.; Wu, W.J.; et al. DPP4/CD26 overexpression in urothelial carcinoma confers an independent prognostic impact and correlates with intrinsic biological aggressiveness. *Oncotarget* **2017**, *8*, 2995–3008. [CrossRef]
160. Abdel-Hamid, A.A.M.; Firgany, A.E.L. Dysplastic urothelial changes accompany empagliflozin administration in urinary bladder of experimental diabetes. *Int. J. Exp. Pathol.* **2019**, *100*, 369–377. [CrossRef]
161. Qi, C.; Bin, L.; Yang, Y.; Yang, Y.; Li, J.; Zhou, Q.; Wen, Y.; Zeng, C.; Zheng, L.; Zhang, Q.; et al. Glipizide suppresses prostate cancer progression in the TRAMP model by inhibiting angiogenesis. *Sci. Rep.* **2016**, *6*, 27819. [CrossRef] [PubMed]
162. Nomiyama, T.; Kawanami, T.; Irie, S.; Hamaguchi, Y.; Terawaki, Y.; Murase, K.; Tsutsumi, Y.; Nagaishi, R.; Tanabe, M.; Morinaga, H.; et al. Exendin-4, a GLP-1 receptor agonist, attenuates prostate cancer growth. *Diabetes* **2014**, *63*, 3891–3905. [CrossRef] [PubMed]
163. Cao, Z.; Liu, L.Z.; Dixon, D.A.; Zheng, J.Z.; Chandran, B.; Jiang, B.H. Insulin-like growth factor-I induces cyclooxygenase-2 expression via PI3K, MAPK and PKC signaling pathways in human ovarian cancer cells. *Cell. Signal.* **2007**, *19*, 1542–1553. [CrossRef]
164. Al-Alem, L.; Southard, R.C.; Kilgore, M.W.; Curry, T.E. Specific thiazolidinediones inhibit ovarian cancer cell line proliferation and cause cell cycle arrest in a PPARγ independent manner. *PLoS ONE* **2011**, *6*, e16179. [CrossRef]
165. Kosowska, A.; Garczorz, W.; Kłych-Ratuszny, A.; Aghdam, M.R.F.; Kimsa-Furdzik, M.; Simka-Lampa, K.; Francuz, T. Sitagliptin Modulates the Response of Ovarian Cancer Cells to Chemotherapeutic Agents. *Int. J. Mol. Sci.* **2020**, *21*, 8976. [CrossRef] [PubMed]

166. Randolph, J.F., Jr.; Kipersztok, S.; Ayers, J.W.; Ansbacher, R.; Peegel, H.; Menon, K.M. The effect of insulin on aromatase activity in isolated human endometrial glands and stroma. *Am. J. Obstet. Gynecol.* **1987**, *157*, 1534–1539. [CrossRef]
167. Kumari, G.K.; Kiran, A.V.V.V.R.; Krishnamurthy, P.T. Preliminary evaluation on the beneficial effects of pioglitazone in the treatment of endometrial cancer. *Med. Oncol.* **2021**, *38*, 71. [CrossRef]
168. Chen, Y.H.; Wang, P.H.; Chen, P.N.; Yang, S.F.; Hsiao, Y.H. Molecular and Cellular Mechanisms of Metformin in Cervical Cancer. *Cancers* **2021**, *13*, 2545. [CrossRef] [PubMed]
169. Rossini, M.; Martini, F.; Torreggiani, E.; Fortini, F.; Aquila, G.; Sega, F.V.D.; Patergnani, S.; Pinton, P.; Maniscalco, P.; Cavallesco, G.; et al. Metformin Induces Apoptosis and Inhibits Notch1 in Malignant Pleural Mesothelioma Cells. *Front. Cell Dev. Biol.* **2021**, *8*, 534499. [CrossRef]
170. Wang, Q.; Lu, P.; Wang, T.; Zheng, Q.; Li, Y.; Leng, S.X.; Meng, X.; Wang, B.; Xie, J.; Zhang, H. Sitagliptin affects gastric cancer cells proliferation by suppressing Melanoma-associated antigen-A3 expression through Yes-associated protein inactivation. *Cancer Med.* **2020**, *9*, 3816–3828. [CrossRef]
171. Alzokaky, A.A.; Al-Karmalawy, A.A.; Saleh, M.A.; Abdo, W.; Farage, A.E.; Belal, A.; Abourehab, M.A.S.; Antar, S.A. Metformin ameliorates doxorubicin-induced cardiotoxicity targeting HMGB1/TLR4/NLRP3 signaling pathway in mice. *Life Sci.* **2023**, *316*, 121390. [CrossRef] [PubMed]
172. Chen, J.; Zhang, S.; Pan, G.; Lin, L.; Liu, D.; Liu, Z.; Mei, S.; Zhang, L.; Hu, Z.; Chen, J.; et al. Modulatory effect of metformin on cardiotoxicity induced by doxorubicin via the MAPK and AMPK pathways. *Life Sci.* **2020**, *249*, 117498. [CrossRef] [PubMed]
173. Kobashigawa, L.C.; Xu, Y.C.; Padbury, J.F.; Tseng, Y.T.; Yano, N. Metformin protects cardiomyocyte from doxorubicin induced cytotoxicity through an AMP-activated protein kinase dependent signaling pathway: An in vitro study. *PLoS ONE* **2014**, *9*, e104888. [CrossRef] [PubMed]
174. Asensio-López, M.C.; Lax, A.; Pascual-Figal, D.A.; Valdés, M.; Sánchez-Más, J. Metformin protects against doxorubicin-induced cardiotoxicity: Involvement of the adiponectin cardiac system. *Free Radic. Biol. Med.* **2011**, *51*, 1861–1871. [CrossRef]
175. Zilinyi, R.; Czompa, A.; Czegledi, A.; Gajtko, A.; Pituk, D.; Lekli, I.; Tosaki, A. The Cardioprotective Effect of Metformin in Doxorubicin-Induced Cardiotoxicity: The Role of Autophagy. *Molecules* **2018**, *23*, 1184. [CrossRef] [PubMed]
176. Van, J.; Hahn, Y.; Silverstein, B.; Li, C.; Cai, F.; Wei, J.; Katiki, L.; Mehta, P.; Livatova, K.; DelPozzo, J.; et al. Metformin Inhibits Autophagy, Mitophagy and Antagonizes Doxorubicin-Induced Cardiomyocyte Death. *Int. J. Drug Discov. Pharm.* **2023**, *2*, 37–51. [CrossRef]
177. Arinno, A.; Maneechote, C.; Khuanjing, T.; Prathumsap, N.; Chunchai, T.; Arunsak, B.; Nawara, W.; Kerdphoo, S.; Shinlapawittayatorn, K.; Chattipakorn, S.C.; et al. Melatonin and metformin ameliorated trastuzumab-induced cardiotoxicity through the modulation of mitochondrial function and dynamics without reducing its anticancer efficacy. *Biochim. Biophys. Acta Mol. Basis. Dis.* **2023**, *1869*, 166618. [CrossRef]
178. Alhowail, A.H.; Aldubayan, M.A. The Impact of Metformin on the Development of Hypothyroidism and Cardiotoxicity Induced by Cyclophosphamide, Methotrexate, and Fluorouracil in Rats. *Pharmaceuticals* **2023**, *16*, 1312. [CrossRef]
179. Sun, M.L.; Chen, W.; Wang, X.H. Reliability of Metformin's protective effects against doxorubicin-induced cardiotoxicity: A meta-analysis of animal studies. *Front. Pharmacol.* **2024**, *15*, 1435866. [CrossRef]
180. Osataphan, N.; Phrommintikul, A.; Leemasawat, K.; Somwangprasert, A.; Apaijai, N.; Suksai, S.; Sirikul, W.; Gunaparn, S.; Chattipakorn, S.C.; Chattipakorn, N. Effects of metformin and donepezil on the prevention of doxorubicin-induced cardiotoxicity in breast cancer: A randomized controlled trial. *Sci. Rep.* **2023**, *13*, 12759. [CrossRef]
181. Furihata, T.; Maekawa, S.; Takada, S.; Kakutani, N.; Nambu, H.; Shirakawa, R.; Yokota, T.; Kinugawa, S. Premedication with pioglitazone prevents doxorubicin-induced left ventricular dysfunction in mice. *BMC Pharmacol. Toxicol.* **2021**, *22*, 27. [CrossRef] [PubMed]
182. Pakravan, G.; Foroughmand, A.M.; Peymani, M.; Ghaedi, K.; Hashemi, M.S.; Hajjari, M.; Nasr-Esfahani, M.H. Downregulation of miR-130a, antagonized doxorubicin-induced cardiotoxicity via increasing the PPARγ expression in mESCs-derived cardiac cells. *Cell Death Dis.* **2018**, *9*, 758. [CrossRef]
183. Pakravan, G.; Peymani, M.; Abedpoor, N.; Safaeinejad, Z.; Yadegari, M.; Derakhshan, M.; Nasr Esfahani, M.H.; Ghaedi, K. Antiapoptotic and anti-inflammatory effects of Pparγ agonist, pioglitazone, reversed Dox-induced cardiotoxicity through mediating of miR-130a downregulation in C57BL/6 mice. *J. Biochem. Mol. Toxicol.* **2022**, *36*, e23041. [CrossRef]
184. Alhowail, A.H. Pioglitazone ameliorates doxorubicin-induced hypothyroidism and cardiotoxicity in rat models. *Eur. Rev. Med. Pharmacol. Sci.* **2023**, *27*, 9388–9395. [CrossRef]
185. Chandra, M.; Miriyala, S.; Panchatcharam, M. PPARγ and Its Role in Cardiovascular Diseases. *PPAR Res.* **2017**, *2017*, 6404638. [CrossRef] [PubMed]
186. Liu, M.L.; Wang, M.L.; Lv, J.J.; Wei, J.; Wan, J. Glibenclamide exacerbates adriamycin-induced cardiotoxicity by activating oxidative stress-induced endoplasmic reticulum stress in rats. *Exp. Ther. Med.* **2018**, *15*, 3425–3431. [CrossRef] [PubMed]
187. El-Agamy, D.S.; Abo-Haded, H.M.; Elkablawy, M.A. Cardioprotective effects of sitagliptin against doxorubicin-induced cardiotoxicity in rats. *Exp. Biol. Med.* **2016**, *241*, 1577–1587. [CrossRef]
188. Aykan, D.A.; Yaman, S.; Eser, N.; Özcan Metin, T.; Seyithanoğlu, M.; Aykan, A.Ç.; Kurt, A.H.; Ergün, Y. Bisoprolol and linagliptin ameliorated electrical and mechanical isometric myocardial contractions in doxorubicin-induced cardiomyopathy in rats. *Pharmacol. Rep.* **2020**, *72*, 867–876. [CrossRef]

189. Abbas, N.A.T.; Kabil, S.L. Liraglutide ameliorates cardiotoxicity induced by doxorubicin in rats through the Akt/GSK-3β signaling pathway. *Naunyn. Schmiedebergs Arch. Pharmacol.* **2017**, *390*, 1145–1153. [CrossRef]
190. Tonon, C.R.; Monte, M.G.; Balin, P.S.; Fujimori, A.S.S.; Ribeiro, A.P.D.; Ferreira, N.F.; Vieira, N.M.; Cabral, R.P.; Okoshi, M.P.; Okoshi, K.; et al. Liraglutide Pretreatment Does Not Improve Acute Doxorubicin-Induced Cardiotoxicity in Rats. *Int. J. Mol. Sci.* **2024**, *25*, 5833. [CrossRef]
191. HamaSalih, R.M. Effects of Semaglutide in Doxorubicin-Induced Cardiac Toxicity in Wistar Albino Rats. *Cancer Manag. Res.* **2024**, *16*, 731–740. [CrossRef] [PubMed]
192. Chen, L.; Chen, X.; Ruan, B.; Yang, H.; Yu, Y. Tirzepatide protects against doxorubicin-induced cardiotoxicity by inhibiting oxidative stress and inflammation via PI3K/Akt signaling. *Peptides* **2024**, *178*, 171245. [CrossRef] [PubMed]
193. Dabour, M.S.; George, M.Y.; Daniel, M.R.; Blaes, A.H.; Zordoky, B.N. The Cardioprotective and Anticancer Effects of SGLT2 Inhibitors: JACC: CardioOncology State-of-the-Art Review. *JACC Cardio Oncol.* **2024**, *6*, 159–182. [CrossRef] [PubMed]
194. Vafa, R.G.; Sabahizadeh, A.; Mofarrah, R. Guarding the heart: How SGLT-2 inhibitors protect against chemotherapy-induced cardiotoxicity: SGLT-2 inhibitors and chemotherapy-induced cardiotoxicity. *Curr. Probl. Cardiol.* **2024**, *49*, 102350. [CrossRef]
195. Hsieh, P.L.; Chu, P.M.; Cheng, H.C.; Huang, Y.T.; Chou, W.C.; Tsai, K.L.; Chan, S.H. Dapagliflozin Mitigates Doxorubicin-Caused Myocardium Damage by Regulating AKT-Mediated Oxidative Stress, Cardiac Remodeling, and Inflammation. *Int. J. Mol. Sci.* **2022**, *23*, 10146. [CrossRef]
196. Ulusan, S.; Gülle, K.; Peynirci, A.; Sevimli, M.; Karaibrahimoglu, A.; Kuyumcu, M.S. Dapagliflozin May Protect Against Doxorubicin-Induced Cardiotoxicity. *Anatol. J. Cardiol.* **2023**, *27*, 339–347. [CrossRef]
197. Sabatino, J.; De Rosa, S.; Tammè, L.; Iaconetti, C.; Sorrentino, S.; Polimeni, A.; Mignogna, C.; Amorosi, A.; Spaccarotella, C.; Yasuda, M.; et al. Empagliflozin prevents doxorubicin-induced myocardial dysfunction. *Cardiovasc. Diabetol.* **2020**, *19*, 66. [CrossRef]
198. Chang, H.Y.; Hsu, H.C.; Fang, Y.H.; Liu, P.Y.; Liu, Y.W. Empagliflozin attenuates doxorubicin-induced cardiotoxicity by inhibiting the JNK signaling pathway. *Biomed. Pharmacother.* **2024**, *176*, 116759. [CrossRef]
199. Ali, F.E.M.; Hassanein, E.H.M.; Abd El-Ghafar, O.A.M.; Rashwan, E.K.; Saleh, F.M.; Atwa, A.M. Exploring the cardioprotective effects of canagliflozin against cisplatin-induced cardiotoxicity: Role of iNOS/NF-κB, Nrf2, and Bax/cytochrome C/Bcl-2 signals. *J. Biochem. Mol. Toxicol.* **2023**, *37*, e23309. [CrossRef]
200. Dabour, M.S.; Abdelgawad, I.Y.; Grant, M.K.O.; El-Sawaf, E.S.; Zordoky, B.N. Canagliflozin mitigates carfilzomib-induced endothelial apoptosis via an AMPK-dependent pathway. *Biomed. Pharmacother.* **2023**, *164*, 114907. [CrossRef]
201. Chen, M. Empagliflozin attenuates doxorubicin-induced cardiotoxicity by activating AMPK/SIRT-1/PGC-1α-mediated mitochondrial biogenesis. *Toxicol. Res.* **2023**, *12*, 216–223. [CrossRef] [PubMed]
202. Quagliariello, V.; De Laurentiis, M.; Rea, D.; Barbieri, A.; Monti, M.G.; Carbone, A.; Paccone, A.; Altucci, L.; Conte, M.; Canale, M.L.; et al. The SGLT-2 inhibitor empagliflozin improves myocardial strain, reduces cardiac fibrosis and pro-inflammatory cytokines in non-diabetic mice treated with doxorubicin. *Cardiovasc. Diabetol.* **2021**, *20*, 150. [CrossRef] [PubMed]
203. Oh, C.M.; Cho, S.; Jang, J.Y.; Kim, H.; Chun, S.; Choi, M.; Park, S.; Ko, Y.G. Cardioprotective Potential of an SGLT2 Inhibitor Against Doxorubicin-Induced Heart Failure. *Korean Circ. J.* **2019**, *49*, 1183–1195. [CrossRef]
204. Gongora, C.A.; Drobni, Z.D.; Quinaglia Araujo Costa Silva, T.; Zafar, A.; Gong, J.; Zlotoff, D.A.; Gilman, H.K.; Hartmann, S.E.; Sama, S.; Nikolaidou, S.; et al. Sodium-Glucose Co-Transporter-2 Inhibitors and Cardiac Outcomes Among Patients Treated With Anthracyclines. *JACC Heart Fail.* **2022**, *10*, 559–567. [CrossRef]
205. Liu, C.H.; Ho, Y.C.; Lee, W.C.; Huang, C.Y.; Lee, Y.K.; Hsieh, C.B.; Huang, N.C.; Wu, C.C.; Nguyen, N.U.N.; Hsu, C.C.; et al. Sodium-Glucose Co-Transporter-2 Inhibitor Empagliflozin Attenuates Sorafenib-Induced Myocardial Inflammation and Toxicity. *Environ. Toxicol.* **2024**, *39*, 4844–4858. [CrossRef]
206. Belger, C.; Abrahams, C.; Imamdin, A.; Lecour, S. Doxorubicin-induced cardiotoxicity and risk factors. *Int. J. Cardiol. Heart Vasc.* **2023**, *50*, 101332. [CrossRef] [PubMed]
207. Li, J.; Gui, Y.; Ren, J.; Liu, X.; Feng, Y.; Zeng, Z.; He, W.; Yang, J.; Dai, C. Metformin Protects Against Cisplatin-Induced Tubular Cell Apoptosis and Acute Kidney Injury via AMPKα-regulated Autophagy Induction. *Sci. Rep.* **2016**, *6*, 23975. [CrossRef]
208. Taghizadeh, F.; Hosseinimehr, S.J.; Zargari, M.; Karimpour Malekshah, A.; Talebpour Amiri, F.B. Gliclazide attenuates cisplatin-induced nephrotoxicity through inhibiting NF-κB and caspase-3 activity. *IUBMB Life* **2020**, *72*, 2024–2033. [CrossRef]
209. Mahmoud, M.F.; El Shazly, S.M. Pioglitazone protects against cisplatin induced nephrotoxicity in rats and potentiates its anticancer activity against human renal adenocarcinoma cell lines. *Food Chem. Toxicol.* **2013**, *51*, 114–122. [CrossRef]
210. Salama, R.M.; Nasr, M.M.; Abdelhakeem, J.I.; Roshdy, O.K.; ElGamal, M.A. Alogliptin attenuates cyclophosphamide-induced nephrotoxicity: A novel therapeutic approach through modulating MAP3K/JNK/SMAD3 signaling cascade. *Drug Chem. Toxicol.* **2022**, *45*, 1254–1263. [CrossRef]
211. Jo, C.H.; Kim, S.; Park, J.S.; Kim, G.H. Anti-Inflammatory Action of Sitagliptin and Linagliptin in Doxorubicin Nephropathy. *Kidney Blood Press. Res.* **2018**, *43*, 987–999. [CrossRef] [PubMed]
212. El-Gohary, R.M.; Ghalwash, A.A.; Awad, M.M.; El-Shaer, R.A.A.; Ibrahim, S.; Eltantawy, A.F.; Elmansy, A.; Okasha, A.H. Novel insights into the augmented effect of curcumin and liraglutide in ameliorating cisplatin-induced nephrotoxicity in rats: Effects on oxidative stress, inflammation, apoptosis and pyroptosis via GSK-3β. *Arch. Biochem. Biophys.* **2023**, *749*, 109801. [CrossRef] [PubMed]

213. Farrokh-Eslamlou, N.; Momtaz, S.; Niknejad, A.; Hosseini, Y.; Mahdaviani, P.; Ghasemnejad-Berenji, M.; Abdolghaffari, A.H. Empagliflozin protective effects against cisplatin-induced acute nephrotoxicity by interfering with oxidative stress and inflammation in Wistar rats. *Naunyn. Schmiedebergs. Arch. Pharmacol.* **2024**, *397*, 7061–7070. [CrossRef] [PubMed]
214. Chang, W.T.; Wu, C.C.; Liao, I.C.; Lin, Y.W.; Chen, Y.C.; Ho, C.H.; Lee, W.C.; Lin, Y.C.; Chen, Z.C.; Shih, J.Y.; et al. Dapagliflozin protects against doxorubicin-induced nephrotoxicity associated with nitric oxide pathway-A translational study. *Free Radic. Biol. Med.* **2023**, *208*, 103–111. [CrossRef]
215. Chen, M.; Yi, Y.; Chen, B.; Zhang, H.; Dong, M.; Yuan, L.; Zhou, H.; Jiang, H.; Ma, Z. Metformin inhibits OCTN1- and OCTN2-mediated hepatic accumulation of doxorubicin and alleviates its hepatotoxicity in mice. *Toxicology* **2024**, *503*, 153757. [CrossRef]
216. Taghizadeh, F.; Hosseinimehr, S.J.; Zargari, M.; Karimpour Malekshah, A.; Mirzaei, M.; Talebpour Amiri, F. Alleviation of cisplatin-induced hepatotoxicity by gliclazide: Involvement of oxidative stress and caspase-3 activity. *Pharmacol. Res. Perspect.* **2021**, *9*, e00788. [CrossRef]
217. Kamel, G.A.M.; Elariny, H.A. Pioglitazone attenuates tamoxifen-induced liver damage in rats via modulating Keap1/Nrf2/HO-1 and SIRT1/Notch1 signaling pathways: In-vivo investigations, and molecular docking analysis. *Mol. Biol. Rep.* **2023**, *50*, 10219–10233. [CrossRef]
218. Satyam, S.M.; Bairy, L.K.; Rehman, A.; Farook, M.; Khan, S.; Nair, A.A.; Binu, N.N.; Yehya, M.; Khan, M.M. Dapagliflozin: A Promising Strategy to Combat Cisplatin-Induced Hepatotoxicity in Wistar Rats. *Biology* **2024**, *13*, 672. [CrossRef]
219. Du, P.; Liu, T.; Luo, P.; Li, H.; Tang, W.; Zong, S.; Xiao, H. SIRT3/GLUT4 signaling activation by metformin protect against cisplatin-induced ototoxicity in vitro. *Arch. Toxicol.* **2023**, *97*, 1147–1162. [CrossRef]
220. Shigematsu, N.; Kawashiri, T.; Kobayashi, D.; Shimizu, S.; Mine, K.; Hiromoto, S.; Uchida, M.; Egashira, N.; Shimazoe, T. Neuroprotective effect of alogliptin on oxaliplatin-induced peripheral neuropathy in vivo and in vitro. *Sci. Rep.* **2020**, *10*, 6734. [CrossRef]
221. Khasabova, I.A.; Khasabov, S.G.; Olson, J.K.; Uhelski, M.L.; Kim, A.H.; Albino-Ramírez, A.M.; Wagner, C.L.; Seybold, V.S.; Simone, D.A. Pioglitazone, a PPARγ agonist, reduces cisplatin-evoked neuropathic pain by protecting against oxidative stress. *Pain* **2019**, *160*, 688–701. [CrossRef] [PubMed]
222. Hussein, S.; Kamel, G.A.M. Pioglitazone ameliorates cisplatin-induced testicular toxicity by attenuating oxidative stress and inflammation via TLR4/MyD88/NF-κB signaling pathway. *J. Trace. Elem. Med. Biol.* **2023**, *80*, 127287. [CrossRef] [PubMed]
223. Alafifi, S.A.; Wahdan, S.A.; Elhemiely, A.A.; Elsherbiny, D.A.; Azab, S.S. Modulatory effect of liraglutide on doxorubicin-induced testicular toxicity and behavioral abnormalities in rats: Role of testicular-brain axis. *Naunyn. Schmiedebergs. Arch. Pharmacol.* **2023**, *396*, 2987–3005. [CrossRef]
224. Cheki, M.; Ghasemi, M.S.; Rezaei Rashnoudi, A.; Erfani Majd, N. Metformin attenuates cisplatin-induced genotoxicity and apoptosis in rat bone marrow cells. *Drug Chem. Toxicol.* **2021**, *44*, 386–393. [CrossRef] [PubMed]
225. Serageldin, M.A.; Kassem, A.B.; El-Kerm, Y.; Helmy, M.W.; El-Mas, M.M.; El-Bassiouny, N.A. The Effect of Metformin on Chemotherapy-Induced Toxicities in Non-diabetic Breast Cancer Patients: A Randomised Controlled Study. *Drug Saf.* **2023**, *46*, 587–599. [CrossRef]
226. Bakry, H.M.; Mansour, N.O.; ElKhodary, T.R.; Soliman, M.M. Efficacy of metformin in prevention of paclitaxel-induced peripheral neuropathy in breast cancer patients: A randomized controlled trial. *Front. Pharmacol.* **2023**, *14*, 1181312. [CrossRef]
227. Iwakura, T.; Fukasawa, H.; Kitamura, A.; Ishibuchi, K.; Yasuda, H.; Furuya, R. Effect of dipeptidyl peptidase-4 inhibitors on cisplatin-induced acute nephrotoxicity in cancer patients with diabetes mellitus: A retrospective study. *PLoS ONE* **2020**, *15*, e0229377. [CrossRef]
228. Galal, M.A.; Al-Rimawi, M.; Hajeer, A.; Dahman, H.; Alouch, S.; Aljada, A. Metformin: A Dual-Role Player in Cancer Treatment and Prevention. *Int. J. Mol. Sci.* **2024**, *25*, 4083. [CrossRef] [PubMed]
229. Zhu, L.; Yang, K.; Ren, Z.; Yin, D.; Zhou, Y. Metformin as anticancer agent and adjuvant in cancer combination therapy: Current progress and future prospect. *Transl. Oncol.* **2024**, *44*, 101945. [CrossRef]
230. Jalali, F.; Fakhari, F.; Sepehr, A.; Zafari, J.; Sarajar, B.O.; Sarihi, P.; Jafarzadeh, E. Synergistic anticancer effects of doxorubicin and metformin combination therapy: A systematic review. *Transl. Oncol.* **2024**, *45*, 101946. [CrossRef]
231. Wen, K.C.; Sung, P.L.; Wu, A.T.H.; Chou, P.C.; Lin, J.H.; Huang, C.F.; Yeung, S.J.; Lee, M.H. Neoadjuvant metformin added to conventional chemotherapy synergizes anti-proliferative effects in ovarian cancer. *J. Ovarian Res.* **2020**, *13*, 95. [CrossRef] [PubMed]
232. Yang, C.; Zhao, N.; Li, D.; Zou, G.; Chen, Y. Metformin improves the sensitivity of ovarian cancer cells to chemotherapeutic agents. *Oncol. Lett.* **2019**, *18*, 2404–2411. [CrossRef] [PubMed]
233. El-Khayat, S.M.; Abouegylah, M.; Abdallah, D.; Geweil, A.G.; Elenbaby, A.M.; Zahra, O.S. The effect of metformin when combined with neoadjuvant chemotherapy in breast cancer patients. *Med. Oncol.* **2021**, *39*, 1. [CrossRef]
234. Kim, S.H.; Kim, S.C.; Ku, J.L. Metformin increases chemo-sensitivity via gene downregulation encoding DNA replication proteins in 5-Fu resistant colorectal cancer cells. *Oncotarget* **2017**, *8*, 56546–56557. [CrossRef]
235. Yip, K.L.; Tsai, T.N.; Yang, I.P.; Miao, Z.F.; Chen, Y.C.; Li, C.C.; Su, W.C.; Chang, T.K.; Huang, C.W.; Tsai, H.L.; et al. Metformin Enhancement of Therapeutic Effects of 5-Fluorouracil and Oxaliplatin in Colon Cancer Cells and Nude Mice. *Biomedicines* **2022**, *10*, 955. [CrossRef]

236. Kordes, S.; Pollak, M.N.; Zwinderman, A.H.; Mathôt, R.A.; Weterman, M.J.; Beeker, A.; Punt, C.J.; Richel, D.J.; Wilmink, J.W. Metformin in patients with advanced pancreatic cancer: A double-blind, randomised, placebo-controlled phase 2 trial. *Lancet Oncol.* **2015**, *16*, 839–847. [CrossRef]
237. Yuan, F.; Cheng, C.; Xiao, F.; Liu, H.; Cao, S.; Zhou, G. Inhibition of mTORC1/P70S6K pathway by Metformin synergistically sensitizes Acute Myeloid Leukemia to Ara-C. *Life Sci.* **2020**, *243*, 117276. [CrossRef] [PubMed]
238. Singh, A.R.; Gu, J.J.; Zhang, Q.; Torka, P.; Sundaram, S.; Mavis, C.; Hernandez-Ilizaliturri, F.J. Metformin sensitizes therapeutic agents and improves outcome in pre-clinical and clinical diffuse large B-cell lymphoma. *Cancer Metab.* **2020**, *8*, 10. [CrossRef] [PubMed]
239. Alqahtani, Q.H.; Alkharashi, L.A.; Alajami, H.; Alkharashi, I.; Alkharashi, L.; Alhinti, S.N. Pioglitazone enhances cisplatin's impact on triple-negative breast cancer: Role of PPARγ in cell apoptosis. *Saudi Pharm. J.* **2024**, *32*, 102059. [CrossRef]
240. Enz, N.; Vliegen, G.; De Meester, I.; Jungraithmayr, W. CD26/DPP4—A potential biomarker and target for cancer therapy. *Pharmacol. Ther.* **2019**, *198*, 135–159. [CrossRef]
241. Klemann, C.; Wagner, L.; Stephan, M.; von Hörsten, S. Cut to the chase: A review of CD26/dipeptidyl peptidase-4's (DPP4) entanglement in the immune system. *Clin. Exp. Immunol.* **2016**, *185*, 1–21. [CrossRef] [PubMed]
242. Kawakita, E.; Koya, D.; Kanasaki, K. CD26/DPP-4: Type 2 Diabetes Drug Target with Potential Influence on Cancer Biology. *Cancers* **2021**, *13*, 2191. [CrossRef] [PubMed]
243. Zheng, X.; Liu, J.; Li, X.; Tian, R.; Shang, K.; Dong, X.; Cao, B. Angiogenesis is promoted by exosomal DPP4 derived from 5-fluorouracil-resistant colon cancer cells. *Cancer Lett.* **2021**, *497*, 190–201. [CrossRef] [PubMed]
244. Wenjing, H.; Shao, Y.; Yu, Y.; Huang, W.; Feng, G.; Li, J. Exendin-4 enhances the sensitivity of prostate cancer to enzalutamide by targeting Akt activation. *Prostate* **2020**, *80*, 367–375. [CrossRef]
245. Fujiyoshi, S.; Honda, S.; Ara, M.; Kondo, T.; Kobayashi, N.; Taketomi, A. SGLT2 is upregulated to acquire cisplatin resistance and SGLT2 inhibition reduces cisplatin resistance in hepatoblastoma. *J. Hepatobiliary Pancreat. Sci.* **2024**, *31*, 223–233. [CrossRef]
246. Zeng, Y.; Jiang, H.; Zhang, X.; Xu, J.; Wu, X.; Xu, Q.; Cai, W.; Ying, H.; Zhou, R.; Ding, Y.; et al. Canagliflozin reduces chemoresistance in hepatocellular carcinoma through PKM2-c-Myc complex-mediated glutamine starvation. *Free Radic. Biol. Med.* **2023**, *208*, 571–586. [CrossRef]
247. Huang, H.; Kung, F.L.; Huang, Y.W.; Hsu, C.C.; Guh, J.H.; Hsu, L.C. Sensitization of cancer cells to paclitaxel-induced apoptosis by canagliflozin. *Biochem. Pharmacol.* **2024**, *223*, 116140. [CrossRef]
248. Karim, S.; Alghanmi, A.N.; Jamal, M.; Alkreathy, H.; Jamal, A.; Alkhatabi, H.A.; Bazuhair, M.; Ahmad, A. A comparative in vitro study on the effect of SGLT2 inhibitors on chemosensitivity to doxorubicin in MCF-7 breast cancer cells. *Oncol. Res.* **2024**, *32*, 817–830. [CrossRef]
249. Okada, J.; Matsumoto, S.; Kaira, K.; Saito, T.; Yamada, E.; Yokoo, H.; Katoh, R.; Kusano, M.; Okada, S.; Yamada, M. Sodium Glucose Cotransporter 2 Inhibition Combined With Cetuximab Significantly Reduced Tumor Size and Carcinoembryonic Antigen Level in Colon Cancer Metastatic to Liver. *Clin. Color. Cancer* **2018**, *17*, e45–e48. [CrossRef]
250. Abdelmoneim, M.; Aboalela, M.A.; Naoe, Y.; Matsumura, S.; Eissa, I.R.; Bustos-Villalobos, I.; Sibal, P.A.; Takido, Y.; Kodera, Y.; Kasuya, H. The Impact of Metformin on Tumor-Infiltrated Immune Cells: Preclinical and Clinical Studies. *Int. J. Mol. Sci.* **2023**, *24*, 13353. [CrossRef]
251. Liu, W.; Wang, Y.; Luo, J.; Liu, M.; Luo, Z. Pleiotropic Effects of Metformin on the Antitumor Efficiency of Immune Checkpoint Inhibitors. *Front. Immunol.* **2021**, *11*, 586760. [CrossRef] [PubMed]
252. Cha, J.H.; Yang, W.H.; Xia, W.; Wei, Y.; Chan, L.C.; Lim, S.O.; Li, C.W.; Kim, T.; Chang, S.S.; Lee, H.H.; et al. Metformin Promotes Antitumor Immunity via Endoplasmic-Reticulum-Associated Degradation of PD-L1. *Mol. Cell* **2018**, *71*, 606–620.e7. [CrossRef] [PubMed]
253. Zhang, J.J.; Zhang, Q.S.; Li, Z.Q.; Zhou, J.W.; Du, J. Metformin attenuates PD-L1 expression through activating Hippo signaling pathway in colorectal cancer cells. *Am. J. Transl. Res.* **2019**, *11*, 6965–6976. [PubMed]
254. Spyrou, N.; Vallianou, N.; Kadillari, J.; Dalamaga, M. The interplay of obesity, gut microbiome and diet in the immune check point inhibitors therapy era. *Semin. Cancer Biol.* **2021**, *73*, 356–376. [CrossRef]
255. Sun, L.; Xie, C.; Wang, G.; Wu, Y.; Wu, Q.; Wang, X.; Liu, J.; Deng, Y.; Xia, J.; Chen, B.; et al. Gut microbiota and intestinal FXR mediate the clinical benefits of metformin. *Nat. Med.* **2018**, *24*, 1919–1929. [CrossRef]
256. Yang, W.; Cong, Y. Gut microbiota-derived metabolites in the regulation of host immune responses and immune-related inflammatory diseases. *Cell. Mol. Immunol.* **2021**, *18*, 866–877. [CrossRef]
257. Chiang, C.H.; Chen, Y.J.; Chiang, C.H.; Chen, C.Y.; Chang, Y.C.; Wang, S.S.; See, X.Y.; Horng, C.S.; Peng, C.Y.; Hsia, Y.P.; et al. Effect of metformin on outcomes of patients treated with immune checkpoint inhibitors: A retrospective cohort study. *Cancer Immunol. Immunother.* **2023**, *72*, 1951–1956. [CrossRef] [PubMed]
258. Wang, J.; Lin, J.; Guo, H.; Wu, W.; Yang, J.; Mao, J.; Fan, W.; Qiao, H.; Wang, Y.; Yan, X.; et al. Prognostic impact of metformin in solid cancer patients receiving immune checkpoint inhibitors: Novel evidences from a multicenter retrospective study. *Front. Pharmacol.* **2024**, *15*, 1419498. [CrossRef]
259. Jia, X.; Qian, J.; Chen, H.; Liu, Q.; Hussain, S.; Jin, J.; Shi, J.; Hou, Y. PPARγ agonist pioglitazone enhances colorectal cancer immunotherapy by inducing PD-L1 autophagic degradation. *Eur. J. Pharmacol.* **2023**, *950*, 175749. [CrossRef]
260. Gao, R.; Liu, Z.; Meng, M.; Song, X.; He, J. Neurogenesis-Associated Protein, a Potential Prognostic Biomarker in Anti-PD-1 Based Kidney Renal Clear Cell Carcinoma Patient Therapeutics. *Pharmaceuticals* **2024**, *17*, 451. [CrossRef]

261. Zuo, B.; Li, T.; Liu, X.; Wang, S.; Cheng, J.; Liu, X.; Cui, W.; Shi, H.; Ling, C. Dipeptidyl peptidase 4 inhibitor reduces tumor-associated macrophages and enhances anti-PD-L1-mediated tumor suppression in non-small cell lung cancer. *Clin. Transl. Oncol.* **2023**, *25*, 3188–3202. [CrossRef] [PubMed]
262. Zhan, Z.T.; Liu, L.; Cheng, M.Z.; Gao, Y.; Zhou, W.J. The Effects of 6 Common Antidiabetic Drugs on Anti-PD1 Immune Checkpoint Inhibitor in Tumor Treatment. *J. Immunol. Res.* **2022**, *2022*, 2651790. [CrossRef]
263. Chen, D.; Liang, H.; Huang, L.; Zhou, H.; Wang, Z. Liraglutide enhances the effect of checkpoint blockade in lung and liver cancers through the inhibition of neutrophil extracellular traps. *FEBS Open Bio* **2024**, *14*, 1365–1377. [CrossRef] [PubMed]
264. De Meo, M.L.; Spicer, J.D. The role of neutrophil extracellular traps in cancer progression and metastasis. *Semin. Immunol.* **2021**, *57*, 101595. [CrossRef] [PubMed]
265. Hahn, A.W.; Venkatesh, N.; Msaouel, P.; McQuade, J.L. The Influence of Obesity on Outcomes with Immune Checkpoint Blockade: Clinical Evidence and Potential Biological Mechanisms. *Cells* **2023**, *12*, 2551. [CrossRef]
266. Manna, P.; Jain, S.K. Obesity, Oxidative Stress, Adipose Tissue Dysfunction, and the Associated Health Risks: Causes and Therapeutic Strategies. *Metab. Syndr. Relat. Disord.* **2015**, *13*, 423–444. [CrossRef]
267. Aurilio, G.; Piva, F.; Santoni, M.; Cimadamore, A.; Sorgentoni, G.; Lopez-Beltran, A.; Cheng, L.; Battelli, N.; Nolè, F.; Montironi, R. The Role of Obesity in Renal Cell Carcinoma Patients: Clinical-Pathological Implications. *Int. J. Mol. Sci.* **2019**, *20*, 5683. [CrossRef]
268. Cortellini, A.; Bersanelli, M.; Buti, S.; Cannita, K.; Santini, D.; Perrone, F.; Giusti, R.; Tiseo, M.; Michiara, M.; Di Marino, P.; et al. A multicenter study of body mass index in cancer patients treated with anti-PD-1/PD-L1 immune checkpoint inhibitors: When overweight becomes favorable. *J. Immunother. Cancer* **2019**, *7*, 57. [CrossRef]
269. Jenkins, B.J.; Blagih, J.; Ponce-Garcia, F.M.; Canavan, M.; Gudgeon, N.; Eastham, S.; Hill, D.; Hanlon, M.M.; Ma, E.H.; Bishop, E.L.; et al. Canagliflozin impairs T cell effector function via metabolic suppression in autoimmunity. *Cell Metab.* **2023**, *35*, 1132–1146.e9. [CrossRef]
270. Ding, L.; Chen, X.; Zhang, W.; Dai, X.; Guo, H.; Pan, X.; Xu, Y.; Feng, J.; Yuan, M.; Gao, X.; et al. Canagliflozin primes antitumor immunity by triggering PD-L1 degradation in endocytic recycling. *J. Clin. Investig.* **2023**, *133*, e154754. [CrossRef]
271. Maurea, N.; Quagliariello, V.; Bonelli, A.; Paccone, A.; Conforti, G.; Caronna, A.; Buccolo, S.; Iovine, M.; Cerrone, F.; Botti, G. TheSGLT-2 inhibitor dapagliflozin enhanced the anticancer activities and exerts cardioprotective effects during exposure to ipilimumab through NLRP3 inflammasome and pro-fibrotic cytokines. *Eur. Heart J.* **2021**, *42* (Suppl. S1), ehab724. [CrossRef]
272. Quagliariello, V.; De Laurentiis, M.; Cocco, S.; Rea, G.; Bonelli, A.; Caronna, A.; Lombari, M.C.; Conforti, G.; Berretta, M.; Botti, G.; et al. NLRP3 as Putative Marker of Ipilimumab-Induced Cardiotoxicity in the Presence of Hyperglycemia in Estrogen-Responsive and Triple-Negative Breast Cancer Cells. *Int. J. Mol. Sci.* **2020**, *21*, 7802. [CrossRef] [PubMed]
273. Perelman, M.G.; Brzezinski, R.Y.; Waissengrin, B.; Leshem, Y.; Bainhoren, O.; Rubinstein, T.A.; Perelman, M.; Rozenbaum, Z.; Havakuk, O.; Topilsky, Y.; et al. Sodium-glucose co-transporter-2 inhibitors in patients treated with immune checkpoint inhibitors. *Cardiooncology* **2024**, *10*, 2. [CrossRef] [PubMed]
274. Corleto, K.A.; Strandmo, J.L.; Giles, E.D. Metformin and Breast Cancer: Current Findings and Future Perspectives from Preclinical and Clinical Studies. *Pharmaceuticals* **2024**, *17*, 396. [CrossRef] [PubMed]
275. Kang, J.; Jeong, S.M.; Shin, D.W.; Cho, M.; Cho, J.H.; Kim, J. The Associations of Aspirin, Statins, and Metformin With Lung Cancer Risk and Related Mortality: A Time-Dependent Analysis of Population-Based Nationally Representative Data. *J. Thorac. Oncol.* **2021**, *16*, 76–88. [CrossRef]
276. Tseng, C.H. Metformin and Risk of Malignant Brain Tumors in Patients with Type 2 Diabetes Mellitus. *Biomolecules* **2021**, *11*, 1226. [CrossRef]
277. Tarhini, Z.; Manceur, K.; Magne, J.; Mathonnet, M.; Jost, J.; Christou, N. The effect of metformin on the survival of colorectal cancer patients with type 2 diabetes mellitus. *Sci. Rep.* **2022**, *12*, 12374. [CrossRef]
278. Kim, J.; Bae, Y.J.; Kang, H.T. Metformin Use May Increase Risk of Pancreatic Cancer in Diabetic Women: An Analysis of the Korean National Health Insurance Service-National Health Screening Cohort Database. *Korean J. Fam. Med.* **2022**, *43*, 327–333. [CrossRef]
279. Tseng, C.H. Use of metformin and risk of kidney cancer in patients with type 2 diabetes. *Eur. J. Cancer* **2016**, *52*, 19–25. [CrossRef]
280. Currie, C.J.; Poole, C.D.; Gale, E.A. The influence of glucose-lowering therapies on cancer risk in type 2 diabetes. *Diabetologia* **2009**, *52*, 1766–1777. [CrossRef]
281. Goossens, M.E.; Buntinx, F.; Zeegers, M.P.; Driessen, J.H.; De Bruin, M.L.; De Vries, F. Influence of metformin intake on the risk of bladder cancer in type 2 diabetes patients. *Br. J. Clin. Pharmacol.* **2015**, *80*, 1464–1472. [CrossRef] [PubMed]
282. El-Arabey, A.A. New insight for metformin against bladder cancer. *Genes Environ.* **2017**, *39*, 13. [CrossRef] [PubMed]
283. Urpilainen, E.; Puistola, U.; Boussios, S.; Karihtala, P. Metformin and ovarian cancer: The evidence. *Ann. Transl. Med.* **2020**, *8*, 1711. [CrossRef] [PubMed]
284. Kaur, P.; Berchuck, A.; Chase, A.; Grout, B.; Deurloo, C.M.; Pearce, L.C.; Pike, M.C.; Richardson, J.; Terry, K.L.; Webb, P.M.; et al. Metformin use and survival in people with ovarian cancer: A population-based cohort study from British Columbia, Canada. *Neoplasia* **2024**, *56*, 101026. [CrossRef] [PubMed]
285. Lee, T.Y.; Martinez-Outschoorn, U.E.; Schilder, R.J.; Kim, C.H.; Richard, S.D.; Rosenblum, N.G.; Johnson, J.M. Metformin as a Therapeutic Target in Endometrial Cancers. *Front. Oncol.* **2018**, *8*, 341. [CrossRef]

286. O'Connor, L.; Bailey-Whyte, M.; Bhattacharya, M.; Butera, G.; Hardell, K.N.L.; Seidenberg, A.B.; Castle, P.E.; Loomans-Kropp, H.A. Association of metformin use and cancer incidence: A systematic review and meta-analysis. *J. Natl. Cancer Inst.* **2024**, *116*, 518–529. [CrossRef]
287. Wang, M.; Noghabaei, G.; Raeisi, T.; Li, D.; Alizadeh, H.; Alizadeh, M. Metformin and risk of hematological cancers in patients with diabetes: A systematic review and meta-analysis. *Ann. Saudi Med.* **2024**, *44*, 126–134. [CrossRef]
288. Mesquita, L.A.; Spiazzi, B.F.; Piccoli, G.F.; Nogara, D.A.; da Natividade, G.R.; Garbin, H.I.; Wayerbacher, L.F.; Wiercinski, V.M.; Baggio, V.A.; Zingano, C.P.; et al. Does metformin reduce the risk of cancer in obesity and diabetes? A systematic review and meta-analysis. *Diabetes Obes. Metab.* **2024**, *26*, 1929–1940. [CrossRef]
289. Xie, H.; Li, M.; Zheng, Y. Associations of metformin therapy treatment with endometrial cancer risk and prognosis: A systematic review and meta-analysis. *Gynecol. Oncol.* **2024**, *182*, 15–23. [CrossRef]
290. Vicentini, M.; Ballotari, P.; Venturelli, F.; Ottone, M.; Manicardi, V.; Gallo, M.; Greci, M.; Pinotti, M.; Pezzarossi, A.; Giorgi Rossi, P. Impact of Insulin Therapies on Cancer Incidence in Type 1 and Type 2 Diabetes: A Population-Based Cohort Study in Reggio Emilia, Italy. *Cancers* **2022**, *14*, 2719. [CrossRef]
291. Chang, C.H.; Lin, J.W.; Wu, L.C.; Lai, M.S.; Chuang, L.M. Oral insulin secretagogues, insulin, and cancer risk in type 2 diabetes mellitus. *J. Clin. Endocrinol. Metab.* **2012**, *97*, E1170–E1175. [CrossRef] [PubMed]
292. Bowker, S.L.; Majumdar, S.R.; Veugelers, P.; Johnson, J.A. Increased cancer-related mortality for patients with type 2 diabetes who use sulfonylureas or insulin. *Diabetes Care* **2006**, *29*, 254–258. [CrossRef]
293. Chen, Y.; Mushashi, F.; Son, S.; Bhatti, P.; Dummer, T.; Murphy, R.A. Diabetes medications and cancer risk associations: A systematic review and meta-analysis of evidence over the past 10 years. *Sci. Rep.* **2023**, *13*, 11844. [CrossRef] [PubMed]
294. Cozma, L.S.; Luzio, S.D.; Dunseath, G.J.; Langendorg, K.W.; Pieber, T.; Owens, D.R. Comparison of the effects of three insulinotropic drugs on plasma insulin levels after a standard meal. *Diabetes Care* **2002**, *25*, 1271–1276. [CrossRef] [PubMed]
295. Tuccori, M.; Filion, K.B.; Yin, H.; Yu, O.H.; Platt, R.W.; Azoulay, L. Pioglitazone use and risk of bladder cancer: Population based cohort study. *BMJ* **2016**, *352*, i1541. [CrossRef]
296. Mehtälä, J.; Khanfir, H.; Bennett, D.; Ye, Y.; Korhonen, P.; Hoti, F. Pioglitazone use and risk of bladder cancer: A systematic literature review and meta-analysis of observational studies. *Diabetol. Int.* **2018**, *10*, 24–36. [CrossRef]
297. Tang, H.; Shi, W.; Fu, S.; Wang, T.; Zhai, S.; Song, Y.; Han, J. Pioglitazone and bladder cancer risk: A systematic review and meta-analysis. *Cancer Med.* **2018**, *7*, 1070–1080. [CrossRef]
298. Dormandy, J.; Bhattacharya, M.; van Troostenburg de Bruyn, A.R. PROactive investigators. Safety and tolerability of pioglitazone in high-risk patients with type 2 diabetes: An overview of data from PROactive. *Drug Saf.* **2009**, *32*, 187–202. [CrossRef]
299. Colmers, I.N.; Bowker, S.L.; Majumdar, S.R.; Johnson, J.A. Use of thiazolidinediones and the risk of bladder cancer among people with type 2 diabetes: A meta-analysis. *CMAJ* **2012**, *184*, 675–683. [CrossRef]
300. Filipova, E.; Uzunova, K.; Kalinov, K.; Vekov, T. Pioglitazone and the Risk of Bladder Cancer: A Meta-Analysis. *Diabetes Ther.* **2017**, *8*, 705–726. [CrossRef]
301. Malhotra, B.; Hiteshi, P.; Khalkho, P.; Malik, R.; Bhadada, S.K.; Bhansali, A.; Shafiq, N.; Malhotra, S.; Kumar, N.; Rajput, R.; et al. Bladder cancer with pioglitazone: A case-control study. *Diabetes Metab. Syndr.* **2022**, *16*, 102637. [CrossRef] [PubMed]
302. Lewis, J.D.; Habel, L.A.; Quesenberry, C.P.; Strom, B.L.; Peng, T.; Hedderson, M.M.; Ehrlich, S.F.; Mamtani, R.; Bilker, W.; Vaughn, D.J.; et al. Pioglitazone Use and Risk of Bladder Cancer and Other Common Cancers in Persons With Diabetes. *JAMA* **2015**, *314*, 265–277. [CrossRef] [PubMed]
303. Cui, H.; Wang, Y.; Yang, S.; He, G.; Jiang, Z.; Gang, X.; Wang, G. Antidiabetic medications and the risk of prostate cancer in patients with diabetes mellitus: A systematic review and meta-analysis. *Pharmacol. Res.* **2022**, *177*, 106094. [CrossRef]
304. Tseng, C.H. Pioglitazone does not affect the risk of kidney cancer in patients with type 2 diabetes. *Metabolism* **2014**, *63*, 1049–1055. [CrossRef]
305. Tseng, C.H. Pioglitazone and breast cancer risk in female patients with type 2 diabetes mellitus: A retrospective cohort analysis. *BMC Cancer* **2022**, *22*, 559. [CrossRef]
306. Keith, R.L.; Blatchford, P.J.; Merrick, D.T.; Bunn, P.A., Jr.; Bagwell, B.; Dwyer-Nield, L.D.; Jackson, M.K.; Geraci, M.W.; Miller, Y.E. A Randomized Phase II Trial of Pioglitazone for Lung Cancer Chemoprevention in High-Risk Current and Former Smokers. *Cancer Prev. Res.* **2019**, *12*, 721–730. [CrossRef]
307. Qiu, Y.Y.; Zhang, J.; Zeng, F.Y.; Zhu, Y.Z. Roles of the peroxisome proliferator-activated receptors (PPARs) in the pathogenesis of nonalcoholic fatty liver disease (NAFLD). *Pharmacol. Res.* **2023**, *192*, 106786. [CrossRef] [PubMed]
308. Lai, S.W.; Lin, C.L.; Liao, K.F. Association of hepatocellular carcinoma with thiazolidinediones use: A population-based case-control study. *Medicine* **2020**, *99*, e19833. [CrossRef]
309. Liu, Y.; Jin, P.P.; Sun, X.C.; Hu, T.T. Thiazolidinediones and risk of colorectal cancer in patients with diabetes mellitus: A meta-analysis. *Saudi J. Gastroenterol.* **2018**, *24*, 75–81. [CrossRef]
310. See, L.C.; Wu, C.Y.; Tsai, C.Y.; Lee, C.C.; Chen, J.J.; Jenq, C.C.; Chen, C.Y.; Chen, Y.C.; Yen, C.L.; Yang, H.Y. PPAR-γ agonist pioglitazone and the risks of malignancy among type2 diabetes mellitus patients. *Acta Diabetol.* **2024**. [CrossRef]
311. Zhao, Z.; He, X.; Sun, Y. Hypoglycemic agents and incidence of pancreatic cancer in diabetic patients: A meta-analysis. *Front. Pharmacol.* **2023**, *14*, 1193610. [CrossRef] [PubMed]
312. Butler, P.C.; Elashoff, M.; Elashoff, R.; Gale, E.A. A critical analysis of the clinical use of incretin-based therapies: Are the GLP-1 therapies safe? *Diabetes Care* **2013**, *36*, 2118–2125. [CrossRef] [PubMed]

313. Dicembrini, I.; Montereggi, C.; Nreu, B.; Mannucci, E.; Monami, M. Pancreatitis and pancreatic cancer in patientes treated with Dipeptidyl Peptidase-4 inhibitors: An extensive and updated meta-analysis of randomized controlled trials. *Diabetes Res. Clin. Pract.* **2020**, *159*, 107981. [CrossRef] [PubMed]
314. Kim, M.K.; Han, K.; Kwon, H.S.; Yoo, S.J. Risk of Pancreatic Cancer and Use of Dipeptidyl Peptidase 4 Inhibitors in Patients with Type 2 Diabetes: A Propensity Score-Matching Analysis. *Endocrinol. Metab.* **2023**, *38*, 426–435. [CrossRef]
315. Ng, L.; Foo, D.C.; Wong, C.K.; Man, A.T.; Lo, O.S.; Law, W.L. Repurposing DPP-4 Inhibitors for Colorectal Cancer: A Retrospective and Single Center Study. *Cancers* **2021**, *13*, 3588. [CrossRef]
316. Pradhan, R.; Yu, O.H.Y.; Platt, R.W.; Azoulay, L. Dipeptidyl peptidase-4 inhibitors and the risk of skin cancer among patients with type 2 diabetes: A UK population-based cohort study. *BMJ Open Diabetes Res. Care* **2023**, *11*, e003550. [CrossRef]
317. Garczorz, W.; Kosowska, A.; Francuz, T. Antidiabetic Drugs in Breast Cancer Patients. *Cancers* **2024**, *16*, 299. [CrossRef]
318. Ali, S.; Fortune, K.; Masur, J.; Viscuse, P.V.; Devitt, M.E.; Dreicer, R.; Skelton, W.P. 4th. Impact of DPP4 Inhibition on Survival in Patients With Metastatic Renal Cell Carcinoma and Type 2 Diabetes Mellitus. *Clin. Genitourin. Cancer* **2024**, *22*, 102173. [CrossRef]
319. Lisco, G.; De Tullio, A.; Disoteo, O.; Piazzolla, G.; Guastamacchia, E.; Sabbà, C.; De Geronimo, V.; Papini, E.; Triggiani, V. Glucagon-like peptide 1 receptor agonists and thyroid cancer: Is it the time to be concerned? *Endocr. Connect.* **2023**, *12*, e230257. [CrossRef]
320. Hegedüs, L.; Moses, A.C.; Zdravkovic, M.; Le Thi, T.; Daniels, G.H. GLP-1 and calcitonin concentration in humans: Lack of evidence of calcitonin release from sequential screening in over 5000 subjects with type 2 diabetes or nondiabetic obese subjects treated with the human GLP-1 analog, liraglutide. *J. Clin. Endocrinol. Metab.* **2011**, *96*, 853–860. [CrossRef]
321. Bethel, M.A.; Patel, R.A.; Thompson, V.P.; Merrill, P.; Reed, S.D.; Li, Y.; Ahmadi, S.; Katona, B.G.; Gustavson, S.M.; Ohman, P.; et al. Changes in Serum Calcitonin Concentrations, Incidence of Medullary Thyroid Carcinoma, and Impact of Routine Calcitonin Concentration Monitoring in the EXenatide Study of Cardiovascular Event Lowering (EXSCEL). *Diabetes Care* **2019**, *42*, 1075–1080. [CrossRef] [PubMed]
322. Bezin, J.; Gouverneur, A.; Pénichon, M.; Mathieu, C.; Garrel, R.; Hillaire-Buys, D.; Pariente, A.; Faillie, J.L. GLP-1 Receptor Agonists and the Risk of Thyroid Cancer. *Diabetes Care* **2023**, *46*, 384–390. [CrossRef] [PubMed]
323. Uppsala Monitoring Centre. Vigibase. Available online: https://who-umc.org/vigibase (accessed on 14 November 2024).
324. Silverii, G.A.; Monami, M.; Gallo, M.; Ragni, A.; Prattichizzo, F.; Renzelli, V.; Ceriello, A.; Mannucci, E. Glucagon-like peptide-1 receptor agonists and risk of thyroid cancer: A systematic review and meta-analysis of randomized controlled trials. *Diabetes Obes. Metab.* **2024**, *26*, 891–900. [CrossRef] [PubMed]
325. Hu, W.; Song, R.; Cheng, R.; Liu, C.; Guo, R.; Tang, W.; Zhang, J.; Zhao, Q.; Li, X.; Liu, J. Use of GLP-1 Receptor Agonists and Occurrence of Thyroid Disorders: A Meta-Analysis of Randomized Controlled Trials. *Front. Endocrinol.* **2022**, *13*, 927859. [CrossRef]
326. Elashoff, M.; Matveyenko, A.V.; Gier, B.; Elashoff, R.; Butler, P.C. Pancreatitis, pancreatic, and thyroid cancer with glucagon-like peptide-1-based therapies. *Gastroenterology* **2011**, *141*, 150–156. [CrossRef]
327. Pinto, L.C.; Falcetta, M.R.; Rados, D.V.; Leitão, C.B.; Gross, J.L. Glucagon-like peptide-1 receptor agonists and pancreatic cancer: A meta-analysis with trial sequential analysis. *Sci. Rep.* **2019**, *9*, 2375. [CrossRef]
328. Nreu, B.; Dicembrini, I.; Tinti, F.; Mannucci, E.; Monami, M. Pancreatitis and pancreatic cancer in patients with type 2 diabetes treated with glucagon-like peptide-1 receptor agonists: An updated meta-analysis of randomized controlled trials. *Minerva Endocrinol.* **2023**, *48*, 206–213. [CrossRef]
329. Ayoub, M.; Faris, C.; Juranovic, T.; Chela, H.; Daglilar, E. The Use of Glucagon-like Peptide-1 Receptor Agonists in Patients with Type 2 Diabetes Mellitus Does Not Increase the Risk of Pancreatic Cancer: A U.S.-Based Cohort Study. *Cancers* **2024**, *16*, 1625. [CrossRef]
330. Dankner, R.; Murad, H.; Agay, N.; Olmer, L.; Freedman, L.S. Glucagon-Like Peptide-1 Receptor Agonists and Pancreatic Cancer Risk in Patients With Type 2 Diabetes. *JAMA Netw. Open* **2024**, *7*, e2350408. [CrossRef]
331. Wang, L.; Berger, N.A.; Kaelber, D.C.; Xu, R. Association of GLP-1 Receptor Agonists and Hepatocellular Carcinoma Incidence and Hepatic Decompensation in Patients With Type 2 Diabetes. *Gastroenterology* **2024**, *167*, 689–703. [CrossRef]
332. Skriver, C.; Friis, S.; Knudsen, L.B.; Catarig, A.M.; Clark, A.J.; Dehlendorff, C.; Mørch, L.S. Potential preventive properties of GLP-1 receptor agonists against prostate cancer: A nationwide cohort study. *Diabetologia* **2023**, *66*, 2007–2016. [CrossRef] [PubMed]
333. Wang, J.; Kim, C.H. Malignancies associated with DPP4 inhibitors and GLP1 receptor agonists: Data from a large real-world database. *J. Clin. Oncol.* **2020**, *38* (Suppl. S15), 1567. [CrossRef]
334. Wang, L.; Xu, R.; Kaelber, D.C.; Berger, N.A. Glucagon-Like Peptide 1 Receptor Agonists and 13 Obesity-Associated Cancers in Patients With Type 2 Diabetes. *JAMA Netw. Open* **2024**, *7*, e2421305. [CrossRef] [PubMed]
335. Popovic, D.S.; Patoulias, D.; Popovic, L.S.; Karakasis, P.; Papanas, N.; Mantzoros, C.S. Tirzepatide use and the risk of cancer among individuals with type 2 diabetes mellitus: A meta-analysis of randomized controlled trials. *Diabetes Res. Clin. Pract.* **2024**, *213*, 111758. [CrossRef] [PubMed]
336. Caruso, I.; Di Gioia, L.; Di Molfetta, S.; Caporusso, M.; Cignarelli, A.; Sorice, G.P.; Laviola, L.; Giorgino, F. The real-world safety profile of tirzepatide: Pharmacovigilance analysis of the FDA Adverse Event Reporting System (FAERS) database. *J. Endocrinol. Investig.* **2024**, *47*, 2671–2678. [CrossRef]

337. Lin, H.W.; Tseng, C.H. A Review on the Relationship between SGLT2 Inhibitors and Cancer. *Int. J. Endocrinol.* **2014**, *2014*, 719578. [CrossRef]
338. Wang, F.; Hendryx, M.; Liu, N.; Bidulescu, A.; Mitra, A.K.; Luo, J. SGLT2 Inhibitor Use and Risk of Breast Cancer Among Adult Women with Type 2 Diabetes. *Drug Saf.* **2024**, *47*, 125–133. [CrossRef]
339. Chung, C.T.; Lakhani, I.; Chou, O.H.I.; Lee, T.T.L.; Dee, E.C.; Ng, K.; Wong, W.T.; Liu, T.; Lee, S.; Zhang, Q.; et al. Sodium-glucose cotransporter 2 inhibitors versus dipeptidyl peptidase 4 inhibitors on new-onset overall cancer in Type 2 diabetes mellitus: A population-based study. *Cancer Med.* **2023**, *12*, 12299–12315. [CrossRef]
340. Pelletier, R.; Ng, K.; Alkabbani, W.; Labib, Y.; Mourad, N.; Gamble, J.M. The association of sodium-glucose cotransporter 2 inhibitors with cancer: An overview of quantitative systematic reviews. *Endocrinol. Diabetes Metab.* **2020**, *3*, e00145. [CrossRef]
341. Kohler, S.; Lee, J.; George, J.T.; Inzucchi, S.E.; Zinman, B. Bladder cancer in the EMPA-REG OUTCOME trial. *Diabetologia* **2017**, *60*, 2534–2535. [CrossRef]
342. Alzenaidi, A.; Alkhalidi, H. PSUN250 Empagliflozin Induced Bladder Cancer in an Adult Patient With Latent Autoimmune Diabetes. *J. Endocr. Soc.* **2022**, *6*, A392. [CrossRef]
343. Spiazzi, B.F.; Naibo, R.A.; Wayerbacher, L.F.; Piccoli, G.F.; Farenzena, L.P.; Londero, T.M.; da Natividade, G.R.; Zoldan, M.; Degobi, N.A.H.; Niches, M.; et al. Sodium-glucose cotransporter-2 inhibitors and cancer outcomes: A systematic review and meta-analysis of randomized controlled trials. *Diabetes Res. Clin. Pract.* **2023**, *198*, 110621. [CrossRef] [PubMed]
344. Lin, Y.; Zhang, Y.; Wang, S.; Cao, L.; Zhao, R.; Ma, X.; Yang, Q.; Zhang, L.; Yang, Q. Pharmacological targets of SGLT2 inhibition on prostate cancer mediated by circulating metabolites: A drug-target Mendelian randomization study. *Front. Pharmacol.* **2024**, *15*, 1443045. [CrossRef] [PubMed]
345. Tang, H.; Dai, Q.; Shi, W.; Zhai, S.; Song, Y.; Han, J. SGLT2 inhibitors and risk of cancer in type 2 diabetes: A systematic review and meta-analysis of randomised controlled trials. *Diabetologia* **2017**, *60*, 1862–1872. [CrossRef] [PubMed]
346. Xu, B.; Kang, B.; Li, S.; Fan, S.; Zhou, J. Sodium-glucose cotransporter 2 inhibitors and cancer: A systematic review and meta-analysis. *J. Endocrinol. Investig.* **2024**, *47*, 2421–2436. [CrossRef]
347. Goodwin, P.J.; Chen, B.E.; Gelmon, K.A.; Whelan, T.J.; Ennis, M.; Lemieux, J.; Ligibel, J.A.; Hershman, D.L.; Mayer, I.A.; Hobday, T.J.; et al. Effect of Metformin vs. Placebo on Invasive Disease-Free Survival in Patients With Breast Cancer: The MA.32 Randomized Clinical Trial. *JAMA* **2022**, *327*, 1963–1973. [CrossRef]
348. Hegedüs, L.; Sherman, S.I.; Tuttle, R.M.; von Scholten, B.J.; Rasmussen, S.; Karsbøl, J.D.; Daniels, G.H. LEADER Publication Committee on behalf of the LEADER Trial Investigators. No Evidence of Increase in Calcitonin Concentrations or Development of C-Cell Malignancy in Response to Liraglutide for Up to 5 Years in the LEADER Trial. *Diabetes Care* **2018**, *41*, 620–622. [CrossRef]

biomolecules

MDPI

Review

Metabolic Syndrome and Biotherapeutic Activity of Dairy (Cow and Buffalo) Milk Proteins and Peptides: Fast Food-Induced Obesity Perspective—A Narrative Review

Kenbon Beyene Abdisa [1], Emőke Szerdahelyi [2], Máté András Molnár [1], László Friedrich [3], Zoltán Lakner [4], András Koris [1], Attila Toth [5,*,†] and Arijit Nath [1,*,†]

1 Department of Food Process Engineering, Institute of Food Science and Technology, Hungarian University of Agriculture and Life Sciences, Ménesi út 44, HU-1118 Budapest, Hungary; kenbonb@gmail.com (K.B.A.)
2 Department of Nutrition, Institute of Food Science and Technology, Hungarian University of Agriculture and Life Sciences, Somlói út 14-16, HU-1118 Budapest, Hungary; nemethne.szerdahelyi.emoke@uni-mate.hu
3 Department of Refrigeration and Livestock Product Technology, Institute of Food Science and Technology, Hungarian University of Agriculture and Life Sciences, Ménesi út 43-45, HU-1118 Budapest, Hungary
4 Department of Agricultural Business and Economics, Institute of Agricultural and Food Economics, Hungarian University of Agriculture and Life Sciences, Villányi út 29-43, HU-1118 Budapest, Hungary
5 Division of Clinical Physiology, Department of Cardiology, Faculty of Medicine, University of Debrecen, Móricz Zsigmond út 22, HU-4032 Debrecen, Hungary
* Correspondence: atitoth@med.unideb.hu (A.T.); arijit.nath@uni-mate.hu or arijit0410@gmail.com (A.N.)
† These authors contributed equally to this work.

Abstract: Metabolic syndrome (MS) is defined by the outcome of interconnected metabolic factors that directly increase the prevalence of obesity and other metabolic diseases. Currently, obesity is considered one of the most relevant topics of discussion because an epidemic heave of the incidence of obesity in both developing and underdeveloped countries has been reached. According to the World Obesity Atlas 2023 report, 38% of the world population are presently either obese or overweight. One of the causes of obesity is an imbalance of energy intake and energy expenditure, where nutritional imbalance due to consumption of high-calorie fast foods play a pivotal role. The dynamic interactions among different risk factors of obesity are highly complex; however, the underpinnings of hyperglycemia and dyslipidemia for obesity incidence are recognized. Fast foods, primarily composed of soluble carbohydrates, non-nutritive artificial sweeteners, saturated fats, and complexes of macronutrients (protein-carbohydrate, starch-lipid, starch-lipid-protein) provide high metabolic calories. Several experimental studies have pointed out that dairy proteins and peptides may modulate the activities of risk factors of obesity. To justify the results precisely, peptides from dairy milk proteins were synthesized under in vitro conditions and their contributions to biomarkers of obesity were assessed. Comprehensive information about the impact of proteins and peptides from dairy milks on fast food-induced obesity is presented in this narrative review article.

Keywords: metabolic syndromes; metabolic factors; hyperglycemia; dyslipidemia; fast food-induced obesity; dairy milk protein; peptides

1. Introduction

Metabolic syndrome (MS) is a consequence of the interplay between different metabolic factors that directly increase the risks of a wide range of diseases. Over the last decade, it has been considered one of the leading causes of morbidity and mortality [1], including premature death [2]. According to the medical classification list of the World Health Organization (WHO), MS has been recognized as the International Statistical Classification of Diseases and Related Health Problems (ICD) with ICD-9 and ICD-10 having codes 277.7 and E88.81, respectively [3]. There are 16 different possible combinations of risk factors that could be diagnosed for MS; however, not all can be considered equally in terms of their impact on risk. It may be supposed that interconnections of some metabolic risk factors, such as insulin resistance, hypertension, glucose intolerance, oxidative stress, inflammation, dyslipidemia (elevated triglycerides (TGs) and low-density lipid-cholesterol (LDL-C), and lower levels of high-density lipid-cholesterol (HDL-C)), thrombophilia, microalbuminuria, and endothelial dysfunction may be considered influencing factors for various diseases, such as diabetes mellitus (DM), cancer, non-alcoholic fatty liver disease (NAFLD), neurological diseases (stroke, depression, and Alzheimer's disease), polycystic ovary syndrome, chronic kidney disease, nonalcoholic steatohepatitis, gout, cardiovascular disease (coronary artery disease, atherosclerosis, heart failure, sleep apnea, arterial thromboembolism, peripheral artery disease), and many more [4]. Thus, MS may certainly be viewed as a risk syndrome, and it is well recognized that patients with MS are more frequently hospitalized and have greater drug expenditure [5]. MS is characterized by some major signs, such as a rise in visceral obesity, dyslipidemia, hyperglycemia, and elevated blood pressure and hypertension. Another characteristic of MS is chronic low-grade inflammation as a consequence of the complex interplay between different risk factors [6]. Different risk factors for MS and associated diseases are presented in Figure 1.

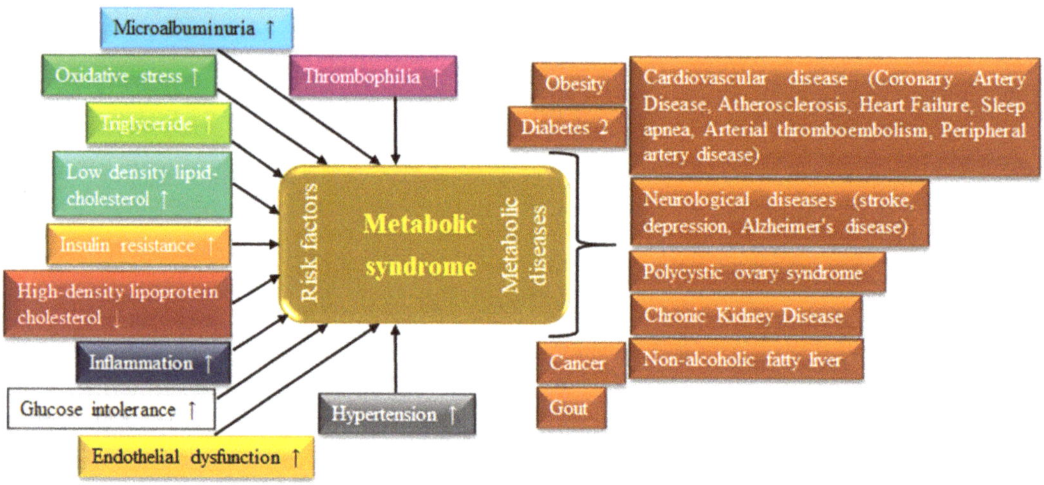

Figure 1. Risk factors of MS and associated metabolic diseases (self-developed, concept was adopted from Mendrick et al., 2018 [7]).

The concept of MS as the association of hypertension, hyperglycemia and gout was demonstrated by Swedish physician Kylin in 1920 [4]. Vague in 1947 described that visceral obesity is usually associated with the metabolic abnormalities found in DM and cardiovascular diseases (CVDs) [8]. Avogaro and Crepaldi described a syndrome which comprised hypertension, hyperglycemia, and obesity at the European Association for the Study of Diabetes annual meeting in 1965 [9]. The term "metabolic syndrome" was first used by Haller and Hanefeld in 1975. They conferred that MS is a combination of simultaneous risk factors of CVDs and diabetes that occur together more frequently than by chance alone [10]. In 1988, Reaven mentioned "a cluster of risk factors for diabetes and cardiovascular disease" and named it "Syndrome X", and insulin resistance as the "driving force" of the syndrome was proposed [11]. Subsequently in 1989, Kaplan renamed the syndrome "The Deadly Quartet" for the combination of upper body obesity, glucose intolerance, hypertriglyceridemia, and hypertension [12]. It was again renamed "The Insulin Resistance Syndrome" in 1992 [13]. Several diagnostic criteria, such as insulin resistance, body weight, blood pressure, glucose and lipid were reported to define MS. The most commonly used criteria for defining MS are from the WHO [14], the International Diabetes Federation (IDF) [15], the European Group for the study of Insulin Resistance (EGIR) [16] and the American Association of Clinical Endocrinologists (AACE) [17]. Diagnostic criteria proposed for defining MS by different authorized bodies are presented in Table 1.

Table 1. Diagnostic criteria proposed for MS (self-developed, concept was adopted from Jha et al., 2023 [6]).

	WHO	IDF	EGIR	AACE
Insulin resistance	Impaired glucose tolerance, Impaired fasting glucose, T2DM, or Lowered insulin sensitivity * plus any 2 of the following.	-	Plasma insulin >75th percentile plus any 2 of the following	Impaired fasting glucose or Impaired fasting glucose plus any of the following based on the clinical judgment
Blood glucose	Impaired glucose tolerance, Impaired fasting glucose, or T2DM	$\geq$100 mg·dL^{-1} (includes diabetes) **	Impaired glucose tolerance or Impaired fasting glucose (but not diabetes)	Impaired fasting glucose or Impaired fasting glucose (but not diabetes)
Dyslipidemia	TG $\geq$ 150 mg·dL^{-1} and/or High density lipoprotein cholesterol < 35 mg·dL^{-1} in men or <39 mg·dL^{-1} in women.	TG $\geq$ 150 mg·dL^{-1} or on receiving treatment for TG. High density lipoprotein cholesterol < 40 mg·dL^{-1} in men or <50 mg·dL^{-1} in women or on receiving treatment for High density lipoprotein cholesterol.	TG $\geq$ 150 mg·dL^{-1} and/or High density lipoprotein cholesterol < 39 mg·dL^{-1} in men or women.	TG $\geq$ 150 mg·dL^{-1} and High density lipoprotein cholesterol < 40 mg·dL^{-1} in men or <50 mg·dL^{-1} in women
Blood pressure	$\geq$140/90 mm Hg	$\geq$130 mm Hg systolic or $\geq$85 mm Hg diastolic or on receiving treatment for hypertension.	$\geq$140/90 mm Hg or on receiving treatment for hypertension.	$\geq$130/85 mm Hg
Body weight	Men: waist-to-hip ratio >0.90; Women: waist-to-hip ratio >0.85 and/or BMI > 30 kg·m^{-2}.	Increased waist circumference (population specific) plus any 2 of the following.	Waist circumference $\geq$94 cm in men or $\geq$80 cm in women.	BMI $\geq$ 25 kg·m^{-2}
Others	Microalbuminuria: Urinary excretion rate of >20 mg·min^{-1} or albumin: creatinine ratio of >30 mg·g^{-1}.	-	-	Other features of insulin resistance ***

* Insulin sensitivity was considered under hyperinsulinemic euglycemic conditions, glucose uptake below lowest quartile for background population under investigation. ** In 2003, the American Diabetes Association (ADA) changed the criteria for impaired fasting glucose tolerance from >110 mg·dL^{-1} to >100 mg·dL^{-1}. *** Includes family history of T2DM, polycystic ovary syndrome, sedentary lifestyle, advancing age, and ethnic groups susceptible to T2DM. Abbreviations: WHO: World Health Organization, IDF: International Diabetes Federation, EGIR: European Group for the study of Insulin Resistance, AACE: American Association of Clinical Endocrinology.

However, each definition holds some common features, there are several parameters that differ according to authorized bodies. Different diagnostic criteria partly influence the estimation of the prevalence of MS. Therefore, it may be realized that accurate estimation of the prevalence of MS is quite difficult. The global prevalence of MS depends on several issues, such as region, urban or rural environment, composition of the population (sex, age, race and ethnicity), no matter which criteria are used to estimate the prevalence of MS [18]. In addition to genetic and epigenetic factors, it was proven that physical inactivity, environmental contaminants, an unhealthy lifestyle and diet are significantly associated with a wide range of metabolic diseases [19]. Therefore, MS comprises several complex biochemical pathways and mechanisms that are not fully comprehended. It has been estimated that the prevalence of MS ranges from around 10% to 30% of the world's adult population, depending on the diagnostic criteria suggested by the Adult Treatment Panel III of the National Cholesterol Education Program (ATP-III) and the Joint Interim Statement (JIS) [20]. There have been several studies on the medical costs of MS and related diseases because the economic burden of MS has been greatly increasing [21,22].

Presently, technological advances have enabled more precise, intrinsic, high-throughput analysis by different types of omics results. Multi-omics concepts, including genetics, transcriptomics, epigenetics, proteomics and metabolomics have been considered as a promising approach to detect, characterize and understand the pathophysiology of different metabolic diseases [23]. However, several risk factors of MS were identified; the pipelines between them are complicated, still unclear and often contrasting results were published [24]. It may be believed that MS is primarily contributed to by overproduction of reactive oxygen species (ROSs), inflammatory biomarkers, dysfunctional adipocytes (accumulation in visceral fat), and insulin resistance [25,26] and prothrombotic state [27]. Furthermore, an unhealthy gut microbiome and environmental contaminants in the food chain modulate risk factors of MS [7]. Several meta-analyses of experimental evidence indicated that the environmental contaminant Dichlorodiphenyltrichloroethane (DDT) and its derivatives, such as Dichlorodiphenyldichloroe-thylene (DDE) in the food chain, modulate mitochondrial dynamics (mitochondrial fusion and fission, mitophagy and mitochondrial biogenesis), mitochondrial function and the prevalence of insulin resistance [28]. Results of metagenomic studies of fecal microbiota and association of circulating trimethylamine N-oxide (TMAO) in pathological conditions reveal the contribution of gut microbiota in the prevalence of MS [29,30].

According to the WHO, IDF, EGIR and AACE, the impacts of two major metabolic diseases, such as dyslipidemia and hyperglycemia, on mortality, morbidity and therapeutic expenditures are rising [10]. Obesity is a multifactorial disorder, influenced by genetic factors, dietary pattern, gut microbiota and environmental factors. Mutations in various genes responsible for controlling appetite and metabolism are primary causes of obesity; however, several investigations documented that lifestyle factors including dietary patterns have contributed to an increased occurrence of hyperglycemia and dyslipidemia, resulting in an increased prevalence of obesity [31,32]. Furthermore, gene-environment interactions play a significant role in MS including obesity. According to epidemiologic studies in 2016, it was estimated that obesity can affect 1.12 billion persons by 2030 [33]. The recent Coronavirus disease 2019 (COVID-19) pandemic also influenced illness severity and obesity. It has been recognized that hyperglycemia in the non-diabetic range is an important risk factor for COVID-19 [34]. However, the underpinnings of the pathogenesis of obesity are not yet fully understood, it was confirmed that the etiology of obesity is the imbalance between energy intake and expenditure, which is related with dietary intake (nutritional imbalance due to mainly empty calorie food intake) and lifestyle [35]. Complex multidirectional interactions, described by the "Interaction Model" in the case of obesity, are presented in Figure 2.

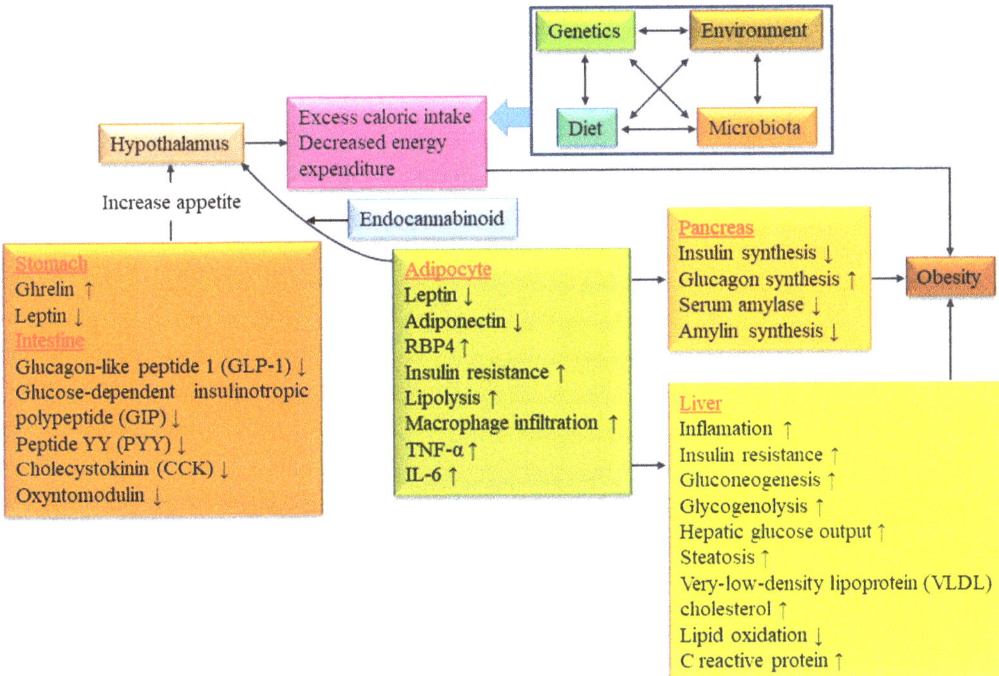

Figure 2. Interconnection between genetic factors, environmental factors, dietary pattern and gut microbiota associated with the incidence of obesity (self-developed, concept was adopted from Ussar et al., 2016 [36]).

Presently, fast foods are being taken into consideration in dietary patterns due to the rapid socio-economic transition, taste preference, the relatively low cost for large portion sizes and quick availability [37]. It has been proven by several investigations that consumption of fast foods is linked with the mentioned two metabolic diseases, which may lead to an increased risk of obesity and a number of non-communicable diseases [38,39]. Typically, high-calorie fast foods are composed with of amounts of soluble carbohydrates, non-nutritive artificial sweeteners, saturated fats, and protein-carbohydrate (Maillard products), starch-lipid and starch-lipid-protein complexes [40]. The nutritional values of foods are lost due to the biochemical modulation of macro- and micro- nutrients during cooking and processing at high temperature [41]. Deep frying, spray drying, microwave heating and extrusion processes are widely used for the preparation of a wide range of fast foods. The oil-food interaction at high temperatures due to the deep frying process leads to physical and chemical changes, such as starch gelatinization, protein denaturation, flavor and color formation, and many more [42]. Edible oils undergo various biochemical modifications during frying at high temperature and massive amounts of toxic active compounds are produced. These products are generally nonvolatile (epoxides, oxidized polymers, carbonyls, and polar dimers) and volatile (aldehydes, alcohols, and hydrocarbons). These toxic compounds are considered potential carcinogenic, mutagenic, genotoxic and teratogenic substances for human health [43]. Due to prolonged heating of cooking oils, TGs begin to break down to free fatty acids (FFAs) and glycerol, and glycerol is quickly dehydrated to acrolein. A wide range of potentially toxic compounds in the class of aldehydes and polycyclic aromatic hydrocarbons were identified in deep fried foods [44]. The lipid peroxidation of fatty acids leads to the formation of lipid peroxides, which first convert to reactive dicarbonyls and finally to advanced glycation end products (AGEs) [45]. Starch-lipid and starch-lipid-protein complexes are produced by a series of noncovalent interactions, including hydrogen bonds, hydrophobic attractions and van der Waals forces among macronutrients during the deep frying process [46]. One common complex biomolecule in fast foods is Maillard products. They are produced by a Maillard reaction, where a covalent bond forms between a free amino group in protein and a carbonyl group of a reducing carbohydrate of polysaccharide due to cooking or processing at high temperature. A series of reactions, such as the formation of a Schiff base, followed by an Amadori rearrangement and subsequently oxidative modifications (glycoxidation) are involved to produce Maillard products [47]. However, Maillard products are produced as a part of the healthy metabolism of carbohydrates, but consumption of excessively high amounts of Maillard products is not appreciable. Apart from Maillard products, several harmful substances and toxic compounds, such as furan, heterocyclic aromatic amines, acrylamide, acrolein and trans fatty acids were identified in food matrices produced by high temperature. They have the ability to promote oxidative stress and inflammation by binding with cell surface receptors or cross-linking with intracellular proteins, and modulate their structure and function [47]. Furthermore, the use of soluble carbohydrates and non-nutritive high-intensity sweeteners, such as aspartame, neotame, acesulfame potassium, advantame, saccharin and sucralose in fast foods has been reported [48]. Non-nutritive sweeteners modulate the secretions of insulin and anorexigenic (appetite suppressing) hormones (glucagon-like peptide 1 (GLP-1), peptide YY (PYY) and cholecystokinin (CCK)), insulin resistance, composition of gut microbiota and postbiotics, mainly short-chain fatty acids (SCFAs) [48].

Obesity is a heterogeneous and multifactorial health disorder; however, it is a well-established fact that the risk of obesity is associated with food intake, appetite and energy balance. Currently, obesity due to consumption of fast foods is considered one of the most relevant topics of discussion in the context of the management of MS because an epidemic heave of the incidence of obesity in both developing and underdeveloped countries has been reached. Furthermore, it has been proven that moderate and higher affinity toward a Western diet is associated with a higher risk of obesity and COVID-19 infection during the pandemic [49]. The primary objective of this article is to make a link between two major metabolic diseases, such as hyperglycemia and dyslipidemia with fast food-induced obesity. Another objective of this review is to highlight the beneficial role of dairy milk (cow and buffalo) proteins and peptides in dietary choice to combat obesity prevalence. Therefore, inclusive information about obesity contributed to by hyperglycemia and dyslipidemia due to consumption of high-calorie fast foods is presented in this narrative review article. Subsequently, the therapeutics and beneficial attributes of dairy milk-based proteins and peptides for fast food-induced obesity are mentioned with a biochemical viewpoint and up-to-date results of clinical trials in this narrative review article.

2. Methodology

A rapid scoping review was conducted. It was conducted to map evidence on the mentioned topic, ascertain information from the available literature, summarize information and concepts, identify prime concepts and theories, and identify the gaps in the existing knowledge with the aid of planning and commissioning future research. The present scoping review fits with the questions of aims and objectives which sit across diverse research. It allowed for a broad search and the review was reported following the PRISMA Scoping Reviews (PRISMA-ScR) Checklist [50].

3. Obesity

Obesity has been recognized as a disease since 1948 by the WHO and in 1997, the WHO recognized this complex and chronic disease is not just constrained to the affluent in Western nations. According to the International Classification of Diseases 11 (ICD-11), obesity is a chronic complex and multifactorial disorder, characterized by an excess of body fat. Obesity is associated with higher consumption of carbohydrates and fat, and a wide range of metabolic dysfunctions, which affect multiple organs of the body and subsequently disrupt their regular function. The epidemic heave of obesity has spread beyond the borders of high-income nations to low- and middle-income countries, predominantly in urban areas [51]. Presently, probable different biomarkers and molecular mechanisms of Western diet-induced obesity and obesity-related carcinogenesis were published. The contribution of insulin resistance, hyperglycemia, hypertriglyceridemia, hypoxia, oxidative stress, mitochondrial dysfunction, dysregulation of glycolysis and lipogenesis, adipokine/cytokine/exosome release, angiogenesis and epithelial to mesenchymal transition (EMT) have been recognized [52].

The prevalence of MS is increasing among adults worldwide and it is largely related to possible factors, such as heredity, epigenetic modifications, socioeconomic status, diets, gut dysbiosis, sedentary lifestyle, medication, medical condition and high body mass index (BMI) [53]. According to the WHO in 2019, more than 1 billion people in the world were obese, including 650 million adults, 340 million adolescents and 39 million children [54]. In another report, there were ~2 billion adults who were overweight, while 650 million were obese. It may be realized that if this rate will not slow down, 2.7 billion adults might be overweight and over 1 billion will be obese by 2025 [55]. According to the World Obesity Atlas 2023 report, 38% of the world population are currently either overweight or obese [56].

It is proven that the central nervous system plays an essential role in the control of food intake and energy homeostasis through the expressions of a wide range of genes in the hypothalamus and pituitary gland. In the early phase of the 20th century, it was assumed that dysfunction of the pituitary/hypothalamus leads to obesity. During the 1940s to 1970s, it was assumed that psychological and psychodynamic aspects are the key factors leading to obesity. In 1956, obesity was introduced as a syndrome. The role of genetic factors in body weight regulation came to the forefront in the 1980s and early 1990s [57]. Recently, it is stated that two factors, such as (a) obesogenic environment triggers obesity-predisposing genes and (b) epigenetic modulation contribute to obesity incidence. The first genome-wide association studies (GWAS) for obesity traits were published in 2007. The first identified obesity-susceptibility locus was the *FTO* gene (fat mass obesity associated) and until now this gene is recognized as the highest contributor to the risk of obesity [58]. Subsequently, many more GWAS were considered and approximately 60 GWAS identified more than 1100 independent loci associated with a range of obesity outcomes [59]. Genomic studies reveal that more than 300 single-nucleotide polymorphisms and 227 genetic variants are related to obesity, although their functional attributions on the obese phenotype are still unclear [60]. Genetic variations, such as single-nucleotide polymorphisms (SNPs), copy number variants (CNVs), and small insertions and deletions of nucleotide sequences contribute to the development of obesity [61,62]. Obesity can be classified into three categories based on genetic contribution. Those are: (a) common polygenic obesity, outcome of the interaction between environmental factors and genetic susceptibilities, (b) syndromic obesity, characterized by obesity combined with developmental delay or dysmorphism and (c) monogenic obesity, outcome of pathogenic variants in single genes, generally involved in the hypothalamic leptin-melanocortin pathway along with the regulation of satiety and hunger [63]. Common polygenic obesity has high prevalence, does not follow the principles of Mendelian inheritance and heritability similar to other multifactorial diseases. In this case, a wide range of variants in several genes interact with environmental factors. Monogenic obesity has a low prevalence, follows a Mendelian inheritance pattern and early onset devastated clinical manifestations. In the case of monogenic obesity, genetic variants in single genes lead to large effects [62]. Furthermore, obesity can be classified into the following categories, such as overweight (BMI = 25 to < 30 kg·m^{-2}), moderate obesity (BMI = 30 to <35 kg·m^{-2}) and severe obesity (BMI ≥ 35 kg·m^{-2}) [64]. Contributions of leptin, leptin receptor, proopiomelanocortin, prohormone convertase 1, melanocortin 4 receptor, single-minded homolog 1, brain-derived neurotrophic factor and neurotrophic tyrosine kinase receptor type 2 genes have been identified as causes for obesity [65,66]; however, more than 500 obesity-related genes were identified [67]. High maternal BMI, high gestational weight gain and diabetes may cause metabolic disorders including obesity in offspring. The epigenetic modifications and imbalance of unhealthy nutritional supply to the fetus in the maternal phase may influence the development of the fetus, fat accumulation and other metabolic complications [68]. Obesity experiences in infancy and childhood phase may predispose to adult obesity. Furthermore, it has been proven that children with obesity are five times more likely to be obese in adulthood [69].

It has been proven that obesity plays an important role in the homeostatic regulation of systemic glucose. Patients with type 1 DM (T1DM) are hardly obese at the time of assessment and diagnosis; whereas patients can be overweight or obese due to insulin resistance and impaired glucose tolerance. Subsequently, it manifests a hyperglycemic condition, characteristic of T2DM. Due to high body fat content, patients with T2DM often experience various types of cardiovascular risk factors, such as hypertension and dyslipidemia (low high-density lipoprotein and/or elevated TGs) [70,71]. On the other hand, a woman may experience hypertension and dyslipidemia prior to GDM [66]. Apart from diabetes and dyslipidemia, obesity is correlated with other metabolic disorders, including NAFLD, CVDs, chronic kidney diseases and cancers [72]. Therefore, the effects of obesity attribute a tremendous financial burden on the health care system. Several investigations indicate that medical costs of obesity management were higher compared to individuals with normal body weight [73]. Obesity is characterized by changes in metabolic pathways and complex metabolic imbalance related to energy balance, glucose, lipid and adipose tissue homeostasis, dysregulation of other physiological processes, including the activity of the central and peripheral nervous system, and their interactions.

3.1. Diabetes

DM has a long history reaching back to antiquity, dated back to 1500 BC. It is a chronic metabolic disease involving all types of macronutrients (carbohydrate, protein and lipid) metabolism. DM is characterized by a physiologically abnormal hyperglycemic condition, represented by continued high levels of blood glucose as a result of a wide range of biochemical pathways, predominantly related to insulin secretion or its action or both. Diabetes with its ever-increasing global prevalence is parallel to the rapid economic growth, urbanization and adoption of modern lifestyle [74]. In the year 2017, diabetes had a global 22.9 million incidences, 476.0 million prevalence, 1.37 million deaths and 67.9 million disability-adjusted life-years. It was projected that in 2025, diabetes may cause 26.6 million incidences, 570.9 million prevalence, 1.59 million deaths and 79.3 million disability-adjusted life-years [75].

Diabetes may be classified into four categories based on both etiology and pathogenesis of the disease. Those are T1DM (accounts for 5–10% of all diabetic cases), T2DM (accounts for 90–95% of all diabetic cases), gestational DM (GDM) (1–14% of all pregnancies) and secondary diabetes caused or associated with certain specific conditions, pathologies, and/or disorders (constitute a smaller percentage) [70,71]. T1DM, also known as insulin-dependent DM (IDDM) or juvenile-onset diabetes, is an autoimmune disorder characterized by T-cell-mediated destruction of β-cells in the pancreas, which results in insulin deficiency and later hyperglycemia [76]. T2DM or adult-onset diabetes is non-insulin-dependent DM (NIDDM) and is characterized by two main insulin-related irregularities, such as insulin resistance and insulin deficiency due to the dysfunction of pancreatic β-cells [77]. GDM is defined as higher fasting and post-prandial blood glucose levels or any degree of glucose intolerance at the onset of or during pregnancy, generally in the second or third trimester [78]. Furthermore, T1DM, T2DM and GDM, secondary diabetes may appear due to some specific conditions including various pathologies and/or several disorders. The majority among this type of diabetes includes maturity-onset diabetes of the young (MODY), neonatal diabetes, lipoatrophic diabetes, drug or chemical-induced diabetes, ketosis-prone diabetes mellitus, and so on [79]. Hyperglycemia and its associated metabolic pathways affect multiple organs of the body and disrupt their normal function, which leads to micro- and macrovascular complications. These complications are manifested in a wide range of body organs and systems, such as the kidney, heart, skin and nerves [70].

One of the major factors responsible for T2DM is the dysfunction of pancreatic β-cells. As an outcome of the dysfunction of pancreatic β-cells and consequently inadequate secretion of insulin, postprandial and subsequently fasting glucose levels are increased. It reduces glucose production in hepatocytes and increases glucose uptake in muscle and hepatocytes. In early diabetic conditions, complications appear to be insufficient responsiveness of pancreatic β-cells to glucose and later it turns to the reduction in pancreatic β-cell mass. The elevated plasma glucose in the case of diabetes might subsidize further disease advancement through glucotoxic effects on pancreatic β-cells and insulin sensitivity [80]. However, the classic pathogenesis of T2DM was considered an insulin deficiency from pancreatic β-cells only, later the contribution of insulin resistance was introduced. Insulin resistance is clinically defined as the incapability of a known amount of insulin (endogenous or exogenous) to escalate glucose uptake and utilization in an individual similar to a usual one [26]. A number of genes are associated with the dysfunction of pancreatic β-cells and insulin resistance in T2DM. Among those genes, the transcription factor TCF7L2 is well studied [81,82]. Glucose homeostasis is maintained by the activation of the transcription factor *FOXO1* in the liver and impaired skeletal muscle GLUT4 translocation. *FOXO1* induces the activities of key enzymes of gluconeogenesis; hence, its upregulation increases the formation of glucose in hepatocytes. In hepatocytes, insulin generally involves phosphorylation and suppression of *FOXO1* activity through the action of protein kinase Akt [83]. Downregulation of GLUT4 translocation to the plasma membrane is the cause of insulin resistance. Interestingly, it has been reported that though GLUT4 level is reduced by 50% in adipose tissue in the case of T2DM, it is unchanged in skeletal muscle [84]. However, non-esterified fatty acids (NEFAs) are critical for normal insulin release, high levels of the formation of NEFA by a lipid infusion-induced pancreatic β-cell dysfunction and insulin resistance in T2DM. The lipotoxicity due to the formation of NEFA in diabetic individuals may cause more deleterious effects, known as 'glucolipotoxicity' [85]. Furthermore, the declined capability of adipocytes to preserve and retain TG in obese individuals, resulting in the accumulation of ectopic fat in hepatocytes and muscle cells, which may cause insulin resistance. Signaling of insulin through Akt in hepatocytes activates fatty acid biosynthesis from glucose and amino acids (de novo lipogenesis (DNL)), which extended the conversion of TG to very-low-density lipoprotein (VLDL) and subsequently, accumulated in peripheral tissues [86,87]. Insulin sensitivity modulates the activity of pancreatic β-cells and it is decreased in the case of obesity. Dysfunction of pancreatic β-cells is noted with increasing fasting glucose level. It is important to mention that fasting hyperinsulinemia without detectable elevations in blood-glucose concentrations is generally noted in some individuals with long-established obesity [88]. Insulin resistance is manifested in both the hepatic and peripheral tissues. Hepatic insulin resistance results in over accumulation of glucose in the basal state, despite the presence of fasting hyperinsulinemia and hyperglycemia. Insulin resistance of peripheral tissues, mainly muscle, reduces insulin-stimulated glucose uptake and postprandial hyperglycemia [89]. Chronic elevated glucose level leads to the generation of ROSs which increase oxidative stress in a wide range of pancreatic cells followed by the dysfunction of pancreatic activity [90]. As a consequence, secretions of insulin and amylin from pancreatic β-cells and amylase from pancreatic acinar cells are reduced. Furthermore, secretion of glucagon from pancreatic α-cells is improved [91]. Dipeptidyl-peptidase-IV (DPP-IV) (EC 3.4.14.5) from epithelial and endothelial cells inhibits the secretion of PYY and oxyntomodulin hormone from intestinal enteroendocrine cells and synthesis of intestinal incretin hormones (glucose-dependent insulinotropic polypeptide (GIP) from K cells of the upper intestine and GLP-1 from L cells of the lower intestine) responsible for postprandial insulin secretion [92]. CCK from the I cells of the gut decreases postprandial glucose levels and increases insulin levels in T2DM. Therefore, absence of CCK causes glucose intolerance and influences T2DM [93]. Deficiency of gastric leptin interrupts glucose homeostasis by modulating hepatic de novo gluconeogenesis and lipogenesis, and secretion of hepatic TG [94] as well as modulating the activity of insulin [95]. Adipose tissue releases glycerol, NEFAs, hormones (adiponectin, leptin and resistin) and proinflammatory cytokines (adipokines) that are involved in the development of insulin resistance [96]. Accumulation of excessive visceral fat may be the cause of dysfunctionality of adipose tissue and consequently increases the secretion of leptin and negatively affects adiponectin. Plasma leptin stimulates oxidative stress, inflammation and insulin resistance [97]. As a consequence of reduced insulin responsiveness in adipocytes, T2DM is associated with lipolysis through the downregulated expression of adipocyte lipid-droplet proteins, such as CIDE proteins [98] and perilipin [99]. Elevated plasma FFAs due to lipolysis may cause insulin resistance in obese individuals, which may lead to cellular uptake of FFA and enhance lipid oxidation. In muscle cells, FFA oxidation reduces insulin-mediated glucose disposal and in hepatic cells, it promotes gluconeogenesis and hepatic glucose output [11]. Postprandial hyperglycemia is associated with the activity of pancreatic α-amylase and α-glucosidase in the brush border of enterocytes of the jejunum in the small intestine. α-amylase is responsible for the initial hydrolysis of the α-D-(1-4) glycosidic bonds of polysaccharides and starch to oligosaccharides; and subsequently, oligosaccharides are hydrolyzed into monosaccharides by α-glucosidase. Therefore, their activity improves the rate of glucose absorption [100]. Insulin resistance encourages individuals to consume high-calorie foods, which is linked with gastric emptying and appetite. Food intake and energy expenditure are associated with lower activity of GLP-1, PYY, CCK and oxyntomodulin from the gut. Furthermore, ghrelin from the stomach, and leptin from the stomach and adipose tissues influence appetite activity [101].

There are a number of cellular stress-responsive metabolic pathways associated with hyperglycemia-induced superoxide synthesis and inflammation. Inflammation in the intestine, hepatocytes, adipocytes and pancreatic cells may influence the synthesis and abnormal regulation of these hormones and peptides, which may contribute to the development of diabetes. It has been proven that circulatory pro-inflammatory molecules, such as tumor necrosis factor (TNF)-α, interferon-gamma (IFN-γ), interleukin (IL)-1, IL-6, IL-8, IL-1β, transforming growth factor (TGF)-β, monocyte chemoattractant protein (MCP)-1, toll-like receptors (TLR)-2 and TLR-4 on the surface of monocytes, leptin, chemerin and plasminogen activator inhibitor (PAI)-1, retinol binding protein-4, C-reactive protein and monocyte chemotactic protein-1 are increased in diabetic and obese individuals. It may be supposed that these pro-inflammatory molecules are associated with systemic insulin resistance and T2DM complications [102]. In a hyperglycemic condition, a wide range of metabolic pathways, namely, the polyol pathway (formation of sorbitol from glucose and conversion to fructose), formation of AGEs pathway (formation of methylglyoxal from glucose and conversion to AGEs), protein kinase C (PKC) isoform pathway (glucose to diacylglycerol (DAG) and stimulation of PKC isoforms, especially PKC-β which subsequently phosphorylate) and hexosamine biosynthetic pathway (glucose to uridine diphosphate N-acetylglucosamine (UDP-GlcNAc) and subsequently N-acetylglucosamine (GlcNAc)) lead to both oxidative and endoplasmic reticulum stress, resulting in chronic inflammation and insulin resistance [103]. These metabolic pathways may cause a wide range of micro- and macro-vascular complications alongside diabetes, such as cardiovascular disease, stroke, peripheral vascular disease, retinopathy, nephropathy, neuropathy and foot ulcers [104]. Inflammation in the intestine is also associated with gut microbiota, which is related to the type of consumed foods and dietary patterns [105]. Dysbiosis reduces the formation of SCFAs and leads to the formation of trimethylamine N-oxide (TMAO). Bacterial translocation and their toxic metabolites lead to inflammation in the gut and reduce insulin sensitivity [106,107]. Putative relationships between different risk factors of diabetes and associated complications are quite complex. Biochemical activities of risk factors and their contribution to the metabolic pathways for diabetes and associated complications, induced by the consumption of high-calorie fast foods are presented in Figure 3.

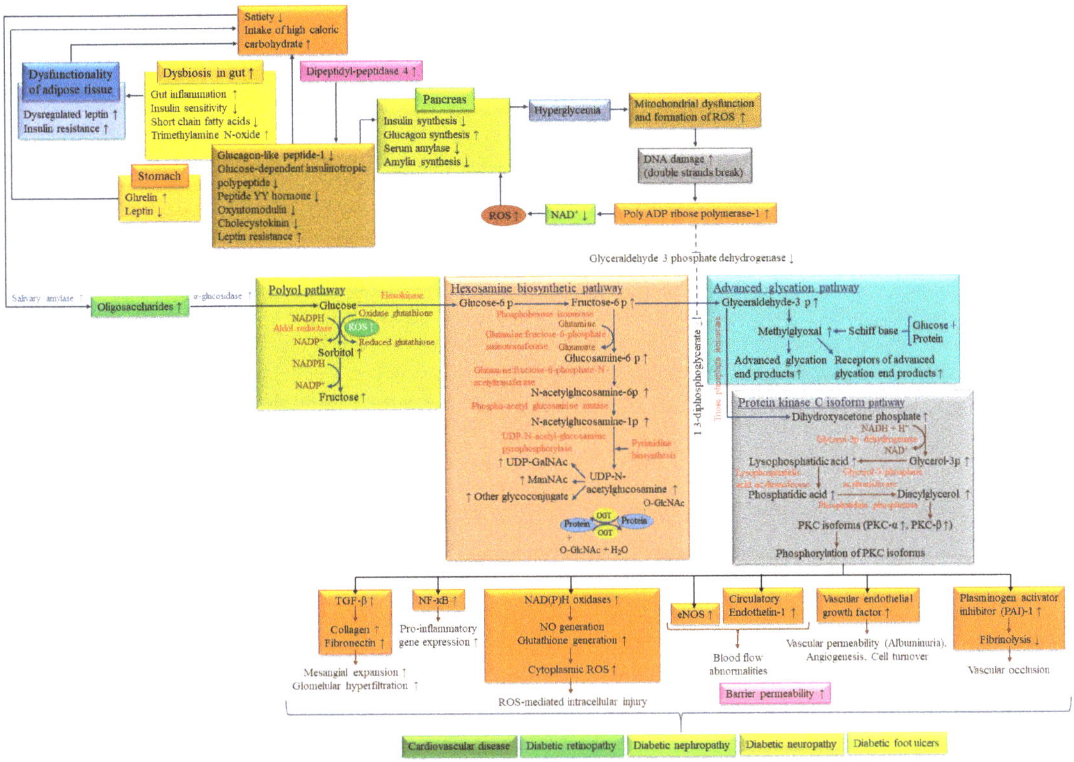

Figure 3. Metabolic pathways for diabetes and associated complications, induced by the consumption of high-calorie fast foods (self-developed, concept was adopted from Kang and Yang, 2020 [108], and Naveen and Baskaran, 2018 [109]).

Reactive carbonyl compounds (RCCs), such as AGEs are produced by non-enzymatic exogenous and endogenous systems and are recognized as key players in the progression of diabetes and diabetes-induced macrovascular and microvascular complications. The pathogenesis of carbonyl stress by AGEs is dependent on two pathways: (a) trapping and cross-linking of cellular or plasma proteins, including laminin, elastin and collagen with AGEs. It induces physiological dysfunction, such as alternation of elasticity and function of tissues, and (b) activating downstream cellular signaling pathways via RAGE receptors to trigger oxidative stress and inflammation [110]. AGEs may exist alone as an amino acid conjugate or with free low-molecular-weight (LMW, <5 kDa) peptides or bound with proteins. Absorption of pyrraline, a sugar derivative of Lys, is performed using peptide transporter 1 (PEPT1) across the intestinal epithelium [111]. The LMW peptide fractions of AGEs are more likely absorbed; while, non-absorbed AGEs are passed through the gastrointestinal tract [112]. AGEs can modulate cellular functions through binding with G-protein-coupled receptors (GPCRs), TLRs, scavenger receptors and pattern recognition receptors; however, the interactions between AGEs with cell surface receptor for advanced glycation end products (RAGE) is well-recognized for T2DM. Interaction between AGEs/RAGE influences an array of signaling events, such as activation of the renin–angiotensin system (RAS)-mediated extracellular signal-regulated kinase (ERK1/2), mitogen-activated protein kinase (MAPK), stress-activated protein kinase/c-Jun N-terminal kinase (SAPK/JNK), phosphoinositide 3-kinase/Protein kinase B (PI3-K/AKT), Janus kinase/signal transducers and activators of transcription (JAK/STAT), Ras homologous GTP-binding proteins (Rho GTPases) Rac-1, Cell division control protein 42 (Cdc42) pathways. Therefore, AGEs promote a wide range of cell-mediated pathophysiological responses, such as pancreatic β-cell toxicity, elevation of cytosolic ROS, activation of nuclear factor kappaB (NF-κB), increased formation of proinflammatory cytokines, stimulation of oxidative and endoplasmic reticulum stress. It may be realized that they are supposed to maintain mechanistic links between the risk factors of DM [110]. More specifically, the activation of JNK promotes the phosphorylation of insulin receptor substrate (IRS-1) at Ser residue, which leads to negative regulation of insulin signal transduction and induces insulin resistance. Activated NF-κB upregulates the expression of various inflammatory cytokines (TNFα, IL-1β, IL-6) and influences insulin resistance [113]. Interestingly, it has been reported that abnormal activation of the ERK1/2 signaling pathway is associated with several diabetogenic factors and adipogenesis [114]. Interaction between AGE and RAGE modulates intracellular cascade of biochemical reactions which inhibits insulin-induced GLUT-4 translocation and induces insulin resistance [115]. RAGE binds with toxic intermediates from amyloid polypeptide (IAPP) and transduces intracellular signals that lead to NADPH oxidase-mediated cellular stress and inflammation. It plays a significant role in pancreatic amyloidosis-induced β-cell proteotoxicity and β-cell apoptosis [116]. The accumulation of AGEs in the endoplasmic reticulum interferes with protein folding. Furthermore, AGEs influence mitochondrial proteins in the respiratory chain and may cause mitochondrial dysfunction (reduces the synthesis of adenosine triphosphate (ATP)) which may cause oxidative stress [117]. When AGEs in the body are relatively high, AGE-R1 typically degrades AGEs in cells and influences receptor-mediated endocytosis. AGE-RI inhibits RAGE-mediated activation of the proinflammatory gene NF-κB through upregulation of SIRT1 and subsequently inhibits oxidative stress [118]. AGE-modified low-density lipoprotein (LDL) contributes to reduced LDL clearance and oxidized LDL via activation of *LOX-1*. Oxidized LDLs exhibit a wide variety of biological and atherogenic properties involving the activation of inflammatory and mitogenic pathways. The biological activity of oxidized LDLs depends on the presence of lipid peroxidation products, such as methylglyoxal. Modified LDL alters the affinity with apoB/E receptor and deviates the metabolism towards scavenger receptor-bearing cells, such as macrophages and smooth muscle cells. It was reported that the formation of atherogenic foam cells from smooth muscle cells with angiotensin II is progressive in T2DM [119]. Activation of *LOX-1* along with oxidized LDL or AGEs can produce ROS through the activation of NADPH oxidase, which subsequently influences the ERK1/2-MAPK and PI3K pathway [120]. Interaction between AGE and RAGE activates PI3K/PDKI along with downstream phosphorylation of mTOR, which triggers the mTOR pathway [121]. Furthermore, *LOX-1* influences cytochrome P450 (CYP450) which modulates the activity of endothelium-derived hyperpolarizing factor (EDHF) and subsequently may cause inflammation [122]. Unabsorbed AGEs in the colon can affect the homeostasis of intestinal microbiota, specifically the loss of butyrate-producing bacteria. The damage to the colonic epithelial barrier and chronic low-grade inflammation due to the loss of butyrate-producing bacteria was correlated with insulin resistance and the pathogenesis of DM [123]. Biochemical pathways and metabolites associated with the consumption of fast foods are presented in Figure 4.

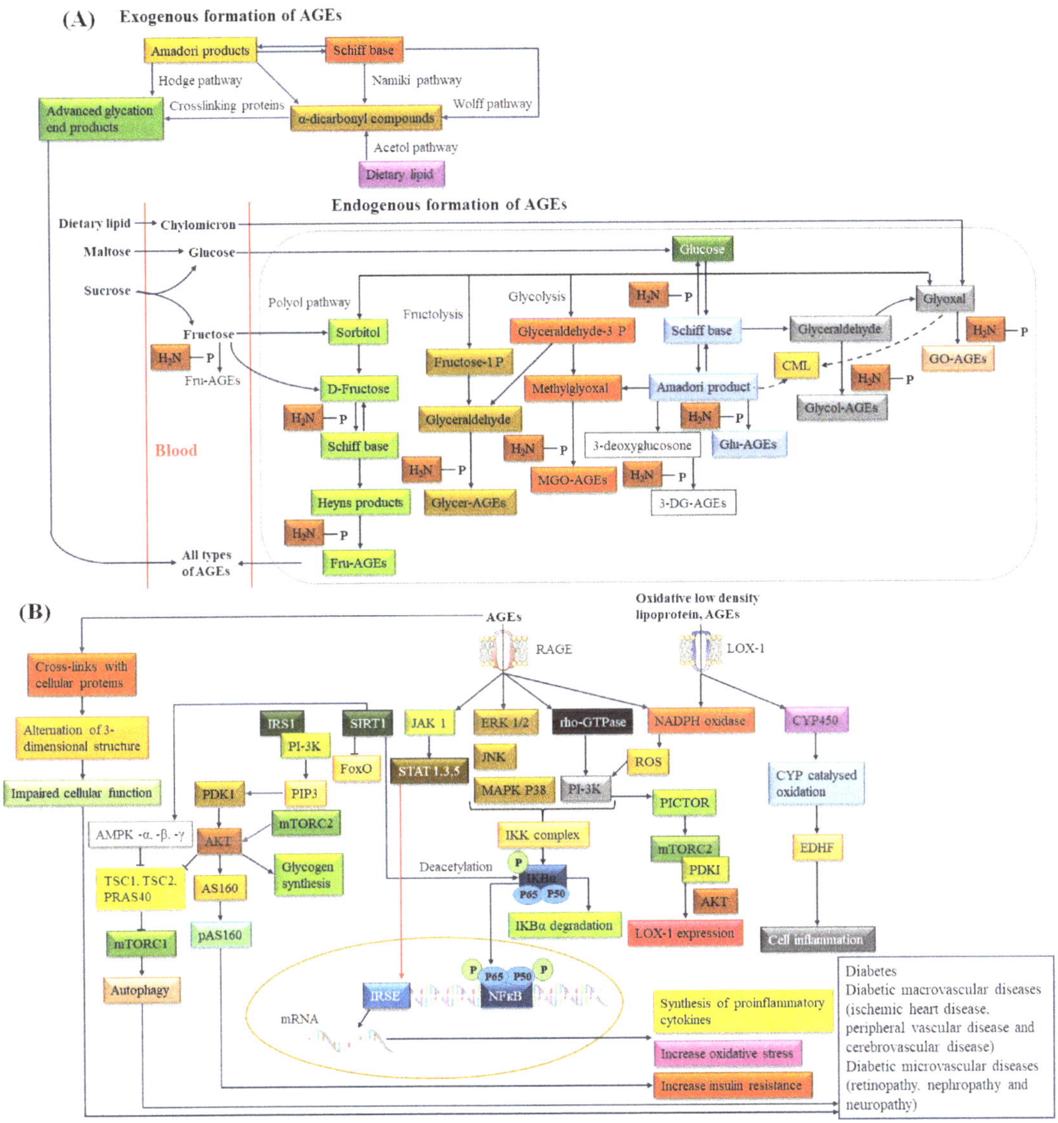

Figure 4. Biochemical pathways of (**A**) formation of exogenous and endogenous AGEs (self-developed, concepts were adopted from Inan-Eroglu et al., 2020 [124] and Takeuchi et al., 2020 [125]), and (**B**) AGE/RAGE-mediated signaling and metabolic pathways for DM (self-developed, concept was adopted from Salazar et al., 2021 [126] and Dong et al., 2023 [127]). Abbreviations: Fru-AGEs: Fructose-derived glycation end products, Glycer-AGEs: Glyceraldehyde-derived glycation end products, MGO-AGEs: Methylglyoxal-derived glycation end products, 3-DG-AGEs: 3-Deoxyglucosone-derived glycation end products, Glu-AGEs: Glucose-derived glycation end products, Glycol-AGEs: Glycol-derived glycation end products, GO-AGEs: Glyoxal-derived glycation end products.

3.2. Dyslipidemia

The metabolic disorder dyslipidemia is the imbalanced formation of elevated levels of TG or LDL-C or lower formation of HDL-C. It may be caused by (a) mutations of a single gene or multiple genes, (b) intake of excessive dietary saturated fats and soluble carbohydrates (glucose, sucrose, fructose) with high caloric index, (c) over consumption of alcohol and cigarette smoking, (d) diseases, including diabetes mellitus, chronic kidney disease, hypothyroidism, HIV infection, nephrotic syndrome, primary biliary cirrhosis and other cholestatic liver diseases, and (e) intake of medications and anabolic steroids [128]. It was shown that the prevalence of dyslipidemia in pediatrics may be associated with high BMI and promotes other cardiometabolic risk factors, such as insulin resistance, high blood pressure and endothelial dysfunction. Therefore, dyslipidemia in childhood is associated with atherosclerosis in adulthood [129].

Dyslipidemia is traditionally classified by patterns of elevation in lipids and lipoproteins in plasma, such as (a) pure or isolated hypercholesterolemia, which is caused by an increase in cholesterol only, (b) pure or isolated hypertriglyceridemia, which is caused by an increase in TGs only, and (c) mixed or combined hyperlipidemias, which is caused by an increase in both cholesterol and TG [130]. Systematically dyslipidemia may be categorized by the Fredrickson phenotype, such as (a) phenotype I where the elevated lipoprotein is chylomicron and the elevated lipid is TG, (b) phenotype IIa* where the elevated lipoprotein is LDL-C and the elevated lipid is cholesterol, (c) phenotype IIb* where the elevated lipoproteins are LDL-C and VLDL, and the elevated lipids are TG and cholesterol, (d) phenotype III where the elevated lipoproteins are VLDL and chylomicron remnants, and the elevated lipids are TG and cholesterol, (e) phenotype IV where the elevated lipoprotein is VLDL and the elevated lipid is TG, and (f) phenotype V where the elevated lipoproteins are chylomicrons and VLDL, and the elevated lipids are TG and cholesterol [131]. Particularly high levels of plasma LDL-C are considered as a major risk factor for CVDs; whereas hypertriglyceridemia associated with other diseases may cause acute pancreatitis and NAFLD. Dyslipidemia could be determined genetically (primary or familial dyslipidemia) or secondary to other conditions including DM, an unhealthy lifestyle, and so on. Hypercholesterolemia with elevated plasma LDL-C is the most common form of dyslipidemia and it increases the risk of CVDs. The global burden of dyslipidemia has significantly increased over the last 30 years. Hypercholesterolemia was considered the 15th leading risk factor for death in 1990 and 8th in 2019 [132].

Primary lipogenic tissues are hepatocytes; however, the contributions of white adipose tissue (WAT) and brown adipose tissue (BAT) are well recognized for TG homeostasis and energy storage [133]. Hepatic TG may be derived from dietary intake of long, medium and free fatty acids, esterification of plasma FFA and hepatic DNL. Fasting and postprandial TG concentrations are significantly increased by hepatic DNL [134]. In hepatocytes and adipocytes, a wide range of transcription factors (liver X receptor (LXR), sterol regulatory element-binding protein-1c (SREBP-1c) and carbohydrate response element binding protein (ChREBP)) and lipogenic enzymes (acetyl-CoA carboxylase, fatty acid synthase, fatty acid elongase 6 and stearoyl-CoA desaturase 1) play a significant role in DNL. The biochemical pathway of hepatic lipogenesis may be characterized by three sequential steps, such as fatty acid synthesis (formation of palmitate), fatty acid elongation (formation of stearate from palmitate)/desaturation (formation of oleate from stearate) and esterification (formation of TG) [135]. Therefore, saturated fatty acids (SFA) through dietary intake are associated with DNL and obesity. The consumption of dietary SFA can promote insulin and leptin resistance, and inflammation via endoplasmic reticulum stress, cellular apoptosis and activation of proinflammatory pathways in the hypothalamus, hepatocytes, adipocytes, skeletal muscle, pancreas and gastrointestinal tract [136]. Interestingly, it was reported that dietary fructose increases insulin resistance and levels of enzymes involved in DNL even more than a high-fat diet [137]. Fructose is absorbed via the portal vein because it does not require insulin for its metabolism, a single phosphorylation step by fructokinase to form fructose-1-phosphate prior to conversion to dihydroxyacetone phosphate (DHAP) and glyceraldehyde (GA) and non-influenced activity of phosphofructokinase by ATP and citrate. Therefore, fructose is delivered to hepatocytes in much higher concentrations, unlike glucose, and induces DNL. Interestingly, it has been reported that prolonged consumption of fructose may lead to hyperinsulinemia and insulin resistance. Furthermore, fructose metabolism is associated with the depletion of ATP, suppression of fatty acid oxidation in mitochondria, formation of uric acid and endoplasmic reticulum stress, which promotes inflammation and the formation of ROSs [138].

DNL is physiologically regulated, where a coordinated series of enzymatic metabolic reactions are involved. For example, during fasting, DNL is downregulated due to high levels of blood glucagon and cellular cAMP. On the other hand, blood glucose and insulin levels are improved after consumption of a carbohydrate diet. Those stimulate DNL through increasing substrate availability, activity of lipogenic enzymes and expressions of lipogenic genes [139]. Glucose and fructose are directly absorbed in the luminal membrane of enterocytes in the intestine via an energy-independent mechanism. Intestinal fructose is transported by glucose transporter 5 (GLUT5) on the lumenal side and glucose transporter 2 (GLUT2) on the basolateral side [140]. Furthermore, enterocytes of the small intestine play a significant role in the absorption of chylomicron from dietary fatty acids which are subsequently transported by carnitine palmitoyltransferase (CPT) and carnitine-acylcarnitine translocase (CACT) in the mitochondrial membrane [141]. The biochemical process of DNL in hepatocytes begins with the formation of DHAP and glyceraldehyde 3-phosphate (GA3P) from soluble dietary carbohydrates. The formation of triose-phosphates (GA, DHAP, GA3P) may concomitantly increase the formation of methylglyoxal (MG) as well as a precursor for AGEs, which have been recognized in the pathogenesis of T2DM [142]. Triose-phosphates are further metabolized to pyruvate and transported to mitochondria by MPC, a mitochondrial pyruvate transporter for energy production. In hepatic mitochondria, pyruvate is oxidized to acetyl-CoA and subsequently used in the tricarboxylic acid (TCA) cycle. When energy accumulation is abundant, citrate is transported from the mitochondria to cytoplasm of hepatocytes by the mitochondrial citrate/isocitrate carrier (CIC). Citrate is converted to acetyl-CoA, which is the first step of endogenous lipogenesis. Palmitate (C16:0) is the primary endogenously synthesized fatty acid; however, it may be elongated to stearate (C18:0) and subsequently to oleate (C18:1). The final step of lipogenesis is the conversion of diacylglycerols into TG, catalyzed by diacylglycerol acyltransferase (DGAT). In hepatocytes, TG is stored as stored energy depots or assembled into VLDL and subsequently transported to the blood. Hepatic DNL induces fasting and postprandial TG concentrations [134]. VLDL in the bloodstream is converted to LDL and lipolytic products via the complex catalytic activity of lipoprotein lipase. LDL enters hepatocytes via LDL receptor (LDLR) and LDLR-related protein-1 (LRP1), and acts as a precursor of TG [143]. Circulating glucose/fructose, fatty acids from dietary chylomicrons, dietary FFAs and VLDL derivatives from hepatocytes are initiators of DNL in adipocytes. Fatty acid derivatives enter adipocytes through fatty acid transporters, such as fatty acid transport protein-1 (FATP1), cluster of differentiation 36 (CD36), Caveolin-1 and FABPpm [144]. DNL in adipose tissue is decreased with increasing insulin resistance [145]. Therefore, in diabetic conditions, adipose-specific GLUT4 is downregulated and TG is formed by lipolytic products from blood in adipocytes by esterification [139]. It has been proven by several clinical investigations that DNL is significantly higher in hepatocytes than adipocytes due to carbohydrate-rich diets [146]. Hydrolysis of TG (lipolysis) by adipose triglyceride lipase (ATGL), hormone-sensitive lipase (HSL) and monoacylglycerol lipase (MGL) in adipose tissues liberates FFA and glycerol, which can be taken up by hepatocytes and converted to VLDL in both fed and fasted states. In the first step, ATGL hydrolyzes TG into diacylglycerol (DAG) and the first molecule of fatty acid. Subsequently, HSL hydrolyzes DAG into monoacylglycerol (MAG) and the second molecule of fatty acid. Finally, MGL hydrolyzes MAG into glycerol and the third molecule of fatty acid. These liberated fatty acids can be oxidized in BAT or muscle and glycerol may be used as a precursor in gluconeogenesis in hepatocytes [147]. Increased FFAs from lipolysis in adipose tissue reduces β-oxidation within the liver and upregulates hepatic DNL [148]. It has been proven by several clinical investigations that a higher proportion of hepatic TG is derived from plasma FFA, produced by lipolysis in adipose tissues than hepatic DNL in subjects with normal liver fat [149,150]. Under pathological conditions, lipolysis in WAT leads to lipotoxicity and insulin resistance [151]. Furthermore, lower lipogenesis in WAT decreases the synthesis of insulin-sensitizing fatty acids and consequently insulin resistance [152]. Due to the consumption of fast foods, the amount of caloric intake exceeds the calorie demand. Adipose tissues accumulate TG and grow both in size and number, resulting in obesity. The accumulation of diacylglycerols in the pathway of DNL induces insulin resistance by activating PKC, the transcription factor of NF-κB and c-Jun N-terminal kinase 1, which is linked with diabetes. Furthermore, hyperinsulinemia promotes lipolysis and the release of FFA from adipocytes, which is the cause of imbalance of adipocytokines (decreased adiponectin and/or increased proinflammatory cytokines) and lipotoxicity, and consequently hepatic steatosis [153]. Interestingly, some recent investigations represent that adipocyte-specific fatty acids improve systemic insulin sensitivity and decrease inflammation [145]. Metabolic pathways related to dyslipidemia due to the consumption of soluble carbohydrates (glucose and fructose) and dietary fats in high-calorie fast foods are presented in Figure 5.

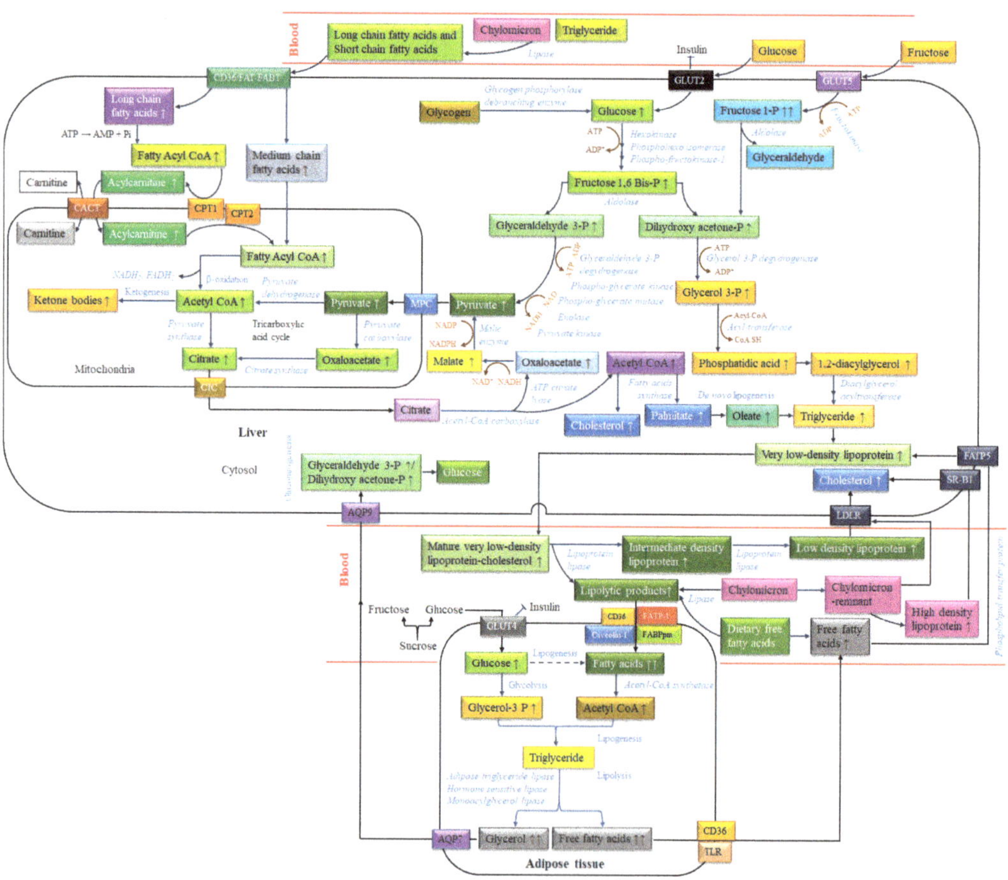

Figure 5. Metabolic pathways for dyslipidemia caused by soluble carbohydrates (glucose and fructose) and dietary fats in high-calorie fast foods (self-developed, concept was adopted from Fukushima et al., 2015 [154], and Mato et al., 2019 [155]).

Dyslipidemia may lead to symptomatic end-organ diseases, including vascular disease (coronary artery disease (CAD), stroke and peripheral arterial disease), acute pancreatitis and hepatosplenomegaly. Very high TG levels may cause dyspnea, paresthesias and confusion [156]. Severe hypertriglyceridemia (>22.6 mmol/L) may cause lipemia retinalis. Severe dyslipidemia may include localized lipid deposits (xanthoma). High levels of LDL-C may cause arcus cornea and xanthelasma palpebrarum. Xanthelasma palpebrarum can also occur in patients with primary biliary cirrhosis. Furthermore, high LDL-C levels may cause tendinous xanthomas at the achilles elbow, and knee tendons and over metacarpophalangeal joints [157].

4. Management of Obesity

Obesity has been acknowledged as a major challenge in the healthcare system and is considered an economic burden around the world. Reduction in the prevalence and risk of obesity involves a combination of lifestyle changes and pharmacological interventions. Reduction in body weight and maintaining the ideal body weight are essential to prevent a wide range of metabolic diseases. The reduction in body weight signifies a loss of 7–10% in baseline body weight within 6–12 months. It is associated with a reduction in caloric intake by 500–1000 calories·day^{-1} [1,158]. Management of obesity includes cooperation between patients and healthcare professionals with different specializations, including dietetics, endocrinologists, medicine specialists, psychologists and physiotherapists in order to achieve the best possible treatment outcome [159]. Three medications, such as sibutramine, orlistat and rimonabant are generally used for weight loss. Sibutramine, a monoamine-reuptake inhibitor, results in average weight losses of 4–5 kg; however, this medication is not suitable for patients with coronary heart disease. Orlistat, a gastrointestinal lipase inhibitor, reduces weight by average 3 kg and decreases development to diabetes in high-risk patients; however, adverse gastrointestinal disorders were noted. Rimonabant, an endocannabinoid receptor antagonist, reduces weight by an average of 4–5 kg and waist circumference. It improves the concentrations of plasma TG and HDL-C; however, an incidence of mood-related disorders can be noted [160]. Unfortunately, the unavailability of a single medication for pharmacotherapy of metabolic diseases including obesity and prolonged use of multiple medications are considered as challenging issues due to the high risk of side effects [161]. Bariatric surgery has been considered an effective treatment for metabolic diseases including obesity due to a significant decrease in calorie intake. Dramatic improvement occurs in the case of glucose intolerance/insulin resistance one year after the surgery in patients [162]; however, some bariatric surgery risks include acid reflux, anesthesia-related risks, dizziness, chronic nausea and vomiting, dilation of the esophagus, intestinal infection, ulcers, bowel obstruction, hernias and low blood sugar [163]. Low-calorie diets and physical activity are generally recommended by dietitians and medical practitioners as a first-line treatment for managing obesity and are approved by most dietary guidelines and scientific societies. For a long time, physical exercise [164,165], intake of healthy diets [166,167], natural products, and herbal medicines were considered for individuals with such risk factors [168,169]. Current guidelines for the management of T2DM recommend progressive resistance training (PRT) to improve muscle mass and glycemic control [170]; however, nutritional/diet management is also considered a central element for the treatment of T2DM and dyslipidemia. Presently, multiple studies have confirmed the beneficial effects of the Mediterranean diet (fruits and vegetables, whole grains, healthy fats, etc.) on obese individuals [171,172].

4.1. Nutraceuticals and MS

Understanding human physiology behind weight and energy gain through food intake is extremely relevant to controlling obesity incidence and other metabolic diseases. Nutritional challenges as well as caloric restriction for the management of metabolic diseases have been reported several times [173]. Therefore, there is growing interest in nutraceuticals/functional foods, management of diet habits, and lifestyle for reducing the risk factors of metabolic diseases. It may be realized that the selection of unique functional foods provides wide ranges of biological activities, which can modulate different metabolic pathways associated with the risk factors of MS [174]. A variety of plant-based nutraceuticals and their contributions to metabolic diseases have been reported. Some examples of them are mentioned herein: curcumin from turmeric (*Curcuma longa*), allicin from garlic (*Allium sativum*), quercetin from onion (*Allium cepa* L.), gingerols, shogaols, and parasols from ginger (*Zingiber officinale*), cuminaldehyde from cumin (*Cuminum cyaminum*), terpenine and cineol from cardamom (*Elettaria cardamomum*), polyphenols from cinnamon (*Cinnamomum verum*), neem oil from neem seed (*Azadirachta indica*), berberine from *Rhizoma coptidis*, bergamot essential oil from bergamot orange (*Citrus bergamia*), resveratrol and 3,5,4′-trihydroxy-*trans*-stilbene from grape seed (*Vitus vinifera* L.), saponins and galactomannan from fenugreek (*Trigonella foenum*), polyunsaturated fatty acids from animal- and plant-based oils, sulforaphrane from broccoli (*Brassica oleracea*), and SCFAs from symbiotic activity of prebiotics and probiotics [158,175].

Some recent investigations indicate that diets rich in protein and vegetables with some dietary restriction, such as consumption of refined carbohydrates can reduce insulin resistance irrespective of the change in body weight [162]. The unique nutritional value of dairy milk has catapulted it to the forefront of the functional food sector. For a long time, dairy milk has been well accepted in diet charts for all ages around the globe. The presence of wide ranges of biomolecules, such as proteins, carbohydrates, fats, minerals, and vitamins in milk offers unique nutritional value to consumers of all ages [176]. Several epidemiological and cohort studies confirmed that the consumption of dairy products decreased the prevalence of MS, while experimental studies pointed to the roles of dairy protein [177] and peptides [178] as dietary components that may modulate the activities of risk factors of MS. Potential beneficial outcomes from milk proteins include a wide range of aspects. Protein generally increases satiety to a greater extent than other primary macromolecules (carbohydrates and fats) and may facilitate a reduction in energy consumption under all dietary conditions. It may be supposed that the metabolic and appetite-suppressing effects of proteins are dependent on the quality of the protein, which is determined by the amino acid composition (non-essential or essential, glucogenic or ketogenic). Diets with higher amounts of protein are associated with increased thermogenesis, which influences the synthesis of satiety hormones, decreasing dietary calorie consumption and improving energy expenditure. Consequently, protein may provide a stimulatory effect on muscle protein anabolism and maintain BMI. Furthermore, peptides with unique functional values are produced by in vivo gastrointestinal digestion of dietary proteins. Their unique characteristics modulate physiological functions in an inclusive way and reduce the risks of a wide range of diseases [179].

4.2. Proteins and Peptides from Milk Proteins

According to the FAO dairy review, ~930 million tons of milk was produced around the globe in 2022, and it is expected to increase by 1.7% annually for almost a decade from now. Cows are the largest global milk source (81%), followed by buffalo (15%). On a global level, milk consumption stands at 112 kg·person^{-1}·year^{-1}, and milk and dairy products contribute about 18–20% of protein consumption in adults [180]. Therefore, peptides from cow and buffalo milk proteins, represented by dairy milk proteins for the management of obesity, are considered in the present review article.

Cow milk and buffalo milk contain ~3.6% and ~4.3% protein by weight, respectively [181], and have been considered excellent sources of all types of amino acids, including essential amino acids (EAAs). The protein digestibility-corrected amino acid score (PDCAAS) and digestible indispensable amino acid score (DIAAS) of milk proteins are higher than those of most other protein sources [182]. Dairy proteins are classified as (a) casein protein (80% (weight basis): αs_1-casein, αs_2-casein, β-casein, κ-casein) and (b) whey protein (20% (weight basis): β-lactoglobulin, α-lactalbumin, immunoglobulins (IgG, IgA, IgM), bovine serum albumin, lactoferrin, lactoperoxidase). The minor proteins in milk are glycoproteins and lipoproteins, which account for less than 2% of total milk proteins. They are produced from distinct genes. Buffalo milk is just like cow milk and contains two major protein fractions, such as casein protein and whey protein. Buffalo milk has a higher protein content than bovine milk. Therefore, it contains higher concentrations of caseins and whey proteins than bovine milk. The characteristics and quantity of milk proteins may change depending on the breed, feeding regime, seasons, and rearing system of animals [181]. Different distributions of caseins (α_{s1}-casein, α_{s2}-casein, β-casein, and κ-casein) and major whey proteins (β-lactoglobulin and α-lactalbumin) in cow milk and buffalo milk are reported in Table 2.

Table 2. Protein profile (g·L^{-1}) of cow and buffalo milk (self-developed, information were adopted from Roy et al., 2020 [183]).

	Casein Protein (g·L^{-1})				**Whey Protein (g·L^{-1})**	
	Total Protein for Cow (24.6–28) and Buffalo (32–40) Milk				**Total Protein for Cow (5.5–7) and Buffalo (6) Milk**	
	α_{s1}-Casein	**α_{s2}-Casein**	**β-Casein**	**κ-Casein**	**β-Lactoglobulin**	**α-Lactalbumin**
Cow	8–10.7	2.8–3.4	8.6–9.3	2.3–3.3	3.2–3.3	1.2–1.3
Buffalo	8.9	5.1	12.6–20.9	4.1–5.4	3.9	1.4

Numerous attempts have been made to characterize milk proteins [184] and their biological activities [185]. Casein contains a higher proportion of different types of EAAs, such as His, Met, Phe, and Val, than whey proteins. Furthermore, casein contains a higher proportion of several non-EAAs, including Arg, Glu, Pro, Ser, and Tyr. On the other hand, whey protein contains a higher proportion of the branched-chain amino acids (BCAAs), such as Leu, Ile, and Val, compared to casein. However, casein and whey proteins contain all EAAs, but their digestion and absorption characteristics are different. Whey proteins are highly soluble in the food matrix and under in vivo conditions. Therefore, after gastrointestinal digestion of whey proteins, their absorption rate is much faster than that of digested casein, which leads to a dramatic short-lived rise in plasma AAs. On the other hand, casein clots in the acidic conditions of the stomach, which leads to delayed gastric emptying and a steady and prolonged release of AAs into the bloodstream [186–188]. Therefore, whey protein is considered to be nutritionally superior to casein in terms of biological value, protein efficiency ratio, and net protein utilization. For example, the biological values for whey proteins and casein are 104 and 77, respectively; the protein efficiency ratio values for whey proteins and casein are 3.2 and 2.5, respectively; and net protein utilization for whey proteins and casein are 92 and 76, respectively [186,189]. Furthermore, the composition of amino acids and protein/food structure may play an important role in protein-stimulated physiological and metabolic effects. For example, casein could attenuate plasma postprandial glucose with nominal stimulation of insulin secretion, whereas the improvement of insulin response after whey causes a greater reduction in plasma postprandial glucose [190,191].

Besides the advantageous outcomes from dairy milk proteins, epitope mapping studies presented the sequential or linear and conformational epitopes within milk proteins. Due to the presence of immunoglobulin E (IgE)- and immunoglobulin G (IgG)-binding epitopes in the protein structure, dairy milk proteins are listed among the "big 8" allergens. Therefore, dairy milk protein allergy can be attributed to IgE-mediated and non-IgE-mediated mechanisms [192]. Dairy milk protein-allergic subjects could be segregated into different phenotypes according to reactivity [193]; however, no specific structure and function of dairy milk proteins account for a major part of the allergenic activity of dairy proteins. Therefore, it may be realized that the heterogeneity of the human IgE response could be attributed to the allergenic potential of any dairy milk protein or protein fragment [194]. Clinical polysensitization (cross-sensitization or cross-reactivity and co-sensitization) to several dairy milk proteins is most often noted. IgEs and IgGs from mice allergic to cow milk are capable of cross-reacting with buffalo milk proteins due to homologies in the composition of amino acids [195]. Some major characteristics of different types of dairy milk proteins are mentioned in Table 3.

Table 3. Characteristics of dairy milk proteins.

Fraction of Protein		Approximate Molecular Weight (kDa) [196,197]	Amino Acids (-) [196,197]	Isoelectric Point (-) [197,198]	Denaturation Temperature (°C) [198,199]	Allergome Name [200]
Casein	α_{s1}-casein	23.6	199	4.9–5.0	-	Bos d 9
	α_{s2}-casein	25.2	207	5.2–5.4	-	Bos d 10
	β-casein	24	209	5.1–5.4	-	Bos d 11
	κ-casein	19	169	5.4–5.6	-	Bos d 12
Whey	β-lactoglobulin	18.3	162	5.3	71.9	Bos d 5
	α-lactalbumin	14.2	123	4.8	64.3	Bos d 4
	Immunoglobulins (IgG, IgA, IgM)	IgG 150–161, IgA 385–417, IgM 1000	-	IgG 5.5–8.3	72	-
	Bovine serum albumin	66.4	583	4.9–5.1	72–74	Bos d 6
	Lactoferrin	80	703	8.7	72–85	-
	Lactoperoxidase	78	612	9.8	70	-

Apart from being an excellent source of EAAs for human nutrition, milk proteins can exhibit a wide range of biological activities by producing peptides due to gastrointestinal digestion. Peptides from dairy milk proteins are recognized as antioxidant, anti-angiotensin, antimicrobial, antidiabetic, anticarcinogenic, anti-inflammatory, antihypertensive, antihypercholesterolemic, immunomodulatory, and many more [185,201]. These advantageous outcomes from dairy milk proteins have encouraged the production of a wide range of dried protein powders, such as milk protein concentrate (MPC), milk protein isolate (MPI), whey protein concentrate (WPC), and whey protein isolate (WPI), as well as regular dairy foods, such as yogurt, kefir, and ice cream. Different physiological functions of dairy milk protein-derived peptides are shown in Figure 6.

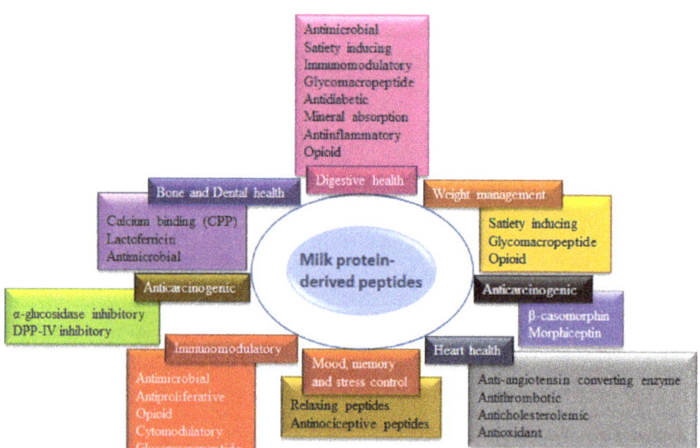

Figure 6. Physiological functions of milk protein-derived peptides (self-developed, concept was adopted from Park and Nam, 2015 [202]).

In the last two decades, the attention of the dairy industry has shifted. Presently, the applications of dairy milk proteins are not limited to the development and fortification of regular dairy foods. The significant income from individual dairy proteins and peptides has been taken into consideration along with the marketing of regular dairy foods [203]. It is well recognized that peptides from milk proteins modulate the synthesis and activities of different risk factors of MS-associated diseases [204,205]. Some milk protein-derived specific peptides offer two or more distinct biological activities (multifunctional activities) [206]. Peptides are protein fragments produced by the action of proteolytic enzymes and generally contain 2–20 amino acid residues and generally have a molecular weight < 6 kDa [207]. Hydrolysis of milk proteins can be performed using suitable acids or alkalis or food-grade proteolytic enzymes. Enzymatic hydrolysis of dairy milk proteins is preferred because acid and alkali hydrolysis of milk proteins is difficult to control. Acid hydrolysis oxidizes Cys and Met and partially destroys Ser and Thr and converts Gln and Asn to Glu and Asp acids, respectively. Furthermore, alkali hydrolysis of milk proteins causes racemization in amino acids. Enzymatic hydrolysis of dairy milk proteins has been recognized as "safe" by the European Food Safety Authority (EFSA) and the FDA [208]. Peptides from both plant- and animal-based proteins can be produced by a wide range of proteases under in vitro conditions, and by fermentation and ripening by microbial proteolytic systems. Furthermore, they can be produced in the gastrointestinal tract (in vivo conditions) by digestive enzymes, such as pepsin, chymotrypsin, and trypsin. Proteolysis of milk proteins through enzymatic routes not only produces peptides with unique functional activities but also reduces the allergenic activity of proteins and oligopeptides [209]. Dairy milk proteins α-casein and β-casein have the potential to liberate more than 20,000 functional peptides each [210]. The first identified food protein-derived peptide was from casein, which had the potential for vitamin D-independent bone calcification in rachitic infants [211]. Intestinal absorption of peptides is supposed to follow any of the following mechanisms: carrier-mediated transporter system, passive diffusion and transcytosis, paracellular pathways through tight junctions, and endocytosis [212]. One of the shortcomings of producing peptides from milk proteins by enzymatic hydrolysis is the generation of peptides with a bitter taste [213]. Some well-recognized peptides produced by in vitro enzymatic hydrolysis of dairy milk proteins and their functional activities against metabolic risk factors are presented in Table 4.

Table 4. Peptides from enzymatic in vitro hydrolysis of cow and buffalo milk proteins, and their biological functions against the risk factors of MS.

Peptide Sequence	Precursor Protein	Source	Hydrolysis Enzymes	Bioactivities	Ref.
FFVAP	α-casein	Cow milk	Trypsin	ACE inhibitory	[214]
KVLPVPQ	β-casein		Proteinase	Anti-hypertensive	[215]
YKVPQL	β-casein		Proteinase	ACE inhibitory	[215]
KVLPVP	β-casein		Carboxypeptidase	ACE inhibitory	[215]
KVLPVP	β-casein		Carboxypeptidase	Antihypertensive	[215]
YGLF	α-lactalbumin		Trypsin, Pepsin	ACE inhibitory	[216]
YLLF	β-lactoglobulin		Trypsin, Pepsin	ACE inhibitory	[216]
KKLGAPSITCVRRAF	Lactoferrin		Pepsin	Anti-inflammatory	[217]
KKLGAPSITCVRRAF	Lactoferrin		Pepsin	Antioxidant	[218]
KKLGAPSITCVRRAF	Lactoferrin		Pepsin	Anticancer	[219]
MAIPPKKNQDK	κ-casein		Chymosin, Trypsin	Antithrombotic	[220]

Table 4. *Cont.*

Peptide Sequence	Precursor Protein	Source	Hydrolysis Enzymes	Bioactivities	Ref.
AVESTVATLEDΣPEVIESPPE, where Σ is Ser(P)	κ-casein		Chymosin	Modulatory of satiety	[221]
TVQVTSTAV, MAIPKKNQDKTEIP	κ-casein		Chymosin, Papain	Anti-obesity	[222]
PGPIPN	β-casein		Trypsin, Chymotrypsin	Immunomodulatory	[220]
YPSYGLNY	κ-casein		Trypsin, Pepsin, Chymotrypsin	Anti-opioid	[220]
YGLF	α-lactalbumin		Trypsin/Pepsin	ACE inhibitory, Anti-opioid	[223]
YLLF	β-lactoglobulin		Trypsin/Pepsin	ACE inhibitory, Anti-opioid	[223]
HIRL	β-lactoglobulin		Trypsin/Pepsin	ACE inhibitory	[223]
RYLGYL, RYLGYLE, YVPFP	α-casein		-	Anticancer	[224]
YPFPGPI, YPFPG	β-casein		-	Anticancer	[224]
FKCRRWQWRMKK, LGAPSITCVRRAF	Lactoferrin		-	Anticancer	[224]
PYPQ, YFYPE, EMPFK, PQSV	Casein		Trypsin, Alcalase	Antioxidant	[225]
YQKFPQYLQY	Casein		Pepsin, Trypsin	Antihypertensive	[226]
LQPE, VAPFPE, TDVEN, VLPVPQ	Milk casein		Neutrase	Cholesterol lowering	[227]
HLPGRG, QNVLPLH, PLMLP, MFE, GPAHCLL, ACGP	Milk casein		Alcalase, Pronase E	Antidiabetic activity (inhibit three diabetic-related enzymes; such as DPP-IV, α-glucosidase and α-amylase)	[228]
RELEELNVPGEIVEΣLΣΣΣEESITRINK	β-casein		Chymotrypsin	Immuno-stimulatory	[229]
LVYPFPGPI	β-casein	Buffalo milk	Proteinase	ACE inhibitory	[230]
FVAPFPE	αs_1-casein		Trypsin	ACE inhibitory	[231]
YQQPVL	β-casein		Fermentation + Pepsin + Trypsin	ACE inhibitory	[232]
FPGPIPK, IPPK, IVPN, QPPQ	β-casein		Papain, Pepsin, or Trypsin	ACE inhibitory	[233]
IPP/VPP	β-casein		-	Anti-diabetic, Antihypertensive	[205]
IPP/VPP	β-casein		Proteases	Antihypertensive	[234]
RNAVPITPTLNR	αs_2-casein		Protease	Antidiabetic/α-glucosidase inhibitory	[235]
TKVIPYVRYL	αs2-casein		Protease	Antidiabetic/α-glucosidase inhibitory	[235]

Table 4. *Cont.*

Peptide Sequence	Precursor Protein	Source	Hydrolysis Enzymes	Bioactivities	Ref.
YLGYLEQLLR	αs_2-casein		Protease	Antidiabetic/α-glucosidase inhibitory	[235]
FALPQYLK	αs_2-casein		Protease	Antidiabetic/α-glucosidase inhibitory	[235]
YVEELKPTPEGDL	β-lactoglobulin		Pepsin	Antioxidant	[236]
VLPVPQK	β-casein		Pepsin + Trypsin	ACE inhibitory	[237]
VLPVPQK	β-casein		Pepsin + Trypsin	Antioxidant	[238]
YPSG, HPFA, KFQ	β-casein		Papain, Pepsin, Trypsin	Antioxidant	[233]
RELEE, TVA, MEDNKQ	Casein		Trypsin, Alcalase	Antioxidant	[225]

Unfortunately, adequate quantities of peptides are not produced through the enzymatic hydrolysis of food proteins, including dairy milk proteins. Therefore, this technique is not always considered economical. Chemical and recombinant DNA technologies came to the forefront to solve the mentioned problem. Chemical synthesis is preferred for producing small and medium-chain-length peptides, usually composed of 5–80 amino acids in sequence, and recombinant DNA technology is preferred for producing larger peptides and proteins. These techniques are usually employed when the peptide sequence is known [239]. Chemical modifications of peptides usually improve membrane permeability, affinity with receptors, and stability during delivery. Furthermore, it has been proven that the bioavailability of peptides is improved by chemical modification. Purification of such synthesized peptides from other products in the reaction mixture is a considerably challenging issue. Gene expression in microorganisms using modern recombinant DNA technology and cloning are advantageous for synthesizing peptides; however, these technologies require intensive research [240]. Bovine β-casomorphin 7 (BCM-7), which has the potential to prevent cardiovascular diseases, type I diabetes, and neurological disorders, was synthesized through this advanced molecular biology technique [241].

Currently, attempts are being made to assess the biological significance of milk protein-derived peptides and their characterization through advanced technologies and concepts, known as peptidomics [242]. In the last few decades, the BIOPEP database [243] has been upgraded. The outstanding development of peptidomics [242] opens a new horizon in science, known as nutrigenetics and nutrigenomics [244]. Recently, they are being considered in dairy foods to understand the biofunctional activities in a comprehensive way [245]. Presently, in silico approaches, including quantitative structure–activity relationship (QSAR), are used to predict the physicochemical information of these peptides and their responsive biological activities [246].

Peptides from food proteins, including dairy milk proteins, have gained attention in both the nutraceutical and pharmacological sectors due to their unique therapeutic potential and biocompatibility without side effects. In some cases, they are used for the fortification of conventional foods to provide unique beneficial outcomes to consumers [247,248]. Dietary products enriched with milk protein-derived peptides are commercially available with therapeutic information [249]. Some of them are used for combating metabolic diseases. Commercial peptides from dairy milk proteins, along with their sequence and appealing health or functional benefits, are mentioned in Table 5.

Table 5. Commercial peptides from dairy milk proteins (self-developed, information were adopted from Carrasco-Castilla et al., 2012 [249]).

Commercial Brand Name	Food Type	Peptide Sequence	Health/Functional Claim	Manufacturers
Ameal bp®, Ameal peptide®	Tablets	VPP, IPP derived from β-casein and κ-casein	Reduction in blood pressure	Calpis Co., Tokyo, Japan
BioZate 1	Hydrolyzed whey protein isolate	Fragments from β-lactoglobulin	Reduction in blood pressure	Davisco Foods International Inc., Savage, MN, USA
BiPRO WPI	WPI	GMP (106–109)	Antithrombic and Anticariogenic	Davisco Foods International Inc., Savage, MN, USA
Calpis	Sour milk	IPP and VPP	Hypotensive	Calpis Co., Tokyo, Japan
Calpico®/Calpis® AMEALs	Fermented milk	IPP and VPP	Hypotensive	Calpis Co., Tokyo, Japan
C12 peptide	Peptide ingredient	FFVAPFPEVFGK	Hypotensive	DMV International, Veghel, The Netherland
Evolus®	Fermented milk	IPP and VPP	Reduction in blood pressure	Valio Ltd., Helsinki, Finland
GC tooth mousse	Water-based creme	Caseinomacropeptide	Anticariogenic	GC Europe N.V., Leuven, Belgium
Glycomacropeptide (GMP)	Fresh cheese, WPI	κ-casein f(106–169)	Anticariogenic and Antithrombotic	Davisco Foods International Inc., Savage, MN, USA
insuVida™	Tablets	Casein hydrolysate	Antidiabetic	DSM Food Specialties, Delft, The Netherland
Lactium®	Milk protein hydrolysate	α_{s1}-casein f(91–100) YLGYLEQLLR	Reduction in stress effects	Ingredia Nutritional, Arras, France
LACPRODAN-DI-3065	Healthy food formula	Whey protein hydrolysate (WPH)	Benefits to sufferers from impaired digestion as a result of surgery, illness or health conditions, such as short bowel syndrome, pancreatic insufficiency and inflammatory bowel disease.	Arla Foods Ingredients Group P/S, Viby, Denmark
LACPRODAN-DI-3091	Healthy food formula	WPH	Supplements for patients with maldigestion or malabsorption.	Arla Foods Ingredients Group P/S, Viby, Denmark
LACPRODAN-DI-3092	Healthy food formula	WPH	Supplements for patients with maldigestion or malabsorption.	Arla Foods Ingredients Group P/S, Viby, Denmark
LACPRODAN-IF-3070	Healthy food formula	WPH	Sustainable gastrointestinal system for infant	Arla Foods Ingredients Group P/S, Viby, Denmark
LACPRODAN-IF-3071	Healthy food formula	WPH	Sustainable gastrointestinal system for infant	Arla Foods Ingredients Group P/S, Viby, Denmark
LACPRODAN-IF-3080	Healthy infant food formula	WPH	Lower allergenic	Arla Foods Ingredients Group P/S, Viby, Denmark

Table 5. *Cont.*

Commercial Brand Name	Food Type	Peptide Sequence	Health/Functional Claim	Manufacturers
LACPRODAN-IF-3090	Healthy infant food formula	WPH	Lower allergenic	Arla Foods Ingredients Group P/S, Viby, Denmark
LACPRODAN-DI-2021	Healthy food formula	Casein phosphopeptides	Increase calcium absorption and bone health	Arla Foods Ingredients Group P/S, Viby, Denmark
LACPRODAN-CGMP-10 and CGMP-20	Healthy food formula	Fragment from α-lactalbumin	Improves sleep and memory	Arla Foods Ingredients Group P/S, Viby, Denmark
MI paste™/MI paste plus™	Toothpaste	Caseinomacropeptide	Anticariogenic	GC America, Alsip, IL, USA
Pep2Dia®	Capsule (vegetable fiber with 350 mg of milk protein hydrolysate/capsule)	RP	Antidiabetic	Ingredia S.A., Arras Cedex, France
PROTARMOR™ 80	Ingredient	Casein hydrolysates	Weight loss	Armor Proteines, Maen Roch, France
Recaldent™	Ingredient	Caseinomacropeptide	Anticariogenic	Cadbury Enterprises Pte. Ltd., Jurong, Singapore
Trident xtra care™	Chewing gum	Caseinomacropeptide	Anticariogenic	Cadbury Adams, East Hanover, NJ, USA

4.2.1. Anti-Diabetic Activity

Both casein and whey protein-derived peptides or amino acids may affect insulin secretion from pancreatic β-cells and the release of incretin hormones, i.e., GIP and GLP-1, from the gut. It is anticipated that insulin sensitivity is related to the characteristics of the amino acid pool (qualitative and quantitative) from dairy proteins in plasma [250]. Whey protein is more effective than micellar casein for the rapid secretion of insulin [187]. It was found that the absorption rate of casein in its native micellar form is lower [251]; however, hydrolysis of casein improves the absorption of amino acids and the secretion of insulin [252]. The insulinotropic activity of WPH might be related to intestinal amino acid absorption and the increased concentration of FAAs (Leu, Ile, Phe, Arg, Tyr, Thr, Val, Ala, and Lys), BCAA-containing dipeptides (IL, LL, and VL), and possibly cyclic dipeptides in plasma [253,254]. Furthermore, whey proteins influence the synthesis of GLP-1 more than casein [255]. Insulin does not have a remarkable contribution in insulin-sensitive tissues exclusively; it has wide-ranging direct and indirect effects on metabolism, including stimulation of glucose uptake, glycogen synthesis, gluconeogenesis, lipid uptake, TG synthesis, lipolysis, protein synthesis, and inhibition of protein breakdown [256]. In addition to stimulating insulin release from pancreatic β-cells, dairy proteins and their hydrolysates may alter tissue glucose uptake in skeletal and muscle cells and suppress the high postprandial blood glucose level [257,258]. Interestingly, it was proven that longer-term consumption of casein and whey protein supplementation decreases fasting blood insulin levels and diminishes the physiological effects of DM [259]. The contribution of milk protein-derived peptides with anti-DPP-IV activity and anti-α-glucosidase and anti-α-amylase activity in hyperglycemic conditions was anticipated by several investigators [260,261]. Both hydrophobic amino acids (Ala, Gly, Ile, Leu, Phe, Pro, Met, Trp, and Val) and hydrophilic amino acids (Thr, His, Gln, Ser, Lys, and Arg) are found within DPP-IV inhibitory peptides. The presence of hydrophobic amino acids in potent DPP-IV inhibitory peptides may enhance the interaction with the active site of DPP-IV (S1 subsite of DPP-IV) [262,263]. Peptides from dairy milk proteins with one or two Pro residues at the N-terminus are highly effective DPP-IV inhibitors; however, tryptophan-containing milk protein-derived dipeptides against DPP-IV were reported [264]. Casein, particularly β-casein, contains a high amount of Pro residues. Therefore, peptides from β-casein are potential DPP-IV inhibitors compared to β-lactoglobulin-derived peptides [265]. It is anticipated that two other enzymes, α-amylase and α-glucosidase, play significant roles in normal glucose homeostasis. α-Amylase competitively interacts with peptide sequences having hydrophobic amino acids, such as Leu, Met, and Pro, at the terminal; while α-glucosidase is inhibited by the hydrophobic amino acids Met, Pro, Phe, and Leu in peptide sequences [228]. It has been proven that antioxidant peptides offer antidiabetic activity and can protect cells from reactive oxygen-induced stress by modulating multiple pathways and biochemical reactions, including inhibition of the synthesis of malondialdehyde (MDA) and protein carbonyls [266], reduction in the activity of lactate dehydrogenase (LDH) [267,268], increase in the activity of oxidative enzymes (catalase, glutathione peroxidase, and superoxide dismutase) [269], and expression of genes in the Keap1-nuclear factor erythroid 2–related factor 2 (Nrf2) signaling pathway [267,270]. Peptides with effective antioxidant capacity are hydrophobic in character and comprise unique amino acids, such as Trp, Tyr, His, and Pro. In addition, the presence of two hydrophobic amino acids, such as Leu and Val, in peptides contributes to antioxidant and lipid peroxidation capacity [266]. Two well-known milk protein-derived peptides, IPP and VPP, demonstrated effective enhancement of insulin signals and anti-inflammation via the NF-κB pathway under TNF stimulation and prevention of insulin resistance. These two peptides offer an insulin-sensitizing effect, which is independent of insulin receptors in adipocytes. Furthermore, they are recognized as potential DPP-IV inhibitors [234,271].

In addition, peptides from casein and whey protein show a satiating effect [272], which is related to the gastrointestinal system and the appetite center of the brain [273]. Casein reduces gastric emptying and improves a slow postprandial increase in amino acids in plasma. Contradictorily, whey proteins improve a fast, high, and transient increase in amino acids in plasma. Some investigations have suggested that whey proteins are more satiating than casein proteins [255,274], and their thermogenesis and influence on body weight differ to some extent [275,276]. Casein hydrolysates activate peripheral opioid and CCK receptors and block the antagonist receptors associated with food consumption [277,278]. Whey protein-derived peptides increase the concentration of postprandial plasma amino acids, CCK, GLP-1, and GIP. The satiating effect of whey protein is mainly due to a high concentration of BCAAs, particularly L-Leu [279]. Therefore, milk protein-derived peptides modulate the release of satiety hormones (CCK, GLP-1, GIP, PYY, an ghrelin), increase diet-induced thermogenesis, and activate opioid receptors [190]. Milk protein-derived small peptides and amino acids act via GPCRs, increasing intracellular Ca^{2+} and/or the concentration of cyclic adenosine monophosphate (cAMP), or via peptide/amino acid transporters, which depolarize the enteroendocrine cell membrane, activating Ca^{2+} influx and stimulating satiety hormones [280,281].

4.2.2. Anti-Dyslipidemic Activity

Dairy milk proteins modulate intrahepatic lipids and circulating TGs through a wide range of biochemical mechanisms, including gastric emptying, insulin secretion, lipogenesis (downregulation), and gluconeogenesis and glycogenesis (upregulation) in hepatocytes. They suppress postprandial lipemia due to insulinotropic effect, modulate metabolic pathways dedicated to lipid metabolism, and offer lipid peroxidation activity [282].

The concentration of postprandial TG is influenced by many factors, such as insulin concentration, secretions of intestinal chylomicrons and hepatic VLDL, transformation of TG-rich lipoproteins to TG-depleted lipoproteins, and tissue uptake of TG-depleted lipoproteins [283]. Insulin is a well-known activator of lipoprotein lipase, but high levels of postprandial insulin inhibit hormone-sensitive lipase [284]. Therefore, consumption of dairy milk proteins may result in lower formation of chylomicrons, and subsequently,

FFA from adipose tissue and lipotoxicity in hepatocytes. Due to the insulinotropic effect of whey proteins, they may enhance the activity of lipoprotein lipase and promote chylomicron clearance. In the chronic state, whey proteins affect intestinal lipid absorption, excretion, and de novo cholesterol biogenesis in hepatocytes [259]. Peptides from dairy milk casein offer cholesterol-lowering activity by diminishing cholesterol micellar solubility and absorption and by reducing mRNA expression of acetyl-CoA acetyltransferase-2 and microsomal triacylglycerols (MTP) in proximal intestinal cells [227].

Dairy milk protein-derived peptides may inhibit hypercholesteremic enzymes and associated genes. They inhibit the activities of cholesterol esterase and pancreatic lipase [285] and expression of the lipogenesis-related genes SCD-1, ACC1, and FASN. They increase the peroxisome proliferator-activated receptor-α (PPAR-α) signaling pathway and expression of the β-oxidation-related genes CPT-1a, PPARα, and ACOX1 [286]. Dairy milk casein-derived peptides enhance trans-intestinal cholesterol excretion (TICE) via regulation of the liver X receptor-α (LXR-α) signaling pathway and upregulation of ABCG5. Lys in dairy milk proteins is known to be involved in the lipid accumulation metabolism of long-chain fatty acids, which are crucial for the endogenous synthesis of carnitine. It was found that Lys-supplemented formula reduces the concentration of TGs in tissues [287]. Moreover, the concentration of triacylglycerols and gene expression of lipogenesis enzymes in hepatocytes are suppressed by Leu and Ala in milk proteins and, consequently, are linked with reduced body fat accumulation [288]. Dairy milk protein-derived peptides induce fibroblast growth factor 15/19 (FGF15/19) from enterocytes, which suppresses hepatic bile acid synthesis involved in adjusting hepatobiliary cholesterol [289].

Metabolic diseases, including obesity, are associated with a proinflammatory state and intracellular redox imbalance and the formation of ROS, which result in mitochondrial dysfunction, lipid oxidation, ROS-related impairment, and many other issues [290]. Milk protein-derived peptides offer lipid peroxidation activity due to the presence of Leu or Val and other hydrophobic amino acids, such as Trp, Tyr, His, and Pro. It has been proven that peptides derived from milk proteins can protect Caco-2 cells from lipid peroxidation (peroxide-induced oxidative stress) by modulating the Keap1-Nrf2 pathway [291], which may be responsible for the suppression of excessive body weight [268]. Peptides from milk proteins inhibit the activity of XO, supposed to be the source of ROS that causes atherosclerosis and cholesterol crystals [292].

4.3. *Biochemical Mechanisms Considering IPP and VPP as Model Peptides*

Two milk protein-derived tripeptides, IPP and VPP, were meticulously studied concerning their effects on biomarkers of metabolic diseases. Hyperglycemia, hyperinsulinemia, and lipotoxicity are influenced by the expression of local RAS components, especially the Ang II-AT1R axis. Ang II impedes adipocyte differentiation, which promotes adipocyte dysfunction and reduces insulin sensitivity through reduced adiponectin and greater proinflammatory adipokine secretion. Ang II further enhances ROS formation and modulates a wide range of metabolic pathways in different tissues. Inflammation in cells, distressed insulin signaling in the endothelium, hepatocytes, muscle, and adipose tissues promote endothelial dysfunction and insulin resistance. Furthermore, stress-responsive damage to pancreatic β-cells is the cause of lower insulin secretion, which enhances fasting blood glucose [293,294]. It may be realized that understanding the pipelines among inflammation, insulin secretion and resistance, and dysregulation of lipid metabolism will be useful to explain how dairy protein-derived peptides IPP and VPP modulate different biochemical pathways and metabolic biomarkers associated with hyperglycemia and dyslipidemia. In Figure 7, the interconnections between RAS-mediated inflammation, oxidative stress, insulin secretion and resistance, and dysregulation of glucose and lipid metabolism are presented.

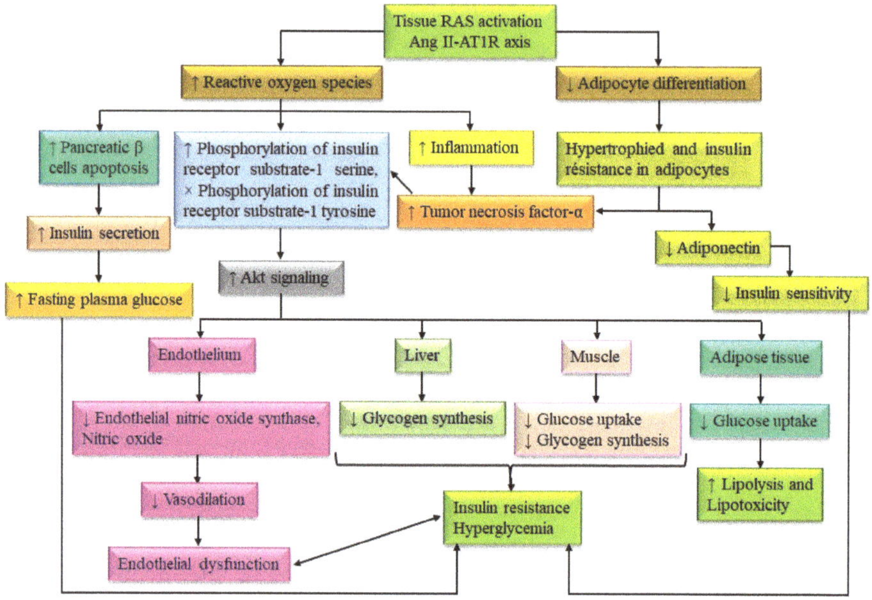

Figure 7. Interconnections between inflammation, oxidative stress, and dysregulation of glucose and lipid metabolism (self-developed; concept was adopted from Putnam et al., 2012 [293]).

Formation of stress-responsive inflammatory cytokines is influenced by several physiological and metabolic factors, such as the activity of the RAS, dysregulation of NADPH oxidase, adipokines, extracellular signal-regulated kinase (ERK), formation of plasminogen activator inhibitor-1 (PAI-1) and C-reactive protein, and activation of the NF-κB pathway, among others [294]. Antioxidant peptides, either direct scavengers or inhibitors of the inflammatory pathways mentioned above, participate in counter-regulatory pathways and act as anti-maladaptation agents in metabolic diseases [295,296]. The ACE inhibitory activity of dairy milk protein-derived IPP and VPP was proven by in vitro studies [297,298], animal models [299–302], and clinical trials [303–306]. Note that the effects were not equivocal regarding their biological efficacy [307,308], but there is a consensus that they are ACE inhibitors. ACE prefers to bind peptides containing a hydrophobic (aromatic or branched side-chain) amino acid residue, such as proline, at the C-terminal position [309]. These hydrophobic residues provide the structural basis for the ACE inhibitory effect; however, the downstream effects are not uniform. While both IPP and VPP inhibited the NF-κB pathway, only VPP antagonized the activation of extracellular signal-regulated kinase (ERK) 1/2 in a rat aortic vascular smooth muscle cell (VSMC) line, A7r5 [310]. It was found that administration of VPP and IPP might be beneficial for preventing atherosclerosis due to the activity of ACE and hypercholesterolemia. Plasma lipid levels and 8-isoprostane, a biomarker of oxidative stress, remained unchanged after 31 weeks of oral feeding of IPP and VPP to mice; however, mRNA expression of inflammatory cytokines, such as IL-6 and IL-1β, oxidized low-density lipoprotein receptor, and transcription regulators were reduced [311]. He antihypertensive and cholesterol-lowering effects of the tripeptides IPP and VPP were proven by a 10-week clinical trial. It was reported that systolic (but not diastolic) blood pressure was significantly decreased. Furthermore, total cholesterol and LDL-C were decreased; however, HDL-C, triacylglycerols, and CRP remained unchanged [312]. Bovine casein-derived tripeptides IPP and VPP from casein hydrolyzate have been reported to prevent obesity-induced chronic adipose inflammation in both animal models and cell cultures. High-fat diet-induced systemic inflammatory factors, hypertrophic white adipocytes, and macrophage infiltration were suppressed in C57BL/6J mice due to casein hydrolyzate feeding for 6 weeks. Casein hydrolyzate was able to mitigate adipocyte dysfunction induced by TNF-α by increasing the expression of CCAAT/enhancer binding protein α (C/EBP-α) rather than PPAR-γ in the 3T3-L1 cell line. Furthermore, casein hydrolyzate suppressed mitogen-activated protein kinase (MAPK)-c-JNK phosphorylation and enhanced phosphorylation of ERK 1/2 in TNF-α-induced 3T3-L1 cells [313]. VPP was tested in high-fat diet-fed C57BL/6J mice for 10 weeks to understand its effect on inflammation in adipose tissues. The results showed that the expression level of CD18 in peripheral blood monocytes was significantly decreased compared with the placebo group. This was particularly prominent in the stromal vascular fractions of the fatty tissue of mice treated with VPP. Moreover, the expression of monocyte chemoattractant protein-1 and IL-6 in adipose tissue was also lower [314]. IPP and VPP were able to promote adipocyte differentiation and inhibit inflammation in the murine preadipocyte cell line 3T3-F442A. The upregulation of PPAR-γ and secretion of the protective lipid hormone adiponectin from these cells were reported as underlying mechanisms for beneficial adipogenic differentiation. IPP and VPP inhibited cytokine-induced changes, such as reduction in adipokine levels and activation of the NF-κB pathway [315]. These findings were supported by another study, suggesting that both IPP and VPP can prevent the activity of the inflammatory mediator NF-κB under TNF stimulation in similar cell lines. Under TNF stimulation, they were able to prevent insulin resistance and enhance the expression of GLUT4 [234]. Furthermore, they can enhance glucose consumption in HepG2 cell lines via upregulated protein expression of p-AKT and GLUT2 and regulate the activities of glucose-metabolizing enzymes in HepG2 cells. IPP also directly interacts with the insulin receptor, activating the insulin/AKT signaling pathway; whereas the contribution of VPP to glucose consumption is not attributed to insulin receptor binding [316]. On the other hand, both peptides promoted glucose uptake via the adenosine monophosphate-activated protein kinase (AMPK) pathway, accompanied by GLUT4 translocation rather than the insulin signaling pathway, in the L6 rat myoblast cell line [317]. Therefore, it may be realized that both IPP and VPP can modulate blood glucose levels and may have an impact on glucose-mediated metabolic disorders.

4.4. Clinical Investigations

Several acute (less than 1 week) and some long-term (4 to 12 weeks) intervention studies were performed with obese and diabetic individuals to understand the influence of dairy proteins on gastric emptying, gastric secretion, amino acid absorption, and enterogastrone response on serum glucose and insulin levels because they are directly related to diabetic conditions; however, the underlying pipelines are not fully comprehended [191]. Furthermore, results of some clinical investigations proved that dyslipidemia in obese or non-obese subjects is associated with insulin resistance and diabetic conditions. Summarizing all the findings, it may be supposed that dairy milk proteins and peptides derived from them can reduce the risks of hyperglycemic and dyslipidemic incidences by stimulating insulin and secretion of incretin hormones (GLP-1, GIP), suppressing appetite, slowing gastric emptying, and modulating glucose and lipid metabolism. It is noteworthy to mention that co-ingestion of a protein hydrolysate or specific amino acids with intact proteins modulates glucose and lipid metabolism [191,318]. There were no experiments performed specifically for fast food-induced obesity, except [319]. Placebo food formulas associated with modulation of biomarkers of hyperglycemia and dyslipidemia may explain fast food-induced obesity. Therefore, it may be realized that results of some clinical investigations about the impacts of milk proteins and peptides on biomarkers of hyperglycemia and dyslipidemia can explain their role in fast food-induced obesity. Results of some clinical investigations about the effects of milk proteins and peptides on outcomes and biomarkers of hyperglycemia and dyslipidemia are summarized in Table 6.

Table 6. Effects of milk proteins and peptides on outcomes and biomarkers of hyperglycemia and dyslipidemia.

Publication Year/Reference	Study Design, Duration	Characteristics and Number of Subjects (n)	Dairy Milk Proteins Formula	Comparison Formula	Effects and Remarks	
					Hyperglycemia	Dyslipidemia
2005/[320]	Randomized, Acute (2 separate occasions with ≥1 week gap between each)	T2DM (14)	Whey protein (27.6 g)	Ham (96 g) and Lactose (5.3 g)	↓ Postprandial plasma glucose by whey protein. ↑ Plasma insulin response and postprandial GIP by whey protein. → GLP-1 by whey protein and ham.	-
2005/[321]	Randomized order and a double-blind (2 trials, separated by a 2 weeks gap between each)	T2DM (10)	No diabetes (n = 10), 0.7 g carbohydrate/kg body weight/h (50% glucose and 50% maltodextrin) with or without 0.35 g/kg body weight/h of a protein hydrolysate and AA mixture (50% casein hydrolysate, 25% free Leu and 25% free phenylalanine) every 15 min until t = 165 min	No diabetes (n = 9), 0.7 g carbohydrate/kg body weight/h (50% glucose and 50% maltodextrin) with or without 0.35 g/kg body weight/h of a protein hydrolysate and AA mixture (50% casein hydrolysate, 25% free Leu and 25% free phenylalanine) every 15 min until t = 165 min	↑ Plasma insulin and ↓ plasma glucose responses by carbohydrate + protein trial than carbohydrate trial in diabetic subjects and matched to control subjects.	-
2006/[322]	Randomized cross-over (4 occasions with 7 days gap between each)	Overweight (19)	Preloads made by water solutions of (a) WPI (55 g), (b) Calcium caseinate (55 g), (c) Glucose (60 g), (d) Lactose (56 g)	-	↓ Acute appetite and energy intake by casein- or whey-, lactose- than glucose-preload. ↓ Postprandial plasma glucose by protein (mean value of casein and WPI)-preload than lactose- and glucose-preload. → Plasma insulin by different preloads. ↑ CCK response and ↓ ghrelin by protein preload.	-

Table 6. *Cont.*

Publication Year/Reference	Study Design, Duration	Characteristics and Number of Subjects (n)	Dairy Milk Proteins Formula	Comparison Formula	Effects and Remarks	
					Hyperglycemia	Dyslipidemia
2006/[323]	Randomized, double-blind, placebo-controlled, cross-over, acute challenge (Participants consumed 3 meals/day in a single 24 h period)	T2DM (11), Healthy (11)	T2DM (11), Casein hydrolysate (0.3 g/kg body weight) + Leu (0.1 g/kg body weight) + meal (64% carbohydrate, 25% fat and 11% protein)	Healthy (11), Flavored water + meal (64% carbohydrate, 25% fat and 11% protein)	↓ Plasma glucose by casein hydrolysate in average 24 h by casein and Leu supplemented meal.	-
2006/[324]	Acute, randomized, double-blind (3 trials, separated by at least 1 week gap between each)	T2DM (10), Healthy (10)	T2DM (10), (a) Carbohydrate (50% glucose and 50% maltodextrin) (0.7 g/kg body weight), (b) Carbohydrate + Casein hydrolysate (0.7 g/kg body weight + 0.3 g/kg body weight), (c) Carbohydrate + Casein hydrolysate + Leu (0.7 g/kg body weight + 0.3 g/kg body weight + 0.1 g/kg body weight)	Healthy (10), Similar diet like diabetic group	↑ Insulin response and ↓ plasma glucose response in subjects with T2DM and control subjects by Carbohydrate + Casein hydrolysate and Carbohydrate + Casein hydrolysate + Leu for both T2DM and healthy subjects (Carbohydrate diet basis).	-

Table 6. *Cont.*

Publication Year/Reference	Study Design, Duration	Characteristics and Number of Subjects (n)	Dairy Milk Proteins Formula	Comparison Formula	Effects and Remarks	
					Hyperglycemia	Dyslipidemia
2007/[325]	Double-blind, randomized, controlled cross-over clinical trial (5 days study with 2 weeks gap between each)	Non-obese, T2DM (12)	Mixed meal (≈31% carbohydrates, ≈17% lipids and ≈52% proteins as total calories) with (a) WPI (fast protein), (b) micellar casein (slow protein), (c) a mixture of FAAs resembling the AA composition of micellar casein.	-	↓ Postprandial plasma glucose by FAA meal than WPI meal and casein meal. ↑ Plasma BCAAs, EAAs, C-peptide, insulin and pro-insulin concentrations by WPI meal than casein meal and similar with FAA meal. ↓ Plasma GLP-1 response by casein meal than WPI meal. ↑ Plasma GIP response by WPI meal and casein meal than FAA meal.	-
2009/[326]	Acute (3 occasions)	T2DM (8)	(a) Pre-meal: 55 g whey in 350 mL beef soup, (b) Main-meal: 55 g whey in potato	No whey in pre-meal and main meal	↑ Plasma insulin, GIP and CCK by whey diet. ↓ Gastric emptying and postprandial glycemia by whey diet.	-
2009/[327]	Randomized crossover (4 separate occasions with 2–5 weeks gap between each)	T2DM (12)	Whey protein (45 g) in meal (80 g fat and 45 g carbohydrate)	(a) Casein (45 g), (b) Cod (45 g), (c) Gluten (45 g) in meal (80 g fat and 45 g carbohydrate)	↑ Postprandial plasma insulin and incretins (GLP-1, GIP) by whey protein. → Glucagon by all types of proteins.	↑ Postprandial lipemia by whey protein than other proteins.

Table 6. *Cont.*

Publication Year/Reference	Study Design, Duration	Characteristics and Number of Subjects (n)	Dairy Milk Proteins Formula	Comparison Formula	Effects and Remarks	
					Hyperglycemia	Dyslipidemia
2009/[328]	Randomized (Energy restriction period of 5–6 weeks followed by a weight maintenance period of 12 weeks with 1 week in-between to change back from liquid to normal food)	Overweight and obese (48)	(a) Casein, (b) Whey protein supplements (2 × 25 g/day)	Maltodextrin	↑ Plasma glucagon. Fasting glucose in normal range by protein diet. → Plasma insulin resistance by casein and whey proteins.	↑ Serum TG by maltodextrin. → Serum total cholesterol, leptin, adiponectin, LDL-C and HDL-C by protein diets.
2010/[329]	Randomized, single blind, three-way crossover design (3 separate intervention days, each preceded by a 1 week washout period)	Overweight or obese (20)	(a) WPI (45 g), (b) Sodium caseinate (45 g)	Glucose (45 g)	↓ Plasma glucose by WPI than sodium caseinate and glucose.	↓ Serum TG-enriched lipoprotein by WPI than casein and glucose. → Serum total cholesterol, LDL-C, HDL-C, NEFA, apolipoprotein B-48, insulin and leptin.
2010/[330]	Randomized, controlled, cross-over, acute challenge trial (4 test meals with intervals of >2 weeks)	T2DM (11)	(a) Casein (45 g) + Carbohydrate (45 g) + Fat (80 g), (b) Casein (45 g) + Fat (80 g), (c) Carbohydrate (45 g) + Fat (80 g)	Fat (80 g)	↑ Plasma insulin and glucagon by Casein + Fat meal and Casein + Carbohydrate + Fat meal than Fat-meal. ↑ GIP response by Casein + Carbohydrate + Fat meal. → GLP-1 response.	→ Serum TG and retinyl palmitate in the chylomicron-rich fraction for all meals. ↑ Retinyl palmitate in the chylomicron-poor fraction by 45 g of protein as casein.
2011/[259]	Randomized parallel (12 weeks)	Overweight and obese (70)	(a) WPI (27 g) in 250 mL of water, (b) Sodium caseinate (27 g) in 250 mL of water. Twice in a day	Glucose (27 g) in 250 mL of water	→ Plasma glucose by WPI than control or casein meal. ↑ Fasting plasma insulin by whey protein.	↓ Fasting serum TG, plasma total cholesterol and LDL-C by WPI.

Table 6. *Cont.*

Publication Year/Reference	Study Design, Duration	Characteristics and Number of Subjects (n)	Dairy Milk Proteins Formula	Comparison Formula	Effects and Remarks	
					Hyperglycemia	Dyslipidemia
2011/[331]	Randomized double-blind partial cross-over (3 out of 4 treatments separated by a week gap between each)	T2DM (36)	(a) Intact casein, (b) Casein hydrolysate, (c) Casein hydrolysate with Leu	Carbohydrate (maltodextrin and glucose monohydrate) without (a) intact casein, (b) unhydrolyzed casein, (c) Casein hydrolysate and Leu	↓ Plasma glucose by casein hydrolysate with or without Leu.	-
2011/[332]	Randomized placebo controlled double blind (Separate 3 days with 1 week gap between each)	T2DM (13)	In a single oral bolus (300 mL) containing 50 g of carbohydrates (50% glucose and 50% maltodextrin and casein hydrolysate (6 g or 12 g)	No casein hydrolysate	→ Plasma insulin and glucose by 6 g of casein hydrolysate. ↑ Post-challenge plasma insulin and ↓ glucose levels by 12 g of casein hydrolysate.	-
2012/[333]	Single blind crossover (4 separate occasions with a washout period ≥2 weeks)	T2DM (12)	Protein supplement ((a) LACPRODAN-ALPHA-10 (45 g), (b) LACPRODAN-DI-9224 (45 g), (c) LACPRODAN CGMP-10 (45 g), (d) LACPRODAN-DI-3065 (45 g)) + Fat (80 g) + Carbohydrate meal (45 g)	-	↑ Plasma glucose, insulin, glucagon and GLP-1 by LACPRODAN-DI-3065 than other proteins in fat-carbohydrate meals.	→ Postprandial serum TG by all dietary proteins. ↑ Retinyl palmitate in the chylomicron-rich fraction by LACPRODAN-DI-3065 than dietary proteins. → FFA and Retinyl palmitate in the chylomicron-poor fraction by all dietary proteins.

Table 6. *Cont.*

Publication Year/Reference	Study Design, Duration	Characteristics and Number of Subjects (n)	Dairy Milk Proteins Formula	Comparison Formula	Effects and Remarks	
					Hyperglycemia	Dyslipidemia
2013/[334]	Randomized crossover (4 different meals in different days with a 2 weeks washout period between each meal)	Obese non-diabetic (11)	WPI + Fat (80 g) + Carbohydrate (45 g)	Carbohydrate (45 g) + Fat (80 g) + (a) Cod or casein (45 g), (b) gluten protein (45 g)	↑ Biomarkers (GLP-1, insulin, glucagon) of postprandial glycemia, ↓ Postprandial GIP by WPI. The larger initial plasma insulin and glucagon response after whey meal did not correlate with the initial GLP-1 or GIP responses.	↓ Postprandial serum TG by WPI than cod and gluten proteins. ↓ NEFA by whey and casein than cod and gluten proteins. → Retinyl palmitate in chylomicron by all types of dietary proteins.
2013/[335]	Double-blind, randomized (Each trial was performed on a distinct day with 3 days intervals between each trial)	T2DM (10)	(a) WPH beverage (0.1 g/kg body weight, 0.2 g/kg body weight and 0.4 g/kg body weight), (b) WPI beverage (0.1 g/kg body weight, 0.2 g/kg body weight and 0.4 g/kg body weight)	Distilled water	↑ Postprandial plasma insulin and ↓ post-challenge plasma glucose by 0.2 g/kg body weight WPH or 0.4 g/kg body weight WPI. ↑ Postprandial plasma insulin and ↓ concomitant glucose level to normal range at 2 h after 0.2 g/kg body weight WPH.	-
2013/[336]	Randomized crossover (4 test days, which were separated by a washout period of at least 3 days)	Obese men (16)	Butter cake (high fat) + (a) 500 mL milk or (b) 500 mL water + milk protein 23.4 g or (c) 500 mL water + calcium 2.3 g	Butter cake + 500 mL water	↑ Plasma insulin and ↓ glucose by milk protein. ↑ Plasma concentrations of total AAs, EAAs and non-EAAs by milk and protein milk.	↑ Serum TG by milk protein. ↓ Apolipoprotein B-48 by calcium meal compared with milk. → NEFA by all drinks.

Table 6. *Cont.*

Publication Year/Reference	Study Design, Duration	Characteristics and Number of Subjects (n)	Dairy Milk Proteins Formula	Comparison Formula	Effects and Remarks	
					Hyperglycemia	Dyslipidemia
2014/[337]	Double-blind randomized and cross-over (3 trials was separated by at least 6 days gap between each)	T2DM (60)	(a) Carbohydrate + Intact casein (0.7 g/kg body weight + 0.3 g/kg body weight), (b) Carbohydrate + Casein hydrolysate (0.7 g/kg body weight + 0.3 g/kg body weight)	Carbohydrate (50% glucose + 50% maltodextrin) (0.7 g/kg body weight)	↑ Plasma insulin and ↓ plasma glucose responses by Carbohydrate + Protein diet than Carbohydrate diet. ↑ Plasma insulin response by Carbohydrate + Casein hydrolysate diet than Carbohydrate + Intact protein diet. ↓ Plasma glucose response by Carbohydrate + Casein hydrolysate diet than Carbohydrate + Intact protein diet.	-
2014/[338]	Randomized, open-label crossover (2 meals in 2 separate days and a at least 2 weeks gap between each)	T2DM (15)	Preload: 50 g whey in 250 mL water	Preload: 250 mL water	↓ Plasma glucose, ↑ plasma insulin, ↑ C-peptide, ↑ GLP-1, and → DPP4 by whey pre-load.	-
2015/[339]	Randomized, cross-over study (4 weeks)	T2DM (7)	Preload: WPI (25 g) + 25 g chocolate-flavor in 100 mL water	Preload: 25 g chocolate-flavor without WPI in 100 mL water	↓ Postprandial plasma blood glucose and peak blood glucose by WPI formula.	-

Table 6. *Cont.*

Publication Year/Reference	Study Design, Duration	Characteristics and Number of Subjects (n)	Dairy Milk Proteins Formula	Comparison Formula	Effects and Remarks	
					Hyperglycemia	Dyslipidemia
2015/[340]	Single-center, randomized, single blind (3 blind challenges on 3 different occasions with a minimum of 2 days gap between them)	Healthy (15), Prediabetic (15)	50 g casein or WPI + 50 g of maltodextrin with dextrose equivalent 19 + 2 g of hydroxyproline + 10 g of lactulose in 300 mL of water.	50 g of maltodextrin with dextrose equivalent 19 + 2 g of hydroxyproline + 10 g of lactulose in 300 mL of water.	↑ Plasma insulin, glucagon, C-peptide, GIP, GLP-1 and satiety by casein and whey proteins than glucose. Plasma glucose by casein and whey proteins than glucose. → All effects two proteins.	-
2015/[341]	Randomized, parallel-controlled, double-blinded (Each test meal with 12 weeks intervention period)	Obese (52)	(a) Whey + Low medium-chain saturated fatty acids, (b) Whey + High medium-chain saturated fatty acids, (c) Casein + Low medium-chain saturated fatty acids, (d) Casein + High medium-chain saturated fatty acids	-	↓ Postprandial apolipoprotein B-48 response after whey compared with casein independently of fatty acid composition. ↑ Postprandial plasma GLP-1 by casein compared with whey. → Postprandial plasma insulin, glucose, glucagon, or GIP among groups.	→ Postprandial serum triacylglycerol and FFA among groups.

Table 6. *Cont.*

Publication Year/Reference	Study Design, Duration	Characteristics and Number of Subjects (n)	Dairy Milk Proteins Formula	Comparison Formula	Effects and Remarks	
					Hyperglycemia	Dyslipidemia
2017/[342]	Randomized open-label parallel-arm (12 weeks, dietary intake on 3 days in each week)	T2DM (56)	(a) Whey protein breakfast: 25% fat + 50% carbohydrates + 25% protein (28 g whey from 42 g total protein). (b) Protein breakfast: 25% fat + 50% carbohydrates + 25% (42 g) protein mainly from eggs (7 g), tuna (20 g), soya (7 g).	Carbohydrate breakfast: 25% fat + 64% carbohydrates + 11% (17 g) soya protein	↓ Overall plasma postprandial incremental area under curve (iAUC) for glucose, ghrelin and hunger scores by whey protein breakfast and protein breakfast than carbohydrate breakfast. ↑ Postprandial plasma overall iAUC for insulin, C-peptide, GLP-1 and satiety scores by whey protein breakfast than protein breakfast and carbohydrate breakfast.	-
2017/[343]	(a) Parallel-armed acute challenge (One serving (21 g) of whey protein) and (b) Crossover design (continuous glucose monitoring (CGM) twice, over 2 consecutive weeks, 3.5 days each week)	n = 18 underwent a challenge test (not crossover design). n = 22 underwent CGM and controlled feeding twice (crossover design), Two consecutive weeks: one week WPI and other week placebo diet.	WPI (21 g protein + 3 g carbohydrate + 0.5 g fat)	Indigestible potato starch (1 g protein + 25 g carbohydrate 20 g fiber + 0.5 g fat)	Acute challenge studies: ↑ Plasma insulin, ↑ GLP-1, ↓ Plasma glucose and ↓ Ghrelin by WPI diet. Placebo diet had no effect.	↓ Hypertriglyceridemia by WPI diet.
2018/[344]	Randomized, double-blind, placebo-controlled (10 weeks)	T2DM (24)	WPI beverage (20 g protein + 10 g carbohydrate + 3 g milk fat)	Without WPI (30 g carbohydrate + 3 g milk fat)	↓ Fasting plasma blood glucose and insulin resistance by whey protein diet.	-

Table 6. *Cont.*

Publication Year/Reference	Study Design, Duration	Characteristics and Number of Subjects (n)	Dairy Milk Proteins Formula	Comparison Formula	Effects and Remarks	
					Hyperglycemia	Dyslipidemia
2018/[345]	Randomized crossover (2 separate test days with 1 week gap between each)	Overweight and obese (9)	Breakfast (74 g carbohydrate, 2.1 g fat and 8.5 g protein in 100 mL orange juice) with 10% (*w*/*v*) solution of a novel casein hydrolysate	Breakfast (74 g carbohydrate, 2.1 g fat and 8.5 g protein in 100 mL orange juice) with 10% (*w*/*v*) solution of sodium caseinate (intact protein).	→ Gastric emptying outcome by intact casein and hydrolysate. ↑ Insulin response and ↓ glucose response by casein hydrolysate compared to intact casein.	-
2018/[346]	Randomized, single-blind crossover (3 separated occasions separated by 7 days gap between each)	T2DM (11)	(a) Intact whey protein (15 g), (b) WPH (15 g) before mixed-macronutrient breakfast and lunch meals, separated by 3 h.	Flavored water	↓ Plasma glucose by WPH (early) and intact whey protein. ↑ Satiety and Plasma insulin by both WPH and intact whey protein.	-
2018/[347]	Acute, randomized, cross-over (Test days separated by a washout period of approximately 1 week gap between each)	Non-diabetic (12), T2DM (12)	Preload: 20 g whey protein in 200 mL water as pre-meal or part of the fat-rich meal.	Preload: 200 mL water	Plasma ↑ insulin, ↑ glucagon, ↑ GIP and ↓ gastric emptying in subjects with and without T2DM by whey protein pre-meal. Plasma ↑ insulin, ↑ glucagon, ↑ GIP and ↓ gastric emptying in subjects with and without T2DM by pre-meal than main meal.	→ Postprandial TG, apolipoprotein B-48 and NEFA in subjects with and without T2DM.

Table 6. *Cont.*

Publication Year/Reference	Study Design, Duration	Characteristics and Number of Subjects (n)	Dairy Milk Proteins Formula	Comparison Formula	Effects and Remarks	
					Hyperglycemia	Dyslipidemia
2018/[348]	Acute, randomized, cross-over (The test days were separated by a wash-out period of approximately 1 week gap between each)	MS (20)	Preload: (a) 10 g whey protein, (b) 20 g whey protein	Preload: No whey protein	↑ Insulin response, ↑ postprandial glucagon, ↓ glucose and ↓ gastric emptying after a pre-meal with 20 g whey protein than 10 g whey protein and placebo. → GIP by whey protein pre-meal.	→ TG, apolipoprotein B-48 and FFA by whey protein pre-meal.
2019/[349]	Randomized, double-blind, placebo-controlled, monocentric, 3-way-cross-over (6 weeks, a wash-out period of 7 days between each study day)	Prediabetic (21)	Milk protein hydrolysate (1.4 g and 2.8 g), 2 dosages/day	Maltodextrin with dextrose equivalent of 9	↓ iAUC of plasma glucose by milk protein hydrolysate in dose-dependent manner. → Insulinotropic properties were insignificant.	-
2020/[350]	Prospective randomized pilot study (45 days)	Obese and insulin resistance (48)	Protein diet (90 g/meal), 5 meals/day for (a) whey protein, (b) vegetable protein (soya, green peas, or cereals), (c) animal protein (meat, fish, egg)	-	↓ Insulin resistance and fasting glycemia by whey protein.	Serum ↓ total cholesterol, LDL-C and TG by whey protein.
2020/[351]	Double-blind randomized clinical trial (13 weeks)	Obese and type 2 (pre-)diabetes (123)	21 g of Leu-enriched whey protein ((3 g total Leu), 9 g carbohydrates, 3 g fat, 800 IU cholecalciferol (Vitamin D3))	Carbohydrate (25 g) and fat (6 g) mix	↓ Insulin resistance and postprandial plasma glucose by Leu-enriched whey protein.	-

Table 6. *Cont.*

Publication Year/Reference	Study Design, Duration	Characteristics and Number of Subjects (n)	Dairy Milk Proteins Formula	Comparison Formula	Effects and Remarks	
					Hyperglycemia	Dyslipidemia
2020/[352]	Single-center randomized, placebo-controlled (3 months)	T2DM (120)	Cys-rich (2.7%) WPI with a standardized lactoferrin in 100 mL of water	5 g casein in 100 mL of water	↓ Fasting plasma glucose by Cys-rich WPI diet.	Serum ↓ total cholesterol, triacylglyceride and LDL-C levels by Cys-rich (2.7%) WPI than placebo.
2020/[319]	Randomized, parallel, placebo-controlled, double-blind study (20 weeks: 1 week pre-study measurement period, a 3 weeks baseline period and a 16 weeks energy restriction intervention period	Overweight or obese	n = 21, Unhealthy western-style eating patterns + 1.5 g total protein (MPI)/kg body weight/day	n = 23, Unhealthy western-style eating patterns + 0.8 g total protein (MPI)/kg body weight/day	→ Fasting plasma insulin and glucose	Serum ↓ fasting TG, → total cholesterol and LDL-C

Legend: ↓: decrease, ↑: increase, →: unchanged or insignificant effect, -: not mentioned, +: and.

4.4.1. Dairy Proteins in Meals and T2DM Subjects

In a clinical investigation, it was found that the glucose response was lower when whey protein was included in breakfast and lunch (test meals) for T2DM subjects. The postprandial plasma GIP response was higher in meals with whey; whereas no differences were found in GLP-1 between the reference (ham and lactose) and test meals [320]. In another investigation, it was found that although gastric emptying and plasma insulin, glucagon, C-peptide, GIP, and GLP-1 were improved by the consumption of bovine casein or whey protein with carbohydrates (co-administration) compared with carbohydrate feeding in healthy and prediabetic adults, their effects were similar. Likewise, venous blood glucose was reduced by the two types of protein diets compared with the carbohydrate diet [340]. Interestingly, it was found that even though whey protein can reduce postprandial TG, FFA, and glucose responses more than casein, cod, and gluten protein in a supplemented fat-rich meal without eliciting plasma insulin, glucagon, GLP-1, and GIP in individuals with T2DM, the retinyl palmitate response in the chylomicron-rich fraction was lower; whereas the retinyl palmitate response in the chylomicron-poor fraction was higher with whey protein than with casein protein and cod protein in the fat-rich meal. A higher retinyl palmitate response in the chylomicron-rich fraction compared to the chylomicron-poor fraction signifies more lipolysis rather than lower hepatic clearance [327]. In another investigation, it was found that casein in a fat meal and a carbohydrate-fat meal increased concentrations of plasma insulin and glucagon more than a fat meal in subjects with T2DM. Casein in a carbohydrate-fat meal increased the plasma GIP response; however, no GLP-1 response was noted. The plasma postprandial glucose concentration was also reduced by the fat meal and casein + fat meal compared to the fat-carbohydrate meal. An increase in chylomicron-rich retinyl palmitate compared to chylomicron-poor retinyl palmitate after the casein + fat meal was noted; however, no significant differences were found in postprandial TG levels between the casein-fat and casein-free meals [330].

4.4.2. Dairy Proteins in Preload Formulas and T2DM Subjects

Several investigations indicate that preloading whey protein plays a significant role in carbohydrate-rich meal consumption and physiological responses in subsequent periods. Consumption of whey protein before a carbohydrate meal suppressed gastric emptying and postprandial glycemia; whereas the secretion of plasma insulin, GIP, and CCK was stimulated in subjects with T2DM [326]. It was found that plasma glucose was reduced and insulin and C-peptide responses and GLP-1 were increased in T2DM subjects due to whey preloading of a high-calorie breakfast [338]. In another investigation with T2DM subjects, the overall postprandial iAUC for plasma glucose, ghrelin, and hunger scores were reduced, and greater postprandial overall AUCs for plasma insulin, C-peptide, GLP-1, and satiety scores were noted for a whey protein-based breakfast than for a high-carbohydrate breakfast [342]. Similar results were also reported by other investigators when T2DM subjects consumed WPI before breakfast [339] and both breakfast and dinner [343].

Likewise, whey protein preloading influences the consumption of fat-rich meals and physiological responses after meal consumption. For example, consumption of whey protein before a fat-rich meal reduced gastric emptying and enhanced plasma insulin, glucagon, and GIP responses but did not influence lipid or glucose responses in subjects with T2DM [347]. In another investigation, pronounced responses of plasma insulin and GLP-1 and lower responses of plasma glucose were noted due to WPI consumption prior to a fat-rich meal; however, responses of GIP, FFA, and appetite were not influenced [348].

Furthermore, a reduction in fasting glucose and insulin resistance due to the consumption of a whey protein beverage for 10 weeks in subjects with T2DM before and after 45 early morning high-intensity mixed-mode interval training (MMIT) sessions was reported [344]. Interestingly, it was found that consumption of WPI and lactoferrin with Cys (non-BCAA) prior to breakfast was more potent than casein in reducing insulin resistance and fasting plasma glucose in T2DM subjects. Positive effects also manifested in blood lipid profiles. Total serum cholesterol, TG, and LDL-C levels were reduced by WPI and lactoferrin with

Cys. Furthermore, it was found that daily supplementation of undenatured Cys-rich WPI for 3 months, even at a low dose (4625 g/day), improved antioxidative biomarkers (superoxide dismutase, glutathione peroxidase, glutathione, and reduced glutathione-to-oxidized glutathione ratio) and suppressed inflammatory biomarkers (high-sensitivity C-reactive protein, IL-6, tumor necrosis factor-α, and malondialdehyde), which play a central role in diabetes complications and risks of other metabolic diseases [352].

4.4.3. Dairy Proteins in Meals and Obese/Overweight Subjects

Positive effects of whey protein on overweight and obese individuals were also noted because it may be supposed that whey proteins reduce insulin resistance and improve fasting insulin levels. WPI was recognized as a better dietary protein than cod, casein, and gluten protein in a high-fat, high-carbohydrate diet [334] and a vegetable protein-supplemented high-fat diet [350] for obese individuals. It was found that there was a significant decrease in postprandial serum TG after consumption of a whey meal compared to a casein meal; however, postprandial total cholesterol, LDL-C, HDL-C, NEFAs, apolipoprotein B-48, plasma insulin, and leptin were unchanged due to the consumption of casein and whey proteins in obese subjects [329]. In another investigation, it was found that serum total cholesterol, fasting TG, and LDL-C were decreased more by whey protein than by casein protein diets when obese subjects consumed them before breakfast and evening meals. Surprisingly, it was found that fasting plasma glucose was improved after the consumption of whey and casein meals [259]. Similar results were reported by other investigators. It was proven that insulin resistance, serum total cholesterol, adiponectin, LDL-C, HDL-C, and leptin were not changed; however, fasting plasma glucose was increased by both casein and whey protein diet compositions [328]. Furthermore, increases in postprandial GLP-1, insulin, and glucagon and reductions in postprandial GIP, total cholesterol, LDL-C, NEFA, and TG by WPI in a ketogenic diet compared with animal- and plant-based proteins in a high-fat diet were proven [334,350]. Likewise, plasma insulin and total AAs were increased when milk or milk protein was consumed with a high-fat meal by obese individuals; however, serum TG was increased by milk protein [336]. There were no significant changes in fasting plasma insulin, glucose, total cholesterol, LDL-C, whole-body mass, lean mass, fat mass, or thigh muscle area with 1.5 g total protein (MPI)/kg body weight/day compared with 0.8 g total protein (MPI)/kg body weight/day, except for fasting serum TG. The reduction in fasting serum TG was more significant with higher MPI consumption than with lower consumption [319]. Low-fat, high-casein, or high-whey protein diets were considered more effective for weight control than low-fat, high-carbohydrate diets for obese subjects [328].

4.4.4. Dairy Proteins in Preload Formulas and Obese/Overweight Subjects

Positive effects of dairy protein consumption before meals were proven in overweight subjects. In an investigation, it was found that acute appetite and energy intake were equally reduced after consumption of casein- or whey-based preloads compared to glucose-based preloads. This might be due to higher CCK and lower ghrelin responses induced by dairy protein preloads. Postprandial plasma glucose was reduced by protein (mean value of casein and WPI)-based preloads compared to lactose- and glucose-based preloads; however, insulin levels did not change among the different preloads [322]. The impacts of a whey protein meal prior to a fat-rich meal on responses of serum TG and apolipoprotein B-48 in subjects with and without T2DM were investigated. Subsequently, efforts were made to understand whether a whey protein pre-meal had a more pronounced effect on lipid responses than whey protein being part of the fat-rich meal. It was found that a whey protein pre-meal before a fat-rich meal improved plasma insulin, glucagon, and GIP responses in subjects with and without T2DM; however, gastric emptying was delayed by the whey protein pre-meal. Plasma insulin, glucagon, and GIP responses were improved in subjects with and without T2DM by the pre-meal compared to whey protein in the fat-rich meal. Interestingly, it was found that gastric emptying was delayed by the whey protein pre-meal compared to whey protein in the fat-rich meal; however, postprandial serum TG, apolipoprotein B-48, and NEFA levels in subjects with and without T2DM were not significantly changed [347]. In another investigation, it was found that a pre-meal with 20 g of whey protein prior to a high-fat meal stimulated the secretion of insulin and glucagon and reduced blood glucose and gastric emptying compared to a pre-meal with 10 g of whey protein and placebo (no whey protein). There were no changes in GIP, TG, apolipoprotein B-48, or FFA levels with the whey protein pre-meal [348].

An interesting investigation was performed by Danish research groups. The interaction between protein and fat in preload diets was examined to understand all gene expressions and postprandial apolipoprotein B-48, GLP-1, GIP, insulin, glucagon, glucose, triacylglycerol, and FFA levels in obese subjects. The fasting gene expressions of lipoprotein lipase, G protein-coupled receptor 120, and CD36 were upregulated; whereas postprandial gene expression of G protein-coupled receptor 120 and CD36 were downregulated after casein with a high amount of medium-chain saturated fatty acids in the preload diet. No postprandial changes in gene expression were observed with casein and a low amount of medium-chain saturated fatty acids in the preload diet. Upregulation of postprandial lipoprotein lipase gene expression was noted after whey with a high amount of medium-chain saturated fatty acids in the preload diet. The fasting gene expressions of G protein-coupled receptor 120 and CD36 were upregulated by whey with a low amount of medium-chain saturated fatty acids in the preload diet; whereas downregulation of postprandial lipoprotein lipase and G protein-coupled receptor 120 gene expression was noted. Interestingly, no modulation of FABP4 or FAS genes in either fasting or postprandial conditions was noted with any of the proteins and medium-chain saturated fatty acid combinations in the preload diets [341].

4.4.5. Dairy Protein Hydrolysates and T2DM Subjects

The insulinotropic effect of WPH was proven in several investigations, and in many cases, the addition of EAAs offered better results. Some results are mentioned herein. Higher glucose, insulin, glucagon, and GLP-1 responses were contributed by LAPRODAN-DI-3065 (WPH) than by LACPRODAN CGMP-10 (caseinoglycomacropeptide), LACPRODAN-ALPHA-10 (α-lactalbumin), and LAC-PRODAN-DI-9224 (whey protein isolate) in fat-carbohydrate meals for T2DM subjects. However, no significant differences were noted for postprandial TG, retinyl palmitate in the chylomicron-poor fraction, and FFA with any of the dietary proteins; higher levels of retinyl palmitate in the chylomicron-rich fraction were contributed by LACPRODAN-DI-3065 than by all other dietary proteins [333]. It was found that plasma insulin concentration was increased significantly by WPH in a dose-dependent manner, which suppressed the plasma glucose level to the normal range at 2 h after the meal [335]. Likewise, the postprandial plasma glucose response in T2DM subjects was decreased by both whey protein and WPH after breakfast and a mixed-macronutrient meal. Similarly, insulin response and satiety were increased by both intact whey protein and WPH after breakfast and lunch [346]. Furthermore, insulin resistance and postprandial plasma glucose were reduced when obese and type 2 (pre-)diabetic subjects consumed whey protein enriched with Leu and vitamin D [351]. Beneficial outcomes from casein hydrolysate were also reported. It was found that intake of 12 g of casein hydrolysate had a positive effect on post-challenge insulin and glucose levels in subjects with T2DM [332]. The insulin response was increased and the glucose response was decreased by casein hydrolysate compared to carbohydrate diet and intact casein [337,345]; however, no significant difference in gastric emptying outcomes was noted between intact casein and hydrolysate [345]. The effects of different concentrations of milk protein hydrolysates containing a bioactive arginine-proline dipeptide with α-glucosidase-inhibiting properties were investigated in prediabetic subjects. Although the iAUC of plasma glucose was significantly reduced by low-dose milk protein hydrolysates compared to placebo, only a minor insulinotropic effect was noted [349].

4.4.6. Co-Consumption of Amino Acids with Dairy Proteins and T2DM Subjects

Furthermore, it was proven that the addition of specific amino acids to protein formulas markedly promoted the secretion of insulin, leading to a concomitant reduction in the post-challenge plasma glucose response. For example, implementation of Leu and casein hydrolysate with a carbohydrate-based diet offered a greater insulin response, reduced postprandial plasma glucose, and improved blood glucose homeostasis in both T2DM [323,324] and healthy [324] subjects. Similar outcomes were reported in other investigations [321,325,331]. The plasma insulin response was higher with a carbohydrate and protein (50% as casein hydrolysate, 25% as free Leu, and 25% as free Phe) diet than with a carbohydrate diet, and concomitant plasma glucose responses were lower with the carbohydrate and protein diet in T2DM subjects [321]. Consumption of insuVida™, an enzymatic hydrolysate of casein with or without the addition of Leu, could significantly lower plasma glucose compared to placebo and intact casein in T2DM subjects due to its insulinotropic effect [331]. In another investigation with non-obese T2DM subjects, concentrations of postprandial plasma glucose were reduced, and total amino acids, BCAAs, EAAs, insulin, C-peptide, and proinsulin were elicited more by whey protein and/or FAA meals than by casein meals, except for GLP-1 [325].

4.4.7. Contradictory Results

Contradictory results were also reported, mentioned herein. Although it was proven by several investigations that predominant insulin responses and suppression of concomitant plasma glucose responses occur due to co-ingestion of a protein hydrolysate with a meal or prior to a meal, unfortunately, in one experiment, it was found that casein hydrolysate with every main meal did not improve glucose homeostasis over a 24 h period in long-standing T2DM [353]. Smaller postprandial suppression of NEFAs after consumption of LACPRODAN-DI-3065 meals compared with other proteins (LACPRODAN-ALPHA-10, LACPRODAN-DI-9224, and LACPRODAN CGMP-10) in fat-carbohydrate meals was noted in obese non-diabetic individuals; however, no significant differences in body weight or fasting concentrations of plasma glucose, glucagon, CCK, ghrelin, GLP-1, and GIP were noted [354]. Although it was proven that dairy milk proteins are superior to many animal (cod, meat, fish, and egg) [327,334,350] and vegetable (gluten, soya, green pea, and cereal) [342,350] proteins in many investigations, lupin protein compared to casein was more potent in improving the LDL-C:HDL-C ratio in obese/hypercholesterolemic subjects. No significant differences in plasma glucose, Arg, Asn, Cys, Gln, Glu, His, Ile, Leu, Phe, Ser, Trp, or Tyr levels were noted when obese subjects consumed casein and lupin protein, except for plasma TG levels. Plasma TG levels and Ala and Gly were reduced in obese subjects due to the consumption of casein [355]. In another investigation, it was found that there were no differences in postprandial and 24 h energy expenditure or appetite regulation when overweight and moderately obese subjects consumed high concentrations of hydrolyzed casein, intact casein, and intact whey. The results were explained by similar absorption rates when proteins were served as high-protein mixed meals. Lipid oxidation and NEFA concentrations were found to be higher after consumption of intact whey than after consumption of intact and hydrolyzed casein during the daytime and after the breakfast meal [356]. Fasting blood glucose and insulin resistance were significantly increased due to 3 months of consumption of bread fortified with 20 g WPC in overweight/obese subjects with T2DM, and no significant difference compared with placebo was noted. Similarly, consumption of the 20 g WPC-fortified bread led to no significant difference in serum lipid (TG, HDL-C, LDL-C) profile compared with the placebo diet. Investigators realized that 20 g of WPC in bread formulation was not sufficient for providing benefits to overweight T2DM subjects [357]. Other investigators also published similar results. For example, 1.5 g total protein (MPI)/kg body weight/day with an unhealthy Western-style eating pattern did not change fasting plasma insulin, glucose, total cholesterol, or LDL-C levels in obese subjects [319]. In older overweight/obese adults with T2DM, regular consumption of whey protein with vitamin D supplementation for 24 weeks did not offer additional benefits to PRT on measures of glycemic control, body composition, muscle strength, or cardiometabolic risk factors. Investigators felt that there might have been an inadequate concentration of whey protein (20 g whey protein each morning on non-training days with an additional 20 g whey protein after exercise on training days) in the diet, which was insufficient to provide additional PRT benefits in adults with T2DM [170].

4.4.8. Limitations of Clinical Investigations

It is necessary to indicate some limitations of the investigations mentioned in Table 6. While the effects of dairy milk proteins on plasma GLP-1 and GIP were investigated in several studies, very little information was published regarding the effects of milk proteins on the physiological roles of ghrelin [322,342,354], CCK [322,326,354], and DPP-4 [338] in subjects with T2DM. Although dairy protein hydrolysates were used in different investigations [321,323–325,331–333,335,337,345,346,349,351], the degree of hydrolysis (DH) and reaction conditions of hydrolysis were not reported, except in [335]. DH of protein is an important issue in protein-based dietary formulas because the absorption of lower-molecular-weight peptide and AAs in the bloodstream and biofunctional activity depend on peptide length and amino acid sequence. Furthermore, AA sequences of peptides in protein hydrolysates were not mentioned by investigators, except in [349]. Different criteria for the consideration of obesity, hyperglycemia, and dyslipidemia were mentioned by different authorized committees (WHO, IDF, EGIR, and AACE) [6]. These criteria were considered to describe obesity, hyperglycemia, and dyslipidemia in different investigations, as mentioned in Table 6. Therefore, results from a particular research group were not directly comparable with those from other investigators.

5. Conclusions

Metabolic diseases, including obesity, are the consequence of a complex interplay between genetic, dietary, gut microbiota, and environmental factors, and they currently stretch across developed and developing countries. One of the major reasons for obesity is an imbalance of energy intake and expenditure, where high calorie fast foods may have a significant contribution. It is associated with a wide range of physiological factors, such as insulin resistance, hyperglycemia, hypertriglyceridemia, hypoxia, oxidative stress, mitochondrial dysfunction, imbalanced glucose and lipid metabolism, release of adipokines/cytokines/exosomes, angiogenesis, and EMT. The underpinnings of the pathogenesis of obesity are not yet fully understood; however, the contributions of hyperglycemia and dyslipidemia in the case of obesity are recognized. Therefore, obesity is considered one of the key drivers of increased healthcare expenditure. Realizing the contribution of high-calorie fast foods to the prevalence of obesity, efforts have been focused on the modulation of dietary perception and components in the context of obesity management corroborated by hyperglycemia and dyslipidemia.

Dairy proteins emerge as pivotal players in the functional food sector, offering bio-therapeutic activities that modulate obesity prevalence. They offer positive effects on postprandial and post-exercise glucose, lipid, and protein metabolism, and may improve metabolic health by reducing body weight and fat mass through enhanced satiety, maintaining the anabolic sensitivity of muscle to nutrition, muscle protein synthesis, and skeletal muscle metabolic function. Diets with higher amounts of dairy protein improve thermogenesis, which influences the balance of dietary calorie consumption and improves energy expenditure. Several epidemiological and cohort studies have confirmed that the consumption of dairy products decreases the prevalence of MS, while experimental studies reveal that dairy proteins and peptides derived from them play a promising role in the modulation of different risk factors of MS. The significance of BCAAs in dairy proteins for the modulation of glucose and lipid metabolism has been recognized. In the present review, possible molecular interactions and biochemical mechanisms related to fast food-induced obesity are mentioned comprehensively. Furthermore, the role of dairy proteins and peptides produced from cow and buffalo milk proteins in the amelioration of fast food-induced obesity has been represented.

Chemical, enzymatic, and recombinant DNA technology have been employed for producing peptides from milk proteins according to the literature survey; however, their production is limited to certain manufacturers, as mentioned in Table 5. This might be due to complicated processing and a lack of awareness of health-promoting benefits and economic factors. The characteristics of dairy protein hydrolysates (FAAs, length, and AA sequence in peptides) are not mentioned by most manufacturers; however, this is an important aspect in protein-based dietary formulations because physiological roles depend on peptide length and amino acid sequence. Therefore, sequencing of peptides produced from dairy milk proteins by liquid chromatography-electrospray ionization quadrupole time-of-flight mass spectrometry (LC-ESI-Q-TOF-MS) might be important for understanding their physiological roles and bio-efficacy comprehensively. Although a lot of scholarly information exists on the synthesis of various milk protein-derived peptides through enzymatic and chemical routes along with their physicochemical information via in silico QSAR approaches, their applications in the food and clinical sectors are hindered to some extent. Therefore, future research efforts should be directed toward evaluating the molecular mechanisms of action and overall possible health-promoting effects in vivo. Production of targeting peptides on an industrial and semi-industrial scale for the development of functional foods could be interesting. Separation of these peptides followed by their encapsulation in order to increase their shelf life and controlled release in the intestinal system would make them astonishing biotherapeutics. Furthermore, masking of the mentioned therapeutic peptides by sweeteners to reduce bittering effect may be investigated by stopped-flow fluorescence, molecular docking, electronic tongue, and clinical trials. The mode of administration (with or without water), time (fed or fasted phase), and inter-individual variability due to sex, age, pathological condition, and ethnicity affect the bioavailability and functional activities of peptides. Therefore, pharmacokinetic information is necessary to understand the effective mode, time, and dosage of administration.

Milk protein hydrolysates have great market value, and their demand is increasing with the progress of time. Presently, some food industries are focusing on marketing individual dairy proteins and peptides along with regular dairy foods; however, this is limited. Protein hydrolysates can be produced from whey, a byproduct of cheese processing. Whey cannot be disposed of directly to the aquatic system due to stricter environmental legislation. It may be supposed that besides the commercialization of regular dairy foods, the production of peptides or protein hydrolysates from whey proteins may bring an economic boon to dairy and biopharmaceutical industries and will be considered a hallmark in the context of zero-waste disposal.

It may be believed that dairy milk protein-derived peptides may be accepted as potential candidates for the treatment of obesity and other metabolic diseases after rectifying the limitations. The multifaceted discussion in this review may open avenues for future investigations and pave the way for impactful therapeutic interventions for the management of fast food-induced obesity.

Author Contributions: Conceptualization, A.T. and A.N.; methodology, A.N., A.T. and E.S.; software, M.A.M. and A.N.; validation, A.N. and A.T.; formal analysis, K.B.A., A.N., A.T. and M.A.M.; data curation, M.A.M., L.F., A.N. and A.K.; writing—original draft preparation, K.B.A., A.N., E.S., A.T. and Z.L.; writing, review, and editing, A.N., A.T., E.S. and Z.L.; supervision, A.T., A.N., A.K. and Z.L.; project administration, A.K., L.F., A.N. and M.A.M.; funding acquisition, L.F., Z.L., A.N. and M.A.M. All authors have read and agreed to the published version of the manuscript.

Funding: This research received no external funding.

Data Availability Statement: Not applicable.

Acknowledgments: K.B.A. acknowledges the Food Science Doctoral School of the Hungarian University of Agriculture and Life Sciences, Hungary.

Conflicts of Interest: The authors declare no conflicts of interest.

References

1. Fahed, G.; Aoun, L.; Bou Zerdan, M.; Allam, S.; Bou Zerdan, M.; Bouferraa, Y.; Assi, H.I. Metabolic Syndrome: Updates on Pathophysiology and Management in 2021. *Int. J. Mol. Sci.* **2022**, *23*, 786. [CrossRef]
2. Wu, M.; Shu, Y.; Wang, L.; Song, L.; Chen, S.; Liu, Y.; Bi, J.; Li, D.; Yang, Y.; Hu, Y.; et al. Visit-to-Visit Variability in the Measurements of Metabolic Syndrome Components and the Risk of All-Cause Mortality, Cardiovascular Disease, and Arterial Stiffness. *Nutr. Metab. Cardiovasc. Dis.* **2021**, *31*, 2895–2903. [CrossRef]
3. Williams, V.F.; Oh, G.-T.; Stahlman, S. Incidence and Prevalence of the Metabolic Syndrome Using ICD-9 and ICD-10 Diagnostic Codes, Active Component, U.S. Armed Forces, 2002–2017. *MSMR* **2018**, *25*, 20–25.
4. Alberti, G. Introduction to the Metabolic Syndrome. *Eur. Heart J. Suppl.* **2005**, *7*, D3–D5. [CrossRef]
5. Marangos, P.J.; Okamoto, L.J.; Caro, J.J. Economic Burden of the Components of the Metabolic Syndrome. In *Handbook of Disease Burdens and Quality of Life Measures*; Preedy, V.R., Watson, R.R., Eds.; Springer: New York, NY, USA, 2010; pp. 1135–1149. ISBN 978-0-387-78665-0.
6. Jha, B.K.; Sherpa, M.L.; Imran, M.; Mohammed, Y.; Jha, L.A.; Paudel, K.R.; Jha, S.K. Progress in Understanding Metabolic Syndrome and Knowledge of Its Complex Pathophysiology. *Diabetology* **2023**, *4*, 134–159. [CrossRef]
7. Mendrick, D.L.; Diehl, A.M.; Topor, L.S.; Dietert, R.R.; Will, Y.; La Merrill, M.A.; Bouret, S.; Varma, V.; Hastings, K.L.; Schug, T.T.; et al. Metabolic Syndrome and Associated Diseases: From the Bench to the Clinic. *Toxicol. Sci.* **2018**, *162*, 36–42. [CrossRef]
8. VAGUE, J. Sexual differentiation, Factor determining forms of obesity. *Presse Med.* **1947**, *55*, 339.
9. Avogaro, P.; Crepaldi, G.; Enzi, G.; Tiengo, A. Metabolic aspects of essential obesity. *Epatologia* **1965**, *11*, 226–238.
10. O'Neill, S.; O'Driscoll, L. Metabolic Syndrome: A Closer Look at the Growing Epidemic and Its Associated Pathologies. *Obes. Rev.* **2015**, *16*, 1–12. [CrossRef]
11. Reaven, G.M. Role of Insulin Resistance in Human Disease. *Diabetes* **1988**, *37*, 1595–1607. [CrossRef]
12. Kaplan, N.M. The Deadly Quartet. Upper-Body Obesity, Glucose Intolerance, Hypertriglyceridemia, and Hypertension. *Arch. Intern. Med.* **1989**, *149*, 1514–1520. [CrossRef]
13. Haffner, S.M.; Valdez, R.A.; Hazuda, H.P.; Mitchell, B.D.; Morales, P.A.; Stern, M.P. Prospective Analysis of the Insulin-Resistance Syndrome (Syndrome X). *Diabetes* **1992**, *41*, 715–722. [CrossRef]
14. Alberti, K.G.M.M.; Zimmet, P.Z. Definition, Diagnosis and Classification of Diabetes Mellitus and Its Complications. Part 1: Diagnosis and Classification of Diabetes Mellitus. Provisional Report of a WHO Consultation. *Diabet. Med.* **1998**, *15*, 539–553. [CrossRef]
15. Alberti, K.G.M.M.; Zimmet, P.; Shaw, J. Metabolic Syndrome—A New World-Wide Definition. A Consensus Statement from the International Diabetes Federation. *Diabet. Med.* **2006**, *23*, 469–480. [CrossRef]
16. Balkau, B.; Charles, M.A. Comment on the Provisional Report from the WHO Consultation. *Diabet. Med.* **1999**, *16*, 442–443. [CrossRef]
17. Einhorn, D.; Reaven, G.M.; Cobin, R.H.; Ford, E.; Ganda, O.P.; Handelsman, Y.; Hellman, R.; Jellinger, P.S.; Kendall, D.; Krauss, R.M.; et al. American College of Endocrinology Position Statement on the Insulin Resistance Syndrome. *Endocr. Pract.* **2003**, *9*, 237–252. [CrossRef]
18. Wang, W.; Hu, M.; Liu, H.; Zhang, X.; Li, H.; Zhou, F.; Liu, Y.-M.; Lei, F.; Qin, J.-J.; Zhao, Y.-C.; et al. Global Burden of Disease Study 2019 Suggests That Metabolic Risk Factors Are the Leading Drivers of the Burden of Ischemic Heart Disease. *Cell Metab.* **2021**, *33*, 1943–1956.e2. [CrossRef]
19. Ghosh, S.; Dhar, S.; Bhattacharjee, S.; Bhattacharjee, P. Contribution of Environmental, Genetic and Epigenetic Factors to Obesity-Related Metabolic Syndrome. *Nucleus* **2023**, *66*, 215–237. [CrossRef]
20. Noubiap, J.J.; Nansseu, J.R.; Lontchi-Yimagou, E.; Nkeck, J.R.; Nyaga, U.F.; Ngouo, A.T.; Tounouga, D.N.; Tianyi, F.-L.; Foka, A.J.; Ndoadoumgue, A.L.; et al. Geographic Distribution of Metabolic Syndrome and Its Components in the General Adult Population: A Meta-Analysis of Global Data from 28 Million Individuals. *Diabetes Res. Clin. Pract.* **2022**, *188*, 109924. [CrossRef]
21. do Vale Moreira, N.C.; Hussain, A.; Bhowmik, B.; Mdala, I.; Siddiquee, T.; Fernandes, V.O.; Montenegro Júnior, R.M.; Meyer, H.E. Prevalence of Metabolic Syndrome by Different Definitions, and Its Association with Type 2 Diabetes, Pre-Diabetes, and Cardiovascular Disease Risk in Brazil. *Diabetes Metab. Syndr. Clin. Res. Rev.* **2020**, *14*, 1217–1224. [CrossRef]
22. Li, W.; Qiu, X.; Ma, H.; Geng, Q. Incidence and Long-Term Specific Mortality Trends of Metabolic Syndrome in the United States. *Front. Endocrinol.* **2022**, *13*, 1029736. [CrossRef] [PubMed]
23. Hu, C.; Jia, W. Multi-Omics Profiling: The Way towards Precision Medicine in Metabolic Diseases. *J. Mol. Cell Biol.* **2021**, *13*, 576–593. [CrossRef] [PubMed]
24. Montgomery, M.K.; Turner, N. Mitochondrial Dysfunction and Insulin Resistance: An Update. *Endocr. Connect.* **2015**, *4*, R1–R15. [CrossRef] [PubMed]
25. Henao-Mejia, J.; Elinav, E.; Jin, C.; Hao, L.; Mehal, W.Z.; Strowig, T.; Thaiss, C.A.; Kau, A.L.; Eisenbarth, S.C.; Jurczak, M.J.; et al. Inflammasome-Mediated Dysbiosis Regulates Progression of NAFLD and Obesity. *Nature* **2012**, *482*, 179–185. [CrossRef]
26. Toubal, A.; Treuter, E.; Clément, K.; Venteclef, N. Genomic and Epigenomic Regulation of Adipose Tissue Inflammation in Obesity. *Trends Endocrinol. Metab.* **2013**, *24*, 625–634. [CrossRef] [PubMed]

27. Nieuwdorp, M.; Stroes, E.S.G.; Meijers, J.C.M.; Büller, H. Hypercoagulability in the Metabolic Syndrome. *Curr. Opin. Pharmacol.* **2005**, *5*, 155–159. [CrossRef] [PubMed]
28. Cano-Sancho, G.; Salmon, A.G.; La Merrill, M.A. Association between Exposure to p,p′-DDT and Its Metabolite p,p′-DDE with Obesity: Integrated Systematic Review and Meta-Analysis. *Environ. Health Perspect.* **2017**, *125*, 96002. [CrossRef] [PubMed]
29. Qin, J.; Li, Y.; Cai, Z.; Li, S.; Zhu, J.; Zhang, F.; Liang, S.; Zhang, W.; Guan, Y.; Shen, D.; et al. A Metagenome-Wide Association Study of Gut Microbiota in Type 2 Diabetes. *Nature* **2012**, *490*, 55–60. [CrossRef] [PubMed]
30. Warrier, M.; Shih, D.M.; Burrows, A.C.; Ferguson, D.; Gromovsky, A.D.; Brown, A.L.; Marshall, S.; McDaniel, A.; Schugar, R.C.; Wang, Z.; et al. The TMAO-Generating Enzyme Flavin Monooxygenase 3 Is a Central Regulator of Cholesterol Balance. *Cell Rep.* **2015**, *10*, 326–338. [CrossRef] [PubMed]
31. Halpern, A.; Mancini, M.C.; Magalhães, M.E.C.; Fisberg, M.; Radominski, R.; Bertolami, M.C.; Bertolami, A.; de Melo, M.E.; Zanella, M.T.; Queiroz, M.S.; et al. Metabolic Syndrome, Dyslipidemia, Hypertension and Type 2 Diabetes in Youth: From Diagnosis to Treatment. *Diabetol. Metab. Syndr.* **2010**, *2*, 55. [CrossRef] [PubMed]
32. Carr, M.C.; Brunzell, J.D. Abdominal Obesity and Dyslipidemia in the Metabolic Syndrome: Importance of Type 2 Diabetes and Familial Combined Hyperlipidemia in Coronary Artery Disease Risk. *J. Clin. Endocrinol. Metab.* **2004**, *89*, 2601–2607. [CrossRef] [PubMed]
33. Sánchez-Alegría, K.; Arias, C. Functional Consequences of Brain Exposure to Saturated Fatty Acids: From Energy Metabolism and Insulin Resistance to Neuronal Damage. *Endocrinol. Diabetes Metab.* **2023**, *6*, e386. [CrossRef] [PubMed]
34. Smith, S.M.; Boppana, A.; Traupman, J.A.; Unson, E.; Maddock, D.A.; Chao, K.; Dobesh, D.P.; Brufsky, A.; Connor, R.I. Impaired Glucose Metabolism in Patients with Diabetes, Prediabetes, and Obesity Is Associated with Severe COVID-19. *J. Med. Virol.* **2021**, *93*, 409–415. [CrossRef] [PubMed]
35. Harlim, A.; Bontong, S.I.; Rahfiludin, M.Z. Relationship between Carbohydrate, Protein, and Fat Consumption with Obesity Incidence in 2014 UKI Faculty of Medicine Students. *Amerta Nutr.* **2022**, *6*, 385–391. [CrossRef]
36. Ussar, S.; Fujisaka, S.; Kahn, C.R. Interactions between Host Genetics and Gut Microbiome in Diabetes and Metabolic Syndrome. *Mol. Metab.* **2016**, *5*, 795–803. [CrossRef] [PubMed]
37. Neumark-Sztainer, D.; Story, M.; Perry, C.; Casey, M.A. Factors Influencing Food Choices of Adolescents: Findings from Focus-Group Discussions with Adolescents. *J. Am. Diet. Assoc.* **1999**, *99*, 929–937. [CrossRef] [PubMed]
38. Pereira, M.A.; Kartashov, A.I.; Ebbeling, C.B.; Van Horn, L.; Slattery, M.L.; Jacobs, D.R.; Ludwig, D.S. Fast-Food Habits, Weight Gain, and Insulin Resistance (the CARDIA Study): 15-Year Prospective Analysis. *Lancet* **2005**, *365*, 36–42. [CrossRef] [PubMed]
39. Rouhani, M.H.; Mirseifinezhad, M.; Omrani, N.; Esmaillzadeh, A.; Azadbakht, L. Fast Food Consumption, Quality of Diet, and Obesity among Isfahanian Adolescent Girls. *J. Obes.* **2012**, *2012*, 597924. [CrossRef] [PubMed]
40. Alexander, L.; Christensen, S.M.; Richardson, L.; Ingersoll, A.B.; Burridge, K.; Golden, A.; Karjoo, S.; Cortez, D.; Shelver, M.; Bays, H.E. Nutrition and Physical Activity: An Obesity Medicine Association (OMA) Clinical Practice Statement 2022. *Obes. Pillars* **2022**, *1*, 100005. [CrossRef] [PubMed]
41. Hu, X.; Li, Z.; Wang, F.; Mu, H.; Guo, L.; Xiao, J.; Liu, Y.; Li, X. Formation of Starch-Lipid Complexes during the Deep-Frying Process and Its Effects on Lipid Oxidation. *Foods* **2022**, *11*, 3083. [CrossRef] [PubMed]
42. Bordin, K.; Kunitake, M.T.; Aracava, K.K.; Silvia, C.; Trindade, F. Changes in Food Caused by Deep Fat Frying—A Review. *Arch. Latinoam. De Nutr.* **2013**, *63*, 5–13.
43. Ganesan, K.; Xu, B. Deep Frying Cooking Oils Promote the High Risk of Metastases in the Breast-A Critical Review. *Food Chem. Toxicol.* **2020**, *144*, 111648. [CrossRef]
44. Choe, E.; Min, D.B. Chemistry of Deep-Fat Frying Oils. *J. Food Sci.* **2007**, *72*, R77–R86. [CrossRef] [PubMed]
45. Pamplona, R. Advanced Lipoxidation End-Products. *Chem. Biol. Interact.* **2011**, *192*, 14–20. [CrossRef] [PubMed]
46. Wang, S.; Chao, C.; Cai, J.; Niu, B.; Copeland, L.; Wang, S. Starch–Lipid and Starch–Lipid–Protein Complexes: A Comprehensive Review. *Compr. Rev. Food Sci. Food Saf.* **2020**, *19*, 1056–1079. [CrossRef]
47. Sharma, C.; Kaur, A.; Thind, S.S.; Singh, B.; Raina, S. Advanced Glycation End-Products (AGEs): An Emerging Concern for Processed Food Industries. *J. Food Sci. Technol.* **2015**, *52*, 7561–7576. [CrossRef]
48. Liauchonak, I.; Qorri, B.; Dawoud, F.; Riat, Y.; Szewczuk, M.R. Non-Nutritive Sweeteners and Their Implications on the Development of Metabolic Syndrome. *Nutrients* **2019**, *11*, 644. [CrossRef] [PubMed]
49. Maharat, M.; Rahimlou, M.; Sioofi, A.; Sajjadi, S.F.; Moosavian, S.P. Association of Major Dietary Patterns with Socioeconomic Status, Obesity, and Contracting COVID-19 among Iranian Adults. *Front. Nutr.* **2024**, *11*, 1301634. [CrossRef] [PubMed]
50. Tricco, A.C.; Lillie, E.; Zarin, W.; O'Brien, K.K.; Colquhoun, H.; Levac, D.; Moher, D.; Peters, M.D.J.; Horsley, T.; Weeks, L.; et al. PRISMA Extension for Scoping Reviews (PRISMA-ScR): Checklist and Explanation. *Ann. Intern. Med.* **2018**, *169*, 467–473. [CrossRef] [PubMed]
51. James, W.P.T. WHO Recognition of the Global Obesity Epidemic. *Int. J. Obes.* **2008**, *32*, S120–S126. [CrossRef] [PubMed]
52. Lathigara, D.; Kaushal, D.; Wilson, R.B. Molecular Mechanisms of Western Diet-Induced Obesity and Obesity-Related Carcinogenesis—A Narrative Review. *Metabolites* **2023**, *13*, 675. [CrossRef] [PubMed]
53. Yang, M.; Liu, S.; Zhang, C. The Related Metabolic Diseases and Treatments of Obesity. *Healthcare* **2022**, *10*, 1616. [CrossRef]
54. Ward, Z.J.; Bleich, S.N.; Cradock, A.L.; Barrett, J.L.; Giles, C.M.; Flax, C.; Long, M.W.; Gortmaker, S.L. Projected U.S. State-Level Prevalence of Adult Obesity and Severe Obesity. *N. Engl. J. Med.* **2019**, *381*, 2440–2450. [CrossRef] [PubMed]

55. Boutari, C.; Mantzoros, C.S. A 2022 Update on the Epidemiology of Obesity and a Call to Action: As Its Twin COVID-19 Pandemic Appears to Be Receding, the Obesity and Dysmetabolism Pandemic Continues to Rage On. *Metabolism* **2022**, *133*, 155217. [CrossRef] [PubMed]
56. Lobstein, T.; Jackson-Leach, R.; Powis, J.; Brinsden, H.; Gray, M. *World Obesity Atlas 2023*; World Obesity Federation: London, UK, 2023; pp. 5–25.
57. Venkatesan, R.; Mohan, V. Obesity—Are We Continuing to Play the Genetic "Blame Game"? *Adv. Genom. Genet.* **2016**, *6*, 11–23. [CrossRef]
58. Scuteri, A.; Sanna, S.; Chen, W.-M.; Uda, M.; Albai, G.; Strait, J.; Najjar, S.; Nagaraja, R.; Orrú, M.; Usala, G.; et al. Genome-Wide Association Scan Shows Genetic Variants in the FTO Gene Are Associated with Obesity-Related Traits. *PLoS Genet.* **2007**, *3*, e115. [CrossRef] [PubMed]
59. Buniello, A.; MacArthur, J.A.L.; Cerezo, M.; Harris, L.W.; Hayhurst, J.; Malangone, C.; McMahon, A.; Morales, J.; Mountjoy, E.; Sollis, E.; et al. The NHGRI-EBI GWAS Catalog of Published Genome-Wide Association Studies, Targeted Arrays and Summary Statistics 2019. *Nucleic Acids Res.* **2019**, *47*, D1005–D1012. [CrossRef] [PubMed]
60. Pigeyre, M.; Yazdi, F.T.; Kaur, Y.; Meyre, D. Recent Progress in Genetics, Epigenetics and Metagenomics Unveils the Pathophysiology of Human Obesity. *Clin. Sci.* **2016**, *130*, 943–986. [CrossRef] [PubMed]
61. van der Klaauw, A.A.; Farooqi, I.S. The Hunger Genes: Pathways to Obesity. *Cell* **2015**, *161*, 119–132. [CrossRef] [PubMed]
62. Loos, R.J.F.; Yeo, G.S.H. The Genetics of Obesity: From Discovery to Biology. *Nat. Rev. Genet.* **2022**, *23*, 120–133. [CrossRef]
63. Hinney, A.; Körner, A.; Fischer-Posovszky, P. The Promise of New Anti-Obesity Therapies Arising from Knowledge of Genetic Obesity Traits. *Nat. Rev. Endocrinol.* **2022**, *18*, 623–637. [CrossRef] [PubMed]
64. Grundy, S.M. Obesity, Metabolic Syndrome, and Cardiovascular Disease. *J. Clin. Endocrinol. Metab.* **2004**, *89*, 2595–2600. [CrossRef] [PubMed]
65. Choquet, H.; Meyre, D. Genetics of Obesity: What Have We Learned? *Curr. Genom.* **2011**, *12*, 169–179. [CrossRef]
66. Frayling, T.M. Genome-Wide Association Studies Provide New Insights into Type 2 Diabetes Aetiology. *Nat. Rev. Genet.* **2007**, *8*, 657–662. [CrossRef] [PubMed]
67. Duis, J.; Butler, M.G. Syndromic and Nonsyndromic Obesity: Underlying Genetic Causes in Humans. *Adv. Biol.* **2022**, *6*, 2101154. [CrossRef] [PubMed]
68. Şanlı, E.; Kabaran, S. Maternal Obesity, Maternal Overnutrition and Fetal Programming: Effects of Epigenetic Mechanisms on the Development of Metabolic Disorders. *Curr. Genom.* **2019**, *20*, 419–427. [CrossRef]
69. Simmonds, M.; Llewellyn, A.; Owen, C.G.; Woolacott, N. Predicting Adult Obesity from Childhood Obesity: A Systematic Review and Meta-Analysis. *Obes. Rev.* **2016**, *17*, 95–107. [CrossRef]
70. Diagnosis and Classification of Diabetes Mellitus. *Diabetes Care* **2014**, *37* (Suppl. S1), S81–S90. [CrossRef] [PubMed]
71. American Diabetes Association. 2. Classification and Diagnosis of Diabetes: Standards of Medical Care in Diabetes-2018. *Diabetes Care* **2018**, *41*, S13–S27. [CrossRef] [PubMed]
72. Sanyaolu, A.; Okorie, C.; Qi, X.; Locke, J.; Rehman, S. Childhood and Adolescent Obesity in the United States: A Public Health Concern. *Glob. Pediatr. Health* **2019**, *6*, 2333794X19891305. [CrossRef] [PubMed]
73. Tsai, A.G.; Williamson, D.F.; Glick, H.A. Direct Medical Cost of Overweight and Obesity in the USA: A Quantitative Systematic Review. *Obes. Rev.* **2011**, *12*, 50–61. [CrossRef] [PubMed]
74. Karamanou, M.; Protogerou, A.; Tsoucalas, G.; Androutsos, G.; Poulakou-Rebelakou, E. Milestones in the History of Diabetes Mellitus: The Main Contributors. *World J. Diabetes* **2016**, *7*, 1–7. [CrossRef] [PubMed]
75. Lin, X.; Xu, Y.; Pan, X.; Xu, J.; Ding, Y.; Sun, X.; Song, X.; Ren, Y.; Shan, P.-F. Global, Regional, and National Burden and Trend of Diabetes in 195 Countries and Territories: An Analysis from 1990 to 2025. *Sci. Rep.* **2020**, *10*, 14790. [CrossRef]
76. Kahaly, G.J.; Hansen, M.P. Type 1 Diabetes Associated Autoimmunity. *Autoimmun. Rev.* **2016**, *15*, 644–648. [CrossRef] [PubMed]
77. Leahy, J.L. Pathogenesis of Type 2 Diabetes Mellitus. *Arch. Med. Res.* **2005**, *36*, 197–209. [CrossRef] [PubMed]
78. Lawrence, J.M.; Contreras, R.; Chen, W.; Sacks, D.A. Trends in the Prevalence of Preexisting Diabetes and Gestational Diabetes Mellitus among a Racially/Ethnically Diverse Population of Pregnant Women, 1999–2005. *Diabetes Care* **2008**, *31*, 899–904. [CrossRef]
79. Grimaldi, A.; Dudek, A.; Otto-Buczkowska, E. Secondary Diabetes. *Med. Des. Mal. Metab.* **2008**, *2*, 571. [CrossRef]
80. Garvey, W.T.; Olefsky, J.M.; Griffin, J.; Hamman, R.F.; Kolterman, O.G. The Effect of Insulin Treatment on Insulin Secretion and Insulin Action in Type II Diabetes Mellitus. *Diabetes* **1985**, *34*, 222–234. [CrossRef] [PubMed]
81. Grant, S.F.A.; Thorleifsson, G.; Reynisdottir, I.; Benediktsson, R.; Manolescu, A.; Sainz, J.; Helgason, A.; Stefansson, H.; Emilsson, V.; Helgadottir, A.; et al. Variant of Transcription Factor 7-like 2 (TCF7L2) Gene Confers Risk of Type 2 Diabetes. *Nat. Genet.* **2006**, *38*, 320–323. [CrossRef]
82. Helgason, A.; Pálsson, S.; Thorleifsson, G.; Grant, S.F.A.; Emilsson, V.; Gunnarsdottir, S.; Adeyemo, A.; Chen, Y.; Chen, G.; Reynisdottir, I.; et al. Refining the Impact of TCF7L2 Gene Variants on Type 2 Diabetes and Adaptive Evolution. *Nat. Genet.* **2007**, *39*, 218–225. [CrossRef] [PubMed]
83. Banday, M.Z.; Sameer, A.S.; Nissar, S. Pathophysiology of Diabetes: An Overview. *Avicenna J. Med.* **2020**, *10*, 174–188. [CrossRef] [PubMed]
84. Shepherd, P.R.; Kahn, B.B. Glucose Transporters and Insulin Action—Implications for Insulin Resistance and Diabetes Mellitus. *N. Engl. J. Med.* **1999**, *341*, 248–257. [CrossRef] [PubMed]

85. Carpentier, A.; Mittelman, S.D.; Lamarche, B.; Bergman, R.N.; Giacca, A.; Lewis, G.F. Acute Enhancement of Insulin Secretion by FFA in Humans Is Lost with Prolonged FFA Elevation. *Am. J. Physiol.* **1999**, *276*, E1055–E1066. [CrossRef] [PubMed]
86. Lotta, L.A.; Gulati, P.; Day, F.R.; Payne, F.; Ongen, H.; van de Bunt, M.; Gaulton, K.J.; Eicher, J.D.; Sharp, S.J.; Luan, J.; et al. Integrative Genomic Analysis Implicates Limited Peripheral Adipose Storage Capacity in the Pathogenesis of Human Insulin Resistance. *Nat. Genet.* **2017**, *49*, 17–26. [CrossRef] [PubMed]
87. Biddinger, S.B.; Hernandez-Ono, A.; Rask-Madsen, C.; Haas, J.T.; Alemán, J.O.; Suzuki, R.; Scapa, E.F.; Agarwal, C.; Carey, M.C.; Stephanopoulos, G.; et al. Hepatic Insulin Resistance Is Sufficient to Produce Dyslipidemia and Susceptibility to Atherosclerosis. *Cell Metab.* **2008**, *7*, 125–134. [CrossRef] [PubMed]
88. Pories, W.J.; Dohm, G.L. Diabetes: Have We Got It All Wrong? Hyperinsulinism as the Culprit: Surgery Provides the Evidence. *Diabetes Care* **2012**, *35*, 2438–2442. [CrossRef] [PubMed]
89. da Silva Rosa, S.C.; Nayak, N.; Caymo, A.M.; Gordon, J.W. Mechanisms of Muscle Insulin Resistance and the Cross-Talk with Liver and Adipose Tissue. *Physiol. Rep.* **2020**, *8*, e14607. [CrossRef] [PubMed]
90. Pickering, R.J.; Rosado, C.J.; Sharma, A.; Buksh, S.; Tate, M.; de Haan, J.B. Recent Novel Approaches to Limit Oxidative Stress and Inflammation in Diabetic Complications. *Clin. Transl. Immunol.* **2018**, *7*, e1016. [CrossRef] [PubMed]
91. Ojha, A.; Ojha, U.; Mohammed, R.; Chandrashekar, A.; Ojha, H. Current Perspective on the Role of Insulin and Glucagon in the Pathogenesis and Treatment of Type 2 Diabetes Mellitus. *Clin. Pharmacol.* **2019**, *11*, 57–65. [CrossRef]
92. Deacon, C.F. Physiology and Pharmacology of DPP-4 in Glucose Homeostasis and the Treatment of Type 2 Diabetes. *Front. Endocrinol.* **2019**, *10*, 440649. [CrossRef]
93. Lo, C.-M.; Obici, S.; Dong, H.H.; Haas, M.; Lou, D.; Kim, D.H.; Liu, M.; D'Alessio, D.; Woods, S.C.; Tso, P. Impaired Insulin Secretion and Enhanced Insulin Sensitivity in Cholecystokinin-Deficient Mice. *Diabetes* **2011**, *60*, 2000–2007. [CrossRef] [PubMed]
94. Hackl, M.T.; Fürnsinn, C.; Schuh, C.M.; Krssak, M.; Carli, F.; Guerra, S.; Freudenthaler, A.; Baumgartner-Parzer, S.; Helbich, T.H.; Luger, A.; et al. Brain Leptin Reduces Liver Lipids by Increasing Hepatic Triglyceride Secretion and Lowering Lipogenesis. *Nat. Commun.* **2019**, *10*, 2717. [CrossRef] [PubMed]
95. Cohen, B.; Novick, D.; Rubinstein, M. Modulation of Insulin Activities by Leptin. *Science* **1996**, *274*, 1185–1188. [CrossRef] [PubMed]
96. Kershaw, E.E.; Flier, J.S. Adipose Tissue as an Endocrine Organ. *J. Clin. Endocrinol. Metab.* **2004**, *89*, 2548–2556. [CrossRef] [PubMed]
97. Katsiki, N.; Mikhailidis, D.P.; Banach, M. Leptin, Cardiovascular Diseases and Type 2 Diabetes Mellitus. *Acta Pharmacol. Sin.* **2018**, *39*, 1176–1188. [CrossRef] [PubMed]
98. Puri, V.; Ranjit, S.; Konda, S.; Nicoloro, S.M.C.; Straubhaar, J.; Chawla, A.; Chouinard, M.; Lin, C.; Burkart, A.; Corvera, S.; et al. Cidea Is Associated with Lipid Droplets and Insulin Sensitivity in Humans. *Proc. Natl. Acad. Sci. USA* **2008**, *105*, 7833–7838. [CrossRef] [PubMed]
99. Kimmel, A.R.; Sztalryd, C. The Perilipins: Major Cytosolic Lipid Droplet-Associated Proteins and Their Roles in Cellular Lipid Storage, Mobilization, and Systemic Homeostasis. *Annu. Rev. Nutr.* **2016**, *36*, 471–509. [CrossRef] [PubMed]
100. Li, X.; Bai, Y.; Jin, Z.; Svensson, B. Food-Derived Non-Phenolic α-Amylase and α-Glucosidase Inhibitors for Controlling Starch Digestion Rate and Guiding Diabetes-Friendly Recipes. *LWT* **2022**, *153*, 112455. [CrossRef]
101. Del Prete, A.; Iadevaia, M.; Loguercio, C. The Role of Gut Hormones in Controlling the Food Intake. What Is Their Role in Emerging Diseases? *Endocrinol. Y Nutr. (Engl. Ed.)* **2012**, *59*, 197–206. [CrossRef]
102. Shoelson, S.E.; Lee, J.; Goldfine, A.B. Inflammation and Insulin Resistance. *J. Clin. Invest.* **2006**, *116*, 1793–1801. [CrossRef] [PubMed]
103. Evans, J.L.; Goldfine, I.D.; Maddux, B.A.; Grodsky, G.M. Oxidative Stress and Stress-Activated Signaling Pathways: A Unifying Hypothesis of Type 2 Diabetes. *Endocr. Rev.* **2002**, *23*, 599–622. [CrossRef] [PubMed]
104. Fowler, M.J. Microvascular and Macrovascular Complications of Diabetes. *Clin. Diabetes* **2008**, *26*, 77–82. [CrossRef]
105. Bielka, W.; Przezak, A.; Pawlik, A. The Role of the Gut Microbiota in the Pathogenesis of Diabetes. *Int. J. Mol. Sci.* **2022**, *23*, 480. [CrossRef] [PubMed]
106. Gradisteanu Pircalabioru, G.; Corcionivoschi, N.; Gundogdu, O.; Chifiriuc, M.-C.; Marutescu, L.G.; Ispas, B.; Savu, O. Dysbiosis in the Development of Type I Diabetes and Associated Complications: From Mechanisms to Targeted Gut Microbes Manipulation Therapies. *Int. J. Mol. Sci.* **2021**, *22*, 2763. [CrossRef] [PubMed]
107. Iatcu, C.O.; Steen, A.; Covasa, M. Gut Microbiota and Complications of Type-2 Diabetes. *Nutrients* **2021**, *14*, 166. [CrossRef] [PubMed]
108. Kang, Q.; Yang, C. Oxidative Stress and Diabetic Retinopathy: Molecular Mechanisms, Pathogenetic Role and Therapeutic Implications. *Redox Biol.* **2020**, *37*, 101799. [CrossRef] [PubMed]
109. Naveen, J.; Baskaran, V. Antidiabetic Plant-Derived Nutraceuticals: A Critical Review. *Eur. J. Nutr.* **2018**, *57*, 1275–1299. [CrossRef] [PubMed]
110. Ott, C.; Jacobs, K.; Haucke, E.; Navarrete Santos, A.; Grune, T.; Simm, A. Role of Advanced Glycation End Products in Cellular Signaling. *Redox Biol.* **2014**, *2*, 411–429. [CrossRef] [PubMed]
111. Geissler, S.; Hellwig, M.; Zwarg, M.; Markwardt, F.; Henle, T.; Brandsch, M. Transport of the Advanced Glycation End Products Alanylpyrraline and Pyrralylalanine by the Human Proton-Coupled Peptide Transporter HPEPT1. *J. Agric. Food Chem.* **2010**, *58*, 2543–2547. [CrossRef]

112. Tessier, F.J.; Niquet-Léridon, C.; Jacolot, P.; Jouquand, C.; Genin, M.; Schmidt, A.-M.; Grossin, N.; Boulanger, E. Quantitative Assessment of Organ Distribution of Dietary Protein-Bound 13C-Labeled Nε-Carboxymethyllysine after a Chronic Oral Exposure in Mice. *Mol. Nutr. Food Res.* **2016**, *60*, 2446–2456. [CrossRef] [PubMed]
113. Sutherland, J.P.; McKinley, B.; Eckel, R.H. The Metabolic Syndrome and Inflammation. *Metab. Syndr. Relat. Disord.* **2004**, *2*, 82–104. [CrossRef] [PubMed]
114. Ozaki, K.; Awazu, M.; Tamiya, M.; Iwasaki, Y.; Harada, A.; Kugisaki, S.; Tanimura, S.; Kohno, M. Targeting the ERK Signaling Pathway as a Potential Treatment for Insulin Resistance and Type 2 Diabetes. *Am. J. Physiol. Metab.* **2016**, *310*, E643–E651. [CrossRef] [PubMed]
115. Sergi, D.; Williams, L.M. Potential Relationship between Dietary Long-Chain Saturated Fatty Acids and Hypothalamic Dysfunction in Obesity. *Nutr. Rev.* **2020**, *78*, 261–277. [CrossRef] [PubMed]
116. Raleigh, D.; Zhang, X.; Hastoy, B.; Clark, A. The β-Cell Assassin: IAPP Cytotoxicity. *J. Mol. Endocrinol.* **2017**, *59*, R121–R140. [CrossRef] [PubMed]
117. Wang, X.; Liu, J.; Yang, Y.; Zhang, X. An Update on the Potential Role of Advanced Glycation End Products in Glycolipid Metabolism. *Life Sci.* **2020**, *245*, 117344. [CrossRef] [PubMed]
118. Li, Y.; Peng, Y.; Shen, Y.; Zhang, Y.; Liu, L.; Yang, X. Dietary Polyphenols: Regulate the Advanced Glycation End Products-RAGE Axis and the Microbiota-Gut-Brain Axis to Prevent Neurodegenerative Diseases. *Crit. Rev. Food Sci. Nutr.* **2023**, *63*, 9816–9842. [CrossRef] [PubMed]
119. Hofnagel, O.; Luechtenborg, B.; Weissen-Plenz, G.; Robenek, H. Statins and Foam Cell Formation: Impact on LDL Oxidation and Uptake of Oxidized Lipoproteins via Scavenger Receptors. *Biochim. Biophys. Acta Mol. Cell Biol. Lipids* **2007**, *1771*, 1117–1124. [CrossRef] [PubMed]
120. Yamamoto, K.; Kakino, A.; Takeshita, H.; Hayashi, N.; Li, L.; Nakano, A.; Hanasaki-Yamamoto, H.; Fujita, Y.; Imaizumi, Y.; Toyama-Yokoyama, S.; et al. Oxidized LDL (OxLDL) Activates the Angiotensin II Type 1 Receptor by Binding to the Lectin-like OxLDL Receptor. *FASEB J.* **2015**, *29*, 3342–3356. [CrossRef] [PubMed]
121. Karar, J.; Maity, A. PI3K/AKT/MTOR Pathway in Angiogenesis. *Front. Mol. Neurosci.* **2011**, *4*, 51. [CrossRef] [PubMed]
122. Arivazhagan, L.; López-Díez, R.; Shekhtman, A.; Ramasamy, R.; Schmidt, A.M. Glycation and a Spark of ALEs (Advanced Lipoxidation End Products)—Igniting RAGE/Diaphanous-1 and Cardiometabolic Disease. *Front. Cardiovasc. Med.* **2022**, *9*, 1–12. [CrossRef] [PubMed]
123. Wang, J.; Cai, W.; Yu, J.; Liu, H.; He, S.; Zhu, L.; Xu, J. Dietary Advanced Glycation End Products Shift the Gut Microbiota Composition and Induce Insulin Resistance in Mice. *Diabetes. Metab. Syndr. Obes.* **2022**, *15*, 427–437. [CrossRef] [PubMed]
124. Inan-Eroglu, E.; Ayaz, A.; Buyuktuncer, Z. Formation of Advanced Glycation Endproducts in Foods during Cooking Process and Underlying Mechanisms: A Comprehensive Review of Experimental Studies. *Nutr. Res. Rev.* **2020**, *33*, 77–89. [CrossRef]
125. Takeuchi, M.; Sakasai-Sakai, A.; Takata, T.; Takino, J.; Koriyama, Y.; Kikuchi, C.; Furukawa, A.; Nagamine, K.; Hori, T.; Matsunaga, T. Intracellular Toxic AGEs (TAGE) Triggers Numerous Types of Cell Damage. *Biomolecules* **2021**, *11*, 387. [CrossRef]
126. Salazar, J.; Navarro, C.; Ortega, Á.; Nava, M.; Morillo, D.; Torres, W.; Hernández, M.; Cabrera, M.; Angarita, L.; Ortiz, R.; et al. Advanced Glycation End Products: New Clinical and Molecular Perspectives. *Int. J. Environ. Res. Public Health* **2021**, *18*, 7236. [CrossRef]
127. Dong, L.; Li, Y.; Chen, Q.; Liu, Y.; Qiao, Z.; Sang, S.; Zhang, J.; Zhan, S.; Wu, Z.; Liu, L. Research Advances of Advanced Glycation End Products in Milk and Dairy Products: Formation, Determination, Control Strategy and Immunometabolism via Gut Microbiota. *Food Chem.* **2023**, *417*, 135861. [CrossRef] [PubMed]
128. Sacks, F.M.; Lichtenstein, A.H.; Wu, J.H.Y.; Appel, L.J.; Creager, M.A.; Kris-Etherton, P.M.; Miller, M.; Rimm, E.B.; Rudel, L.L.; Robinson, J.G.; et al. Dietary Fats and Cardiovascular Disease: A Presidential Advisory from the American Heart Association. *Circulation* **2017**, *136*, e1–e23. [CrossRef] [PubMed]
129. Dybiec, J.; Baran, W.; Dąbek, B.; Fularski, P.; Młynarska, E.; Radzioch, E.; Rysz, J.; Franczyk, B. Advances in Treatment of Dyslipidemia. *Int. J. Mol. Sci.* **2023**, *24*, 13288. [CrossRef] [PubMed]
130. Arvanitis, M.; Lowenstein, C.J. Dyslipidemia. *Ann. Intern. Med.* **2023**, *176*, ITC81–ITC96. [CrossRef] [PubMed]
131. Rosenson, R.S.; Najera, S.D.; Hegele, R.A. Heterozygous Familial Hypercholesterolemia Presenting as Chylomicronemia Syndrome. *J. Clin. Lipidol.* **2017**, *11*, 294–296. [CrossRef] [PubMed]
132. Pirillo, A.; Casula, M.; Olmastroni, E.; Norata, G.D.; Catapano, A.L. Global Epidemiology of Dyslipidaemias. *Nat. Rev. Cardiol.* **2021**, *18*, 689–700. [CrossRef]
133. Schwarz, J.-M.; Linfoot, P.; Dare, D.; Aghajanian, K. Hepatic de Novo Lipogenesis in Normoinsulinemic and Hyperinsulinemic Subjects Consuming High-Fat, Low-Carbohydrate and Low-Fat, High-Carbohydrate Isoenergetic Diets123. *Am. J. Clin. Nutr.* **2003**, *77*, 43–50. [CrossRef] [PubMed]
134. Stanhope, K.L.; Schwarz, J.M.; Keim, N.L.; Griffen, S.C.; Bremer, A.A.; Graham, J.L.; Hatcher, B.; Cox, C.L.; Dyachenko, A.; Zhang, W.; et al. Consuming Fructose-Sweetened, Not Glucose-Sweetened, Beverages Increases Visceral Adiposity and Lipids and Decreases Insulin Sensitivity in Overweight/Obese Humans. *J. Clin. Invest.* **2009**, *119*, 1322–1334. [CrossRef] [PubMed]
135. Strable, M.S.; Ntambi, J.M. Genetic Control of de Novo Lipogenesis: Role in Diet-Induced Obesity. *Crit. Rev. Biochem. Mol. Biol.* **2010**, *45*, 199–214. [CrossRef] [PubMed]
136. Sears, B.; Perry, M. The Role of Fatty Acids in Insulin Resistance. *Lipids Health Dis.* **2015**, *14*, 121. [CrossRef] [PubMed]

137. Kennedy, A.R.; Pissios, P.; Otu, H.; Roberson, R.; Xue, B.; Asakura, K.; Furukawa, N.; Marino, F.E.; Liu, F.-F.; Kahn, B.B.; et al. A High-Fat, Ketogenic Diet Induces a Unique Metabolic State in Mice. *Am. J. Physiol. Endocrinol. Metab.* **2007**, *292*, E1724–E1739. [CrossRef] [PubMed]
138. Pereira, R.M.; Botezelli, J.D.; da Cruz Rodrigues, K.C.; Mekary, R.A.; Cintra, D.E.; Pauli, J.R.; da Silva, A.S.R.; Ropelle, E.R.; de Moura, L.P. Fructose Consumption in the Development of Obesity and the Effects of Different Protocols of Physical Exercise on the Hepatic Metabolism. *Nutrients* **2017**, *9*, 405. [CrossRef] [PubMed]
139. Ameer, F.; Scandiuzzi, L.; Hasnain, S.; Kalbacher, H.; Zaidi, N. De Novo Lipogenesis in Health and Disease. *Metabolism* **2014**, *63*, 895–902. [CrossRef] [PubMed]
140. Herman, M.A.; Birnbaum, M.J. Molecular Aspects of Fructose Metabolism and Metabolic Disease. *Cell Metab.* **2021**, *33*, 2329–2354. [CrossRef] [PubMed]
141. Fraser, F.; Corstorphine, C.G.; Zammit, V.A. Topology of Carnitine Palmitoyltransferase I in the Mitochondrial Outer Membrane. *Biochem. J.* **1997**, *323 Pt. 3*, 711–718. [CrossRef]
142. Nowotny, K.; Jung, T.; Höhn, A.; Weber, D.; Grune, T. Advanced Glycation End Products and Oxidative Stress in Type 2 Diabetes Mellitus. *Biomolecules* **2015**, *5*, 194–222. [CrossRef]
143. Ganda, O.P. Triglyceride-Rich Lipoproteins, Remnant-Cholesterol, and Atherosclerotic Cardiovascular Disease. *Curr. Opin. Lipidol.* **2023**, *34*, 105–113. [CrossRef] [PubMed]
144. He, Q.; Chen, Y.; Wang, Z.; He, H.; Yu, P. Cellular Uptake, Metabolism and Sensing of Long-Chain Fatty Acids. *Front. Biosci. Landmark* **2023**, *28*, 10. [CrossRef] [PubMed]
145. Vijayakumar, A.; Aryal, P.; Wen, J.; Syed, I.; Vazirani, R.P.; Moraes-Vieira, P.M.; Camporez, J.P.; Gallop, M.R.; Perry, R.J.; Peroni, O.D.; et al. Absence of Carbohydrate Response Element Binding Protein in Adipocytes Causes Systemic Insulin Resistance and Impairs Glucose Transport. *Cell Rep.* **2017**, *21*, 1021–1035. [CrossRef] [PubMed]
146. Aarsland, A.; Chinkes, D.; Wolfe, R.R. Hepatic and Whole-Body Fat Synthesis in Humans during Carbohydrate Overfeeding. *Am. J. Clin. Nutr.* **1997**, *65*, 1774–1782. [CrossRef] [PubMed]
147. Bolsoni-Lopes, A.; Alonso-Vale, M.I.C. Lipolysis and Lipases in White Adipose Tissue—An Update. *Arch. Endocrinol. Metab.* **2015**, *59*, 335–342. [CrossRef] [PubMed]
148. Barrows, B.R.; Parks, E.J. Contributions of Different Fatty Acid Sources to Very Low-Density Lipoprotein-Triacylglycerol in the Fasted and Fed States. *J. Clin. Endocrinol. Metab.* **2006**, *91*, 1446–1452. [CrossRef] [PubMed]
149. Donnelly, K.L.; Smith, C.I.; Schwarzenberg, S.J.; Jessurun, J.; Boldt, M.D.; Parks, E.J. Sources of Fatty Acids Stored in Liver and Secreted via Lipoproteins in Patients with Nonalcoholic Fatty Liver Disease. *J. Clin. Invest.* **2005**, *115*, 1343–1351. [CrossRef] [PubMed]
150. Lambert, J.E.; Ramos-Roman, M.A.; Browning, J.D.; Parks, E.J. Increased de Novo Lipogenesis Is a Distinct Characteristic of Individuals with Nonalcoholic Fatty Liver Disease. *Gastroenterology* **2014**, *146*, 726–735. [CrossRef]
151. Kim, J.Y.; Nasr, A.; Tfayli, H.; Bacha, F.; Michaliszyn, S.F.; Arslanian, S. Increased Lipolysis, Diminished Adipose Tissue Insulin Sensitivity, and Impaired β-Cell Function Relative to Adipose Tissue Insulin Sensitivity in Obese Youth With Impaired Glucose Tolerance. *Diabetes* **2017**, *66*, 3085–3090. [CrossRef] [PubMed]
152. Morigny, P.; Houssier, M.; Mouisel, E.; Langin, D. Adipocyte Lipolysis and Insulin Resistance. *Biochimie* **2016**, *125*, 259–266. [CrossRef] [PubMed]
153. Roden, M. Mechanisms of Disease: Hepatic Steatosis in Type 2 Diabetes—Pathogenesis and Clinical Relevance. *Nat. Clin. Pract. Endocrinol. Metab.* **2006**, *2*, 335–348. [CrossRef] [PubMed]
154. Fukushima, A.; Milner, K.; Gupta, A.; Lopaschuk, D.G. Myocardial Energy Substrate Metabolism in Heart Failure: From Pathways to Therapeutic Targets. *Curr. Pharm. Des.* **2015**, *21*, 3654–3664. [CrossRef] [PubMed]
155. Mato, J.M.; Alonso, C.; Noureddin, M.; Lu, S.C. Biomarkers and Subtypes of Deranged Lipid Metabolism in Non-Alcoholic Fatty Liver Disease. *World J. Gastroenterol.* **2019**, *25*, 3009–3020. [CrossRef] [PubMed]
156. Mach, F.; Baigent, C.; Catapano, A.L.; Koskinas, K.C.; Casula, M.; Badimon, L.; Chapman, M.J.; De Backer, G.G.; Delgado, V.; Ference, B.A.; et al. 2019 ESC/EAS Guidelines for the Management of Dyslipidaemias: Lipid Modification to Reduce Cardiovascular Risk. *Eur. Heart J.* **2020**, *41*, 111–188. [CrossRef]
157. Nair, P.A.; Patel, C.R.; Ganjiwale, J.D.; Diwan, N.G.; Jivani, N.B. Xanthelasma Palpebrarum with Arcus Cornea: A Clinical and Biochemical Study. *Indian J. Dermatol.* **2016**, *61*, 295–300. [CrossRef] [PubMed]
158. Rochlani, Y.; Pothineni, N.V.; Kovelamudi, S.; Mehta, J.L. Metabolic Syndrome: Pathophysiology, Management, and Modulation by Natural Compounds. *Ther. Adv. Cardiovasc. Dis.* **2017**, *11*, 215–225. [CrossRef]
159. Dobrowolski, P.; Prejbisz, A.; Kuryłowicz, A.; Baska, A.; Burchardt, P.; Chlebus, K.; Dzida, G.; Jankowski, P.; Jaroszewicz, J.; Jaworski, P.; et al. Metabolic Syndrome—A New Definition and Management Guidelines. A joint position paper by the Polish Society of Hypertension, Polish Society for the Treatment of Obesity, Polish Lipid Association, Polish Association for Study of Liver, Polish Society of Family Medicine, Polish Society of Lifestyle Medicine, Division of Prevention and Epidemiology Polish Cardiac Society, "Club 30" Polish Cardiac Society, and Division of Metabolic and Bariatric Surgery Society of Polish Surgeons. *Arch. Med. Sci.* **2022**, *18*, 1133–1156. [CrossRef] [PubMed]
160. Padwal, R.S.; Majumdar, S.R. Drug Treatments for Obesity: Orlistat, Sibutramine, and Rimonabant. *Lancet* **2007**, *369*, 71–77. [CrossRef] [PubMed]

161. Lillich, F.F.; Imig, J.D.; Proschak, E. Multi-Target Approaches in Metabolic Syndrome. *Front. Pharmacol.* **2021**, *11*, 1–18. [CrossRef] [PubMed]
162. Lopez-Jimenez, F. Metabolic Syndrome—A Common and Dangerous Health Problem. *US Cardiol.* **2007**, *4*, 37–40. [CrossRef]
163. Lim, R.; Beekley, A.; Johnson, D.C.; Davis, K.A. Early and Late Complications of Bariatric Operation. *Trauma Surg. Acute Care Open* **2018**, *3*, e000219. [CrossRef] [PubMed]
164. Rai, R.H.; Singh, R.B.; Mehta, V.; Sakshi; Asif, M.; Goyal, K.; Balodhi, A.; Manglik, P.; Sharma, A.; Chahal, A. Impact of Exercise Training Duration on Obesity and Cardiometabolic Biomarkers: A Systematic Review. *J. Diabetes Metab. Disord.* **2023**, *22*, 155–174. [CrossRef] [PubMed]
165. Moschonis, G.; Trakman, G.L. Overweight and Obesity: The Interplay of Eating Habits and Physical Activity. *Nutrients* **2023**, *15*, 2896. [CrossRef] [PubMed]
166. Parmar, R.M.; Can, A.S. *Dietary Approaches to Obesity Treatment*; StatPearls Publishing: Treasure Island, FL, USA, 2024.
167. Lee, V. Introduction to the Dietary Management of Obesity in Adults. *Clin. Med. J. R. Coll. Physicians Lond.* **2023**, *23*, 304–310. [CrossRef] [PubMed]
168. Shaik Mohamed Sayed, U.F.; Moshawih, S.; Goh, H.P.; Kifli, N.; Gupta, G.; Singh, S.K.; Chellappan, D.K.; Dua, K.; Hermansyah, A.; Ser, H.L.; et al. Natural Products as Novel Anti-Obesity Agents: Insights into Mechanisms of Action and Potential for Therapeutic Management. *Front. Pharmacol.* **2023**, *14*, 1182937. [CrossRef] [PubMed]
169. Chan, Y.; Ng, S.W.; Tan, J.Z.X.; Gupta, G.; Negi, P.; Thangavelu, L.; Balusamy, S.R.; Perumalsamy, H.; Yap, W.H.; Singh, S.K.; et al. Natural Products in the Management of Obesity: Fundamental Mechanisms and Pharmacotherapy. *S. Afr. J. Bot.* **2021**, *143*, 176–197. [CrossRef]
170. Miller, E.G.; Nowson, C.A.; Dunstan, D.W.; Kerr, D.A.; Menzies, D.; Daly, R.M. Effects of Whey Protein plus Vitamin D Supplementation Combined with Progressive Resistance Training on Glycaemic Control, Body Composition, Muscle Function and Cardiometabolic Risk Factors in Middle-Aged and Older Overweight/Obese Adults with Type 2 Dia. *Diabetes. Obes. Metab.* **2021**, *23*, 938–949. [CrossRef] [PubMed]
171. Dominguez, L.J.; Veronese, N.; Di Bella, G.; Cusumano, C.; Parisi, A.; Tagliaferri, F.; Ciriminna, S.; Barbagallo, M. Mediterranean Diet in the Management and Prevention of Obesity. *Exp. Gerontol.* **2023**, *174*, 112121. [CrossRef] [PubMed]
172. Del Bo', C.; Perna, S.; Allehdan, S.; Rafique, A.; Saad, S.; AlGhareeb, F.; Rondanelli, M.; Tayyem, R.F.; Marino, M.; Martini, D.; et al. Does the Mediterranean Diet Have Any Effect on Lipid Profile, Central Obesity and Liver Enzymes in Non-Alcoholic Fatty Liver Disease (NAFLD) Subjects? A Systematic Review and Meta-Analysis of Randomized Control Trials. *Nutrients* **2023**, *15*, 2250. [CrossRef] [PubMed]
173. Hoyas, I.; Leon-Sanz, M. Nutritional Challenges in Metabolic Syndrome. *J. Clin. Med.* **2019**, *8*, 1301. [CrossRef] [PubMed]
174. Castro-Barquero, S.; Ruiz-León, A.M.; Sierra-Pérez, M.; Estruch, R.; Casas, R. Dietary Strategies for Metabolic Syndrome: A Comprehensive Review. *Nutrients* **2020**, *12*, 2983. [CrossRef] [PubMed]
175. Mohamed, S.M.; Shalaby, M.A.; El-Shiekh, R.A.; El-Banna, H.A.; Emam, S.R.; Bakr, A.F. Metabolic Syndrome: Risk Factors, Diagnosis, Pathogenesis, and Management with Natural Approaches. *Food Chem. Adv.* **2023**, *3*, 100335. [CrossRef]
176. Hsieh, C.C.; Hernández-Ledesma, B.; Fernández-Tomé, S.; Weinborn, V.; Barile, D.; De Moura Bell, J.M.L.N. Milk Proteins, Peptides, and Oligosaccharides: Effects against the 21st Century Disorders. *Biomed. Res. Int.* **2015**, *2015*. [CrossRef]
177. McGregor, R.A.; Poppitt, S.D. Milk Protein for Improved Metabolic Health: A Review of the Evidence. *Nutr. Metab.* **2013**, *10*, 46. [CrossRef] [PubMed]
178. Ricci-Cabello, I.; Olalla Herrera, M.; Artacho, R. Possible Role of Milk-Derived Bioactive Peptides in the Treatment and Prevention of Metabolic Syndrome. *Nutr. Rev.* **2012**, *70*, 241–255. [CrossRef] [PubMed]
179. Paddon-Jones, D.; Westman, E.; Mattes, R.D.; Wolfe, R.R.; Astrup, A.; Westerterp-Plantenga, M. Protein, Weight Management, and Satiety. *Am. J. Clin. Nutr.* **2008**, *87*, 1558S–1561S. [CrossRef] [PubMed]
180. Achaglinkame, M.A.; Dari, L.; Mörlein, D. A Review of Dairy Production and Utilization in Ghana and Benin. *Discov. Food* **2023**, *3*, 13. [CrossRef]
181. Garau, V.; Manis, C.; Scano, P.; Caboni, P. Compositional Characteristics of Mediterranean Buffalo Milk and Whey. *Dairy* **2021**, *2*, 469–488. [CrossRef]
182. Mathai, J.K.; Liu, Y.; Stein, H.H. Values for Digestible Indispensable Amino Acid Scores (DIAAS) for Some Dairy and Plant Proteins May Better Describe Protein Quality than Values Calculated Using the Concept for Protein Digestibility-Corrected Amino Acid Scores (PDCAAS). *Br. J. Nutr.* **2017**, *117*, 490–499. [CrossRef]
183. Roy, D.; Ye, A.; Moughan, P.J.; Singh, H. Composition, Structure, and Digestive Dynamics of Milk From Different Species—A Review. *Front. Nutr.* **2020**, *7*. [CrossRef]
184. Jenness, R. *Composition of Milk BT—Fundamentals of Dairy Chemistry*; Wong, N.P., Jenness, R., Keeney, M., Marth, E.H., Eds.; Springer US: Boston, MA, USA, 1988; pp. 1–38; ISBN 978-1-4615-7050-9.
185. Hossein, S.; Shahbazi, R.; Esmaeili, S.; Sohrabvandi, S. Health-Related Aspects of Milk Proteins. *Iran. J. Pharm. Res.* **2016**, *15*, 573–591.
186. Sindayikengera, S.; Xia, W. Nutritional Evaluation of Caseins and Whey Proteins and Their Hydrolysates from Protamex. *J. Zhejiang Univ. Sci. B* **2006**, *7*, 90–98. [CrossRef] [PubMed]
187. Boirie, Y.; Dangin, M.; Gachon, P.; Vasson, M.-P.; Maubois, J.-L.; Beaufrère, B. Slow and Fast Dietary Proteins Differently Modulate Postprandial Protein Accretion. *Proc. Natl. Acad. Sci. USA* **1997**, *94*, 14930–14935. [CrossRef] [PubMed]

188. Paul, G.L. The Rationale for Consuming Protein Blends in Sports Nutrition. *J. Am. Coll. Nutr.* **2009**, *28* (Suppl. S4), 464S–472S. [CrossRef] [PubMed]
189. Hoffman, J.R.; Falvo, M.J. Protein—Which Is Best? *J. Sports Sci. Med.* **2004**, *3*, 118–130. [PubMed]
190. Bendtsen, L.Q.; Lorenzen, J.K.; Bendsen, N.T.; Rasmussen, C.; Astrup, A. Effect of Dairy Proteins on Appetite, Energy Expenditure, Body Weight, and Composition: A Review of the Evidence from Controlled Clinical Trials. *Adv. Nutr.* **2013**, *4*, 418–438. [CrossRef]
191. Hidayat, K.; Du, X.; Shi, B.-M. Milk in the Prevention and Management of Type 2 Diabetes: The Potential Role of Milk Proteins. *Diabetes. Metab. Res. Rev.* **2019**, *35*, e3187. [CrossRef] [PubMed]
192. Sackesen, C.; Altintas, D.U.; Bingol, A.; Bingol, G.; Buyuktiryaki, B.; Demir, E.; Kansu, A.; Kuloglu, Z.; Tamay, Z.; Sekerel, B.E. Current Trends in Tolerance Induction in Cow's Milk Allergy: From Passive to Proactive Strategies. *Front. Pediatr.* **2019**, *7*, 372. [CrossRef] [PubMed]
193. Cuomo, B.; Indirli, G.C.; Bianchi, A.; Arasi, S.; Caimmi, D.; Dondi, A.; La Grutta, S.; Panetta, V.; Verga, M.C.; Calvani, M. Specific IgE and Skin Prick Tests to Diagnose Allergy to Fresh and Baked Cow's Milk According to Age: A Systematic Review. *Ital. J. Pediatr.* **2017**, *43*, 93. [CrossRef] [PubMed]
194. Wal, J.-M. Cow's Milk Proteins/Allergens. *Ann. Allergy Asthma Immunol.* **2002**, *89*, 3–10. [CrossRef] [PubMed]
195. Shandilya, U.K.; Panwar, V.; Haq, M.R.; Kapila, R.; Kansal, V.K. IgE and IgG Cross Reactivity between Cow and Buffalo Milk Proteins in Swiss Albino Mice Model. *Milchwissenschaft* **2012**, *67*, 235–238.
196. Villa, C.; Costa, J.; Oliveira, M.B.P.P.; Mafra, I. Bovine Milk Allergens: A Comprehensive Review. *Compr. Rev. Food Sci. Food Saf.* **2018**, *17*, 137–164. [CrossRef] [PubMed]
197. Restani, P.; Ballabio, C.; Di Lorenzo, C.; Tripodi, S.; Fiocchi, A. Molecular Aspects of Milk Allergens and Their Role in Clinical Events. *Anal. Bioanal. Chem.* **2009**, *395*, 47–56. [CrossRef] [PubMed]
198. Rao, Q.; Klaassen Kamdar, A.; Labuza, T.P. Storage Stability of Food Protein Hydrolysates—A Review. *Crit. Rev. Food Sci. Nutr.* **2016**, *56*, 1169–1192. [CrossRef] [PubMed]
199. Ramos, O.L.; Pereira, R.N.; Martins, A.; Rodrigues, R.; Fuciños, C.; Teixeira, J.A.; Pastrana, L.; Malcata, F.X.; Vicente, A.A. Design of Whey Protein Nanostructures for Incorporation and Release of Nutraceutical Compounds in Food. *Crit. Rev. Food Sci. Nutr.* **2017**, *57*, 1377–1393. [CrossRef] [PubMed]
200. Alessandri, C.; Sforza, S.; Palazzo, P.; Lambertini, F.; Paolella, S.; Zennaro, D.; Rafaiani, C.; Ferrara, R.; Bernardi, M.L.; Santoro, M.; et al. Tolerability of a Fully Maturated Cheese in Cow's Milk Allergic Children: Biochemical, Immunochemical, and Clinical Aspects. *PLoS ONE* **2012**, *7*, e40945. [CrossRef] [PubMed]
201. Manson, W. Nutritional Aspects of Milk Protein. *Int. J. Dairy Technol.* **1975**, *28*, 203–209. [CrossRef]
202. Park, Y.W.; Nam, M.S. Bioactive Peptides in Milk and Dairy Products: A Review. *Korean J. Food Sci. Anim. Resour.* **2015**, *35*, 831–840. [CrossRef] [PubMed]
203. Singh, B.P.; Aluko, R.E.; Hati, S.; Solanki, D. Bioactive Peptides in the Management of Lifestyle-Related Diseases: Current Trends and Future Perspectives. *Crit. Rev. Food Sci. Nutr.* **2022**, *62*, 4593–4606. [CrossRef] [PubMed]
204. Nongonierma, A.B.; FitzGerald, R.J. The Scientific Evidence for the Role of Milk Protein-Derived Bioactive Peptides in Humans: A Review. *J. Funct. Foods* **2015**, *17*, 640–656. [CrossRef]
205. Koirala, P.; Dahal, M.; Rai, S.; Dhakal, M.; Nirmal, N.P.; Maqsood, S.; Al-Asmari, F.; Buranasompob, A. Dairy Milk Protein–Derived Bioactive Peptides: Avengers Against Metabolic Syndrome. *Curr. Nutr. Rep.* **2023**, *12*, 299–317. [CrossRef] [PubMed]
206. Meisel, H.; Bockelmann, W. Bioactive Peptides Encrypted in Milk Proteins: Proteolytic Activation and Thropho-Functional Properties. *Antonie Van Leeuwenhoek* **1999**, *76*, 207–215. [CrossRef] [PubMed]
207. Zaky, A.A.; Simal-Gandara, J.; Eun, J.B.; Shim, J.H.; Abd El-Aty, A.M. Bioactivities, Applications, Safety, and Health Benefits of Bioactive Peptides From Food and By-Products: A Review. *Front. Nutr.* **2022**, *8*, 1–18. [CrossRef]
208. Abd El-Salam, M.H.; El-Shibiny, S. Preparation, Properties, and Uses of Enzymatic Milk Protein Hydrolysates. *Crit. Rev. Food Sci. Nutr.* **2017**, *57*, 1119–1132. [CrossRef] [PubMed]
209. Abd El-Salam, M.H.; El-Shibiny, S. Reduction of Milk Protein Antigenicity by Enzymatic Hydrolysis and Fermentation. A Review. *Food Rev. Int.* **2021**, *37*, 276–295. [CrossRef]
210. Meisel, H. Biochemical Properties of Regulatory Peptides Derived from Milk Proteins. *Biopolym. Pept. Sci. Sect.* **1997**, *43*, 119–128. [CrossRef]
211. Mellander, O. The Physiological Importance of the Casein Phosphopeptide Calcium Salts. II. Peroral Calcium Dosage of Infants. *Acta Soc. Med. Ups.* **1950**, *55*, 247–255. [PubMed]
212. Xu, Q.; Yan, X.; Zhang, Y.; Wu, J. Current Understanding of Transport and Bioavailability of Bioactive Peptides Derived from Dairy Proteins: A Review. *Int. J. Food Sci. Technol.* **2019**, *54*, 1930–1941. [CrossRef]
213. Nath, A.; Eren, B.A.; Zinia Zaukuu, J.-L.; Koris, A.; Pásztorné-Huszár, K.; Szerdahelyi, E.; Kovacs, Z. Detecting the Bitterness of Milk-Protein-Derived Peptides Using an Electronic Tongue. *Chemosens.* **2022**, *10*, 215. [CrossRef]
214. Maruyama, S.; Nakagomi, K.; Tomizuka, N.; Suzuki, H. Angiotensin I-Converting Enzyme Inhibitor Derived from an Enzymatic Hydrolysate of Casein. Ii. Isolation and Bradykinin-Potentiating Activity on the Uterus and the Ileum of Rats. *Agric. Biol. Chem.* **1985**, *49*, 1405–1409. [CrossRef]
215. Maeno, M.; Yamamoto, N.; Takano, T. Identification of an Antihypertensive Peptide from Casein Hydrolysate Produced by a Proteinase from Lactobacillus Helveticus CP790. *J. Dairy Sci.* **1996**, *79*, 1316–1321. [CrossRef] [PubMed]

216. Mullally, M.M.; Meisel, H.; Fitzgerald, R.J. Angiotensin-I-Converting Enzyme Inhibitory Activities of Gastric and Pancreatic Proteinase Digests of Whey Proteins. *Int. Dairy J.* **1997**, *7*, 299–303. [CrossRef]
217. Yan, D.; Chen, D.; Shen, J.; Xiao, G.; van Wijnen, A.J.; Im, H.J. Bovine Lactoferricin Is Anti-Inflammatory and Anti-Catabolic in Human Articular Cartilage and Synovium. *J. Cell. Physiol.* **2013**, *228*, 447–456. [CrossRef] [PubMed]
218. Kanwar, R.K.; Kanwar, J.R. Immunomodulatory Lactoferrin in the Regulation of Apoptosis Modulatory Proteins in Cancer. *Protein Pept. Lett.* **2013**, *20*, 450–458. [PubMed]
219. Furlong, S.J.; Mader, J.S.; Hoskin, D.W. Bovine Lactoferricin Induces Caspase-Independent Apoptosis in Human B-Lymphoma Cells and Extends the Survival of Immune-de Fi Cient Mice Bearing B-Lymphoma Xenografts. *Exp. Mol. Pathol.* **2010**, *88*, 371–375. [CrossRef] [PubMed]
220. Meisel, H. Biochemical Properties of Peptides Encrypted in Bovine Milk Proteins. *Curr. Med. Chem.* **2005**, *12*, 1905–1919. [CrossRef] [PubMed]
221. Yvon, M.; Beucher, S.; Guilloteau, P.; Le Huerou-Luron, I.; Corring, T. Effects of Caseinomacropeptide (CMP) on Digestion Regulation. *Reprod. Nutr. Dev.* **1994**, *34*, 527–537. [CrossRef] [PubMed]
222. Córdova-Dávalos, L.E.; Jiménez, M.; Salinas, E. Glycomacropeptide Bioactivity and Health: A Review Highlighting Action Mechanisms and Signaling Pathways. *Nutrients* **2019**, *11*, 598. [CrossRef] [PubMed]
223. FitzGerald, R.J.; Meisel, H. Lactokinins: Whey Protein-Derived ACE Inhibitory Peptides. *Nahr. Food* **1999**, *43*, 165–167. [CrossRef]
224. Pepe, G.; Tenore, G.C.; Mastrocinque, R.; Stusio, P.; Campiglia, P. Potential Anticarcinogenic Peptides from Bovine Milk. *J. Amino Acids* **2013**, *2013*, 939804. [CrossRef] [PubMed]
225. Shazly, A.B.; He, Z.; El-Aziz, M.A.; Zeng, M.; Zhang, S.; Qin, F.; Chen, J. Fractionation and Identification of Novel Antioxidant Peptides from Buffalo and Bovine Casein Hydrolysates. *Food Chem.* **2017**, *232*, 753–762. [CrossRef] [PubMed]
226. Xue, L.; Wang, X.; Hu, Z.; Wu, Z.; Wang, L.; Wang, H.; Yang, M. Identification and Characterization of an Angiotensin-Converting Enzyme Inhibitory Peptide Derived from Bovine Casein. *Peptides* **2018**, *99*, 161–168. [CrossRef] [PubMed]
227. Jiang, X.; Pan, D.; Zhang, T.; Liu, C.; Zhang, J.; Su, M.; Wu, Z.; Zeng, X.; Sun, Y.; Guo, Y. Novel Milk Casein-Derived Peptides Decrease Cholesterol Micellar Solubility and Cholesterol Intestinal Absorption in Caco-2 Cells. *J. Dairy Sci.* **2020**, *103*, 3924–3936. [CrossRef] [PubMed]
228. Mudgil, P.; Kamal, H.; Priya Kilari, B.; Mohd Salim, M.A.S.; Gan, C.-Y.; Maqsood, S. Simulated Gastrointestinal Digestion of Camel and Bovine Casein Hydrolysates: Identification and Characterization of Novel Anti-Diabetic Bioactive Peptides. *Food Chem.* **2021**, *353*, 129374. [CrossRef] [PubMed]
229. Fitzgerald, M. Biofunctional Peptides from Milk Proteins: Mineral Binding and Cytomodulatory Effects. *Curr. Pharm. Des.* **2003**, *9*, 1289–1295. [CrossRef]
230. Minervini, F.; Algaron, F.; Rizzello, C.G.; Fox, P.F.; Monnet, V.; Gobbetti, M. Angiotensin I-Converting-Enzyme-Inhibitory and Antibacterial Peptides from Lactobacillus Helveticus PR4 Proteinase-Hydrolyzed Caseins of Milk from Six Species. *Appl. Environ. Microbiol.* **2003**, *69*, 5297–5305. [CrossRef] [PubMed]
231. Hern, B.; Miralles, B.; Amigo, L.; Ramos, M. Identification of Antioxidant and ACE-Inhibitory Peptides in Fermented Milk. *J. Sci. Food Agric.* **2005**, *1048*, 1041–1048. [CrossRef]
232. Meisel, H. Overview on Milk Protein-Derived Peptides. *Int. Dairy J.* **1998**, *8*, 363–373. [CrossRef]
233. Abdel-Hamid, M.; Otte, J.; De Gobba, C.; Osman, A.; Hamad, E. Angiotensin I-Converting Enzyme Inhibitory Activity and Antioxidant Capacity of Bioactive Peptides Derived from Enzymatic Hydrolysis of Buffalo Milk Proteins. *Int. Dairy J.* **2017**, *66*, 91–98. [CrossRef]
234. Chakrabarti, S.; Jahandideh, F.; Davidge, S.T.; Wu, J. Milk-Derived Tripeptides IPP (Ile-Pro-Pro) and VPP (Val-Pro-Pro) Enhance Insulin Sensitivity and Prevent Insulin Resistance in 3T3-F442A Preadipocytes. *J. Agric. Food Chem.* **2018**, *66*, 10179–10187. [CrossRef]
235. Zhao, Q.; Wei, G.; Li, K.; Duan, S.; Ye, R.; Huang, A. Identification and Molecular Docking of Novel α-Glucosidase Inhibitory Peptides from Hydrolysates of Binglangjiang Buffalo Casein. *LWT* **2022**, *156*, 113062. [CrossRef]
236. Basilicata, M.G.; Pepe, G.; Adesso, S.; Ostacolo, C.; Sala, M.; Sommella, E.; Scala, M.C.; Messore, A.; Autore, G.; Marzocco, S.; et al. Antioxidant Properties of Buffalo-Milk Dairy Products: A β-Lg Peptide Released after Gastrointestinal Digestion of Buffalo Ricotta Cheese Reduces Oxidative Stress in Intestinal Epithelial Cells. *Int. J. Mol. Sci.* **2018**, *19*, 1955. [CrossRef] [PubMed]
237. Shanmugam, V.P.; Kapila, S.; Kemgang, T.S.; Reddi, S.; Kapila, R.; Muthukumar, S.; Rajesh, D. Isolation and Characterization of Angiotensin Converting Enzyme Inhibitory Peptide from Buffalo Casein. *Int. J. Pept. Res. Ther.* **2021**, *27*, 1481–1491. [CrossRef]
238. Shanmugam, V.P.; Kapila, S.; Sonfack, T.K.; Kapila, R. Antioxidative Peptide Derived from Enzymatic Digestion of Buffalo Casein. *Int. Dairy J.* **2015**, *42*, 1–5. [CrossRef]
239. Kimmerlin, T.; Seebach, D.; Hilvert, D. Synthesis of Β3-Peptides and Mixed α/Β3-Peptides by Thioligation. *Helv. Chim. Acta* **2002**, *85*, 1812–1826. [CrossRef]
240. Guzmán, F.; Barberis, S.; Illanes, A. Peptide Synthesis: Chemical or Enzymatic. *Electron. J. Biotechnol.* **2007**, *10*, 279–314. [CrossRef]
241. Alpine Roads. Nobell Foods Food Compositions Comprising Milk Proteins Produced in Transgenic Plants. U.S. Patent Application No. 16/862,011, 2021.
242. Dallas, D.; Nielsen, S.D. Milk Peptidomics to Identify Functional Peptides and for Quality Control of Dairy Products. *Methods Mol. Biol.* **2018**, *1719*, 223–240. [CrossRef] [PubMed]

243. Nielsen, S.D.; Beverly, R.L.; Qu, Y.; Dallas, D.C. Milk Bioactive Peptide Database: A Comprehensive Database of Milk Protein-Derived Bioactive Peptides and Novel Visualization. *Food Chem.* **2017**, *232*, 673–682. [CrossRef] [PubMed]
244. Kussmann, M.; Affolter, M. Proteomics at the Center of Nutrigenomics: Comprehensive Molecular Understanding of Dietary Health Effects. *Nutrition* **2009**, *25*, 1085–1093. [CrossRef] [PubMed]
245. Comerford, K.B.; Pasin, G. Gene-Dairy Food Interactions and Health Outcomes: A Review of Nutrigenetic Studies. *Nutrients* **2017**, *9*, 710. [CrossRef]
246. FitzGerald, R.J.; Cermeño, M.; Khalesi, M.; Kleekayai, T.; Amigo-Benavent, M. Application of in Silico Approaches for the Generation of Milk Protein-Derived Bioactive Peptides. *J. Funct. Foods* **2020**, *64*, 103636. [CrossRef]
247. Iwaniak, A.; Mogut, D. Metabolic Syndrome-Preventive Peptides Derived from Milk Proteins and Their Presence in Cheeses: A Review. *Appl. Sci.* **2020**, *10*, 2772. [CrossRef]
248. Chen, G.-C.; Szeto, I.M.Y.; Chen, L.-H.; Han, S.-F.; Li, Y.-J.; van Hekezen, R.; Qin, L.-Q. Dairy Products Consumption and Metabolic Syndrome in Adults: Systematic Review and Meta-Analysis of Observational Studies. *Sci. Rep.* **2015**, *5*, 14606. [CrossRef] [PubMed]
249. Carrasco-Castilla, J.; Hernández-Álvarez, A.J.; Jiménez-Martínez, C.; Gutiérrez-López, G.F.; Dávila-Ortiz, G. Use of Proteomics and Peptidomics Methods in Food Bioactive Peptide Science and Engineering. *Food Eng. Rev.* **2012**, *4*, 224–243. [CrossRef]
250. Nilsson, M.; Holst, J.J.; Björck, I.M.E. Metabolic Effects of Amino Acid Mixtures and Whey Protein in Healthy Subjects: Studies Using Glucose-Equivalent Drinks. *Am. J. Clin. Nutr.* **2007**, *85*, 996–1004. [CrossRef] [PubMed]
251. Dangin, M.; Boirie, Y.; Garcia-Rodenas, C.; Gachon, P.; Fauquant, J.; Callier, P.; Ballèvre, O.; Beaufrère, B. The Digestion Rate of Protein Is an Independent Regulating Factor of Postprandial Protein Retention. *Am. J. Physiol. Metab.* **2001**, *280*, E340–E348. [CrossRef] [PubMed]
252. Calbet, J.A.L.; Holst, J.J. Gastric Emptying, Gastric Secretion and Enterogastrone Response after Administration of Milk Proteins or Their Peptide Hydrolysates in Humans. *Eur. J. Nutr.* **2004**, *43*, 127–139. [CrossRef] [PubMed]
253. Koopman, R.; Crombach, N.; Gijsen, A.P.; Walrand, S.; Fauquant, J.; Kies, A.K.; Lemosquet, S.; Saris, W.H.M.; Boirie, Y.; van Loon, L.J.C. Ingestion of a Protein Hydrolysate Is Accompanied by an Accelerated in Vivo Digestion and Absorption Rate When Compared with Its Intact Protein. *Am. J. Clin. Nutr.* **2009**, *90*, 106–115. [CrossRef] [PubMed]
254. Morifuji, M.; Ishizaka, M.; Baba, S.; Fukuda, K.; Matsumoto, H.; Koga, J.; Kanegae, M.; Higuchi, M. Comparison of Different Sources and Degrees of Hydrolysis of Dietary Protein: Effect on Plasma Amino Acids, Dipeptides, and Insulin Responses in Human Subjects. *J. Agric. Food Chem.* **2010**, *58*, 8788–8797. [CrossRef] [PubMed]
255. Veldhorst, M.A.B.; Nieuwenhuizen, A.G.; Hochstenbach-Waelen, A.; van Vught, A.J.A.H.; Westerterp, K.R.; Engelen, M.P.K.J.; Brummer, R.-J.M.; Deutz, N.E.P.; Westerterp-Plantenga, M.S. Dose-Dependent Satiating Effect of Whey Relative to Casein or Soy. *Physiol. Behav.* **2009**, *96*, 675–682. [CrossRef] [PubMed]
256. Pistrosch, F.; Natali, A.; Hanefeld, M. Is Hyperglycemia a Cardiovascular Risk Factor? *Diabetes Care* **2011**, *34*, 10–13. [CrossRef] [PubMed]
257. Claessens, M.; Saris, W.H.M.; van Baak, M.A. Glucagon and Insulin Responses after Ingestion of Different Amounts of Intact and Hydrolysed Proteins. *Br. J. Nutr.* **2008**, *100*, 61–69. [CrossRef] [PubMed]
258. Gerich, J.E. Clinical Significance, Pathogenesis, and Management of Postprandial Hyperglycemia. *Arch. Intern. Med.* **2003**, *163*, 1306–1316. [CrossRef] [PubMed]
259. Pal, S.; Ellis, V.; Dhaliwal, S. Effects of Whey Protein Isolate on Body Composition, Lipids, Insulin and Glucose in Overweight and Obese Individuals. *Br. J. Nutr.* **2010**, *104*, 716–723. [CrossRef] [PubMed]
260. Patil, P.; Mandal, S.; Tomar, S.K.; Anand, S. Food Protein-Derived Bioactive Peptides in Management of Type 2 Diabetes. *Eur. J. Nutr.* **2015**, *54*, 863–880. [CrossRef] [PubMed]
261. Nongonierma, A.B.; FitzGerald, R.J. Features of Dipeptidyl Peptidase IV (DPP-IV) Inhibitory Peptides from Dietary Proteins. *J. Food Biochem.* **2019**, *43*, e12451. [CrossRef] [PubMed]
262. Engel, M.; Hoffmann, T.; Wagner, L.; Wermann, M.; Heiser, U.; Kiefersauer, R.; Huber, R.; Bode, W.; Demuth, H.-U.; Brandstetter, H. The Crystal Structure of Dipeptidyl Peptidase IV (CD26) Reveals Its Functional Regulation and Enzymatic Mechanism. *Proc. Natl. Acad. Sci. USA* **2003**, *100*, 5063–5068. [CrossRef]
263. Guasch, L.; Ojeda, M.J.; González-Abuín, N.; Sala, E.; Cereto-Massagué, A.; Mulero, M.; Valls, C.; Pinent, M.; Ardévol, A.; Garcia-Vallvé, S.; et al. Identification of Novel Human Dipeptidyl Peptidase-IV Inhibitors of Natural Origin (Part I): Virtual Screening and Activity Assays. *PLoS ONE* **2012**, *7*, e44971. [CrossRef] [PubMed]
264. Nongonierma, A.B.; FitzGerald, R.J. Inhibition of Dipeptidyl Peptidase IV (DPP-IV) by Tryptophan Containing Dipeptides. *Food Funct.* **2013**, *4*, 1843–1849. [CrossRef]
265. Nongonierma, A.B.; FitzGerald, R.J. Inhibition of Dipeptidyl Peptidase IV (DPP-IV) by Proline Containing Casein-Derived Peptides. *J. Funct. Foods* **2013**, *5*, 1909–1917. [CrossRef]
266. Sowmya, K.; Mala, D.; Bhat, M.I.; Kumar, N.; Bajaj, R.K.; Kapila, S.; Kapila, R. Bio-Accessible Milk Casein Derived Tripeptide (LLY) Mediates Overlapping Anti- Inflammatory and Anti-Oxidative Effects under Cellular (Caco-2) and in Vivo Milieu. *J. Nutr. Biochem.* **2018**, *62*, 167–180. [CrossRef] [PubMed]
267. Devi, S.; Kumar, N.; Kapila, S.; Mada, S.B.; Reddi, S.; Vij, R.; Kapila, R. Buffalo Casein Derived Peptide Can Alleviates H2O2 Induced Cellular Damage and Necrosis in Fibroblast Cells. *Exp. Toxicol. Pathol.* **2017**, *69*, 485–495. [CrossRef] [PubMed]

268. Mada, S.B.; Reddi, S.; Kumar, N.; Kumar, R.; Kapila, S.; Kapila, R.; Trivedi, R.; Karvande, A.; Ahmad, N. Antioxidative Peptide from Milk Exhibits Antiosteopenic Effects through Inhibition of Oxidative Damage and Bone-Resorbing Cytokines in Ovariectomized Rats. *Nutrition* **2017**, *43–44*, 21–31. [CrossRef] [PubMed]
269. Wen, C.; Zhang, J.; Zhang, H.; Duan, Y.; Ma, H. Plant Protein-Derived Antioxidant Peptides: Isolation, Identification, Mechanism of Action and Application in Food Systems: A Review. *Trends Food Sci. Technol.* **2020**, *105*, 308–322. [CrossRef]
270. Kumar, N.; Devi, S.; Mada, S.B.; Reddi, S.; Kapila, R.; Kapila, S. Anti-Apoptotic Effect of Buffalo Milk Casein Derived Bioactive Peptide by Directing Nrf2 Regulation in Starving Fibroblasts. *Food Biosci.* **2020**, *35*, 100566. [CrossRef]
271. Iwasa, M.; Takezoe, S.; Kitaura, N.; Sutani, T.; Miyazaki, H.; Aoi, W. A Milk Casein Hydrolysate-Derived Peptide Enhances Glucose Uptake through the AMP-Activated Protein Kinase Signalling Pathway in Skeletal Muscle Cells. *Exp. Physiol.* **2021**, *106*, 496–505. [CrossRef] [PubMed]
272. Kondrashina, A.; Brodkorb, A.; Giblin, L. Dairy-Derived Peptides for Satiety. *J. Funct. Foods* **2020**, *66*, 103801. [CrossRef]
273. Kaelberer, M.M.; Buchanan, K.L.; Klein, M.E.; Barth, B.B.; Montoya, M.M.; Shen, X.; Bohórquez, D. V A Gut-Brain Neural Circuit for Nutrient Sensory Transduction. *Science* **2018**, *361*, eaat5236. [CrossRef] [PubMed]
274. Hall, W.L.; Millward, D.J.; Long, S.J.; Morgan, L.M. Casein and Whey Exert Different Effects on Plasma Amino Acid Profiles, Gastrointestinal Hormone Secretion and Appetite. *Br. J. Nutr.* **2003**, *89*, 239–248. [CrossRef]
275. Demling, R.H.; DeSanti, L. Effect of a Hypocaloric Diet, Increased Protein Intake and Resistance Training on Lean Mass Gains and Fat Mass Loss in Overweight Police Officers. *Ann. Nutr. Metab.* **2000**, *44*, 21–29. [CrossRef]
276. Acheson, K.J.; Blondel-Lubrano, A.; Oguey-Araymon, S.; Beaumont, M.; Emady-Azar, S.; Ammon-Zufferey, C.; Monnard, I.; Pinaud, S.; Nielsen-Moennoz, C.; Bovetto, L. Protein Choices Targeting Thermogenesis and Metabolism. *Am. J. Clin. Nutr.* **2011**, *93*, 525–534. [CrossRef] [PubMed]
277. Froetschel, M.A.; Azain, M.J.; Edwards, G.L.; Barb, C.R.; Amos, H.E. Opioid and Cholecystokinin Antagonists Alleviate Gastric Inhibition of Food Intake by Premeal Loads of Casein in Meal-Fed Rats. *J. Nutr.* **2001**, *131*, 3270–3276. [CrossRef] [PubMed]
278. Pupovac, J.; Anderson, G.H. Dietary Peptides Induce Satiety via Cholecystokinin-A and Peripheral Opioid Receptors in Rats. *J. Nutr.* **2002**, *132*, 2775–2780. [CrossRef] [PubMed]
279. Toedebusch, R.G.; Childs, T.E.; Hamilton, S.R.; Crowley, J.R.; Booth, F.W.; Roberts, M.D. Postprandial Leucine and Insulin Responses and Toxicological Effects of a Novel Whey Protein Hydrolysate-Based Supplement in Rats. *J. Int. Soc. Sports Nutr.* **2012**, *9*, 24. [CrossRef] [PubMed]
280. Tolhurst, G.; Reimann, F.; Gribble, F.M. Intestinal Sensing of Nutrients. In *Appetite Control. Handbook of Experimental Pharmacology*; Springer: Berlin/Heidelberg, Germany, 2012; pp. 309–335. [CrossRef]
281. Santos-Hernández, M.; Miralles, B.; Amigo, L.; Recio, I. Intestinal Signaling of Proteins and Digestion-Derived Products Relevant to Satiety. *J. Agric. Food Chem.* **2018**, *66*, 10123–10131. [CrossRef] [PubMed]
282. Udenigwe, C.C.; Rouvinen-Watt, K. The Role of Food Peptides in Lipid Metabolism during Dyslipidemia and Associated Health Conditions. *Int. J. Mol. Sci.* **2015**, *16*, 9303–9313. [CrossRef]
283. Karpe, F. Postprandial Lipoprotein Metabolism and Atherosclerosis. *J. Intern. Med.* **1999**, *246*, 341–355. [CrossRef] [PubMed]
284. Kersten, S. Physiological Regulation of Lipoprotein Lipase. *Biochim. Biophys. Acta* **2014**, *1841*, 919–933. [CrossRef]
285. Mudgil, P.; Baba, W.N.; Kamal, H.; FitzGerald, R.J.; Hassan, H.M.; Ayoub, M.A.; Gan, C.-Y.; Maqsood, S. A Comparative Investigation into Novel Cholesterol Esterase and Pancreatic Lipase Inhibitory Peptides from Cow and Camel Casein Hydrolysates Generated upon Enzymatic Hydrolysis and In-Vitro Digestion. *Food Chem.* **2022**, *367*, 130661. [CrossRef] [PubMed]
286. Chen, H.; Qi, X.; Guan, K.; Gu, Y.; Wang, R.; Li, Q.; Ma, Y. Peptides Released from Bovine α-Lactalbumin by Simulated Digestion Alleviated Free Fatty Acids-Induced Lipid Accumulation in HepG2 Cells. *J. Funct. Foods* **2021**, *85*, 104618. [CrossRef]
287. Hlais, S.; Reslan, D.R.A.; Sarieddine, H.K.; Nasreddine, L.; Taan, G.; Azar, S.; Obeid, O.A. Effect of Lysine, Vitamin B(6), and Carnitine Supplementation on the Lipid Profile of Male Patients with Hypertriglyceridemia: A 12-Week, Open-Label, Randomized, Placebo-Controlled Trial. *Clin. Ther.* **2012**, *34*, 1674–1682. [CrossRef] [PubMed]
288. Freudenberg, A.; Petzke, K.J.; Klaus, S. Dietary L-Leucine and L-Alanine Supplementation Have Similar Acute Effects in the Prevention of High-Fat Diet-Induced Obesity. *Amino Acids* **2013**, *44*, 519–528. [CrossRef] [PubMed]
289. Lee, S.; Youn, B. Hypolipidemic Roles of Casein-Derived Peptides by Regulation of Trans-Intestinal Cholesterol Excretion and Bile Acid Synthesis. *Nutrients* **2020**, *12*, 3058. [CrossRef]
290. Vona, R.; Gambardella, L.; Cittadini, C.; Straface, E.; Pietraforte, D. Biomarkers of Oxidative Stress in Metabolic Syndrome and Associated Diseases. *Oxid. Med. Cell. Longev.* **2019**, *2019*, 8267234. [CrossRef] [PubMed]
291. Tonolo, F.; Folda, A.; Cesaro, L.; Scalcon, V.; Marin, O.; Ferro, S.; Bindoli, A.; Rigobello, M.P. Milk-Derived Bioactive Peptides Exhibit Antioxidant Activity through the Keap1-Nrf2 Signaling Pathway. *J. Funct. Foods* **2020**, *64*, 103696. [CrossRef]
292. Qi, X.; Chen, H.; Guan, K.; Sun, Y.; Wang, R.; Li, Q.; Ma, Y. Novel Xanthine Oxidase Inhibitory Peptides Derived from Whey Protein: Identification, in Vitro Inhibition Mechanism and in Vivo Activity Validation. *Bioorg. Chem.* **2022**, *128*, 106097. [CrossRef] [PubMed]
293. Putnam, K.; Shoemaker, R.; Yiannikouris, F.; Cassis, L.A. The Renin-Angiotensin System: A Target of and Contributor to Dyslipidemias, Altered Glucose Homeostasis, and Hypertension of the Metabolic Syndrome. *Am. J. Physiol. Circ. Physiol.* **2012**, *302*, H1219–H1230. [CrossRef]
294. Tangvarasittichai, S. Oxidative Stress, Insulin Resistance, Dyslipidemia and Type 2 Diabetes Mellitus. *World J. Diabetes* **2015**, *6*, 456. [CrossRef] [PubMed]

295. Fernández-Sánchez, A.; Madrigal-Santillán, E.; Bautista, M.; Esquivel-Soto, J.; Morales-González, A.; Esquivel-Chirino, C.; Durante-Montiel, I.; Sánchez-Rivera, G.; Valadez-Vega, C.; Morales-González, J.A. Inflammation, Oxidative Stress, and Obesity. *Int. J. Mol. Sci.* **2011**, *12*, 3117–3132. [CrossRef] [PubMed]
296. Ruiz-Ojeda, F.J.; Olza, J.; Gil, Á.; Aguilera, C.M. Chapter 1—Oxidative Stress and Inflammation in Obesity and Metabolic Syndrome. In *Obesity*; del Moral, A.M., Aguilera García, C.M.B.T.-O., Eds.; Academic Press: Cambridge, MA, USA, 2018; pp. 1–15. ISBN 978-0-12-812504-5.
297. Meyer, J.; Bütikofer, U.; Walther, B.; Wechsler, D.; Sieber, R. Hot Topic: Changes in Angiotensin-Converting Enzyme Inhibition and Concentrations of the Tripeptides Val-Pro-Pro and Ile-Pro-Pro during Ripening of Different Swiss Cheese Varieties. *J. Dairy Sci.* **2009**, *92*, 826–836. [CrossRef] [PubMed]
298. Chen, Y.; Li, C.; Xue, J.; Kwok, L.Y.; Yang, J.; Zhang, H.; Menghe, B. Characterization of Angiotensin-Converting Enzyme Inhibitory Activity of Fermented Milk Produced by Lactobacillus Helveticus. *J. Dairy Sci.* **2015**, *98*, 5113–5124. [CrossRef] [PubMed]
299. Sipola, M.; Finckenberg, P.; Korpela, R.; Vapaatalo, H.; Nurminen, M.-L. Effect of Long-Term Intake of Milk Products on Blood Pressure in Hypertensive Rats. *J. Dairy Res.* **2002**, *69*, 103–111. [CrossRef] [PubMed]
300. Yamaguchi, N.; Kawaguchi, K.; Yamamoto, N. Study of the Mechanism of Antihypertensive Peptides VPP and IPP in Spontaneously Hypertensive Rats by DNA Microarray Analysis. *Eur. J. Pharmacol.* **2009**, *620*, 71–77. [CrossRef] [PubMed]
301. Watanabe, M.; Kurihara, J.; Suzuki, S.; Nagashima, K.; Hosono, H.; Itagaki, F. The Influence of Dietary Peptide Inhibitors of Angiotensin-Converting Enzyme on the Hypotensive Effects of Enalapril. *J. Pharm. Health Care Sci.* **2015**, *1*, 17. [CrossRef] [PubMed]
302. Kim, S.M.; Kim, J.Y.; Choue, R.W. Effects of IPP and VPP Supplementation on Blood Pressure, ACE Activity and NO Production in Spontaneously Hypertensive Rats. *FASEB J.* **2008**, *22*, 1112.3. [CrossRef]
303. Hirota, T.; Ohki, K.; Kawagishi, R.; Kajimoto, Y.; Mizuno, S.; Nakamura, Y.; Kitakaze, M. Casein Hydrolysate Containing the Antihypertensive Tripeptides Val-Pro-Pro and Ile-Pro-Pro Improves Vascular Endothelial Function Independent of Blood Pressure-Lowering Effects: Contribution of the Inhibitory Action of Angiotensin-Converting Enzyme. *Hypertens. Res.* **2007**, *30*, 489–496. [CrossRef] [PubMed]
304. Fekete, Á.A.; Givens, D.I.; Lovegrove, J.A. Casein-Derived Lactotripeptides Reduce Systolic and Diastolic Blood Pressure in a Meta-Analysis of Randomised Clinical Trials. *Nutrients* **2015**, *7*, 659–681. [CrossRef] [PubMed]
305. de Leeuw, P.W.; van der Zander, K.; Kroon, A.A.; Rennenberg, R.M.W.; Koning, M.M.G. Dose-dependent Lowering of Blood Pressure by Dairy Peptides in Mildly Hypertensive Subjects. *Blood Press.* **2009**, *18*, 44–50. [CrossRef] [PubMed]
306. Cicero, A.F.G.; Aubin, F.; Azais-Braesco, V.; Borghi, C. Do the Lactotripeptides Isoleucine–Proline–Proline and Valine–Proline–Proline Reduce Systolic Blood Pressure in European Subjects? A Meta-Analysis of Randomized Controlled Trials. *Am. J. Hypertens.* **2013**, *26*, 442–449. [CrossRef] [PubMed]
307. Engberink, M.F.; Schouten, E.G.; Kok, F.J.; van Mierlo, L.A.J.; Brouwer, I.A.; Geleijnse, J.M. Lactotripeptides Show No Effect on Human Blood Pressure. *Hypertension* **2008**, *51*, 399–405. [CrossRef]
308. van Mierlo, L.A.J.; Koning, M.M.G.; van der Zander, K.; Draijer, R. Lactotripeptides Do Not Lower Ambulatory Blood Pressure in Untreated Whites: Results from 2 Controlled Multicenter Crossover Studies2. *Am. J. Clin. Nutr.* **2009**, *89*, 617–623. [CrossRef] [PubMed]
309. FitzGerald, R.J.; Murray, B.A.; Walsh, D.J. Hypotensive Peptides from Milk Proteins. *J. Nutr.* **2004**, *134*, 980S–988S. [CrossRef] [PubMed]
310. Chakrabarti, S.; Liao, W.; Davidge, S.T.; Wu, J. Milk-Derived Tripeptides IPP (Ile-Pro-Pro) and VPP (Val-Pro-Pro) Differentially Modulate Angiotensin II Effects on Vascular Smooth Muscle Cells. *J. Funct. Foods* **2017**, *30*, 151–158. [CrossRef]
311. Nakamura, T.; Hirota, T.; Mizushima, K.; Ohki, K.; Naito, Y.; Yamamoto, N.; Yoshikawa, T. Milk-Derived Peptides, Val-Pro-Pro and Ile-Pro-Pro, Attenuate Atherosclerosis Development in Apolipoprotein e-Deficient Mice: A Preliminary Study. *J. Med. Food* **2013**, *16*, 396–403. [CrossRef] [PubMed]
312. Turpeinen, A.M.; Kumpu, M.; Rönnback, M.; Seppo, L.; Kautiainen, H.; Jauhiainen, T.; Vapaatalo, H.; Korpela, R. Antihypertensive and Cholesterol-Lowering Effects of a Spread Containing Bioactive Peptides IPP and VPP and Plant Sterols. *J. Funct. Foods* **2009**, *1*, 260–265. [CrossRef]
313. Liu, L.; Yu, S.; Bu, T.; He, G.; Li, S.; Wu, J. Casein Hydrolysate Alleviates Adipose Chronic Inflammation in High Fat-Diet Induced Obese C57BL/6J Mice through MAPK Pathway. *Nutrients* **2023**, *15*, 1813. [CrossRef] [PubMed]
314. Aihara, K.; Osaka, M.; Yoshida, M. Oral Administration of the Milk Casein-Derived Tripeptide Val-Pro-Pro Attenuates High-Fat Diet-Induced Adipose Tissue Inflammation in Mice. *Br. J. Nutr.* **2014**, *112*, 513–519. [CrossRef] [PubMed]
315. Chakrabarti, S.; Wu, J. Milk-Derived Tripeptides IPP (Ile-Pro-Pro) and VPP (Val-Pro-Pro) Promote Adipocyte Differentiation and Inhibit Inflammation in 3T3-F442A Cells. *PLoS ONE* **2015**, *10*, e0117492. [CrossRef] [PubMed]
316. Wang, C.; Zheng, L.; Zhao, M. Molecular Targets and Mechanisms of Casein-Derived Tripeptides Ile-Pro-Pro and Val-Pro-Pro on Hepatic Glucose Metabolism. *J. Agric. Food Chem.* **2023**, *71*, 18802–18814. [CrossRef] [PubMed]
317. Liu, L.; Zheng, J.; Zhou, M.; Li, S.; He, G.; Wu, J. Peptide Analogues of VPP and IPP with Improved Glucose Uptake Activity in L6 Myotubes Can Be Released from Cereal Proteins. *J. Agric. Food Chem.* **2021**, *69*, 2875–2883. [CrossRef] [PubMed]
318. van Meijl, L.E.C.; Vrolix, R.; Mensink, R.P. Dairy Product Consumption and the Metabolic Syndrome. *Nutr. Res. Rev.* **2008**, *21*, 148–157. [CrossRef] [PubMed]

319. Hudson, J.L.; Zhou, J.; Kim, J.E.; Campbell, W.W. Incorporating Milk Protein Isolate into an Energy-Restricted Western-Style Eating Pattern Augments Improvements in Blood Pressure and Triglycerides, but Not Body Composition Changes in Adults Classified as Overweight or Obese: A Randomized Controlled Tria. *Nutrients* **2020**, *12*, 851. [CrossRef] [PubMed]
320. Frid, A.H.; Nilsson, M.; Holst, J.J.; Björck, I.M.E. Effect of Whey on Blood Glucose and Insulin Responses to Composite Breakfast and Lunch Meals in Type 2 Diabetic Subjects. *Am. J. Clin. Nutr.* **2005**, *82*, 69–75. [CrossRef] [PubMed]
321. Manders, R.J.F.; Wagenmakers, A.J.M.; Koopman, R.; Zorenc, A.H.G.; Menheere, P.P.C.A.; Schaper, N.C.; Saris, W.H.M.; van Loon, L.J.C. Co-Ingestion of a Protein Hydrolysate and Amino Acid Mixture with Carbohydrate Improves Plasma Glucose Disposal in Patients with Type 2 Diabetes. *Am. J. Clin. Nutr.* **2005**, *82*, 76–83. [CrossRef] [PubMed]
322. Bowen, J.; Noakes, M.; Trenerry, C.; Clifton, P.M. Energy Intake, Ghrelin, and Cholecystokinin after Different Carbohydrate and Protein Preloads in Overweight Men. *J. Clin. Endocrinol. Metab.* **2006**, *91*, 1477–1483. [CrossRef] [PubMed]
323. Manders, R.J.F.; Praet, S.F.E.; Meex, R.C.R.; Koopman, R.; de Roos, A.L.; Wagenmakers, A.J.M.; Saris, W.H.M.; van Loon, L.J.C. Protein Hydrolysate/Leucine Co-Ingestion Reduces the Prevalence of Hyperglycemia in Type 2 Diabetic Patients. *Diabetes Care* **2006**, *29*, 2721–2722. [CrossRef] [PubMed]
324. Manders, R.J.; Koopman, R.; Sluijsmans, W.E.; van den Berg, R.; Verbeek, K.; Saris, W.H.; Wagenmakers, A.J.; van Loon, L.J. Co-Ingestion of a Protein Hydrolysate with or without Additional Leucine Effectively Reduces Postprandial Blood Glucose Excursions in Type 2 Diabetic Men. *J. Nutr.* **2006**, *136*, 1294–1299. [CrossRef] [PubMed]
325. Tessari, P.; Kiwanuka, E.; Cristini, M.; Zaramella, M.; Enslen, M.; Zurlo, C.; Garcia-Rodenas, C. Slow versus Fast Proteins in the Stimulation of Beta-Cell Response and the Activation of the Entero-Insular Axis in Type 2 Diabetes. *Diabetes. Metab. Res. Rev.* **2007**, *23*, 378–385. [CrossRef] [PubMed]
326. Ma, J.; Stevens, J.E.; Cukier, K.; Maddox, A.F.; Wishart, J.M.; Jones, K.L.; Clifton, P.M.; Horowitz, M.; Rayner, C.K. Effects of a Protein Preload on Gastric Emptying, Glycemia, and Gut Hormones After a Carbohydrate Meal in Diet-Controlled Type 2 Diabetes. *Diabetes Care* **2009**, *32*, 1600–1602. [CrossRef] [PubMed]
327. Mortensen, L.S.; Hartvigsen, M.L.; Brader, L.J.; Astrup, A.; Schrezenmeir, J.; Holst, J.J.; Thomsen, C.; Hermansen, K. Differential Effects of Protein Quality on Postprandial Lipemia in Response to a Fat-Rich Meal in Type 2 Diabetes: Comparison of Whey, Casein, Gluten, and Cod Protein. *Am. J. Clin. Nutr.* **2009**, *90*, 41–48. [CrossRef] [PubMed]
328. Claessens, M.; Van Baak, M.A.; Monsheimer, S.; Saris, W.H.M. The Effect of a Low-Fat, High-Protein or High-Carbohydrate Ad Libitum Diet on Weight Loss Maintenance and Metabolic Risk Factors. *Int. J. Obes.* **2009**, *33*, 296–304. [CrossRef] [PubMed]
329. Pal, S.; Ellis, V.; Ho, S. Acute Effects of Whey Protein Isolate on Cardiovascular Risk Factors in Overweight, Post-Menopausal Women. *Atherosclerosis* **2010**, *212*, 339–344. [CrossRef] [PubMed]
330. Brader, L.; Holm, L.; Mortensen, L.; Thomsen, C.; Astrup, A.; Holst, J.J.; de Vrese, M.; Schrezenmeir, J.; Hermansen, K. Acute Effects of Casein on Postprandial Lipemia and Incretin Responses in Type 2 Diabetic Subjects. *Nutr. Metab. Cardiovasc. Dis.* **2010**, *20*, 101–109. [CrossRef] [PubMed]
331. Geerts, B.F.; Van Dongen, M.G.J.; Flameling, B.; Moerland, M.M.; Kam, M.L.D.; Cohen, A.F.; Romijn, J.A.; Gerhardt, C.C.; Kloek, J.; Burggraaf, J. Hydrolyzed Casein Decreases Postprandial Glucose Concentrations in T2DM Patients Irrespective of Leucine Content. *J. Diet. Suppl.* **2011**, *8*, 280–292. [CrossRef] [PubMed]
332. Jonker, J.T.; Wijngaarden, M.A.; Kloek, J.; Groeneveld, Y.; Gerhardt, C.; Brand, R.; Kies, A.K.; Romijn, J.A.; Smit, J.W.A. Effects of Low Doses of Casein Hydrolysate on Post-Challenge Glucose and Insulin Levels. *Eur. J. Intern. Med.* **2011**, *22*, 245–248. [CrossRef] [PubMed]
333. Mortensen, L.S.; Holmer-Jensen, J.; Hartvigsen, M.L.; Jensen, V.K.; Astrup, A.; De Vrese, M.; Holst, J.J.; Thomsen, C.; Hermansen, K. Effects of Different Fractions of Whey Protein on Postprandial Lipid and Hormone Responses in Type 2 Diabetes. *Eur. J. Clin. Nutr.* **2012**, *66*, 799–805. [CrossRef] [PubMed]
334. Holmer-Jensen, J.; Mortensen, L.S.; Astrup, A.; de Vrese, M.; Holst, J.J.; Thomsen, C.; Hermansen, K. Acute Differential Effects of Dietary Protein Quality on Postprandial Lipemia in Obese Non-Diabetic Subjects. *Nutr. Res.* **2013**, *33*, 34–40. [CrossRef] [PubMed]
335. Goudarzi, M.; Madadlou, A. Influence of Whey Protein and Its Hydrolysate on Prehypertension and Postprandial Hyperglycaemia in Adult Men. *Int. Dairy J.* **2013**, *33*, 62–66. [CrossRef]
336. Van Meijl, L.E.C.; Mensink, R.P. Effects of Milk and Milk Constituents on Postprandial Lipid and Glucose Metabolism in Overweight and Obese Men. *Br. J. Nutr.* **2013**, *110*, 413–419. [CrossRef] [PubMed]
337. Manders, R.J.F.; Hansen, D.; Zorenc, A.H.G.; Dendale, P.; Kloek, J.; Saris, W.H.M.; van Loon, L.J.C. Protein Co-Ingestion Strongly Increases Postprandial Insulin Secretion in Type 2 Diabetes Patients. *J. Med. Food* **2014**, *17*, 758–763. [CrossRef] [PubMed]
338. Jakubowicz, D.; Froy, O.; Ahrén, B.; Boaz, M.; Landau, Z.; Bar-Dayan, Y.; Ganz, T.; Barnea, M.; Wainstein, J. Incretin, Insulinotropic and Glucose-Lowering Effects of Whey Protein Pre-Load in Type 2 Diabetes: A Randomised Clinical Trial. *Diabetologia* **2014**, *57*, 1807–1811. [CrossRef] [PubMed]
339. Ma, J.; Jesudason, D.R.; Stevens, J.E.; Keogh, J.B.; Jones, K.L.; Clifton, P.M.; Horowitz, M.; Rayner, C.K. Sustained Effects of a Protein 'Preload' on Glycaemia and Gastric Emptying over 4 Weeks in Patients with Type 2 Diabetes: A Randomized Clinical Trial. *Diabetes Res. Clin. Pract.* **2015**, *108*, e31–e34. [CrossRef] [PubMed]
340. Hoefle, A.S.; Bangert, A.M.; Stamfort, A.; Gedrich, K.; Rist, M.J.; Lee, Y.-M.; Skurk, T.; Daniel, H. Metabolic Responses of Healthy or Prediabetic Adults to Bovine Whey Protein and Sodium Caseinate Do Not Differ. *J. Nutr.* **2015**, *145*, 467–475. [CrossRef] [PubMed]

341. Bohl, M.; Bjørnshave, A.; Rasmussen, K.V.; Schioldan, A.G.; Amer, B.; Larsen, M.K.; Dalsgaard, T.K.; Holst, J.J.; Herrmann, A.; O'Neill, S.; et al. Dairy Proteins, Dairy Lipids, and Postprandial Lipemia in Persons with Abdominal Obesity (DairyHealth): A 12-Wk, Randomized, Parallel-Controlled, Double-Blinded, Diet Intervention Study. *Am. J. Clin. Nutr.* **2015**, *101*, 870–878. [CrossRef] [PubMed]
342. Jakubowicz, D.; Wainstein, J.; Landau, Z.; Ahren, B.; Barnea, M.; Bar-Dayan, Y.; Froy, O. High-Energy Breakfast Based on Whey Protein Reduces Body Weight, Postprandial Glycemia and HbA1C in Type 2 Diabetes. *J. Nutr. Biochem.* **2017**, *49*, 1–7. [CrossRef]
343. Almario, R.U.; Buchaxn, W.M.; Rocke, D.M.; Karakas, S.E. Glucose-Lowering Effect of Whey Protein Depends upon Clinical Characteristics of Patients with Type 2 Diabetes. *BMJ Open Diabetes Res. Care* **2017**, *5*, e000420. [CrossRef] [PubMed]
344. Gaffney, K.A.; Lucero, A.; Stoner, L.; Faulkner, J.; Whitfield, P.; Krebs, J.; Rowlands, D.S. Nil Whey Protein Effect on Glycemic Control after Intense Mixed-Mode Training in Type 2 Diabetes. *Med. Sci. Sports Exerc.* **2018**, *50*, 11–17. [CrossRef] [PubMed]
345. Horner, K.; Drummond, E.; O'Sullivan, V.; S C Sri Harsha, P.; Brennan, L. Effects of a Casein Hydrolysate versus Intact Casein on Gastric Emptying and Amino Acid Responses. *Eur. J. Nutr.* **2019**, *58*, 955–964. [CrossRef] [PubMed]
346. King, D.G.; Walker, M.; Campbell, M.D.; Breen, L.; Stevenson, E.J.; West, D.J. A Small Dose of Whey Protein Co-Ingested with Mixed-Macronutrient Breakfast and Lunch Meals Improves Postprandial Glycemia and Suppresses Appetite in Men with Type 2 Diabetes: A Randomized Controlled Trial. *Am. J. Clin. Nutr.* **2018**, *107*, 550–557. [CrossRef] [PubMed]
347. Bjørnshave, A.; Holst, J.J.; Hermansen, K. Pre-Meal Effect of Whey Proteins on Metabolic Parameters in Subjects with and without Type 2 Diabetes: A Randomized, Crossover Trial. *Nutrients* **2018**, *10*, 122. [CrossRef]
348. Bjørnshave, A.; Holst, J.J.; Hermansen, K. A Pre-Meal of Whey Proteins Induces Differential Effects on Glucose and Lipid Metabolism in Subjects with the Metabolic Syndrome: A Randomised Cross-over Trial. *Eur. J. Nutr.* **2019**, *58*, 755–764. [CrossRef] [PubMed]
349. Sartorius, T.; Weidner, A.; Dharsono, T.; Boulier, A.; Wilhelm, M.; Schön, C. Postprandial Effects of a Proprietary Milk Protein Hydrolysate Containing Bioactive Peptides in Prediabetic Subjects. *Nutrients* **2019**, *11*, 1700. [CrossRef]
350. Basciani, S.; Camajani, E.; Contini, S.; Persichetti, A.; Risi, R.; Bertoldi, L.; Strigari, L.; Prossomariti, G.; Watanabe, M.; Mariani, S.; et al. Very-Low-Calorie Ketogenic Diets With Whey, Vegetable, or Animal Protein in Patients With Obesity: A Randomized Pilot Study. *J. Clin. Endocrinol. Metab.* **2020**, *105*, 2939–2949. [CrossRef] [PubMed]
351. Memelink, R.G.; Pasman, W.J.; Bongers, A.; Tump, A.; van Ginkel, A.; Tromp, W.; Wopereis, S.; Verlaan, S.; de Vogel-van den Bosch, J.; Weijs, P.J.M. Effect of an Enriched Protein Drink on Muscle Mass and Glycemic Control during Combined Lifestyle Intervention in Older Adults with Obesity and Type 2 Diabetes: A Double-Blind RCT. *Nutrients* **2020**, *13*, 64. [CrossRef] [PubMed]
352. Derosa, G.; D'Angelo, A.; Maffioli, P. Change of Some Oxidative Stress Parameters after Supplementation with Whey Protein Isolate in Patients with Type 2 Diabetes. *Nutrition* **2020**, *73*, 110700. [CrossRef] [PubMed]
353. Manders, R.J.F.; Praet, S.F.E.; Vikström, M.H.; Saris, W.H.M.; van Loon, L.J.C. Protein Hydrolysate Co-Ingestion Does Not Modulate 24 h Glycemic Control in Long-Standing Type 2 Diabetes Patients. *Eur. J. Clin. Nutr.* **2009**, *63*, 121–126. [CrossRef] [PubMed]
354. Holmer-Jensen, J.; Hartvigsen, M.L.; Mortensen, L.S.; Astrup, A.; de Vrese, M.; Holst, J.J.; Thomsen, C.; Hermansen, K. Acute Differential Effects of Milk-Derived Dietary Proteins on Postprandial Lipaemia in Obese Non-Diabetic Subjects. *Eur. J. Clin. Nutr.* **2012**, *66*, 32–38. [CrossRef] [PubMed]
355. Weisse, K.; Brandsch, C.; Zernsdorf, B.; Nkengfack Nembongwe, G.S.; Hofmann, K.; Eder, K.; Stangl, G.I. Lupin Protein Compared to Casein Lowers the LDL Cholesterol: HDL Cholesterol-Ratio of Hypercholesterolemic Adults. *Eur. J. Nutr.* **2010**, *49*, 65–71. [CrossRef] [PubMed]
356. Bendtsen, L.Q.; Lorenzen, J.K.; Gomes, S.; Liaset, B.; Holst, J.J.; Ritz, C.; Reitelseder, S.; Sjödin, A.; Astrup, A. Effects of Hydrolysed Casein, Intact Casein and Intact Whey Protein on Energy Expenditure and Appetite Regulation: A Randomised, Controlled, Cross-over Study. *Br. J. Nutr.* **2014**, *112*, 1412–1422. [CrossRef] [PubMed]
357. Nouri, M.; Tarighat-Esfanjani, A.; Sadra, V.; Ghasempour, Z.; Jafarabadi, M.A.; Tajfar, P.; Gargari, B.P. Effects of Whey Protein Concentrate on Glycemic Status, Lipid Profile, and Blood Pressure in Overweight/Obese Women with Type 2 Diabetes Mellitus: A Randomized Placebo-Controlled Clinical Trial. *Turk. J. Endocrinol. Metab.* **2022**, *26*, 148–155. [CrossRef]

biomolecules

MDPI

Review

Polyphenols: Role in Modulating Immune Function and Obesity

Md Abdullah Al Mamun, Ahmed Rakib, Mousumi Mandal, Santosh Kumar, Bhupesh Singla and Udai P. Singh *

Department of Pharmaceutical Sciences, College of Pharmacy, The University of Tennessee Health Science Center, 881 Madison Avenue, Memphis, TN 38163, USA; mmamun3@uthsc.edu (M.A.A.M.); arakib@uthsc.edu (A.R.); mmandal1@uthsc.edu (M.M.); ksantosh@uthsc.edu (S.K.); bsingla@uthsc.edu (B.S.)

* Correspondence: usingh1@uthsc.edu

Abstract: Polyphenols, long-used components of medicinal plants, have drawn great interest in recent years as potential therapeutic agents because of their safety, efficacy, and wide range of biological effects. Approximately 75% of the world's population still use plant-based medicinal compounds, indicating the ongoing significance of phytochemicals for human health. This study emphasizes the growing body of research investigating the anti-adipogenic and anti-obesity functions of polyphenols. The functions of polyphenols, including phenylpropanoids, flavonoids, terpenoids, alkaloids, glycosides, and phenolic acids, are distinct due to changes in chemical diversity and structural characteristics. This review methodically investigates the mechanisms by which naturally occurring polyphenols mediate obesity and metabolic function in immunomodulation. To this end, hormonal control of hunger has the potential to inhibit pro-obesity enzymes such as pancreatic lipase, the promotion of energy expenditure, and the modulation of adipocytokine production. Specifically, polyphenols affect insulin, a hormone that is essential for regulating blood sugar, and they also play a role, in part, in a complex web of factors that affect the progression of obesity. This review also explores the immunomodulatory properties of polyphenols, providing insight into their ability to improve immune function and the effects of polyphenols on gut health, improving the number of commensal bacteria, cytokine production suppression, and immune cell mediation, including natural killer cells and macrophages. Taken together, continuous studies are required to understand the prudent and precise mechanisms underlying polyphenols' therapeutic potential in obesity and immunomodulation. In the interim, this review emphasizes a holistic approach to health and promotes the consumption of a wide range of foods and drinks high in polyphenols. This review lays the groundwork for future developments, indicating that the components of polyphenols and their derivatives may provide the answer to urgent worldwide health issues. This compilation of the body of knowledge paves the way for future discoveries in the global treatment of pressing health concerns in obesity and metabolic diseases.

Keywords: polyphenols; obesity; metabolic disorders; gut health; immunomodulation

1. Introduction

Obesity is associated with an increased level of body fat mass caused by any specific or combination of contributing factors such as genetics, environmental factors, dietary habits, lifestyle, or multiple pathophysiological clinical conditions [1,2]. Obesity is a consequence of an energy imbalance between caloric input and output, and when this phenomenon continues for a long time, it results in metabolic disorders [3]. Obesity has been emerging as a major health concern worldwide in recent years, resulting in incalculable social costs [4]. Recently, the World Health Organization (WHO) declared obesity as an epidemic hazard worldwide. Currently, more than 1 billion people are affected by obesity and this prevalence is increasing day by day [5]. The multifactorial nature of obesity drives researchers to various treatment approaches, as the management of obesity is a slow process, including managing through physical exercise and diet control. Thus, medications, either natural or synthetic, are preferable to patients for effective outcomes during obesity [6]. There are several effective medications available nowadays, but even a few years back, phentermine and orlistat were the only drugs approved by the FDA as anti-obesity drugs [7]. Currently, orlistat, phentermine/topiramate, naltrexone/bupropion, and liraglutide are the anti-obesity drugs that have been FDA-approved for chronic weight management [8]. However, significant side effects of orlistat include gastrointestinal issues, liver damage, allergic reactions, and aberrant endocrine system responses shown during treatment [9]. Additionally, existing therapeutics for obesity are expensive, and some people may experience serious negative side effects.

Obesity has been shown to affect both innate and adaptive immune systems [10]. The emergence of numerous obesity-related issues brought about by altered innate and adaptive immune responses induces chronic adipose tissue (AT) inflammation [11]. Epidemiological data from various studies have demonstrated that obese people have a higher incidence and severity of different types of infections than lean participants [12]. However, the etiology of altered innate immune responses in obesity is still an intensely debated issue to date. Studying the relationship between immune response and metabolism has become more popular recently, and uncovering this role may shed light on the defective innate immunity during obesity. According to research on the subject of "immunometabolism", immune cell function is linked to a certain state of cellular metabolism [13]. The significance of metabolism in enabling functional alterations to immune cells by blocking glycolysis in activated macrophages to stop inflammatory cytokine release has been shown [14]. Further, research has revealed that, when endogenous and external stimuli activate macrophages and other innate immune cells, these cells undergo unique metabolic rewiring for appropriate functional responses [15]. Thus, it is crucial to understand the mutual interaction that exists between immunity and obesity.

To this end, plant products have served as the most ancient source of medication from the start of civilization [16]. According to recent data, around three quarters of the total global population still rely on plant-derived therapeutic agents [17]. For example, many plants are used traditionally for treating diabetes, and the safety and efficacy of these plant-derived substances have been validated through many successful clinical trials. At present, the vast resources of phytochemicals are being explored intensely to unmask their antiadipogenic and anti-obesity properties [18], because phytochemicals are considered to be safe compared to synthetic compounds. Polyphenols are beneficial plant compounds with highly antioxidant properties [19]. Mounting evidence reports the role of polyphenolic compounds as being highly effective with lower side effects for the management of obesity-related complications. In this review, we will describe the mechanisms of how polyphenols are associated with ameliorating obesity conditions.

2. Diversity of Polyphenols and Their Effects

Approximately 8000 distinct polyphenols exist today, and they all have an aromatic ring with hydroxyl groups as part of their common phenolic structure. Despite being classified chemically as substances with phenolic structural characteristics, polyphenols are a wide class of natural products that include many subgroups of phenolic substances. Polyphenols can be found in abundance in fruits, vegetables, whole grains, and other forms of food and drink, including tea, chocolate, and wine [20]. Mounting evidence suggests that dietary polyphenols help to prevent obesity by influencing the brain's different neurohormones that regulate satiety and food intake. Experimental studies indicate that polyphenols may play a part in the neurohormones that regulate energy balance and calorie intake in obese people [21]. Further, curcumin treatment resulted in the downregulation of the insulin-like growth factor (IGF) pathway in medulloblastoma cells [22]. Polyphenols can be categorized into various groups based on the number of aromatic rings and structural moiety connected to them [23]. The majority of them are produced by plants' secondary metabolite shikimate pathway as a defense strategy [24]. There is enough evidence to conclude that the appropriate consumption of polyphenols has several positive health effects, even though most diseases are notlinked to inadequate or absent polyphenol intake [25]. Most of these advantageous traits of polyphenols are thought to be a result of their capacity to scavenge free radicals, create stable complexes, and obstruct subsequent chemical events. Additionally, they can scavenge hydrogen peroxide or limit its synthesis, protecting against oxidative stress, which might modulate immunological responses [26]. Depending on factors like the quantity of aromatic rings, their distribution in nature, and other factors, polyphenols can be categorized in many ways. Below is a basic description of the main groups of polyphenols.

2.1. Phenolic Acids

Benzoic acids and cinnamic acids and their derivatives are the two classes into which phenolic acids can be classified. Benzoic acids are the most basic phenolic acids found in nature, containing seven carbon atoms (C6–C1). While cinnamic acids can have nine carbon atoms (C6–C3), plants are the most prevalent source of these acids, which have seven. These compounds are identified by the presence of one or more hydroxyl and/or methoxyl groups, a carboxylic group, and a benzenic ring in the molecule [27]. While bound phenolic acids are connected to the cell walls, free phenolic acids are present in the pericarp, testa, and aleurone, the outer layers of the kernel. The majority of sorghum's phenolic chemicals are found bound and in the bran. In sorghum, ferulic acid is the most prevalent bound phenolic acid; nevertheless, other, more abundant phenolic acids include syringic, protocatechuic, caffeic, p-coumaric, and sinapic [28]. Considering these different phenolic acids, for example, ferulic acid is reported to reduce the activity of hepatic lipogenic enzymes, including glucose-6-phosphate dehydrogenase (G6PD), malic enzyme (ME), and fatty acid synthase (FAS), which are in charge of synthesizing fatty acids and cholesterol [29]. Moreover, it has been reported that ferulic acid effectively reduces high fat diet (HFD)-induced visceral obesity and weight gain by modulating the peptide hormones associated with food regulation such as ghrelin, leptin, and insulin (Table 1) [30].

Table 1. Summary of polyphenolic compounds and their functions in obesity.

Compound	Structure	Function	References
Ferulic acid	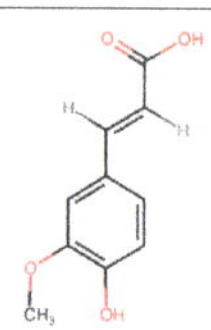	Reduces lipogenic enzyme activity Prevents HFD-induced visceral adiposity	[29,30]
Quercetin	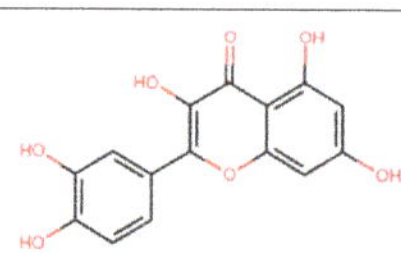	Suppresses adipogenic factors Reduces TG synthesis enzymes	[31,32]
Cyanidin-3-glucoside	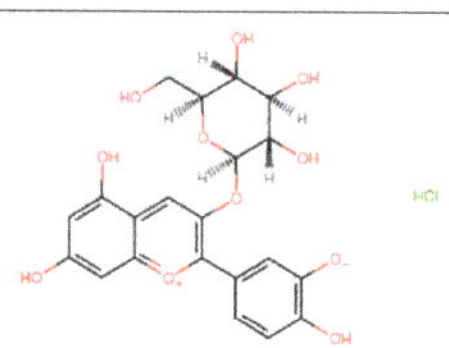	Increase energy expenditure Increase thermogenesis	[33]

Table 1. *Cont.*

Compound	Structure	Function	References
7,8-dihydroxy-3-(4-methylphenyl) coumarin	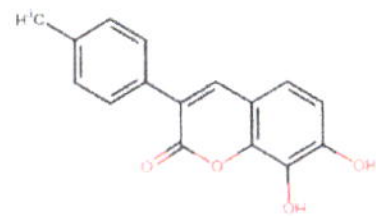	Lower serum cholesterol	[34,35]
Curcumin	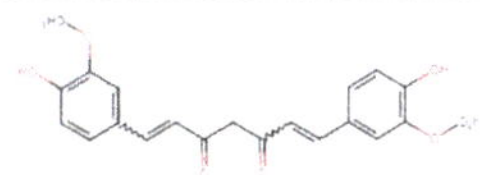	Decrease adipocyte size Reduce M1 macrophage marker	[36]
Resveratrol	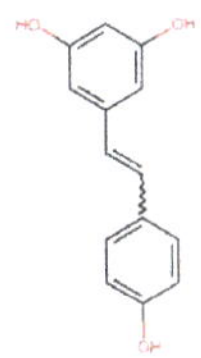	Inhibit NF-κB activation Increase Tregs	[37,38]

2.2. Flavonoids

Flavonoids are mostly studied and found in large quantities in foods and drinks that people consume daily, such as fruits, vegetables, tea, chocolate, and wine [39]. Flavonoids are found to reduce the risk associated with type 2 diabetes [40,41], cardiovascular diseases [42], obesity, and non-alcoholic fatty liver disease [43,44]. The beneficial effects of flavonoids have been thoroughly studied at the molecular level, with AMP-activated protein kinase (AMPK) activation being a common focal point. An essential function of AMPK is to regulate adipogenesis and lipid metabolism. Anabolic pathways like fatty acid production and gluconeogenesis are suppressed, while catabolic processes like fatty acid oxidation (FAO), glucose absorption, and glycolysis are encouraged through the phosphorylation and activation of AMPK [45]. A recent study showed that quercetin, a flavonoid, suppressed the key adipogenic factors, including CCAAT-enhancer-binding proteinα (C/EBPα), C/EBPβ, peroxisome proliferator activated receptor γ (PPARγ), and triglycerides (TG) synthesis enzymes, such as diacylglycerol acyltransferase-1, lipin1 (Table 1) [31]. In addition, quercetin is reported to inhibit the MAPK signaling factors, including extracellular signal-regulated kinases1/2 (ERK1/2) and c-Jun N-terminal kinase in adipocytes and macrophages [31]. Moreover, certain doses of quercetin have been shown to downregulate different inflammatory signaling pathways, including Nuclear factor kappa B (NF-κB) and cyclooxygenase-2 (COX-2) [32].

2.3. Tannins

Dietary tannic acid is a naturally occurring polyphenolic substance that is frequently present in plants and has long been associated with negative nutritional effects. Advances in structure identification and separation technologies have allowed researchers to isolate and identify thousands of tannin monomers from plants in the last few decades. Proanthocyanidins, gallotannins, and ellagitannins are widely reported tannins that have many pharmacological effects. A great deal of pharmacological research has also been carried out, including on anti-inflammatory, antioxidant, anti-aging, hypoglycemic, lipid-lowering, anticancer, antibacterial, and antiviral properties [46–49]. Tannins can interact with a variety of viral targets. Herpes simplex radiolabelled virus particles have shown that the antiviral properties of galloylated and hydrolyzable condensed tannins stem from their ability to impede virus adsorption [50]. Certain anti-nutritional characteristics of hydrolysable tannins have been demonstrated, particularly in the nutrition of monogastric animals by building complexes with proteins, carbs, and starches. When hydrolysable tannins were used in varying dosages, the degree of these beneficial and detrimental effects on digestion was seen; however, in animal testing, smaller doses did not significantly affect the digestibility of crude protein, organic matter, or ash [51]. The intake of hydrolysable tannins in higher amounts decreases weight gain and daily food intake [52,53]. In a recent study, You et al. reported that cyanidin-3-glucoside, a most abundant anthocyanin in plants, mitigates obesity by increasing energy expenditure and thermogenic properties in brown AT (BAT) (Table 1) [33].

2.4. Coumarins

Coumarins are molecules with both natural and synthetic origins that are among the most researched natural substances concerning human health. They belong to a family of heterocyclic compounds and have been thoroughly investigated in the domains of biochemistry and pharmacology [54]. As lipid-lowering agents, coumarins with various heterocycles based on the cyclization of 2-ethoxy-3-phenylpropanoic acid and 2-benzylmalonic acid have been studied to prevent the buildup of TGs and cholesterol in the walls of arteries or blood vessels, which causes atheroma and lowers the incidence of cardiovascular disease [55]. In a mouse model of hypercholesterolemia, the lipid-lowering properties of coumarin 7,8-dihydroxy-3-(4-methylphenyl) coumarin were evaluated, and it resulted in a considerable drop in serum cholesterol levels (Table 1) [34]. Reports showed that a HFD supplemented with 0·05% coumarin for eight weeks dramatically lowered blood lipid levels, body weight, and body fat [35]. The antiadipogenic properties of coumarins have also been demonstrated and found to suppress lipogenic gene expressions and lipid accumulation in 3T3-L1 adipocyte cells, potentially via the PPARγ pathway [56,57].

3. Polyphenols Alter Obesity

Anti-obesity phytochemicals are broadly classified into phenylpropanoids, flavonoids, terpenoids, alkaloids, glycosides, and phenolic acids according to their chemical structure [58]. Numerous studies have been conducted to determine the exact mechanisms associated with improving obesity and obesity-related complications using polyphenols. The inhibition of transcription factors involved in adipogenesis, such as C/EBP and PPAR, has been hypothesized as the method by which natural products alter obesity state [59]. One of the proposed mechanisms is to prevent the activation of C/EBP and PPAR signals, and the Wnt/β-catenin signaling pathway can prevent preadipocytes from differentiating into adipogenesis, which has an anti-obesity impact [60]. Several phytochemicals have been demonstrated to date in the literature to support uncoupled protein 1 (UCP1)-related adipose differentiation and thermogenesis capacity, boosting energy expenditure and reducing obesity and its side effects [61]. For a better understanding of the mechanism of polyphenols against obesity, a pictorial representation is shown in Figure 1.

3.1. Hormonal Regulation of Food Intake and Satiety

Mounting evidence suggests that dietary polyphenols prevent obesity by influencing the brain's neurohormones that regulate satiety and food intake. Studies conducted in vitro and in vivo indicate that polyphenols may play a part in the neurohormones that regulate energy balance and calorie intake in obese people. Insulin is a vital hormone that controls blood sugar levels and cues for adipocytes to store energy. To date, research has not been able to adequately address the complicated and contentious relationships between insulin and obesity. However, the pathophysiology and development of obesity have been linked to the pancreas' hypersecretion of insulin [21]. In the past, several studies have been conducted to determine how insulin affects the onset of obesity. For instance, a study found that, over a 15-year follow-up period, persons who hypersecreted insulin in response to an intravenous glucose tolerance test experienced abnormal weight gain [62]. Further, the effects of polyphenols on insulin highlight their potential significance in obesity, because insulin is a crucial neurohormone in the etiology of obesity. It has been shown that long-term resveratrol intracerebroventricular infusion alleviated hyperinsulinemia and corrected hyperglycemia in obese mice [63,64]. When curcumin was administered, the IGF pathway was found to be downregulated in medulloblastoma cells, and it has been shown that curcumin includes polyphenol metabolites that affect how the central nervous system regulates neurohormones like insulin [65]. Exciting studies performed in ob/ob and db/db mouse models of obesity led to this discovery, and profoundly influential studies in obese humans stimulated interest in leptin as a rational therapy for obesity and its numerous associated disorders. This leptin will forever be known as a potent regulator of feeding behavior and body weight [66]. Further, the daily intake of 200 mg/kg of resveratrol reestablished leptin sensitivity in obese rats and lowered their overall body weight [67]. Anthocyanins were reported to inhibit neuropeptide Y, reducing obesity in HFD-fed rats [68]. Polyphenols have also been proven to exert anti-obesity benefits by directly influencing neuropeptides in food intake [69].

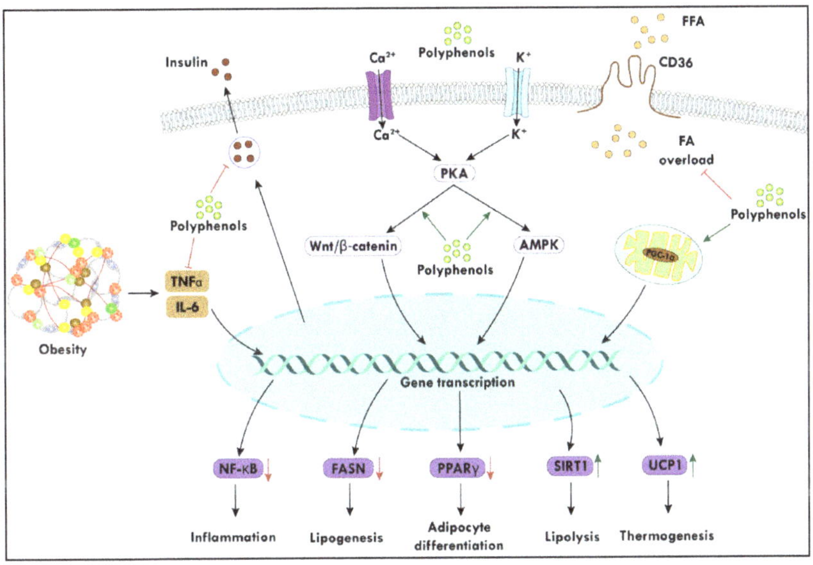

Figure 1. Polyphenols play a role in modifying obesity. They hinder the production of free fatty acids and mitigate the hypersecretion of insulin associated with obesity. Natural polyphenols facilitate BAT thermogenesis and lipolysis, aiding in the regulation of adiposity by enhancing the expression of UCP1 and SIRT1. Furthermore, polyphenols downregulate gene transcription of FAS and PPARγ, thereby inhibiting lipogenesis and adipocyte differentiation. Additionally, they prevent TNF-α and IL-6-mediated NF- κB expression, subsequently averting inflammation.

3.2. *Polyphenols Amend Pro-Obesity Enzymes*

The primary enzyme responsible for breaking down dietary lipids in the digestive system is pancreatic lipase. Numerous clinical and experimental studies have demonstrated that lipase inhibitors can improve lipid metabolism in obese people. It has been shown that preventing the absorption of fatty acids lowers serum LDL (low-density lipoprotein) levels and raises HDL (high-density lipoprotein) levels [70]. Therefore, the creation of pancreatic lipase inhibitors is a crucial goal for the treatment of obesity. Pancreatic lipase activity has been inhibited both in vitro and in vivo by polyphenol-rich extracts from a variety of plants [71]. When fat synthesis outpaces fat oxidation, it leads to the development of obesity, and the body's ability to store fat is closely controlled by lipogenesis and lipolysis. Lipolysis hydrolyzes TGs to produce free fatty acids, mono- or diacylglycerol, or free glycerol, whereas lipogenesis transforms simple sugars and other substrates into fatty acids and finally TGs. Thus, FAS is a crucial component of animal de novo lipogenesis. The processes in the body's synthesis of endogenous lipids have the potential to become excessive and cause obesity. Towards this, polyphenols can disrupt the lipogenic pathway and inhibit the activity of the FAS enzyme to alter obesity. In HFD-induced obese rats, polyphenol treatment dramatically decreased the obesity index by preventing FAS activity [72]. Further, inhibiting FAS activity in 3T3-L1 cells, chokeberry-derived polyphenols drastically reduced body weight and blood TG levels [73]. A powerful natural FAS inhibitor, epigallocatechin-3-gallate has been shown to limit the enzyme's activity in prostate cancer cells, lowering endogenous lipid production [74]. Taken together, uncovering better mechanisms of polyphenols in mediating obesity and metabolic syndrome has great potential to develop therapeutics for obesity.

3.3. *Enhancement of Energy Expenditure through Polyphenols*

Physical activity and thermogenesis are well-reported categories of total daily energy expenditure to date. Typically, an increase in TG lipolysis and FAO occurs simultaneously with the creation of heat. Hence, it will be helpful to stimulate thermogenesis for altering obesity conditions [75]. The two primary ATs in mammals, white AT (WAT) and BAT, are noticed. To control adiposity, BAT has a brown hue because of its strong vascularization and high mitochondrion concentration, which releases surplus energy through non-shivering thermogenesis. The inner membrane of the mitochondria uses UCP1 to assist BAT in using and dispersing the energy from lipids to produce heat [76]. Further, UCP1 is also reported to control BAT's thermogenesis by reducing the proton gradient and separating oxidation from ATP production [77]. There is some distribution of WAT in the visceral region, such as in the omental, mesenteric, mediastinal, and epicardial regions, but the majority of WAT is stored in the subcutaneous region in the deep and superficial abdominal sections, as well as the gluteal-femoral regions [78]. WAT's primary purpose is to store excess energy as TGs, which is counter to BAT's ability to release energy by generating heat and warming the blood supply to essential organs [79]. Several animal studies have proven the role of polyphenolic compounds in the browning of AT, which is the key regulator of energy expenditure. By interfering with AMPK, sirtuin 1 (SIRT1), proliferator-activated-receptor-gamma-coactivator-1, catechol-O-methyltransferase, and sympathetic nervous system, which are key players in the transcriptional regulation and physiology of AT, phenolic compounds such as curcumin, quercetin, resveratrol, and isoflavones play crucial roles in thermogenesis and consequently lowering obesity [80]. The potential effect may be able to increase metabolic performance by efficiently controlling energy metabolism, as well as by boosting glucose uptake and mitochondrial activity, according to many preclinical studies [81]. Here, it becomes clear that polyphenolic substances such as gingerol, icariin, and resveratrol can affect skeletal muscle in preclinical metabolic syndrome models to control energy metabolism and enhance mitochondrial function [82]. In 3T3-L1 adipocytes, a polyphenolic extract of mango reduced adipogenesis and boosted thermogenesis, as demonstrated by increased expressions of thermogenic markers (UCP1, SIRT1, and AMPK) [83]. Expressions of C/EBPα and PPARγ, which are two important transcriptional factors for adipocyte differentiation, were also found to be suppressed in rodent models treated with polyphenolic extracts isolated from mango [84].

3.4. *Regulation of Expression of Adipokines*

Recent research has demonstrated that adipocytes have a secretory role and can release a variety of physiologic active compounds known as adipocytokines. Leptin and adiponectin are examples of adipokines that are selectively expressed in AT. Tumor necrosis factor (TNF)-α and interleukin-6 (IL-6) are examples of adipokines that are non-specifically expressed in AT. Propolis and its extract have been proven in studies to have an anti-obesity effect by altering adipokine secretion [85]. By influencing the hypothalamus's weight regulation region, leptin reduces fat storage and prevents body weight gain by increasing energy intake and decreasing appetite. Brazilian green propolis was found to dramatically boost leptin expression when administered in vitro to the 3T3-L1 cell line and in vivo to C57BL/6 mice [86]. Adiponectin is another advantageous adipokine that controls an organism's energy homeostasis, glucose metabolism, and fat metabolism [87]. An increase in adiponectin has been associated with enhanced AT and overall body energy metabolism [88]. Key metabolic processes are carried out by adiponectin in the liver and skeletal muscles. PPARα and AMPK are the mechanisms by which adiponectin's effects on insulin sensitivity are mediated in muscle. Adiponectin stimulates FAO and reduces inflammation through the PPARα pathway, but it stimulates glucose transport and inhibits gluconeogenesis in the liver via AMPK [89]. Numerous plant-derived polyphenols, including resveratrol from red grapes, quercetin from fruits, vegetables, and grains, genistein from many plants, including soybeans, epigallocatechin gallate from green tea, berberine from *Coptis chinensis*, and curcumin from *Curcuma longa*, have been found to activate AMPK [90]. Since many of these substances are known to limit mitochondrial ATP generation, it appears that the mechanisms by which AMPK is activated necessitate the raising of AMP levels. Epigallocatechin-3-gallate, curcumin, resveratrol, and quercetin specifically target and block the mitochondrial F1F0-ATPase/ATP synthase [91,92]. Further evidence for the molecular basis of resveratrol and quercetin's inability to activate AMPK in cells expressing the AMP-insensitive (R531G) AMPK2 subunit is provided by their failure to do so [93]. Taken together, it has been well established that polyphenols alter the levels of adipokines.

3.5. Regulation of Lipid Metabolism by Polyphenols

Numerous studies have investigated the role of polyphenols in regulating lipid metabolism, particularly in the context of obesity and cardiovascular health. Towards this, resveratrol, a polyphenol found in red wine, has been shown to affect lipid metabolism by enhancing the expression of genes involved in lipid oxidation and inhibiting lipogenesis, the process of fat storage [94,95]. Similarly, green tea catechins, such as epigallocatechin gallate, have demonstrated lipid-lowering effects by increasing the oxidation of fatty acids and reducing fat absorption in the intestines [96,97]. Further, quercetin, another polyphenol abundant in fruits and vegetables, has been reported to modulate lipid metabolism by reducing inflammation and oxidative stress, which are known contributors to dyslipidemia and atherosclerosis [98]. In addition, curcumin, a polyphenol from turmeric, has shown the potential to improve lipid profiles by increasing the expression of genes involved in cholesterol metabolism and reducing the formation of lipid plaques in blood vessels [99,100]. These studies collectively suggest that polyphenols can impact lipid metabolism through various mechanisms, including the regulation of gene expression, enhancement of lipid oxidation, inhibition of lipogenesis, and reduction in inflammation and oxidative stress. Thus, together, these effects have important implications for the prevention and management of conditions related to dyslipidemia, such as obesity and cardiovascular diseases.

4. Polyphenols Modulate Immune Response

4.1. Gut Health Reinforces the Immune Response

The increasing market size of immunity booster food and supplements indicates the concern of people worldwide regarding immunomodulation and protection from various diseases. It is always better to prevent infection and other disease conditions by boosting the natural immune system. Food contains relatively low levels of bioactive chemicals, yet their effects on health were constantly studied in the past. The mucosal layer, epithelium, and lamina propria are the three defense mechanisms of the intestinal innate immune system. The mucosal layer is the host's initial line of defense against foreign pathogens in the intestinal tract [101]. Numerous investigations on the modulatory effects of polyphenols on intestinal immune function have produced compelling data that required more mechanistic studies. The nutritional protection of polyphenol-induced abnormal crypt lesions reduction may be a crucial step in the prevention of gastrointestinal tract tumors [102]. The bioactive substances known as polyphenols improve gut health by controlling mucosal immunity and inflammation. It has been demonstrated that polyphenols boost intestinal mucosal immunity in vivo after boosting the number of intraepithelial T cells and mucosal eosinophils in pigs infected with *Ascaris suum* [103]. The composition of the microflora populations may be modulated and subject to fluctuations by the phenolic substrates provided to the gut bacteria through varying dietary patterns and the aromatic metabolites generated, which have been shown to have selective prebiotic effects and antimicrobial activities against gut pathogenic bacteria [104,105]. Several studies have shown the critical function that gut bacteria plays in controlling the development of antigen-presenting cells [106]. The studies have shown that the monocolonization of germ-free (GF) mice with *Escherichia coli* was sufficient to recruit dendritic cells (DCs) to the intestines, and GF animals showed a decreased number of intestinal but not systemic DCs [107]. Furthermore, it has recently been demonstrated that microbe-derived ATP stimulates a subset of DCs that produce CD70 and CX3CR1 on their surface, which, in turn, causes T helper 17 (Th17) cells to differentiate [108]. Epigallocatechin-3-gallate, epicatechin-3-gallate, and epigallocatechin are examples of polyphenols that are said to increase IL-10 production by human white blood cells. Thus, they cause the activity of proinflammatory cytokines released by macrophages to decrease and increase the activity of anti-inflammatory cytokines [109]. The study targeted various types of immune cells, such as primary macrophages, to find out the potential targets [110]. Further, nitric oxide (NO) generation in healthy peripheral blood mononuclear cells (PBMC) was used as a model showing that red wine might cause human monocytes to produce NO and that the subsequently released NO's vasodilatory properties could prevent atherosclerosis [111]. It also reduces the secretion of IL-6, TNF-α, and IL-1β from PBMCs [112]. Animal studies have shown that epigallocatechin-3-gallate reduces the signs and symptoms of autoimmune disorders. Mice given epigallocatechin-3-gallate had significantly more regulatory T (Treg) cells in their lymph nodes and spleens, and their T-cell response was reduced [113]. IL-10, transforming growth factor-beta-1 (TGF-β1), IL-6, and IL-17 are found to be balanced by polyphenols as a result of the regulation of Th17 and Tregs. Additionally, polyphenols block NF-κB activation, which prevents the development of dextran sulfate sodium-induced colitis [114]. A graphical representation of the roles of polyphenols in protecting gut health is shown in Figure 2.

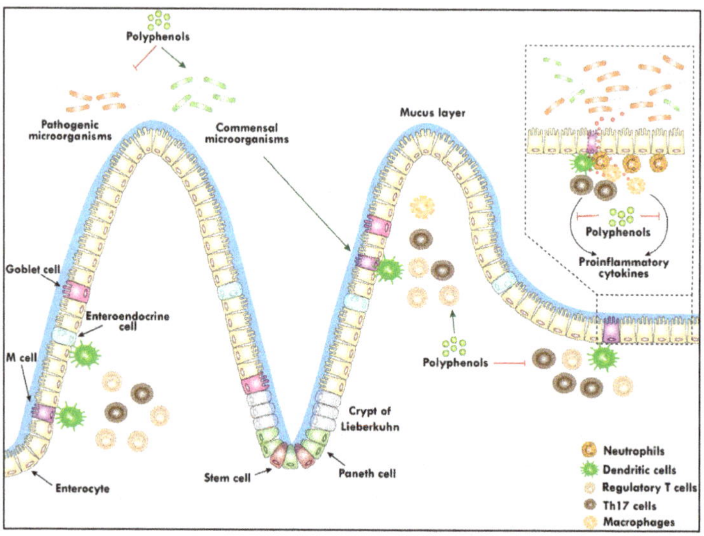

Figure 2. Polyphenols enhance the well-being of the digestive system by promoting a healthy community of beneficial bacteria and preserving an equilibrium between Th17 and Treg. Harmful microorganisms can harm the mucosal lining of the gut, allowing the invasion of antigens and triggering the activation of antigen-presenting cells, such as dendritic cells. Polyphenols intervene in this process, obstructing the pathway and averting subsequent inflammation driven by proinflammatory cytokines such as IL-10, IL-6, and IL-17.

4.2. Polyphenols Modulate Macrophage Functions and Inflammation

Macrophages are phagocytes that differentiate from transient monocytes, eliminate foreign substances, and initiate immune response. Macrophages like dendritic cells (DCs) serve as antigen-presenting cells, activating immature T cells into effector T cells in the presence of an antigen [115]. Macrophages are crucial for tissue healing, host defense, and controlling inflammation, and also play a harmful role in several chronic disorders, such as rheumatoid arthritis, inflammatory bowel disease, asthma, and atherosclerosis. The traditional classification of macrophages includes the traditional inflammatory M1 and immunosuppressive M2 phenotypes. The stimulation of interferon (IFN) and the activation of toll-like receptors (TLRs) by bacterial lipopolysaccharides (LPS) cause M1 differentiation to begin, whereas IL-4 causes M2 polarization to begin [116]. Certain polyphenols from cinnamon are reported to activate macrophages and have an impact on lowering inflammation and enhancing immunological performance [117,118]. Reactive oxygen species (ROS) govern a wide range of complex biological activities, including angiogenesis, inflammation, differentiation, and proliferation. The nicotinamide adenine dinucleotide phosphate (NADPH) oxidase family consists of seven members with tissue- and cell-type-specific expression profiles. The primary purpose of all family members is to produce controlled levels of ROS [119]. Flavanols can inhibit transcription factors (such as NF-κB), resulting in a decrease in NADPH oxidase activity [120]. The polarization of macrophages and the role of polyphenols are shown graphically in Figure 3.

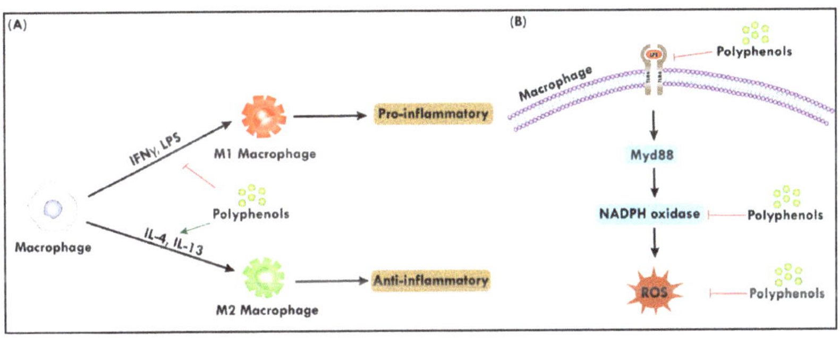

Figure 3. (**A**) The initiation of M1 differentiation is prompted by the activation of toll-like receptors (TLRs) and the stimulation of IFN through bacterial LPS. Conversely, M2 polarization is initiated by IL-4 and IL-13. Polyphenols promote the conversion of macrophages to anti-inflammatory M2 phenotype. (**B**) The generation of ROS is facilitated by the NADPH oxidase family. Polyphenols, particularly flavonols, can reduce NADPH activity, leading to a subsequent reduction in ROS production and inflammation.

In colitis, dietary polyphenols were able to encourage the phenotypic conversion of M1 into M2 macrophages and suppress colitis [121]. It has been shown that punicalagin and ellagic acid from pomegranate peel decreased TLR4 mRNA and protein expression levels in a dose-dependent manner, and inhibited LPS-induced intracellular ROS generation. Further, the inhibition of LPS-induced phosphorylation and the nuclear translocation of p65 were also facilitated by the anti-inflammatory mechanism [122]. In macrophage cell lines J774, it has been shown that pomegranate polyphenols dose-dependently reduced the macrophage response to M1 proinflammatory activation [123]. Accordingly, studies conducted both in vivo and in vitro have shown that one of the primary effects of polyphenols on macrophages is the inhibition of important inflammatory response regulators, with the most consistent effect being the repression of COX-2, inducible nitric oxide synthase (iNOS), and the cytokines TNF-α, IL-1β, and IL-6 [124]. When Chinese propolis was added to mouse RAW 264.7 macrophages challenged with LPS, a dose-dependent reduction in iNOS, IL-1β, and IL-6 mRNA expression was seen, followed by an increase in NO, IL-1β, and IL-6 production [125]. Some studies have described the implication of MAPK pathways for cell signaling implicated in these outcomes. The flavonoid procyanidin C1 was able to suppress the expression of TLR4 and COX-2, as well as phosphorylate p38 and ERK-1/2, and reduce the secretion of TNF-α, IL-1β, and IL-6 in bone-marrow-derived macrophages [126].

4.3. Modifying the Function of Natural Killer Cells by Polyphenols

The use of cytotoxic immune cells in the prevention and treatment of cancer is a growing field of research known as immune-cell-mediated cancer therapy. Natural killer (NK) cells, which also attack other microorganisms and alter body cells, are one of the promising cells for combating cancer. About 10–15% of blood lymphocytes are NK cells, also referred to as cytotoxic lymphocytes of the innate immune system. NK cells detect 'stressed' cells like tumor- or virus-infected cells, and then they eliminate those cells on their accord [127]. The impact of different plant-derived compounds on the ability of NK cells to fight against malignant illnesses has long been the subject of much research. Numerous fruits and vegetables naturally contain flavonoids, which are plentiful phytonutrients, and the subgroup called quercetin significantly affects cytotoxic immune cells [128]. Both endogenous and exogenous immune-modulating substances can either strengthen or weaken immunological responses and inflammation. Many secondary metabolites from plants, such as flavonoids like quercetin, have stimulating effects on NK cells and increase the ability of NK cells to destroy YAC-1 target cells [129]. Further, it has been shown that resveratrol modulates direct and indirect effects on NK cell function, which is consistent with the immune-modulating abilities of nutrition-derived substances [130]. By altering the expression of activating cell surface receptors such as NKG2D on NK cells or by stimulating the production of their corresponding ligands on cancerous cells, resveratrol appears to improve immune responses [131]. When viewed in light of the additional chemical characteristics of this natural substance, this immune modulation is intriguing. In distinct human hepatoblastoma cells, resveratrol was found to have an inhibitory effect on the classical histone deacetylases (class I, II, and IV). It has been shown that HDAC inhibition caused a dose-dependent decrease in the growth of cancer cells [132]. Further, the enhanced expression of various NKG2D ligands made leukemia K562 and gastric cancer SNU1 and SNU-C4 cells more susceptible to NK cell-mediated lysis [133]. Increased tumor resistance to treatment, whether it be chemotherapy, radiation, or any of the targeted therapies, is one of the most significant unsolved issues in cancer therapy. As a general rule, defects in apoptosis have a direct impact on this enhanced resistance of cancer therapy, and the solution to this issue might be autophagy, a different type of cell death. Numerous research studies have shown that polyphenolic substances, including rottlerin, genistein, quercetin, curcumin, and resveratrol, can mediate both canonical and non-canonical autophagy via multiple pathways and may offer novel therapeutic approaches in the treatment of cancer to address the serious issue of drug resistance in cancer therapy [134].

5. Role of Polyphenols in Obesity-Associated Immunomodulation

Obesity, which is characterized by the intricate activation of different inflammatory pathways, is crucial for contributing to insulin resistance, as well as other metabolic dysfunctions. Importantly, AT comprises both immune cells from the innate and adaptive immune systems. Immune cells from the innate and adaptive immune systems crosstalk with adipocytes during obesity-related inflammation. This crosstalk between adipocytes and other immune cells results in a vicious cycle of inflammatory responses, which causes more recruitment of immune cells such as monocytes/macrophages, neutrophils, and T cells, which exacerbates the obesity condition [135]. Therefore, therapeutic agents that can suppress the inflammatory behaviors of different immune cells may have the potential to ameliorate obesity and associated metabolic diseases.

It has been well established that polyphenols possess several pharmacological activities, including anti-inflammatory, antioxidant, and anticancer activities. As obesity-induced inflammation potentially increases the proinflammatory mediators and accumulation of immune cells, polyphenolic compounds might have therapeutic potential for treating obesity-related inflammatory conditions. It has been shown that curcumin suppresses the migration of macrophages in the mesenteric AT and significantly reduces the secretion of monocyte chemoattractant protein-1 (MCP-1) from RAW264.7 macrophages that are treated with the conditioned media from mesenteric AT [136]. In addition, curcumin supplementation has been found to reduce the expression of several pro-inflammatory M1 macrophage markers like CD11c, CD38, and CD80 and decrease adipocyte size compared to HFD-fed obese mice, as well as causing reduced macrophage infiltration in AT [36]. Moreover, curcumin, along with another polyphenolic compound, resveratrol, has been shown to inhibit NF-κB activation in adipocytes [37]. To this end, it has been shown that resveratrol suppresses the phosphorylation of the p65 subunit of NF-κB signaling in adipocyte cell lines, which is also well reported in several inflammatory signaling by canonical NF-κB signaling pathways [137]. Resveratrol also increases the accumulation of Tregs, as well as maintains glucose homeostasis by regulating SIRT1 signaling in HFD-fed mice [38]. Furthermore, capsaicin inhibits the p65 subunit of the NF-κB in adipocytes, as well as the mesenteric AT-conditioned medium and MCP-1-induced macrophage migration. Additionally, capsaicin significantly suppresses the activation of macrophages to produce inflammatory mediators like NO, TNF-α, and MCP-1, which are induced by the obese-mouse mesenteric AT-conditioned medium [138]. The polyphenol butein, which is an active component of *Toxicodendron vernicifluum*, is also reported to inhibit the NF-κB signaling activated by TNF-α in adipocytes, and butein prevents inflammation in the suppression of the NF-κB and MAPK signaling pathway in macrophages. [139]. The upregulation of pro-inflammatory genes, such as iNOS, is associated with the activation of MAPKs in macrophages [140]. 6-gingerol, available from ginger, has been reported to reduce the expression of adipocyte differentiation genes, such as C/EBPα,and PPARγ after HFD feeding. Moreover, animals fed an HFD along with 6-gingerol showed a reduced expression of the macrophage marker F4/80 in their WAT compared to mice fed an HFD alone, indicating a possible immunoregulatory role for 6-gingerol in the WAT [141]. Another polyphenolic compound from ginger, gingerenone A, was also reported to suppress adipocyte differentiation markers and the chemotaxis of macrophages, and, interestingly, promotes M2 macrophages in HFD-fed mice [142].

Further, polyphenols obtained from different plant extracts showed a notable improvement in ameliorating obesity conditions with inflammation. Polyphenol-rich fractions obtained from table grapes have significantly decreased the body weight and inflammatory gene expression in visceral fat in obese mice fed on an HFD [143]. Polyphenol-rich grape powder supplementation increases the glucose tolerance in HFD-fed mice compared to mice fed only with an HFD. Importantly, quercetin 3-O-glucoside, a polyphenol obtained from grape powder, decreased the expression of inflammatory markers like TNF-α, IL-6, and CD11c in the serum and AT of HFD-fed mice and MCP-1 and IL-1β in human primary adipocytes [144]. Moreover, polyphenols from tea suppressed obesity-related fatty liver and inflammatory cytokine expression in HFD-fed dogs [145]. These promising studies suggest that polyphenols have a beneficial effect in the prevention of AT inflammation under obese conditions.

6. Conclusions

In conclusion, polyphenols play a crucial role in immunomodulation and the fight against obesity, and have a lot of potential for enhancing human health. The health benefits of polyphenols present in a wide range of foods and beverages have been the subject of detailed research. The mechanisms by which polyphenols combat obesity are multifaceted, but the neurohormones in the brain that control insulin linked to hunger and fullness are the most studied field. Polyphenols such as resveratrol and curcumin have been demonstrated in animal models to reduce hyperinsulinemia and correct hyperglycemia, reducing inflammation and cancer, suggesting that they may be useful in the treatment of obesity. Furthermore, polyphenols can inhibit the lipogenic pathway and pro-obesity enzymes like pancreatic lipase, which lowers the synthesis and storage of fat. Polyphenols also increase thermogenesis, which increases energy expenditure and aids in weight control and calorie burning. The hormones leptin and adiponectin control the expression of adipocytokines, which are essential for maintaining energy homeostasis and metabolic processes. Polyphenols also influence mucosal immunity and inflammation to mediate the gut immune system and overall health. Polyphenols have been found to impact immune cells, such as T cells, macrophages, and NK cells. Polyphenols promote the production of anti-inflammatory cytokines while inhibiting proinflammatory cytokines, enhancing NK cell function, and making them valuable in preventing inflammation-related diseases. Taken together, these studies highlight the possibility that polyphenols support general health by reducing obesity and altering the immune system and inflammation. However, further studies are required to understand the precise mechanisms of how polyphenols, as well as their derivatives, enhance the immune system, reduce AT inflammation, and manage obesity, by utilizing the potential of some of these and developing tailored therapeutics in humans.

Author Contributions: M.A.A.M. and A.R.: wrote the manuscript, and made all the figures; M.M.: edited the review and helped in making figures; S.K. and B.S.; edited the manuscript; U.P.S.: conceived the idea, designed and edited the review. All authors have read and agreed to the published version of the manuscript.

Funding: This study was supported by grants from NIH (NIAID R01 AI140405) to the U.S., and the Intramural Research Program at UTHSC in Memphis, TN.

Institutional Review Board Statement: Not applicable.

Informed Consent Statement: Not applicable.

Data Availability Statement: No new data were created or analyzed in this study. Data sharing is not applicable in the review article.

Acknowledgments: The authors are thankful to the PubChem database, ChemDraw, and RCSB Chemical Sketch Tool for the chemical structures used in this review article.

Conflicts of Interest: The authors declare that they do not have any competing interests that might influence this study or personal relationships that could have appeared to influence the work reported in this paper.

References

1. Kumar, V.; Singh, D.D.; Lakhawat, S.S.; Yasmeen, N.; Pandey, A.; Singla, R.K. Biogenic phytochemicals modulating obesity: From molecular mechanism to preventive and therapeutic approaches. *Evid.-Based Complement. Altern. Med.* **2022**, *2022*, 6852276. [CrossRef]
2. Mohamed, G.A.; Ibrahim, S.R.; Elkhayat, E.S.; El Dine, R.S. Natural anti-obesity agents. *Bull. Fac. Pharm. Cairo Univ.* **2014**, *52*, 269–284. [CrossRef]
3. Liu, J.; Wang, H.; Zeng, D.; Xiong, J.; Luo, J.; Chen, X.; Chen, T.; Xi, Q.; Sun, J.; Ren, X. The novel importance of miR-143 in obesity regulation. *Int. J. Obes.* **2023**, *47*, 100–108. [CrossRef]
4. González-Castejón, M.; Rodriguez-Casado, A. Dietary phytochemicals and their potential effects on obesity: A review. *Pharmacol. Res.* **2011**, *64*, 438–455. [CrossRef]
5. Vizmanos, B.; Cascales, A.I.; Rodríguez-Martín, M.; Salmerón, D.; Morales, E.; Aragón-Alonso, A.; Scheer, F.A.; Garaulet, M. Lifestyle mediators of associations among siestas, obesity, and metabolic health. *Obesity* **2023**, *31*, 1227–1239. [CrossRef]
6. Bhardwaj, M.; Yadav, P.; Vashishth, D.; Sharma, K.; Kumar, A.; Chahal, J.; Dalal, S.; Kataria, S.K. A review on obesity management through natural compounds and a green nanomedicine-based approach. *Molecules* **2021**, *26*, 3278. [CrossRef] [PubMed]
7. Patel, D. Pharmacotherapy for the management of obesity. *Metabolism* **2015**, *64*, 1376–1385. [CrossRef] [PubMed]
8. Tak, Y.J.; Lee, S.Y. Long-term efficacy and safety of anti-obesity treatment: Where do we stand? *Curr. Obes. Rep.* **2021**, *10*, 14–30. [CrossRef] [PubMed]
9. Kang, J.G.; Park, C.-Y. Anti-obesity drugs: A review about their effects and safety. *Diabetes Metab. J.* **2012**, *36*, 13–25. [CrossRef]
10. Herrera-Martínez, A.D.; Herrero-Aguayo, V.; Pérez-Gómez, J.M.; Gahete, M.D.; Luque, R.M. Inflammasomes: Cause or consequence of obesity-associated comorbidities in humans. *Obesity* **2022**, *30*, 2351–2362. [CrossRef] [PubMed]
11. Zatterale, F.; Longo, M.; Naderi, J.; Raciti, G.A.; Desiderio, A.; Miele, C.; Beguinot, F. Chronic adipose tissue inflammation linking obesity to insulin resistance and type 2 diabetes. *Front. Physiol.* **2020**, *10*, 1607. [CrossRef] [PubMed]
12. Dhurandhar, N.; Bailey, D.; Thomas, D. Interaction of obesity and infections. *Obes. Rev.* **2015**, *16*, 1017–1029. [CrossRef] [PubMed]

13. O'Neill, L.A.; Pearce, E.J. Immunometabolism governs dendritic cell and macrophage function. *J. Exp. Med.* **2016**, *213*, 15–23. [CrossRef] [PubMed]
14. Palsson-McDermott, E.M.; Curtis, A.M.; Goel, G.; Lauterbach, M.A.; Sheedy, F.J.; Gleeson, L.E.; van den Bosch, M.W.; Quinn, S.R.; Domingo-Fernandez, R.; Johnston, D.G. Pyruvate kinase M2 regulates Hif-1α activity and IL-1β induction and is a critical determinant of the warburg effect in LPS-activated macrophages. *Cell Metab.* **2015**, *21*, 65–80. [CrossRef]
15. Ratter, J.M.; Tack, C.J.; Netea, M.G.; Stienstra, R. Environmental signals influencing myeloid cell metabolism and function in diabetes. *Trends Endocrinol. Metab.* **2018**, *29*, 468–480. [CrossRef]
16. Rakib, A.; Mandal, M.; Showkat, A.; Kiran, S.; Mazumdar, S.; Singla, B.; Bajwa, A.; Kumar, S.; Park, F.; Singh, U.P. Piceatannol induces regulatory T cells and modulates the inflammatory response and adipogenesis. *Biomed. Pharmacother.* **2023**, *161*, 114514. [CrossRef]
17. Gilani, A.H. Trends in ethnopharmacology. *J. Ethnopharmacol.* **2005**, *100*, 43–49. [CrossRef]
18. Sun, N.-N.; Wu, T.-Y.; Chau, C.-F. Natural dietary and herbal products in anti-obesity treatment. *Molecules* **2016**, *21*, 1351. [CrossRef]
19. Abbas, M.; Saeed, F.; Anjum, F.M.; Afzaal, M.; Tufail, T.; Bashir, M.S.; Ishtiaq, A.; Hussain, S.; Suleria, H.A.R. Natural polyphenols: An overview. *Int. J. Food Prop.* **2017**, *20*, 1689–1699. [CrossRef]
20. Boccellino, M.; D'Angelo, S. Anti-obesity effects of polyphenol intake: Current status and future possibilities. *Int. J. Mol. Sci.* **2020**, *21*, 5642. [CrossRef] [PubMed]
21. Isganaitis, E.; Lustig, R.H. Fast food, central nervous system insulin resistance, and obesity. *Arterioscler. Thromb. Vasc. Biol.* **2005**, *25*, 2451–2462. [CrossRef]
22. Hosseini, S.A.; Zand, H.; Cheraghpour, M. The influence of curcumin on the downregulation of MYC, insulin and IGF-1 receptors: A possible mechanism underlying the anti-growth and anti-migration in chemoresistant colorectal cancer cells. *Medicina* **2019**, *55*, 90. [CrossRef] [PubMed]
23. Pandey, K.B.; Rizvi, S.I. Plant polyphenols as dietary antioxidants in human health and disease. *Oxidative Med. Cell. Longev.* **2009**, *2*, 270–278. [CrossRef] [PubMed]
24. Tsao, R. Chemistry and biochemistry of dietary polyphenols. *Nutrients* **2010**, *2*, 1231–1246. [CrossRef]
25. Fraga, C.G.; Croft, K.D.; Kennedy, D.O.; Tomás-Barberán, F.A. The effects of polyphenols and other bioactives on human health. *Food Funct.* **2019**, *10*, 514–528. [CrossRef]
26. Sroka, Z.; Cisowski, W. Hydrogen peroxide scavenging, antioxidant and anti-radical activity of some phenolic acids. *Food Chem. Toxicol.* **2003**, *41*, 753–758. [CrossRef]
27. Yang, C.S.; Landau, J.M.; Huang, M.-T.; Newmark, H.L. Inhibition of carcinogenesis by dietary polyphenolic compounds. *Annu. Rev. Nutr.* **2001**, *21*, 381–406. [CrossRef]
28. Van Hung, P. Phenolic compounds of cereals and their antioxidant capacity. *Crit. Rev. Food Sci. Nutr.* **2016**, *56*, 25–35. [CrossRef]
29. Son, M.J.; Rico, C.W.; Nam, S.H.; Kang, M.Y. Influence of oryzanol and ferulic acid on the lipid metabolism and antioxidative status in high fat-fed mice. *J. Clin. Biochem. Nutr.* **2010**, *46*, 150–156. [CrossRef] [PubMed]
30. De Melo, T.; Lima, P.; Carvalho, K.; Fontenele, T.; Solon, F.; Tomé, A.; De Lemos, T.; da Cruz Fonseca, S.; Santos, F.; Rao, V.S. Ferulic acid lowers body weight and visceral fat accumulation via modulation of enzymatic, hormonal and inflammatory changes in a mouse model of high-fat diet-induced obesity. *Braz. J. Med. Biol. Res.* **2017**, *50*, e5630. [CrossRef] [PubMed]
31. Seo, M.-J.; Lee, Y.-J.; Hwang, J.-H.; Kim, K.-J.; Lee, B.-Y. The inhibitory effects of quercetin on obesity and obesity-induced inflammation by regulation of MAPK signaling. *J. Nutr. Biochem.* **2015**, *26*, 1308–1316. [CrossRef]
32. Chen, S.; Jiang, H.; Wu, X.; Fang, J. Therapeutic effects of quercetin on inflammation, obesity, and type 2 diabetes. *Mediat. Inflamm.* **2016**, *2016*, 9340637. [CrossRef]
33. You, Y.; Han, X.; Guo, J.; Guo, Y.; Yin, M.; Liu, G.; Huang, W.; Zhan, J. Cyanidin-3-glucoside attenuates high-fat and high-fructose diet-induced obesity by promoting the thermogenic capacity of brown adipose tissue. *J. Funct. Foods* **2018**, *41*, 62–71. [CrossRef]
34. Yuce, B.; Danis, O.; Ogan, A.; Sener, G.; Bulut, M.; Yarat, A. Antioxidative and lipid lowering effects of 7, 8-dihydroxy-3-(4-methylphenyl) coumarin in hyperlipidemic rats. *Arzneimittelforschung* **2009**, *59*, 129–134. [CrossRef]
35. Um, M.Y.; Moon, M.K.; Ahn, J.; Ha, T.Y. Coumarin attenuates hepatic steatosis by down-regulating lipogenic gene expression in mice fed a high-fat diet. *Br. J. Nutr.* **2013**, *109*, 1590–1597. [CrossRef] [PubMed]
36. Islam, T.; Koboziev, I.; Albracht-Schulte, K.; Mistretta, B.; Scoggin, S.; Yosofvand, M.; Moussa, H.; Zabet-Moghaddam, M.; Ramalingam, L.; Gunaratne, P.H. Curcumin Reduces Adipose Tissue Inflammation and Alters Gut Microbiota in Diet-Induced Obese Male Mice. *Mol. Nutr. Food Res.* **2021**, *65*, 2100274. [CrossRef] [PubMed]
37. Gonzales, A.M.; Orlando, R.A. Curcumin and resveratrol inhibit nuclear factor-kappaB-mediated cytokine expression in adipocytes. *Nutr. Metab.* **2008**, *5*, 17. [CrossRef]
38. Wang, B.; Sun, J.; Li, L.; Zheng, J.; Shi, Y.; Le, G. Regulatory effects of resveratrol on glucose metabolism and T-lymphocyte subsets in the development of high-fat diet-induced obesity in C57BL/6 mice. *Food Funct.* **2014**, *5*, 1452–1463. [CrossRef] [PubMed]
39. Panche, A.N.; Diwan, A.D.; Chandra, S.R. Flavonoids: An overview. *J. Nutr. Sci.* **2016**, *5*, e47. [CrossRef] [PubMed]
40. Al-Ishaq, R.K.; Abotaleb, M.; Kubatka, P.; Kajo, K.; Büsselberg, D. Flavonoids and their anti-diabetic effects: Cellular mechanisms and effects to improve blood sugar levels. *Biomolecules* **2019**, *9*, 430. [CrossRef] [PubMed]
41. Caro-Ordieres, T.; Marín-Royo, G.; Opazo-Ríos, L.; Jiménez-Castilla, L.; Moreno, J.A.; Gómez-Guerrero, C.; Egido, J. The coming age of flavonoids in the treatment of diabetic complications. *J. Clin. Med.* **2020**, *9*, 346. [CrossRef]

42. Rees, A.; Dodd, G.F.; Spencer, J.P. The effects of flavonoids on cardiovascular health: A review of human intervention trials and implications for cerebrovascular function. *Nutrients* **2018**, *10*, 1852. [CrossRef]
43. Salomone, F.; Godos, J.; Zelber-Sagi, S. Natural antioxidants for non-alcoholic fatty liver disease: Molecular targets and clinical perspectives. *Liver Int.* **2016**, *36*, 5–20. [CrossRef]
44. Kawser Hossain, M.; Abdal Dayem, A.; Han, J.; Yin, Y.; Kim, K.; Kumar Saha, S.; Yang, G.-M.; Choi, H.Y.; Cho, S.-G. Molecular mechanisms of the anti-obesity and anti-diabetic properties of flavonoids. *Int. J. Mol. Sci.* **2016**, *17*, 569. [CrossRef]
45. Garcia, D.; Shaw, R.J. AMPK: Mechanisms of cellular energy sensing and restoration of metabolic balance. *Mol. Cell* **2017**, *66*, 789–800. [CrossRef] [PubMed]
46. Tong, Z.; He, W.; Fan, X.; Guo, A. Biological function of plant tannin and its application in animal health. *Front. Vet. Sci.* **2022**, *8*, 803657. [CrossRef] [PubMed]
47. Ajebli, M.; Eddouks, M. The promising role of plant tannins as bioactive antidiabetic agents. *Curr. Med. Chem.* **2019**, *26*, 4852–4884. [CrossRef] [PubMed]
48. Sp, N.; Kang, D.Y.; Kim, D.H.; Yoo, J.-S.; Jo, E.S.; Rugamba, A.; Jang, K.-J.; Yang, Y.M. Tannic acid inhibits non-small cell lung cancer (NSCLC) stemness by inducing G0/G1 cell cycle arrest and intrinsic apoptosis. *Anticancer Res.* **2020**, *40*, 3209–3220. [CrossRef] [PubMed]
49. Buzzini, P.; Arapitsas, P.; Goretti, M.; Branda, E.; Turchetti, B.; Pinelli, P.; Ieri, F.; Romani, A. Antimicrobial and antiviral activity of hydrolysable tannins. *Mini-Rev. Med. Chem.* **2008**, *8*, 1179. [CrossRef] [PubMed]
50. Fukuchi, K.; Sakagami, H.; Okuda, T.; Hatano, T.; Tanuma, S.-i.; Kitajima, K.; Inoue, Y.; Inoue, S.; Ichikawa, S.; Nonoyama, M. Inhibition of herpes simplex virus infection by tannins and related compounds. *Antivir. Res.* **1989**, *11*, 285–297. [CrossRef] [PubMed]
51. Addisu, S. Effect of dietary tannin source feeds on ruminal fermentation and production of cattle; A review. *Online J. Anim. Feed Res.* **2016**, *6*, 45–56.
52. Schiavone, A.; Guo, K.; Tassone, S.; Gasco, L.; Hernandez, E.; Denti, R.; Zoccarato, I. Effects of a natural extract of chestnut wood on digestibility, performance traits, and nitrogen balance of broiler chicks. *Poult. Sci.* **2008**, *87*, 521–527. [CrossRef] [PubMed]
53. Jamroz, D.; Wiliczkiewicz, A.; Skorupińska, J.; Orda, J.; Kuryszko, J.; Tschirch, H. Effect of sweet chestnut tannin (SCT) on the performance, microbial status of intestine and histological characteristics of intestine wall in chickens. *Br. Poult. Sci.* **2009**, *50*, 687–699. [CrossRef]
54. Gao, W.; Li, Q.; Chen, J.; Wang, Z.; Hua, C. Total synthesis of six 3, 4-unsubstituted coumarins. *Molecules* **2013**, *18*, 15613–15623. [CrossRef] [PubMed]
55. Madhavan, G.R.; Balraju, V.; Mallesham, B.; Chakrabarti, R.; Lohray, V.B. Novel coumarin derivatives of heterocyclic compounds as lipid-lowering agents. *Bioorganic Med. Chem. Lett.* **2003**, *13*, 2547–2551. [CrossRef]
56. Shin, E.; Choi, K.-M.; Yoo, H.-S.; Lee, C.-K.; Hwang, B.Y.; Lee, M.K. Inhibitory effects of coumarins from the stem barks of *Fraxinus rhynchophylla* on adipocyte differentiation in 3T3-L1 cells. *Biol. Pharm. Bull.* **2010**, *33*, 1610–1614. [CrossRef] [PubMed]
57. Taira, N.; Nugara, R.N.; Inafuku, M.; Takara, K.; Ogi, T.; Ichiba, T.; Iwasaki, H.; Okabe, T.; Oku, H. In vivo and in vitro anti-obesity activities of dihydropyranocoumarins derivatives from Peucedanum japonicum Thunb. *J. Funct. Foods* **2017**, *29*, 19–28. [CrossRef]
58. Luo, J.; Yu, Z.; Tovar, J.; Nilsson, A.; Xu, B. Critical review on anti-obesity effects of phytochemicals through Wnt/β-catenin signaling pathway. *Pharmacol. Res.* **2022**, *184*, 106461. [CrossRef]
59. Jakab, J.; Miškić, B.; Mikšić, Š.; Juranić, B.; Ćosić, V.; Schwarz, D.; Včev, A. Adipogenesis as a potential anti-obesity target: A review of pharmacological treatment and natural products. *Diabetes Metab. Syndr. Obes.* **2021**, *14*, 67–83. [CrossRef]
60. Shen, Y.; Honma, N.; Kobayashi, K.; Jia, L.N.; Hosono, T.; Shindo, K.; Ariga, T.; Seki, T. Cinnamon extract enhances glucose uptake in 3T3-L1 adipocytes and C2C12 myocytes by inducing LKB1-AMP-activated protein kinase signaling. *PLoS ONE* **2014**, *9*, e87894. [CrossRef]
61. Li, H.; Qi, J.; Li, L. Phytochemicals as potential candidates to combat obesity via adipose non-shivering thermogenesis. *Pharmacol. Res.* **2019**, *147*, 104393. [CrossRef] [PubMed]
62. Sigal, R.J.; El-Hashimy, M.; Martin, B.C.; Soeldner, J.S.; Krolewski, A.S.; Warram, J.H. Acute postchallenge hyperinsulinemia predicts weight gain: A prospective study. *Diabetes* **1997**, *46*, 1025–1029. [CrossRef] [PubMed]
63. Leontieva, O.; Paszkiewicz, G.; Demidenko, Z.; Blagosklonny, M. Resveratrol potentiates rapamycin to prevent hyperinsulinemia and obesity in male mice on high fat diet. *Cell Death Dis.* **2013**, *4*, e472. [CrossRef]
64. Ghadiri Soufi, F.; Arbabi-Aval, E.; Rezaei Kanavi, M.; Ahmadieh, H. Anti-inflammatory properties of resveratrol in the retinas of type 2 diabetic rats. *Clin. Exp. Pharmacol. Physiol.* **2015**, *42*, 63–68. [CrossRef] [PubMed]
65. Lim, K.J.; Bisht, S.; Bar, E.E.; Maitra, A.; Eberhart, C.G. A polymeric nanoparticle formulation of curcumin inhibits growth, clonogenicity and stem-like fraction in malignant brain tumors. *Cancer Biol. Ther.* **2011**, *11*, 464–473. [CrossRef] [PubMed]
66. DePaoli, A.M. 20 years of leptin: Leptin in common obesity and associated disorders of metabolism. *J. Endocrinol.* **2014**, *223*, T71–T81. [CrossRef]
67. Ardid-Ruiz, A.; Ibars, M.; Mena, P.; Del Rio, D.; Muguerza, B.; Bladé, C.; Arola, L.; Aragonès, G.; Suárez, M. Potential involvement of peripheral leptin/STAT3 signaling in the effects of resveratrol and its metabolites on reducing body fat accumulation. *Nutrients* **2018**, *10*, 1757. [CrossRef]
68. Sivamaruthi, B.S.; Kesika, P.; Chaiyasut, C. The influence of supplementation of anthocyanins on obesity-associated comorbidities: A concise review. *Foods* **2020**, *9*, 687. [CrossRef]

69. Badshah, H.; Ullah, I.; Kim, S.E.; Kim, T.-h.; Lee, H.Y.; Kim, M.O. Anthocyanins attenuate body weight gain via modulating neuropeptide Y and GABAB1 receptor in rats hypothalamus. *Neuropeptides* **2013**, *47*, 347–353. [CrossRef]
70. Liu, T.-T.; Liu, X.-T.; Chen, Q.-X.; Shi, Y. Lipase inhibitors for obesity: A review. *Biomed. Pharmacother.* **2020**, *128*, 110314. [CrossRef]
71. Buchholz, T.; Melzig, M.F. Polyphenolic compounds as pancreatic lipase inhibitors. *Planta Medica* **2015**, *81*, 771–783. [CrossRef]
72. Othman, Z.A.; Wan Ghazali, W.S.; Noordin, L.; Mohd. Yusof, N.A.; Mohamed, M. Phenolic compounds and the anti-atherogenic effect of bee bread in high-fat diet-induced obese rats. *Antioxidants* **2019**, *9*, 33. [CrossRef]
73. Kim, N.-H.; Jegal, J.; Kim, Y.N.; Heo, J.-D.; Rho, J.-R.; Yang, M.H.; Jeong, E.J. Chokeberry extract and its active polyphenols suppress adipogenesis in 3T3-L1 adipocytes and modulates fat accumulation and insulin resistance in diet-induced obese mice. *Nutrients* **2018**, *10*, 1734. [CrossRef]
74. Brusselmans, K.; De Schrijver, E.; Heyns, W.; Verhoeven, G.; Swinnen, J.V. Epigallocatechin-3-gallate is a potent natural inhibitor of fatty acid synthase in intact cells and selectively induces apoptosis in prostate cancer cells. *Int. J. Cancer* **2003**, *106*, 856–862. [CrossRef]
75. Pan, R.; Zhu, X.; Maretich, P.; Chen, Y. Combating obesity with thermogenic fat: Current challenges and advancements. *Front. Endocrinol.* **2020**, *11*, 185. [CrossRef]
76. Cannon, B.; Nedergaard, J. Metabolic consequences of the presence or absence of the thermogenic capacity of brown adipose tissue in mice (and probably in humans). *Int. J. Obes.* **2010**, *34*, S7–S16. [CrossRef]
77. Panic, V. Regulation of Glucose Flux and Metabolism in Brown Adipose Tissue. Ph.D. Dissertation, The University of Utah, Salt Lake City, UT, USA, 2020.
78. Shiffman, M.A.; Di Giuseppe, A.; Bassetto, F. *Stem Cells in Aesthetic Procedures: Art, Science, and Clinical Techniques*; Springer: Berlin/Heidelberg, Germany, 2014.
79. Saely, C.H.; Geiger, K.; Drexel, H. Brown versus white adipose tissue: A mini-review. *Gerontology* **2011**, *58*, 15–23. [CrossRef]
80. Mele, L.; Bidault, G.; Mena, P.; Crozier, A.; Brighenti, F.; Vidal-Puig, A.; Del Rio, D. Dietary (Poly) phenols, brown adipose tissue activation, and energy expenditure: A narrative review. *Adv. Nutr.* **2017**, *8*, 694–704. [CrossRef] [PubMed]
81. Hanhineva, K.; Törrönen, R.; Bondia-Pons, I.; Pekkinen, J.; Kolehmainen, M.; Mykkänen, H.; Poutanen, K. Impact of dietary polyphenols on carbohydrate metabolism. *Int. J. Mol. Sci.* **2010**, *11*, 1365–1402. [CrossRef] [PubMed]
82. Mthembu, S.X.; Dludla, P.V.; Ziqubu, K.; Nyambuya, T.M.; Kappo, A.P.; Madoroba, E.; Nyawo, T.A.; Nkambule, B.B.; Silvestri, S.; Muller, C.J. The potential role of polyphenols in modulating mitochondrial bioenergetics within the skeletal muscle: A systematic review of preclinical models. *Molecules* **2021**, *26*, 2791. [CrossRef] [PubMed]
83. Singh, M.; Thrimawithana, T.; Shukla, R.; Adhikari, B. Managing obesity through natural polyphenols: A review. *Future Foods* **2020**, *1*, 100002. [CrossRef]
84. Fang, C.; Kim, H.; Noratto, G.; Sun, Y.; Talcott, S.T.; Mertens-Talcott, S.U. Gallotannin derivatives from mango (*Mangifera indica* L.) suppress adipogenesis and increase thermogenesis in 3T3-L1 adipocytes in part through the AMPK pathway. *J. Funct. Foods* **2018**, *46*, 101–109. [CrossRef]
85. Kitamura, H. Effects of propolis extract and propolis-derived compounds on obesity and diabetes: Knowledge from cellular and animal models. *Molecules* **2019**, *24*, 4394. [CrossRef] [PubMed]
86. Washio, K.; Shimamoto, Y.; Kitamura, H. Brazilian propolis extract increases leptin expression in mouse adipocytes. *Biomed. Res.* **2015**, *36*, 343–346. [CrossRef] [PubMed]
87. Berg, A.H.; Combs, T.P.; Scherer, P.E. ACRP30/adiponectin: An adipokine regulating glucose and lipid metabolism. *Trends Endocrinol. Metab.* **2002**, *13*, 84–89. [CrossRef] [PubMed]
88. De Rosa, A.; Monaco, M.L.; Capasso, M.; Forestieri, P.; Pilone, V.; Nardelli, C.; Buono, P.; Daniele, A. Adiponectin oligomers as potential indicators of adipose tissue improvement in obese subjects. *Eur. J. Endocrinol.* **2013**, *169*, 37–43. [CrossRef] [PubMed]
89. Nigro, E.; Scudiero, O.; Monaco, M.L.; Palmieri, A.; Mazzarella, G.; Costagliola, C.; Bianco, A.; Daniele, A. New insight into adiponectin role in obesity and obesity-related diseases. *BioMed Res. Int.* **2014**, *2014*, 658913. [CrossRef]
90. Kim, J.; Yang, G.; Kim, Y.; Kim, J.; Ha, J. AMPK activators: Mechanisms of action and physiological activities. *Exp. Mol. Med.* **2016**, *48*, e224. [CrossRef]
91. Gledhill, J.R.; Montgomery, M.G.; Leslie, A.G.; Walker, J.E. Mechanism of inhibition of bovine F1-ATPase by resveratrol and related polyphenols. *Proc. Natl. Acad. Sci. USA* **2007**, *104*, 13632–13637. [CrossRef]
92. Zheng, J.; Ramirez, V.D. Inhibition of mitochondrial proton F0F1-ATPase/ATP synthase by polyphenolic phytochemicals. *Br. J. Pharmacol.* **2000**, *130*, 1115–1123. [CrossRef]
93. Hawley, S.A.; Ross, F.A.; Chevtzoff, C.; Green, K.A.; Evans, A.; Fogarty, S.; Towler, M.C.; Brown, L.J.; Ogunbayo, O.A.; Evans, A.M. Use of cells expressing γ subunit variants to identify diverse mechanisms of AMPK activation. *Cell Metab.* **2010**, *11*, 554–565. [CrossRef] [PubMed]
94. Timmers, S.; Konings, E.; Bilet, L.; Houtkooper, R.H.; van de Weijer, T.; Goossens, G.H.; Hoeks, J.; van der Krieken, S.; Ryu, D.; Kersten, S. Calorie restriction-like effects of 30 days of resveratrol supplementation on energy metabolism and metabolic profile in obese humans. *Cell Metab.* **2011**, *14*, 612–622. [CrossRef] [PubMed]
95. Jimenez-Gomez, Y.; Mattison, J.A.; Pearson, K.J.; Martin-Montalvo, A.; Palacios, H.H.; Sossong, A.M.; Ward, T.M.; Younts, C.M.; Lewis, K.; Allard, J.S. Resveratrol improves adipose insulin signaling and reduces the inflammatory response in adipose tissue of rhesus monkeys on high-fat, high-sugar diet. *Cell Metab.* **2013**, *18*, 533–545. [CrossRef] [PubMed]

96. Wolfram, S.; Raederstorff, D.; Preller, M.; Wang, Y.; Teixeira, S.R.; Riegger, C.; Weber, P. Epigallocatechin gallate supplementation alleviates diabetes in rodents. *J. Nutr.* **2006**, *136*, 2512–2518. [CrossRef] [PubMed]
97. Incalza, M.A.; D'Oria, R.; Natalicchio, A.; Perrini, S.; Laviola, L.; Giorgino, F. Oxidative stress and reactive oxygen species in endothelial dysfunction associated with cardiovascular and metabolic diseases. *Vasc. Pharmacol.* **2018**, *100*, 1–19. [CrossRef]
98. Egert, S.; Bosy-Westphal, A.; Seiberl, J.; Kürbitz, C.; Settler, U.; Plachta-Danielzik, S.; Wagner, A.E.; Frank, J.; Schrezenmeir, J.; Rimbach, G. Quercetin reduces systolic blood pressure and plasma oxidised low-density lipoprotein concentrations in overweight subjects with a high-cardiovascular disease risk phenotype: A double-blinded, placebo-controlled cross-over study. *Br. J. Nutr.* **2009**, *102*, 1065–1074. [CrossRef]
99. Ji, X.; Shi, S.; Liu, B.; Shan, M.; Tang, D.; Zhang, W.; Zhang, Y.; Zhang, L.; Zhang, H.; Lu, C. Bioactive compounds from herbal medicines to manage dyslipidemia. *Biomed. Pharmacother.* **2019**, *118*, 109338. [CrossRef]
100. Sahebkar, A. Are curcuminoids effective C-reactive protein-lowering agents in clinical practice? Evidence from a meta-analysis. *Phytother. Res.* **2014**, *28*, 633–642. [CrossRef]
101. Xu, X.-R.; Liu, C.-Q.; Feng, B.-S.; Liu, Z.-J. Dysregulation of mucosal immune response in pathogenesis of inflammatory bowel disease. *World J. Gastroenterol. WJG* **2014**, *20*, 3255. [CrossRef]
102. Banerjee, N.; Kim, H.; Talcott, S.T.; Turner, N.D.; Byrne, D.H.; Mertens-Talcott, S.U. Plum polyphenols inhibit colorectal aberrant crypt foci formation in rats: Potential role of the miR-143/protein kinase B/mammalian target of rapamycin axis. *Nutr. Res.* **2016**, *36*, 1105–1113. [CrossRef]
103. Maheshwari, S.; Kumar, V.; Bhadauria, G.; Mishra, A. Immunomodulatory potential of phytochemicals and other bioactive compounds of fruits: A review. *Food Front.* **2022**, *3*, 221–238. [CrossRef]
104. Lee, H.C.; Jenner, A.M.; Low, C.S.; Lee, Y.K. Effect of tea phenolics and their aromatic fecal bacterial metabolites on intestinal microbiota. *Res. Microbiol.* **2006**, *157*, 876–884. [CrossRef]
105. Queipo-Ortuño, M.I.; Boto-Ordóñez, M.; Murri, M.; Gomez-Zumaquero, J.M.; Clemente-Postigo, M.; Estruch, R.; Cardona Diaz, F.; Andres-Lacueva, C.; Tinahones, F.J. Influence of red wine polyphenols and ethanol on the gut microbiota ecology and biochemical biomarkers. *Am. J. Clin. Nutr.* **2012**, *95*, 1323–1334. [CrossRef] [PubMed]
106. Gaudino, S.J.; Kumar, P. Cross-talk between antigen presenting cells and T cells impacts intestinal homeostasis, bacterial infections, and tumorigenesis. *Front. Immunol.* **2019**, *10*, 360. [CrossRef] [PubMed]
107. Haverson, K.; Rehakova, Z.; Sinkora, J.; Sver, L.; Bailey, M. Immune development in jejunal mucosa after colonization with selected commensal gut bacteria: A study in germ-free pigs. *Vet. Immunol. Immunopathol.* **2007**, *119*, 243–253. [CrossRef] [PubMed]
108. Atarashi, K.; Nishimura, J.; Shima, T.; Umesaki, Y.; Yamamoto, M.; Onoue, M.; Yagita, H.; Ishii, N.; Evans, R.; Honda, K. ATP drives lamina propria TH17 cell differentiation. *Nature* **2008**, *455*, 808–812. [CrossRef] [PubMed]
109. Crouvezier, S.; Powell, B.; Keir, D.; Yaqoob, P. The effects of phenolic components of tea on the production of pro-and anti-inflammatory cytokines by human leukocytes in vitro. *Cytokine* **2001**, *13*, 280–286. [CrossRef] [PubMed]
110. Liu, G.; Yu, L.; Fang, J.; Hu, C.-A.A.; Yin, J.; Ni, H.; Ren, W.; Duraipandiyan, V.; Chen, S.; Al-Dhabi, N.A. Methionine restriction on oxidative stress and immune response in dss-induced colitis mice. *Oncotarget* **2017**, *8*, 44511. [CrossRef] [PubMed]
111. Loke, W.M.; Hodgson, J.M.; Proudfoot, J.M.; McKinley, A.J.; Puddey, I.B.; Croft, K.D. Pure dietary flavonoids quercetin and (−)-epicatechin augment nitric oxide products and reduce endothelin-1 acutely in healthy men. *Am. J. Clin. Nutr.* **2008**, *88*, 1018–1025. [CrossRef] [PubMed]
112. Urpi-Sarda, M.; Monagas, M.; Khan, N.; Lamuela-Raventos, R.M.; Santos-Buelga, C.; Sacanella, E.; Castell, M.; Permanyer, J.; Andres-Lacueva, C. Epicatechin, procyanidins, and phenolic microbial metabolites after cocoa intake in humans and rats. *Anal. Bioanal. Chem.* **2009**, *394*, 1545–1556. [CrossRef]
113. Pae, M.; Meydani, S.N.; Wu, D. The role of nutrition in enhancing immunity in aging. *Aging Dis.* **2012**, *3*, 91.
114. Gairola, K.; Gururani, S.; Dubey, S.K. Polyphenols and its effect on the immune system. In *Nutraceuticals and Functional Foods in Immunomodulators*; Springer: Singapore, 2023; pp. 121–140.
115. Hachimura, S.; Totsuka, M.; Hosono, A. Immunomodulation by food: Impact on gut immunity and immune cell function. *Biosci. Biotechnol. Biochem.* **2018**, *82*, 584–599. [CrossRef] [PubMed]
116. Dugo, L.; Belluomo, M.G.; Fanali, C.; Russo, M.; Cacciola, F.; Maccarrone, M.; Sardanelli, A.M. Effect of cocoa polyphenolic extract on macrophage polarization from proinflammatory M1 to anti-inflammatory M2 state. *Oxidative Med. Cell. Longev.* **2017**, *2017*, 6293740. [CrossRef]
117. Song, Y.; Jung, Y.S.; Park, S.; Park, H.S.; Lee, S.J.; Maeng, S.; Kim, H.; Kim, D.O.; Park, K.W.; Kang, H. Anti-Inflammatory Effects and Macrophage Activation Induced by Bioavailable Cinnamon Polyphenols in Mice. *Mol. Nutr. Food Res.* **2023**, *67*, 2200768. [CrossRef]
118. Ben Lagha, A.; Azelmat, J.; Vaillancourt, K.; Grenier, D. A polyphenolic cinnamon fraction exhibits anti-inflammatory properties in a monocyte/macrophage model. *PLoS ONE* **2021**, *16*, e0244805. [CrossRef]
119. Ushio-Fukai, M.; Nakamura, Y. Reactive oxygen species and angiogenesis: NADPH oxidase as target for cancer therapy. *Cancer Lett.* **2008**, *266*, 37–52. [CrossRef]
120. Kim, J.M.; Lee, E.K.; Kim, D.H.; Yu, B.P.; Chung, H.Y. Kaempferol modulates pro-inflammatory NF-κB activation by suppressing advanced glycation endproducts-induced NADPH oxidase. *Age* **2010**, *32*, 197–208. [CrossRef]

121. Mohammadi, A.; Blesso, C.N.; Barreto, G.E.; Banach, M.; Majeed, M.; Sahebkar, A. Macrophage plasticity, polarization and function in response to curcumin, a diet-derived polyphenol, as an immunomodulatory agent. *J. Nutr. Biochem.* **2019**, *66*, 1–16. [CrossRef] [PubMed]
122. Du, L.; Li, J.; Zhang, X.; Wang, L.; Zhang, W.; Yang, M.; Hou, C. Pomegranate peel polyphenols inhibits inflammation in LPS-induced RAW264. 7 macrophages via the suppression of TLR4/NF-κB pathway activation. *Food Nutr. Res.* **2019**, *63*, 3392. [CrossRef] [PubMed]
123. Aharoni, S.; Lati, Y.; Aviram, M.; Fuhrman, B. Pomegranate juice polyphenols induce a phenotypic switch in macrophage polarization favoring a M 2 anti-inflammatory state. *BioFactors* **2015**, *41*, 44–51. [CrossRef] [PubMed]
124. González, R.; Ballester, I.; López-Posadas, R.; Suárez, M.; Zarzuelo, A.; Martínez-Augustin, O.; Medina, F.S.D. Effects of flavonoids and other polyphenols on inflammation. *Crit. Rev. Food Sci. Nutr.* **2011**, *51*, 331–362. [CrossRef] [PubMed]
125. Wang, K.; Ping, S.; Huang, S.; Hu, L.; Xuan, H.; Zhang, C.; Hu, F. Molecular mechanisms underlying the in vitro anti-inflammatory effects of a flavonoid-rich ethanol extract from Chinese propolis (poplar type). *Evid.-Based Complement. Altern. Med.* **2013**, *2013*, 127672.
126. Byun, E.-B.; Sung, N.-Y.; Byun, E.-H.; Song, D.-S.; Kim, J.-K.; Park, J.-H.; Song, B.-S.; Park, S.-H.; Lee, J.-W.; Byun, M.-W. The procyanidin trimer C1 inhibits LPS-induced MAPK and NF-κB signaling through TLR4 in macrophages. *Int. Immunopharmacol.* **2013**, *15*, 450–456. [CrossRef] [PubMed]
127. Vivier, E.; Tomasello, E.; Baratin, M.; Walzer, T.; Ugolini, S. Functions of natural killer cells. *Nat. Immunol.* **2008**, *9*, 503–510. [CrossRef]
128. Lu, H.-Y.; Peng, T.-S.; Hu, X.-D.; Li, S.-J.; Luo, M.; He, Y.-H.; Nie, T. Quercetin potentiates the effect of γδ T cells via modulating the expressions of Granzyme B, perforin and IFN-γ and also regulates the Wnt/β-catenin signalling pathway in human colon cancer cells. *Bangladesh J. Pharmacol.* **2015**, *10*, 251–259. [CrossRef]
129. Yu, C.S.; Lai, K.C.; Yang, J.S.; Chiang, J.H.; Lu, C.C.; Wu, C.L.; Lin, J.P.; Liao, C.L.; Tang, N.Y.; Wood, W.G. Quercetin inhibited murine leukemia WEHI-3 cells in vivo and promoted immune response. *Phytother. Res.* **2010**, *24*, 163–168. [CrossRef]
130. Lee, Y.; Shin, H.; Kim, J. In vivo anti-cancer effects of resveratrol mediated by NK cell activation. *J. Innate Immun.* **2021**, *13*, 94–106. [CrossRef]
131. Mu, Q.; Najafi, M. Resveratrol for targeting the tumor microenvironment and its interactions with cancer cells. *Int. Immunopharmacol.* **2021**, *98*, 107895. [CrossRef]
132. Venturelli, S.; Berger, A.; Böcker, A.; Busch, C.; Weiland, T.; Noor, S.; Leischner, C.; Schleicher, S.; Mayer, M.; Weiss, T.S. Resveratrol as a pan-HDAC inhibitor alters the acetylation status of jistone proteins in human-derived hepatoblastoma cells. *PLoS ONE* **2013**, *8*, e73097. [CrossRef]
133. Bae, J.-H.; Kim, J.-Y.; Kim, M.-J.; Chang, S.-H.; Park, Y.-S.; Son, C.-H.; Park, S.-J.; Chung, J.-S.; Lee, E.-Y.; Kim, S.-H. Quercetin enhances susceptibility to NK cell-mediated lysis of tumor cells through induction of NKG2D ligands and suppression of HSP70. *J. Immunother.* **2010**, *33*, 391–401. [CrossRef] [PubMed]
134. Hasima, N.; Ozpolat, B. Regulation of autophagy by polyphenolic compounds as a potential therapeutic strategy for cancer. *Cell Death Dis.* **2014**, *5*, e1509. [CrossRef]
135. Kiran, S.; Kumar, V.; Murphy, E.A.; Enos, R.T.; Singh, U.P. High fat diet-induced CD8+ T cells in adipose tissue mediate macrophages to sustain low-grade chronic inflammation. *Front. Immunol.* **2021**, *12*, 680944. [CrossRef]
136. Woo, H.-M.; Kang, J.-H.; Kawada, T.; Yoo, H.; Sung, M.-K.; Yu, R. Active spice-derived components can inhibit inflammatory responses of adipose tissue in obesity by suppressing inflammatory actions of macrophages and release of monocyte chemoattractant protein-1 from adipocytes. *Life Sci.* **2007**, *80*, 926–931. [CrossRef]
137. Kang, L.; Heng, W.; Yuan, A.; Baolin, L.; Fang, H. Resveratrol modulates adipokine expression and improves insulin sensitivity in adipocytes: Relative to inhibition of inflammatory responses. *Biochimie* **2010**, *92*, 789–796. [CrossRef] [PubMed]
138. Kang, J.-H.; Kim, C.-S.; Han, I.-S.; Kawada, T.; Yu, R. Capsaicin, a spicy component of hot peppers, modulates adipokine gene expression and protein release from obese-mouse adipose tissues and isolated adipocytes, and suppresses the inflammatory responses of adipose tissue macrophages. *FEBS Lett.* **2007**, *581*, 4389–4396. [CrossRef] [PubMed]
139. Wang, Z.; Lee, Y.; Eun, J.S.; Bae, E.J. Inhibition of adipocyte inflammation and macrophage chemotaxis by butein. *Eur. J. Pharmacol.* **2014**, *738*, 40–48. [CrossRef] [PubMed]
140. Chan, E.D.; Riches, D.W. IFN-γ+ LPS induction of iNOS is modulated by ERK, JNK/SAPK, and p38 mapk in a mouse macrophage cell line. *Am. J. Physiol.-Cell Physiol.* **2001**, *280*, C441–C450. [CrossRef] [PubMed]
141. Hong, K.H.; Um, M.Y.; Ahn, J.; Ha, T.Y. 6-Gingerol ameliorates adiposity and inflammation in adipose tissue in high fat diet-induced obese mice: Association with regulating of adipokines. *Nutrients* **2023**, *15*, 3457. [CrossRef] [PubMed]
142. Suk, S.; Kwon, G.T.; Lee, E.; Jang, W.J.; Yang, H.; Kim, J.H.; Thimmegowda, N.; Chung, M.Y.; Kwon, J.Y.; Yang, S. Gingerenone A, a polyphenol present in ginger, suppresses obesity and adipose tissue inflammation in high-fat diet-fed mice. *Mol. Nutr. Food Res.* **2017**, *61*, 1700139. [CrossRef] [PubMed]
143. Collins, B.; Hoffman, J.; Martinez, K.; Grace, M.; Lila, M.A.; Cockrell, C.; Nadimpalli, A.; Chang, E.; Chuang, C.-C.; Zhong, W. A polyphenol-rich fraction obtained from table grapes decreases adiposity, insulin resistance and markers of inflammation and impacts gut microbiota in high-fat-fed mice. *J. Nutr. Biochem.* **2016**, *31*, 150–165. [CrossRef] [PubMed]

144. Chuang, C.-C.; Shen, W.; Chen, H.; Xie, G.; Jia, W.; Chung, S.; McIntosh, M.K. Differential effects of grape powder and its extract on glucose tolerance and chronic inflammation in high-fat-fed obese mice. *J. Agric. Food Chem.* **2012**, *60*, 12458–12468. [CrossRef] [PubMed]
145. Rahman, S.U.; Huang, Y.; Zhu, L.; Chu, X.; Junejo, S.A.; Zhang, Y.; Khan, I.M.; Li, Y.; Feng, S.; Wu, J. Tea polyphenols attenuate liver inflammation by modulating obesity-related genes and down-regulating COX-2 and iNOS expression in high fat-fed dogs. *BMC Vet. Res.* **2020**, *16*, 234. [CrossRef] [PubMed]

biomolecules

Review

Trace Amine-Associated Receptors and Monoamine-Mediated Regulation of Insulin Secretion in Pancreatic Islets

Anastasia N. Vaganova [1,2], Taisiia S. Shemyakova [1], Karina V. Lenskaia [3], Roman N. Rodionov [4], Charlotte Steenblock [4] and Raul R. Gainetdinov [1,2,*]

1 Institute of Translational Biomedicine, St. Petersburg State University, 199034 St. Petersburg, Russia; a.n.vaganova@spbu.ru (A.N.V.); st035112@student.spbu.ru (T.S.S.)
2 St. Petersburg State University Hospital, St. Petersburg State University, 199034 St. Petersburg, Russia
3 Department of Medicine, St. Petersburg State University, Universitetskaya nab. 7/9, 199034 St. Petersburg, Russia; k.lenskaya@spbu.ru
4 Department of Internal Medicine III, University Hospital Carl Gustav Carus, Technische Universität Dresden, 01307 Dresden, Germany; roman.rodionov@ukdd.de (R.N.R.); charlotte.steenblock@ukdd.de (C.S.)
* Correspondence: r.gainetdinov@spbu.ru

Abstract: Currently, metabolic syndrome treatment includes predominantly pharmacological symptom relief and complex lifestyle changes. Trace amines and their receptor systems modulate signaling pathways of dopamine, norepinephrine, and serotonin, which are involved in the pathogenesis of this disorder. Trace amine-associated receptor 1 (TAAR1) is expressed in endocrine organs, and it was revealed that TAAR1 may regulate insulin secretion in pancreatic islet β-cells. For instance, accumulating data demonstrate the positive effect of TAAR1 agonists on the dynamics of metabolic syndrome progression and MetS-associated disease development. The role of other TAARs (TAAR2, TAAR5, TAAR6, TAAR8, and TAAR9) in the islet's function is much less studied. In this review, we summarize the evidence of TAARs' contribution to the metabolic syndrome pathogenesis and regulation of insulin secretion in pancreatic islets. Additionally, by the analysis of public transcriptomic data, we demonstrate that TAAR1 and other TAAR receptors are expressed in the pancreatic islets. We also explore associations between the expression of TAARs mRNA and other genes in studied samples and demonstrate the deregulation of TAARs' functional associations in patients with metabolic diseases compared to healthy donors.

Keywords: trace amines; trace amine-associated receptors; TAAR; dopamine; pancreatic islets; insulin; transcriptomic; GPCR

1. Introduction

Trace amines (TA) are a group of biogenic amines that were identified in mammalian brains in extremely low concentrations and were long considered neuromodulators of classical monoamine signaling [1,2]. The receptors, known as trace amine-associated receptors (TAARs), recognize both these and some other amine compounds, belong to the family of G protein-coupled receptors, and were discovered only in 2001 [3,4]. Humans possess six functional TAAR types, including TAAR1, TAAR2, TAAR5, TAAR6, TAAR8, and TAAR9 [1,2]. TAAR1 and TAAR2 sense primary amines and TAAR5–9 is tuned toward diamines, polyamines, or tertiary amines [1]. These receptors were identified at first in the olfactory epithelium, except TAAR1, which was detected primarily in the brain. Later, all TAARs were also found in the brain structures and outside the nervous system, in the immune cells, gastrointestinal tract, thyroid, and other organs and tissues [1,2,5], and their involvement in regulating immune response and metabolism is the subject of current research.

The prevalence of cardiovascular diseases (CVD), obesity, and type 2 diabetes mellitus (T2DM) is reaching epidemic proportions worldwide. One of the most significant reasons for this phenomenon is the Western diet, i.e., increased calorie intake with a predominance of saturated and trans fatty acids and insufficient physical activity. Such a kind of lifestyle leads to visceral fat accumulation and insulin resistance, which are the primary pathophysiological signs of metabolic syndrome (MetS) [6].

The term "metabolic syndrome" was introduced in 1998 by the World Health Organization to determine the condition precedent of T2DM, obesity, and related CVD. Besides excessive weight and insulin resistance, dyslipidemia, hypertension, hyperglycemia, and microalbuminuria are also associated with this state. These disorders occur together and seem to have common roots, which allows us to consider them as a syndrome [7].

MetS diagnosis is essential for public health and disease prevention. It is believed that this complex condition increases the risk of CVD, T2DM, non-alcoholic fatty liver disease, and cancer by 2–5 times compared to its individual components [8]. However, the mechanisms of MetS are not yet understood, and the search for a therapy for MetS is continuing. Currently, MetS therapy comprises lifestyle changes and pharmacological treatment of its components. Diet, physical exercise, and rejection of unhealthy habits are effective in combating MetS [9]. Metformin, statins, angiotensin-converting enzyme (ACE) inhibitors, and angiotensin receptor blockers may be prescribed to improve one or more components of MetS [10].

Dopamine (DA) is well studied as a neurotransmitter involved in movement and cognitive function regulation. It is also suggested to regulate insulin sensitivity, and lipid and carbohydrate metabolism. Recent studies have revealed the implication of DA in the pathogenesis of MetS [11–13]. A promising way to regulate DA signaling is its indirect modulation by TAAR1, which acts as an endogenous rheostat of DA signaling [14] and is implicated in immunity and energy homeostasis regulatory pathways [15].

TAAR1 interactions with norepinephrine (NE) [16] and serotonin (5-HT) [17] signaling were also identified. The contribution of these signaling systems in metabolic syndrome progression was demonstrated [18–21]. Pancreatic islet dysfunction is one of the metabolic syndrome development drivers [22]. Currently, the impact of DA [23,24], NE [25,26], and 5-HT [27,28] signaling on pancreatic islet function is confirmed. This review synthesizes the current evidence for the involvement of DA, NE, 5-HT, and trace amine (TA) signaling systems in regulating pancreatic islet function. Assuming that TAAR1 is the prospective therapeutic target for diseases associated with monoamine signaling disorders, we believe that TA signaling pathways should be kept in mind when developing new pharmaceutical approaches to managing metabolic syndrome.

2. Metabolic Syndrome and Insulin

The prevalence of MetS in the world is growing every year, varying from 9.8% to 54% depending on the criterion used [29]. However, despite the obvious social significance of MetS, it is still difficult to diagnose and treat because of the lack of a unified definition of this state. The most recognized MetS criteria are obesity, dyslipidemia, hypertension, microalbuminuria, and, especially, insulin resistance [7].

It is believed that insulin resistance is the central mechanism of the onset of MetS and related disorders, and obesity is the main manifestation of these conditions [30]. Insulin resistance is associated with a decrease in muscle and adipose cell sensitivity to insulin, while its production by the β-cells of the pancreas does not change or increase. Hyperinsulinemia is considered a compensatory attempt to overcome insulin resistance and maintain normal glucose transport into cells. However, it is contributing to metabolic, hemodynamic, and organ disorders such as T2DM and CVD, and the impairment of insulin-sensitive tissues [31].

The effects of insulin on the organs and tissues are multifaceted, which explains the heterogeneity of the MetS manifestations. The key function of insulin is the promotion of intracellular glucose accumulation by increasing cell membrane permeability, gluconeogenesis and glycogenolysis suppression, glycogenesis activation, and enhancing the synthesis of fats and proteins [32,33]. Insulin is recognized by the insulin receptor (IR), which is expressed in various organs, including the liver, heart, lung, placenta, muscle, spleen, kidney, adipose tissue, and central nervous system (CNS) [34]. IR tyrosine kinase domain autophosphorylation leads to activation of the Akt-cascade through phosphoinositol 3-kinase (PI3K) phosphorylation, which results in downstream insulin effects like decrease in gluconeogenesis and increases in glycogenesis, lipolysis suppression, and uptake of circulating glucose into cells [35]. The interaction of insulin with its receptor in the brain can activate the ERK signaling pathway [36]. Insulin together with aldosterone participates in regulating vascular stiffness, and controlling the delivery of nutrients so that insulin sensitivity failure could reduce the bioavailability of nitric oxide, and initiation of atherogenic processes like the decrease of blood flow rate and glucose removal [37].

In MetS, the transport of insulin through the blood–brain barrier is disrupted, and the level of insulin in the brain decreases at the same time as the decline of its ability to reg.ulate peripheral metabolism [38]. Because of insulin resistance, glycogen synthesis and glucose transport decrease, and lipids, free fatty acids, atherogenic triglycerides, and very low-density lipoproteins accumulate, increasing the lipolysis rate [39].

Thus, insulin resistance may be considered the most crucial prerequisite for MetS, and the prospective therapy of this condition should be targeted first at this disorder.

3. G Protein-Coupled Receptors and Insulin Secretion Regulation

The β-cells seem to be electrically excitable and depolarize in response to glucose and generate action potentials [40]. Commonly, β-cells uptake glucose, and when its concentration grows, the activation of glycolysis increases ATP, which mediates the inhibition of ATP-sensitive K+ channels, the opening of voltage-dependent Ca^{2+} channels, and increases cytosolic Ca^{2+} concentration with consequent insulin release [41,42]. This effect is known as glucose-stimulated insulin secretion (GSIS). Likewise, insulin secretion depends on the signals of multiple ligand–receptor systems integrated by β-cells [43]. Thus, β-cells may be classified into high or low glucose responders. The proportion of high responders depends on physiological conditions and, for example, increases during pregnancy [44].

G protein-coupled receptors (GPCRs) bind extracellular peptides or proteins, small organic molecules like free fatty acids or amino acids and their derivatives, monatomic ions, and large biological macromolecules and transmit signals to intracellular guanine nucleotide-binding proteins (G proteins). Human islets express at least 293 GPCRs, and 99 of them are the targets of either approved drugs or drugs undergoing clinical trials [41]. These receptors are critical players in GSIS paracrine regulation in β-cells by binding the glucagon, glucagon-like peptide 1, and acetylcholine derived from the α-cells which potentiate GSIS, or somatostatin (SST) originating from δ-cells and inhibiting insulin secretion [45]. Insulin, islet amyloid polypeptide (IAPP), zinc ions, and other signaling molecules are involved in GPCR-mediated autocrine regulation of GSIS in β-cells [46]. There is a plethora of other GPCR ligands that regulate islet hormone secretion like oxytocin, neurotensin, bradykinin, secretin, ATP, orexins, bile acids, fructose, fatty acids, amino acids, melanin-concentrating hormones, corticotropin-releasing hormones, and histamine [26,41,45,47,48].

The repertoire of GPCRs on pancreatic cells varies and depends on individual peculiarities and circumstances. For example, it has been demonstrated that the alpha-2C adrenergic receptor (ADRA2C) expression in mouse β-cells regularly varies depending on the circadian rhythm [49].

Pancreatic islets comprise functionally different cell types, mainly β-cells (54% of islet cells), α-cells (35%), and δ-cells (11%). The γ-cells, polypeptides (PP) cells, and ε-cells are minor cell populations in the pancreatic islets [25]. The cytoarchitecture of human islets differs somewhat from mouse islets. β-cells, α-cells, and δ-cells are scattered throughout the human islet and aligned along blood vessels. In contrast, in rodents, δ-cells are localized peripherally and β-cells are clustered together and demonstrate synchronized oscillation of membranous potential. The proportion of α-cells is significantly lower in mice compared to human islets, and their physiological impact is weaker. Together with the discrepancy in the GPCR repertoire, these differences complicate the translation of model animal studies in clinical practice [50].

Even though islets account for only 1–2% of the pancreas, they receive nearly 10% of the blood supply in this organ. The vascularization allows islets to rapidly sense the changes in nutritional or hormonal status and coordinate the insulin and glucagon secretion in the circulation. Hormone production also depends on complex autocrine and paracrine regulation, which involves not only insulin and glucagon but also δ-cell-derived SST, ε-cell-derived ghrelin, and PP synthesized and secreted in PP cells [51]. Serotonin, urocortin 3, GABA, and zinc modulate regulatory molecule secretion in α- and δ-cells [26]. All islet cell populations are thought to contribute to the function of islets through integrating systemic signals, and disturbances in these interactions could cause the development of T2DM [25].

The elucidation of monoamine-binding GPCRs' effect on GSIS and glucose homeostasis regulation will be described below.

4. DA-Mediated Regulation of Insulin Secretion

4.1. DA Modulation of GSIS in Islets

Ingestion of a standard meal increases plasma levels of DA and its precursor L-DOPA [43]. DA infusion stimulates glucagon release and downregulates insulin secretion in healthy humans [52]. DA is found both in the endocrine and exocrine parts of the pancreas [53]. The pancreas is the main source of dopamine, which is secreted with duodenal juice and protects duodenal epithelium from damage by digestive enzymes and other harmful agents [54,55]. Recently, it was demonstrated that pancreatic acinar cells can synthesize L-DOPA, but not DA; thus, DA likely originates from nerves or endocrine cells [56].

Paracrine DA coordinates the secretory activity of the major cell groups in pancreatic islets, i.e., α-, β- and δ-cells [25]. In islets, both α- and β-cells express the catecholamine biosynthetic and signaling machinery. These cells express the components of the DA synthesis pathway, including tyrosine hydroxylase, which is the rate-limiting enzyme of this process, and aromatic amino acid decarboxylase (AADC). DA degradation enzymes monoamine oxidases MAO-A and MAO-B, catechol-O-methyltransferase (COMT), vesicular monoamine transporter, VMAT2, and the DA precursor L-DOPA were also found in islet cells [25,54,57].

In the β-cells, DA is packaged in the insulin granules and is released simultaneously with insulin secretion [25,42,43]. Notably, it is not only a regulatory signaling compound co-secreted with insulin. For example, insulin granules contain high concentrations of Ca^{2+}, which act at neighboring β-cells to amplify insulin output via the activation of a calcium-sensing GPCR named CASR [41]. The main sources of DA in islets are summarized in Figure 1a.

The DA effect on GSIS is context dependent. DA dose-dependently inhibits insulin secretion in isolated islets in high-glucose conditions but not in a low-glucose-containing medium [58,59]. It also was demonstrated that DA regulates insulin secretion but does not modulate its production and storage in β-cells [59]. It can also suppress cell growth and initiate apoptosis in β-cells in vitro [60].

DA releases its physiological effects through interaction with two families of GPCRs, D1-like receptors (includes types 1 and 5 of DA receptors (D1R, D5R) and D2-like receptors (includes types 2, 3, and 4 of DA receptors (D2R, D3R, D4R). D1-like receptors activate G proteins of the $G_{\alpha\,s/olf}$ family, which stimulates adenylate cyclase. The D2-like receptors are coupled with G proteins of the $G_{\alpha\,i/o}$ family and inhibit adenylate cyclase [61]. In the CNS, D1-like receptors are localized mainly on the postsynaptic membranes of cells sensitive to DA, while D2-like receptors are also found on the presynaptic membranes of dopaminergic neurons, having the ability to autoregulate the release of DA [62]. DA receptors are heterogeneously expressed in different islet cell populations. In particular, D1R receptors are expressed in β -cells, D2R in α-, β-, δ-, and PP cells, D4R in β- and PP cells, and D5R in α- and δ-cells [23,24].

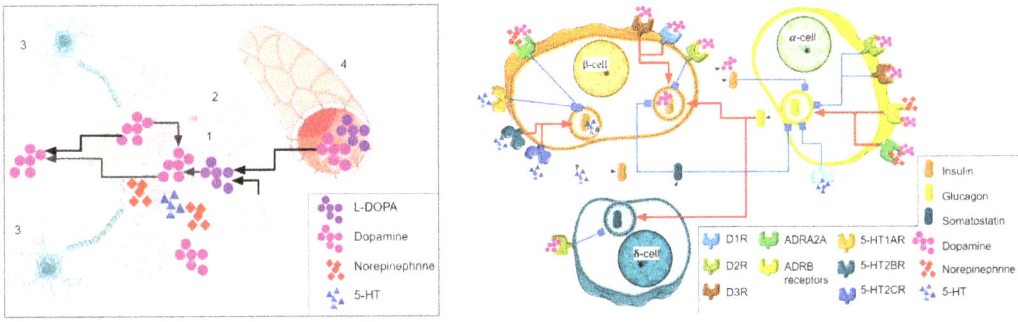

(**a**) (**b**)

Figure 1. The role of monoamine neurotransmitters in the insulin secretion of β-cells. (**a**) The main sources of monoamine neurotransmitters in pancreatic islets; all monoamine neurotransmitters may be synthesized in islets (1), the dopamine synthesis in islets needs L-DOPA which is realized from acini (2) or other sources, dopamine or norepinephrine are realized from the nerve endings (3), or acquired from the circulation (4). (**b**) Monoamines and their receptors are involved in the pancreatic islet hormone production. The activation effect is marked by red arrows, inhibition is marked by blue blunt arrows, and the dotted line indicates bioactive molecules' release. Parts of the figure were drawn by using pictures from Servier Medical Art. Servier Medical Art by Servier is licensed under a Creative Commons Attribution 3.0 Unported License.

The D1R receptor mediates GSIS activation in β-cells, and its expression seems to prevail over the D2R expression in this cell population [52]. A D1-like receptor agonist (SKF-38393, 10 μM) stimulates GSIS, but the mechanism by which the D1 receptor regulates this process needs further investigation [23].

In contrast, D2 family activation provides an inhibitory effect on GSIS by cell depolarization [58,63]. D2R function in β-cells is not limited to participating in GSIS modulation. In vitro, the D2R antagonist domperidone protects β-cells from dedifferentiation and apoptosis and promotes the proliferation of the differentiated β-cells [64]. Recently, it was shown that D2R is involved in circadian β-cell regulation [49]. It should be noted that in the rat insulinoma cells INS-1E, D2R localization is atypical. No D2R was detected in either the cytoplasmic or the plasma membrane fractions, but it was associated with insulin granules. It remains unknown whether D2R on the granule surface is activated by extracellular dopamine only at the time of its transient contact with the cellular membrane or whether it interacts with DA in some other way [58].

Higher DA concentrations may activate ADRA2A adrenergic receptors, which also inhibit insulin secretion [25]. It is also possible that the DA effect on β-cells is indirect and mediated by its interaction with D2R on δ-cells with consequent downregulation of SST secretion. In such a way, DA activates insulin secretion [23,42]. Controversial results of some studies highlight the complexity of D2R involvement in GSIS. Knockout of D2 receptors in the insulinoma cell line INS-1 832/13 resulted in increased insulin secretion, but the global D2R knockout (KO) in mice impaired insulin secretion and caused glucose intolerance [43,65].

D3R and D4R involvement in GSIS regulation was shown in hamster β-cells. GSIS is downregulated by D3 and D4 antagonists, whereas a D3 agonist increases it [48].

Pronounced species-specific differences mark the DA-mediated regulation of α-cell glucagon secretion. DA enhances glucagon release from mouse islets, but in human islets, its effect is biphasic. DA concentrations below 1 μM reduce glucagon secretion, while higher DA concentrations enhance glucagon release. Such dualistic effects are determined by a complex repertoire of DA-binding receptors in human α-cells. D2R and D3R enable the inhibitory DA effect even in low concentrations. Higher DA concentrations lead to activation of β-adrenergic receptors, resulting in glucagon release. In mouse α-cells, β-adrenergic receptors are significantly more abundant than in humans, so the DA effect almost entirely depends on these receptors and becomes monophasic [25]. In δ-cells, D2R expression is significantly higher than in other islet cell populations and these cells seem to be the main link in dopamine-mediated GSIS regulation [23,25]. DA's effect on pancreatic hormone secretion is briefly summarized in Figure 1b.

4.2. DA Signaling in CNS in the Aspect of Metabolic Syndrome

The assumption of a connection between DA homeostasis and metabolic syndrome arose when the increased risk of weight gain, hyperinsulinemia, hyperlipidemia, and diabetes mellitus, was found in patients receiving antipsychotics and other DA metabolism-targeting drugs [66]. The administration of DA receptor agonists sometimes has an ameliorative effect on the dynamics of metabolic disease recovery [67].

DA receptors are heterogeneously distributed in the mammalian CNS. D1R and D2R are commonly expressed in the striatum, nucleus accumbens, olfactory bulb, amygdala, hippocampus, substantia nigra (SN), ventral tegmental area (VTA), hypothalamus, and frontal cortex; D3R was detected in the nucleus accumbens, striatum, and cortex; D4R is expressed in the frontal cortex, amygdala, hypothalamus, and nucleus accumbens; D5R was detected in the cortex, the SN, and the hypothalamus [68,69].

D2-like receptors play the most important role in postsynaptic receptor-mediated behavioral and extrapyramidal activity [70]. Eating behavior is regulated by homeostatic signals and non-homeostatic hedonic cues both in animals and humans [71,72]. D2R contributes to the rewarding effects secondary to food intake [59], since food consumption leads to an increase in the level of DA in the brain, associated with a decrease in the expression of D2R in the nucleus accumbens in the forebrain and the ventral tegmental area in the midbrain [73]. Experimental data on mice and rats confirm the DA neurotransmission deficiency and the downregulation of D2R, DA transporter (DAT), and DA synthetic enzyme tyrosine hydroxylase (TH) under conditions of a high-calorie obesity-inducing diet [74,75]. In addition, depletion of striatal D2Rs mediates the onset of compulsive-like food seeking in rats [76,77].

The role of D2R-mediated signaling in eating behavior regulation is not restricted to its central position in the reward system. DA is released by the hypothalamic cells and tonically inhibits prolactin secretion from the pituitary through binding with D2R [78,79]. Thus, D2R antagonists may contribute to hyperinsulinemia and obesity via the control of the pituitary hormone prolactin [78].

D1R implication in the MetS phenotype is also not excluded [80]. This assumption is supported by the noticeable role of D1R in reward processing [68,69]. Furthermore, mice deficient in the SLC35D3 transporter protein exhibit MetS and obesity. SLC35D3 is expressed only in striatonigral medium spiny neurons and provides D1R transfer from the endoplasmic reticulum to the membrane [81]. The basic points of the interactions between insulin and dopamine signaling in brain structures are summarized in Figure 2a.

In humans, DA system hypofunction may be a predictor of obesity. A postmortem study of obese and non-obese patients revealed an obesity-associated decrease in DAT and TH expression without significant changes in the D2R expression level [82]. Thus, DA itself and its receptors, transporters, and synthesis enzymes are involved in eating behavior and energy balance, and their coordinated functioning breakdown may lead to metabolic disruptions.

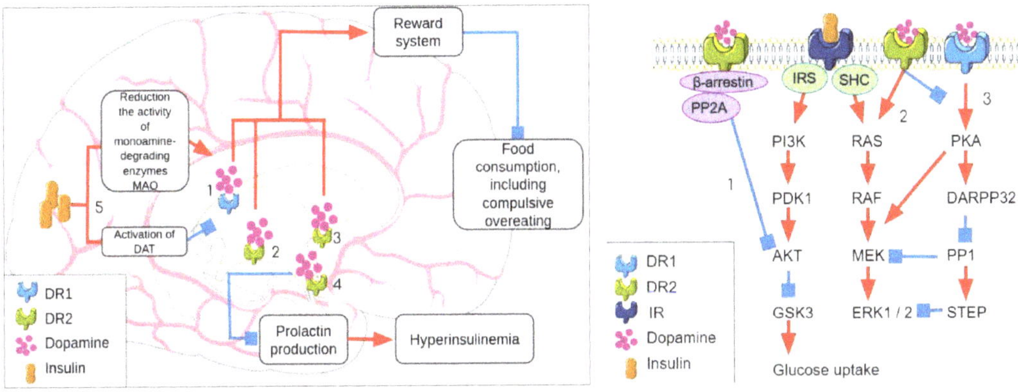

(a) (b)

Figure 2. The interactions between dopamine and insulin signaling. (**a**) The crosstalk between insulin signaling and dopamine signaling in the brain where the D1R and D2R receptors in the nucleus accumbens and VTA and dopamine receptors D1R in striatonigral pathway neurons regulate the reward system, which contributes to the normalization of eating behavior (1, 2, 3). The dopamine interaction with D2R receptor in the hypothalamus prevents prolactin production and consequent hyperinsulinemia (4). At the same time, insulin dualistically affects dopamine transmission by the activation of dopamine reuptake by DAT and prevention of dopamine degradation by MAO enzymes. (**b**) Intracellular interaction of insulin signaling and dopamine signaling resulting from common downstream targets, D2R inhibits Akt signaling by the β-arrestin-mediated signaling cascade (1) and activates MEK signaling (2) while D1R stimulation activates MEK via PKA-DARPP32 pathway (3) instead. Parts of the figure were drawn by using pictures from Servier Medical Art. Servier Medical Art by Servier is licensed under a Creative Commons Attribution 3.0 Unported License. DR1—dopamine receptor 1, DR2—dopamine receptor 2, IR—insulin receptor, IRS—insulin receptor substrate, SHC—*Src* homology 2 domain containing, PP2A—protein phosphatase 2A, PI3K—phosphatidylinositol 3-kinase, PDK1—pyruvate dehydrogenase kinase 1, GSK3—glycogen synthase kinase 3, MEK—mitogen-activated protein kinase kinase, ERK 1/2—extracellular signal-regulated kinase 1/2, PKA—protein kinase A, DARPP32—dopamine- and cAMP-regulated phosphoprotein with an apparent Mr of 32,000, PP1—protein phosphatase 1, STEP—striatal-enriched protein tyrosine phosphatase.

4.3. Systemic Interaction of Insulin and DA Signaling Pathways

The regulatory influence of insulin extends beyond the control of peripheral glucose levels. Insulin, along with the DA, is involved in regulating eating behavior and reward. In addition, insulin affects DA signaling, modulating the duration of DA action by regulating MAOs and DAT expression [83].

Insulin activates tyrosine-kinase insulin receptor IR, and its downstream signaling includes two main pathways. The insulin receptor substrate (IRS) node transduces the signal to the phosphatidylinositol 3-kinase (PI3K, a lipid kinase)/Akt (also known as PKB or protein kinase B) and the Raf/Ras/MEK/MAPK (mitogen-activated protein kinase, also known as ERK or extracellular signal-regulated kinase) pathways. Additionally, the Src homology 2 domain-containing (Shc) adapter is involved in Raf/Ras/MEK/MAPK pathway activation [84].

Overlapping insulin and DA signal transduction reflects the contribution of common downstream components in both pathways. The key point, where these pathways have crossed, is the Akt signaling. Akt is involved in a range of processes, including the control of glucose homeostasis by insulin and regulating DAT expression. Glycogen synthase kinase 3 (GSK3), which influences cell survival, circadian rhythms, mood, and cognition, is one of Akt's downstream targets [60]. Insulin activates the Akt-signaling pathway [85]. In contrast, D2R activation blocks Akt activity through a G protein-independent beta-arrestin 2-mediated mechanism [68,86].

D2R can also trigger the extracellular signal-regulated kinase (ERK) pathway, which is involved in insulin signaling. A short-term effect of DA on D2R leads to G protein-dependent inhibition of cAMP, Ca^{2+} influx into the cell, and suppression of PKA activity. Reduction in PKA activity removes the inhibitory effect of 32 kDa DA and cAMP-regulated phosphoprotein (DARPP-32) on protein phosphatase 1 (PP1) activity, which leads to the activation of ERK. The ERK pathway mediates changes in gene expression, including DAT [68,83,87]. These interactions of two signaling pathways are represented in Figure 2b.

Metabolic diseases, in particular, T2DM, are often associated with CNS disorders. Insulin deficiency-associated disturbances in the DA system may be manifested as eating disorders, depression, and cognitive dysfunction. Insulin receptor gene KO in mice leads to the deregulation of MAO-A and MAO-B expression in the striatum and consequent increase in DA clearance time [88]. DA neuron damage in the suprachiasmatic nucleus in rats causes MetS with characteristic manifestations, including weight gain, hypertension, and insulin resistance [89].

The reduced secretory functions of pancreatic β-cells, together with complete or partial lack of insulin sensitivity, are implicated in the pathogenesis of MetS. Blockade of pancreatic D2R in mice causes glucose intolerance by decreasing insulin production and decreasing β-cells mass. The impaired β-cells replicative capacity was demonstrated in 2-month-old D2R-KO mice [65]. DA-mediated inhibition of GSIS seems to depend on the combined effect of D2 and D3 members of the D2-like receptor family and the L-DOPA transporter LAT. This transporter is expressed in islet cells as two isoforms, LAT1 and LAT2. Its concerted functioning provides effective absorption of L-DOPA in response to stimulation by the glucose supplied with food. In summary, the bioavailability of DA synthesized from dietary L-DOPA may contribute to the inhibition of GSIS [90].

Human data also support a link between DA and insulin signaling. Historically, insulin coma therapy was used in schizophrenia patients [91]. Examination of postmortem brain tissue samples from individuals with psychiatric or neurodegenerative diseases demonstrated a coordinated decrease of DA metabolism-related genes (i.e., D1R, D2R, TH, MAOB) and insulin receptor expression [92].

Considering the reciprocal interaction of the DA and insulin systems, prevention and therapy of MetS by targeting the DA metabolism may be a promising approach. Thus, D2R agonist bromocriptine has become the prospective medication to manage MetS, obesity, and T2DM. It improves glycemic parameters such as plasma cholesterol and being overweight [93]. The bromocriptine effect is mediated by both central and peripheral processes modulation. As demonstrated in animal models, it impacts the imbalance of DA signaling in the adipose tissue. Bromocriptine administration stimulates D1R and TH expression in white adipose tissue and the liver activates lipid oxidation. Additionally, bromocriptine increases insulin receptor expression levels in adipose tissue and contributes to the improvement of fasting glucose levels [94].

5. Norepinephrine Signaling in GSIS Regulation

Islets are innervated by adrenergic nerves projecting from the celiac ganglion or the paravertebral sympathetic ganglia. Thus, NE is released in pancreatic islets from sympathetic inputs. In addition, the murine α-cell line, alpha TC1 clone 6, may also produce NE [25]. NE reduces insulin and SST secretion and stimulates glucagon and PP secretion in islets [25,26].

Different adrenergic receptors are involved in GSIS regulation. The inhibitory effect of NE is mediated by ADRA adrenergic receptors (Figure 1a), whose blockade counteracts the inhibition of GSIS by electrical sympathetic nerve activation. GSIS inhibition by ADRA2A activation is accompanied by hyperpolarization of the β-cells and reduction in the cytoplasmic concentration of Ca^{2+} and cAMP [95]. It was found that rs553668 polymorphism in the 3′ UTR region of *ADRA2A* may be linked to impaired insulin secretion and GSIS, and an increased risk of T2DM [96].

The stimulatory action of ADRB receptors is mediated by the increased formation of cAMP in β-cells [51]. ADRB2-KO mice are glucose intolerant because of impaired GSIS; they have significantly higher blood glucose both upon fasting and under random feeding conditions and decreased fasting serum insulin levels compared to their wild-type littermates. It also was demonstrated that in 20-month-old mice, ADRB2 expression was significantly lower compared to young animals; thus, it was postulated that the reduced ADRB2 expression might contribute to the age-related decline of glucose tolerance in mice [97]. In an experimental model of T2D, the ADRB2 ligand ephedrine showed a hypoglycemic effect together with facilitating the regeneration of the pancreatic islets after their atrophy [98].

Prolonged NE exposure leads to the upregulation of insulin secretion and profound changes in the GPCR repertoire in islet cells. After chronic NE treatment of islets, the levels of $G_{\alpha s}$, $G_{\alpha z}$, $G_{\beta 1}$, and $G_{\beta 2}$ proteins were 42.8%, 19.4%, 24.8%, and 16.9% lower compared to the untreated islets, respectively. The negative regulator of insulin secretion, Ucp2, was also reduced in the NE-exposed β-cells [57].

A study of genetic factors responsible for the development of insulin resistance showed that mutations in ADRB3 may be involved in the development of tissue insulin sensitivity disorders [99]. However, this association seems to be related to NE signaling from the pancreatic islets, in particular, in visceral adipose tissue. Previously, a relationship between obesity and its associated metabolic complications, as well as increased visceral fat ADRB3 sensitivity, was proven [100].

6. 5-HT Participation in β-cell Modulation

5-HT regulates several functions in the CNS and outside the brain, including colon motility, immune function, and blood flow. In β-cell, 5-HT regulates GSIS (Figure 1b) and proliferation, particularly in the adaptation to hyperglycemia, a high-fat diet, or pregnancy [101]. Basal 5-HT production in β-cells is required for physiological insulin release [27]. These cells express all proteins required to synthesize and package 5-HT, which is stored in insulin granules and co-released with insulin after glucose stimulation [28,45,102,103]. Islet cells also express SERT, which allows the re-uptake of secreted 5-HT [101].

5-HT receptors are structurally different. 5-HT3Rs are ligand-gated cation channels, but all the other 5-HT receptors are GPCRs. These receptors are expressed in different populations of islet cells. β-cells express 5-HT1AR, 5-HT2CR, 5-HT1DR, 5-HT2BR, and 5-HT3Rs. 5-HT1FRs were detected in α-cells, and 5-HT5ARs are specific to δ-cells [104].

5-HT1AR's agonists significantly reduced GSIS in rats [105]. In contrast, 5-HT2BR and 5-HT2CR receptors increased insulin secretion from pancreatic β-cells [43,99] through the activation of the phospholipase C pathway, with a subsequent increase in intracellular Ca^{2+} [27]. When β-cells grow in medium supplemented with 5-HT2B receptor antagonist SB204741 or high (30 mmol/l) glucose concentrations, the expression of this receptor becomes reduced [106]. It was also demonstrated that this receptor is involved in GSIS activation and β-cells proliferation in pregnancy [27,102]. The 5-HT2B receptor is a critical modulator of physiological insulin secretion and its inactivation blunts GSIS [27].

5-HT2CR expression in β-cells and its effect on GSIS activity depends on its mRNA editing. Adenosine to inosine (A-to-I) RNA editing increases the receptor 5-HT potency, agonist binding affinity, constitutive activities, and G protein coupling activity. The expression of this 5-HT2CR also depends on the feed consumption induced by an insulin resistance-initiating diet. The expression of the enzyme ADAR2 involved in the 5-HT2CR mRNA editing is also upregulated by the same diet [107].

GSIS is accompanied by 5-HT release from the β cells, and then 5-HT provides negative paracrine and autocrine feedback loops for the insulin secretion. 5-HT1ARs mediate autocrine GSIS downregulation by 5-HT on β-cells, and paracrine downregulation is implemented by the binding with 5-HT1FR on α-cells and the consequent inhibition of glucagon secretion [101,104]. In contrast, activation of 5-HT2 and 5-HT3 receptors potentiate GSIS in β-cells [101].

Downstream insulin effects may also be modulated by 5-HT. It is expected that the 5-HT2 antagonist ketanserin impairs insulin sensitivity by suppressing 5-HT2A receptor-mediated glucose uptake in skeletal muscle [108].

Several effects of 5-HT in pancreatic cells are bypassing receptors and mediated by serotonylation of proteins. In β-cells, protein serotonylation is a meaningful mechanism, thus enhancing insulin granule exocytosis by activating small GTPases [27,28].

7. Trace Amine Signaling and TAARs in Pancreatic Islets

7.1. TAARs-Dependent Regulation Mechanisms

TAs are a group of endogenous and exogenous compounds that structurally resemble biogenic monoamines. In the mammalian CNS, their concentration is extremely low and does not exceed 50 ng/g (500 nM), so the TAs were named according to their trace concentrations [1]. TAs are structurally and metabolically related to classical monoamine neurotransmitters. In low concentrations, these molecules become modulators of monoaminergic activity, especially the activities of the DA or 5-HT systems. This property determines their influence on mood and emotional background. At high concentrations, TAs demonstrate an amphetamine-like effect on the release, reuptake, and biosynthesis of catecholamines and indolamines. The ability of TA signaling to modulate monoaminergic systems attracted interest as a potential therapeutic target for the management of mental and neurodegenerative diseases [109].

The most-studied TAs are β-phenylethylamine (PEA), tyramine (TYR), tryptamine (TRP), 3-iodothyronamine (T1AM), and octopamine (OCT). These and many other less-studied mammalian TAs are TAAR agonists at their physiological concentrations. Despite 20 years of TAAR research, the progress regarding their study is relatively slow, reflecting some limitations in its exploration. First, the TAARs investigation is limited because of the low expression of these proteins. Other constraints are associated with intracellular TAAR localization and, last but not least, the wide diversity of TAAR ligands [2,110].

As is currently known, TAARs are members of the rhodopsin A-like GPCR family with seven transmembrane domains. The ligand–receptor interaction leads to cAMP accumulation, modulation of PKA and PKC signaling, and, possibly, β-arrestin 2-dependent pathway activation (at least after TAAR1 binding with its ligands). The influence of TAAR1 on the opening of internally (inwardly) rectifying K+ GirK channels was also revealed [17]. At least 26 TAAR subtypes belonging to 9 families were described in vertebrates, and their distribution in different species varies significantly. The TAAR gene family members which were identified in human and several animal models following the Ensembl database [111] are summarized in Table 1. Although TAAR expression was initially found in pancreatic β-cells, the studies of these receptors were for a long time focused specifically on their function in the CNS. This interest was attributed to the TAAR gene localization on the human chromosome 6q23.2 locus associated with susceptibility to schizophrenia [3].

Table 1. TAAR genes diversity in human and animal models.

Species	Number of Genes	Identified TAAR Genes
Homo sapiens	6	*TAAR1, TAAR2, TAAR5, TAAR6, TAAR8, TAAR9*
Pan trogloditus	3	*TAAR1, TAAR5, TAAR6*
Chlorocebus sabaeus	4	*TAAR1, TAAR2, TAAR5, TAAR6*
Mus musculus	15	*Taar1, Taar2, Taar3, Taar4, Taar5, Taar6, Taar7a, Taar7b, Taar7d, Taar7e, Taar7f, Taar8a, Taar8b, Taar8c, Taar9*
Rattus norvegicus	18	*Taar1, Taar2, Taar3, Taar4, Taar5, Taar6, Taar7a, Taar7b, Taar7c, Taar7d, Taar7e, Taar7f, Taar7h, Taar7g, Taar8a, Taar8b, Taar8c, Taar9*

The first discovered and most studied of the TAARs is TAAR1. In the brain, its expression is concentrated in the monoaminergic systems, including the VTA, the dorsal raphe, and their projections in the striatum, amygdala, hypothalamus, and frontal cortex, suggesting a TAAR1 role in modulating DA, 5-HT, and possibly NE neurotransmission [1,110,112].

All TAARs except TAAR1 are represented in the olfactory epithelium, where they are involved in the detection of various innate olfactory stimuli mediated by volatile amines [113,114]. However, growing evidence confirms that TAARs are also present in the CNS, particularly in limbic brain areas receiving olfactory input [5], and can be involved in adult neurogenesis, as was demonstrated in TAAR5-KO mice [115,116]. TAAR5 was found in the amygdala, arcuate nucleus, ventromedial hypothalamus, frontal cortex, hippocampus, and SN [2,117]. Systemic transcriptomic analysis demonstrated a wide representation of TAAR5 in the adult human brain, especially in the limbic brain regions associated with the processing of olfactory signals and control of emotions [117]. A TAAR6 transcript was found in the amygdala, frontal cortex, hippocampus, striatum, and SN [2,118,119].

TAAR expression has also been demonstrated outside the CNS, in the mammalian digestive, endocrine, cardiovascular, and immune systems. The expression of TAAR1, TAAR2, and TAAR6 was shown in the cells of the gastrointestinal tract. TAAR1 is predominately expressed in the mouse, rat, or human stomach and small intestine. TAAR2, TAAR6, and TAAR9 expression were detected in the mucous cells of the duodenum in mice and rats [2,120]. Human leukocytes express TAAR1, TAAR2, TAAR5, TAAR6, TAAR8, and TAAR9 [121]. Of particular interest is the TAAR expression in the islet β-cells. TAAR1 co-localization with insulin in pancreatic β-cells, as well as the modulating effects of TAAR1 on DA transmission, may underlie the relationship between TA-mediated signaling pathways and metabolic diseases, in particular, T2DM [122].

7.2. Relationship between the Trace Amine System and the Monoaminergic Systems

Given the involvement of the DA system in the pathogenesis of MetS, it is important to understand the possibilities of its targeting in metabolic diseases. It is known that TAAR1 modulates the DA neurotransmission regulating the rate of dopaminergic neurons' excitation, the sensitivity of D2R to ligands, and the activity of the DAT [123]. The mechanisms underlying the interaction of the DA and TA systems will be considered. TAAR1 and D2R have a functional and structural relationship. Both receptors belong to the GPCR family and could form a heterodimer on the surface of the cell membrane [124,125]. The interaction of two receptors in the complex impacts its properties. After heterodimerization with D2R, TAAR1, which is predominantly intracellular in the monomeric form, migrates to the cell surface. In parallel, D2R signaling activity shifts from the β-arrestin 2 signaling pathway to Gi activation. The β-arrestin 2 affinities to D2-like receptors decrease, which leads to the suppression of GSK3β signaling [1,126,127]. Furthermore, after the heterodimerization, D2R and TAAR1 become co-internalized upon any of the receptors' stimulation [124,125].

Currently, the evidence suggests an antagonistic relationship between the DA and TA systems. D2R blockade selectively enhances TAAR1 signaling [128], while TAARs' agonists, such as PEA and TYR, inhibit the excitation of DA neurons in the VTA, SN, and nucleus accumbens. The supposed mechanism of DA neuron inhibition is related to alterations in D2R autoreceptor activity [129]. In contrast, in TAAR1-deficient mice, the DA metabolite, homovanillic acid accumulates extracellularly in the dorsal striatum and nucleus accumbens, which suggests the increased DA release in these structures. DA release is also stimulated by TAAR1 antagonists and inhibited by TAAR1 agonists in wild-type mice. However, TAAR1 insufficiency does not affect the kinetics of DA uptake and DAT function [128,130].

Thus, the TA system, in particular TAAR1, apparently functions as a negative control mechanism for DA release and exerts an inhibitory effect on DA neurons.

There is some evidence for the interaction of TA and 5-HT signaling. For example, TA may also indirectly cause increased susceptibility to oxidative stress in tissue through their modulation of 5-HT. The absence of TAAR1 in model animals is associated with increased

DA and 5-HT signaling [131,132]. TAAR5-KO mice also showed a deregulated 5-HT and DA system [5,115]. Since it was suggested that TAAR1 could act as a rheostat of 5-HT signaling [1,131], the relation between TAARs and 5-HT signaling components is expected.

7.3. Background and Prospects for the Clinical Use of Trace Amines and their Receptors

The involvement of the TA system in regulating biogenic amine-mediated processes suggests its role in MetS pathogenesis. Besides the modulating effect of TAAR1 activation on DA transmission, TAARs may be implicated in the control of energy metabolism, nutrition, eating behavior, body weight, and insulin sensitivity [2]. TAAR expression outside the CNS is interesting in terms of MetS pathogenesis and therapy. TAAR1 expression was detected in the stomach and intestinal neuroendocrine cells, and in pancreatic β-cells which could impact insulin production and systemic function of the endocrine pancreas [122,133]. It was suggested that TAAR1 activation in β-cells by ractopamine can stimulate GSIS [134]. Also, it was demonstrated that the acute TAAR1 activation in vivo by the selective small molecule agonist RO5166017 stimulates insulin secretion, but in sub-chronic treatment by supplying RO5166017 with food for 7 days improves insulin sensitivity and reduces insulin plasma levels in mice [122].

The gradually accumulating evidence for TAARs' involvement in energy metabolism makes it possible to consider these receptors as the targets in MetS therapy. In the aspect of MetS, it is important to consider the TAARs' role in regulating eating behavior, the violation of which may underlie the etiology of obesity, which is an important factor in MetS development. A TAAR1 agonist, RO5256390, was demonstrated to prevent the compulsive eating of delicious food. The proposed background of this effect, as with drug abuse and addiction, is the modulation of DA projections in the medial prefrontal cortex. In conditions associated with disturbed eating behavior, i.e., overeating and obesity, an increased DA release during food intake occurs. Rats receiving the TAAR1 agonist RO5256390 exhibited a selective reduction in sugary food consumption without effect on either baseline intake or food restriction-induced overeating on the standard chow diet [135].

TAARs expressed outside the CNS also may be potential medication targets for MetS, as TAAR-expressing cells outside the brain also may be involved in the pathogenesis of metabolic diseases. This perspective is attractive since such TAAR-targeted compounds may release their therapeutic potential without impacting the cognitive or behavioral functions if they have a limited capacity to cross the blood–brain barrier [2]. For example, TAAR1 expression in the SST-producing D-cells in the stomach was found [136]. SST released from D-cells has a paracrine effect on the ghrelin production in the stomach and consequent downregulation of the ghrelin synthesis [137]. Ghrelin is known as the "hunger hormone" and its depletion leads to decreased food consumption. The TAAR1 agonist RO5166017 stimulates SST production in stomach D-cells and this effect seems to be TAAR1 dependent [136]. Also, it was identified that TAAR1 gene knockout in mice is associated with an increase in catabolic reaction, although the basis of this association is not described yet [138].

TAAR1, TAAR2, and TAAR9 receptors are expressed in the mucosal layer of the duodenum in mice. These receptors are suggested to be involved in hormonal secretion, or motility regulation [2,120]. In the duodenum, TAARs are co-localized with the cells producing the neuroendocrine marker chromogranin A or hormones involved in appetite regulation and glucose level control as GLP1 and PPY [122].

The inhibition of certain TAARs may improve MetS severity. Two TAAR9-KO rat strains have lower low-density lipoprotein cholesterol levels in the plasma compared to their wild-type counterparts [139]. A high LDL-C level is a significant risk factor for CVD development. The extremely low LDL cholesterol levels may also be related to adverse effects like hemorrhagic strokes, dementia, depression, hematuria, and cancers [140]. However, TAAR9-KO rats had LDL plasma levels only twice lower than the wild-type animals, so this TAAR9-KO rats feature seems to be safe for health. No health-damaging consequences of TAAR9 deletion were found in both rat knockout lines tested [139]. The identified associations between TAARs and the regulation of mechanisms involved in metabolic syndrome development are represented in Figure 3.

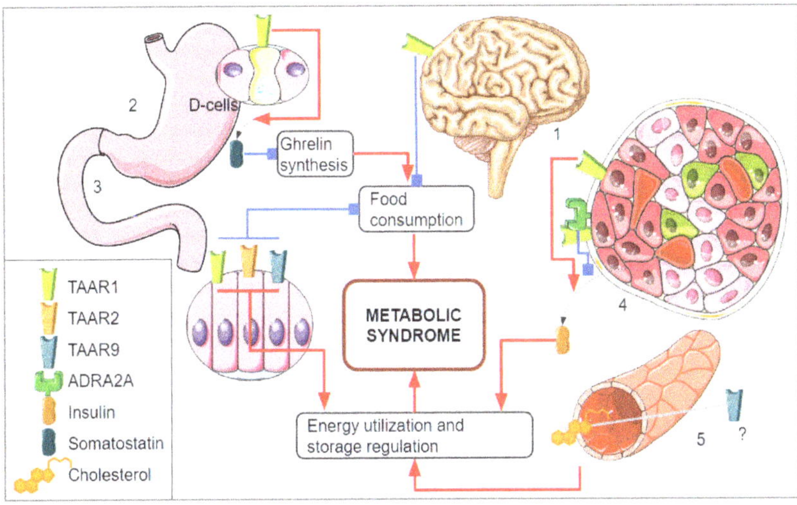

Figure 3. The associations between TAAR-mediated signaling and mechanisms involved in metabolic syndrome development. (1) TAAR1 expression was identified in brain structure involved in the reward system regulation which is involved in food consumption regulation; (2) TAAR1 was identified in the stomach epithelium, its activation on the D-cells leads to the somatostatin realizing, downregulation of ghrelin production, and reduces the feeling of hunger; (3) in the duodenum, TAAR1, TAAR2, and TAAR9 were identified; it is suggested that these receptors are involved in the regulation of neuroendocrine secretory cells that control appetite and glucose metabolism; (4) TAAR1 is expressed in β-cells and regulates insulin secretion, TAAR1 agonists could stimulate insulin secretion, but if TAAR1 dimerizes with ADRA2A, the effect of its activation on insulin secretion becomes inhibitory; (5) TAAR9 seems to be involved in the cholesterol metabolism regulation, but this regulatory mechanism is not clearly understood yet. Parts of the figure were drawn using pictures from Servier Medical Art. Servier Medical Art by Servier is licensed under a Creative Commons Attribution 3.0 Unported License.

Recently, TAAR1 missense mutations, S49L, and I171L were found in obese/overweight patients with slightly impaired glucose homeostasis, and R23C substitution was identified in a patient with a complete loss of insulin production. These alleles are rare, thus their impact on glucose metabolism currently has not been fully investigated. An in vitro assay revealed a partial loss of function for S49L and a complete loss of function for R23C [141].

However, it was demonstrated that the overrepresentation of tryptamine and phenethylamine producers in the enteric microbial community may play a pathogenic role in insulin resistance development. It was evidenced that the identified effect of dysbiosis is mediated by the TAAR1 activation that impairs insulin signaling in the metabolic tissues, including white adipose tissue (WAT), liver, and skeletal muscle. Thus, the TAARs, at least TAAR1, impact on the metabolic disorders seems to be dualistic [142]. Also, TAAR1 agonist N-methyltyramine provokes insulin resistance in adipocytes in vitro, despite its weak insulin-like effect in these cells in insulin-free conditions [143].

Because of the complex MetS nature, involving the disturbance of glucose and lipid metabolism accompanied by hypertension, activation inflammatory pathways, and different organ complications, multi-component treatment regimens are administered in this condition. This resulted in low patient compliance with the difficult-to-apply regimen and additional undesirable drug–drug interactions. The hopeful way to overcome these disadvantages is the development of multi-targeted therapeutic compounds that could improve different individual MetS symptoms at the same time [144,145]. The commonly known examples of such medications are PPAR agonists because PPARγ plays a significant role in the regulation of lipid metabolism, glucose homeostasis, adipogenesis, and several inflammatory processes [144].

The TAAR1 interaction with D2R with downstream behavioral effects seems to be only the "tip of the iceberg" in terms of its complex function. TAAR1 is suggested to be constitutively active or tonically activated in the central nervous system and modulate monoamine neurotransmission activity [142]. TAAR1 was found in cells involved in innate and adaptive immune response, thrombocytes, lung tissue, the esophagus and stomach, uterine cervix, prostate, kidneys, liver, and others. The function of TAARs in these cells largely remains unknown [1,15,146,147]. So, TAARs seem to be an attractive drug target as they are modulators of multiple processes in different organs and tissues.

7.4. TAAR Activity in the Pancreatic Islets as Part of the Systemic Metabolism Regulation by TAs

First, the idea of TA implication in metabolic processes arose when it was found that the TAAR1 agonist RO5263397 exerted potential antipsychotic activity with no metabolic side effects and minimized the weight gain in rodents undergoing D2R-blocking antipsychotic treatment [148]. Recent Phase II clinical trials with the first-in-class antipsychotic not affecting D2R, TAAR1 agonist with 5-HT1A activity SEP-363856 (Ulotaront), supported these observations [149]. In a 6-month open-label study, Ulotaront demonstrated significant antipsychotic activity without extrapyramidal-related adverse effects, a low liability for adverse weight and metabolic effects, and no effect on prolactin levels [150].

TAAR1 is expressed in the β-cells [133,151]. TAAR activation affects insulin secretion through cAMP signaling pathways. It was demonstrated that TAAR1, TAAR3, and TAAR4 ligands increase the level of cAMP and GSIS in the INS-1 mouse β-cell line [151]. Furthermore, TAAR family gene expression in islets is downregulated by glucolipotoxicity [151,152]. In β-cells, in glucolipotoxic conditions, TAAR1 expression loss is accompanied by the downregulation of stimulatory $G_{\alpha s}$ protein, whereas inhibitory Gαi protein becomes overexpressed. These expression changes are suggested to be associated with the suppression of GPCR-mediated GSIS upregulation [152].

The effect of TAAR1 activation in β-cells is the most studied TAAR-dependent process in islets. This receptor mediates GSIS potentiation by membrane depolarization, increases vesicle priming and fusion at docking sites, and stimulates phosphorylation of the key transcription factor CREB that promotes β-cell functioning. Also, TAAR1 activation could improve β-cell mass in vitro [153]. Furthermore, TAAR1 agonist RO5256390 induces potentiation of GSIS, possibly by the activation of calcium channels on the plasma membrane, which was demonstrated in the Ins-1 β-cell line in vitro [151].

TAAR1, TAAR2, and TAAR5 ligand T1AM and, especially, its synthetic analog SG2 with high affinity to TAAR1, affect energy metabolism, inducing ameliorative effects on insulin secretion and glucose homeostasis [154]. SG2 is also considered a promising medication in the therapy and prevention of MetS and demonstrates more suitable pharmacological properties than T1AM [150]. In contrast, GSIS stimulation by T1AM through TAAR1 binding may be demonstrated only in insulinoma cell lines [133,155]. Simultaneously, T1AM is an agonist of other receptors, including TAARs, such as TAAR5 and TAAR8, so its influence on insulin secretion may be mediated by some other targets as well [153]. On the other hand, T1AM injection stimulates hyperglycemia [156] and increases plasma glucagon levels in mice, but these effects seem to be rather dependent on the Gi/o coupled Alpha-2A adrenergic receptor (ADRA2A) than TAAR1, which is coupled with coupled Gs [157]. Another TAAR1 agonist, TYR, activates islets' pericytes, decreases blood flow, lowers plasma insulin levels, and increases plasma glucagon levels [158]. The effect of TYR is considered to be associated with its sympathomimetic properties [154].

8. Evidence for TAARs Involvement in Islets Functioning from Transcriptomic Data

8.1. Whole Islet Transcriptomic Data

Publicly available transcriptomic data seem to be a prospective source of information about expression patterns and their systemic analysis. In this context, the Gene Expression Omnibus [159] database was searched for RNA-seq-generated transcriptomic data for human pancreatic islets. The relevant whole-tissue datasets included in the analysis are listed in Table 2. For the estimation of *TAAR* expression, all data were CPM-normalized by the edgeR R package [160] version 3.42.4. A value of 0.5 CPM was applied as the threshold for positive expression. As a result, all *TAARs*, except *TAAR2*, were detected in human islet whole-tissue samples, at least in one of the studied datasets.

Table 2. Whole tissue sample transcriptome datasets included in analysis.

Accesion	Title	*n*	TAAR1	TAAR2	TAAR5	TAAR6	TAAR8	TAAR9
GSE108072	Global transcriptomic analysis of human pancreatic islets (RNAseq)	88	0.03–1.78 *	0.01–0.06	0.01–0.07	0.01–0.06	0.01–0.05	0.01–0.06
GSE165121	Transcriptional analysis of islets of Langerhans from organ donors of different ages	30	0.8–28,6	0–0.1	0–18.9	0–10.5	0.5–18	0–10.6
GSE50244	Global transcriptomic analysis of human pancreatic islets reveals novel genes influencing glucose metabolism (RNA-seq)	89	0–0.03	NA	NA	NA	NA	NA
GSE164416	Multi-omics profiling of living human pancreatic islet donors reveals heterogeneous beta cell trajectories toward type 2 diabetes	133 **	0–4.8	0–0.2	0–0.9	0–0.2	0–0.3	0–0.9
GSE86468	Single-cell transcriptomics defines human islet cell signatures and reveals cell-type-specific expression changes in type 2 diabetes (bulk)	24	0	0	0	0	0	0

* All expression values are represented in CPM. ** The study includes three groups, i.e., 18 healthy controls, 41 patients with impaired glucose tolerance, 35 type 3c diabetic patients, and 39 type 2 diabetic patients.

Because the dataset GSE165121 was the most informative for *TAAR* expression in the pancreatic islets, we selected it for the study of the functional significance of *TAAR* expression in these structures. In this dataset, all human TAARs' mRNA except TAAR2 were identified in most studied islet samples (Table 2, Figure 4a,b). *TAARs'* co-expressed genes in this dataset were further analyzed to explain the significance of TAAR expression in healthy human pancreatic islets.

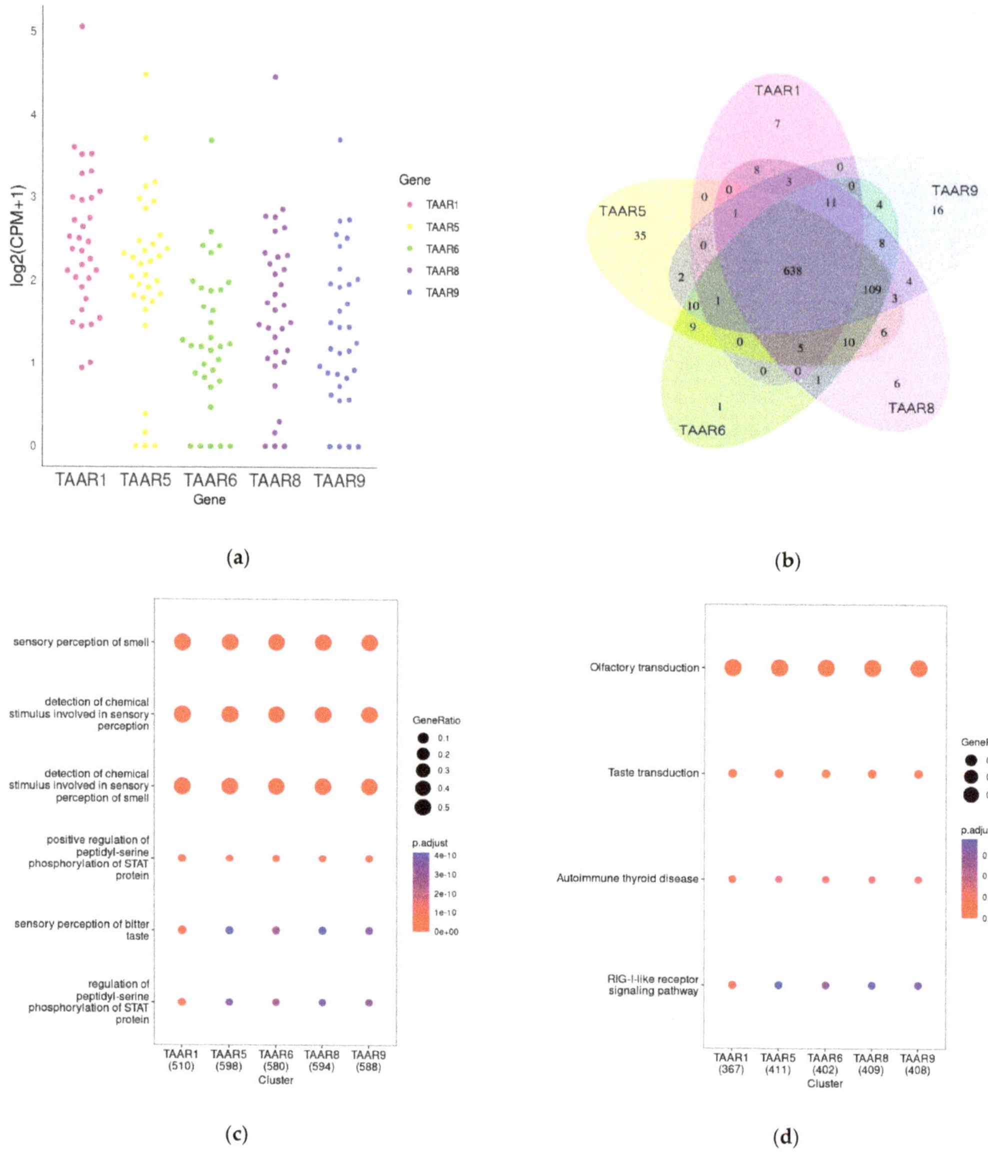

Figure 4. *TAARs* and *TAAR* co-expressed gene clusters in pancreatic islets. (**a**) Expression levels of *TAAR1*, *TAAR5*, *TAAR6*, *TAAR8*, and *TAAR9* in pancreatic islets isolated from healthy donors; (**b**) Venn diagram representing overlaps between *TAAR1*, *TAAR5*, *TAAR6*, *TAAR8*, and *TAAR9* co-expressed gene clusters; (**c**) analysis of Gene Ontology (GO) enrichment of *TAAR* co-expressed gene clusters; (**d**) analysis of KEGG pathway enrichment of *TAAR* co-expressed gene clusters.

TAARs' co-expressed genes were selected by Pearson correlation analysis ($r > 0.7$, $p < 0.05$). Intriguingly, the expression levels of all detected TAARs were correlated with the expression of the same genes. *TAAR*-associated gene clusters were analyzed by the Gene Ontology (GO) terms enrichment test (i.e., identification of the gene's functional groups that are over-represented in the gene set) and the KEGG pathway enrichment test.

The analysis was performed by the clusterProfiler R package [161] version 4.8.3. Biologic process (BP) ontology was applied for the GO terms enrichment analysis.

Both applied methods reveal the co-expression of *TAARs* with genes implicated in other GPCR-mediated biological processes, particularly with olfactory and taste receptor activity, and perception of other chemical stimuli (Figure 4c,d). The co-expression of TAARs with other chemosensory receptors may mirror their implication in the islet hormone secretion regulation in the response to chemical stimuli. For example, it was demonstrated that ectopic odorant receptors (ORs) in pancreatic islets regulate insulin and glucagon secretion, fatty acid oxidation, lipogenesis, thermogenesis, and other processes. At least 47 ORs have been proven to be expressed in human pancreatic islets and the pancreatic β-cell line MIN6 [162], where these receptors regulate insulin secretion and inflammation [163–167]. The olfactory marker protein (OMP) is also expressed in pancreatic islets, especially in α-cells, in which it participates in glucagon secretion regulation. Taste receptors are identified in pancreatic islets [168,169]. Sweet taste receptors TAS1R1, TAS1R2, and TAS1R3 stimulate insulin secretion from β-cells [170–173].

8.2. TAAR1 Co-Expression Cluster Changes in Metabolic Diseases

The GSE164416 dataset includes gene expression data for islets of healthy donors ($n = 18$) and patients with pre-diabetic conditions ($n = 41$), T2DM ($n = 39$), or type 3c diabetes (T3cD, pancreatogenic diabetes, $n = 35$). *TAAR1* expression was detected in all study groups (Figure 5a), while other *TAARs* were not expressed or expressed below the cut-off (this discrepancy with data discussed above may be because of the methodological aspects that impact sequencing sensitivity).

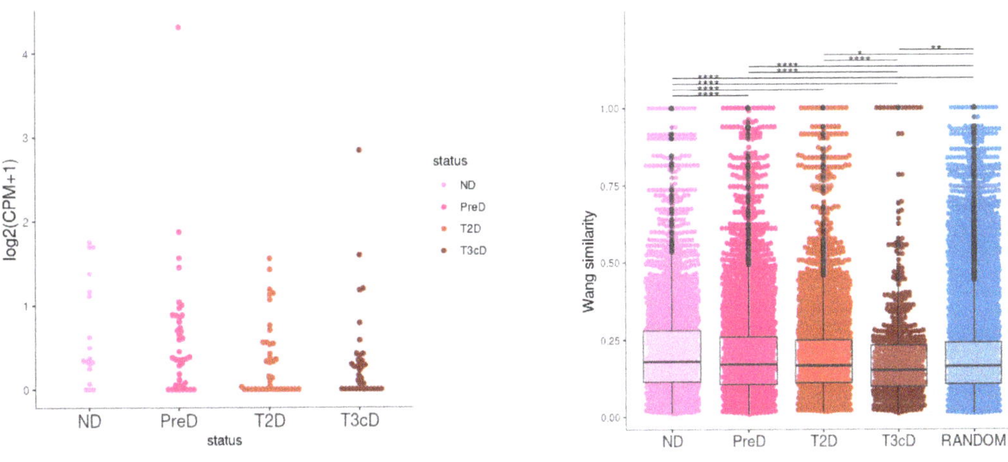

(a) (b)

Figure 5. *TAAR1* and *TAAR1* co-expressed gene clusters in pancreatic islets isolated from healthy donors and patients with metabolic diseases. (**a**) Expression levels of *TAAR1* healthy donors and patients with metabolic diseases; (**b**) functional similarity of *TRAA1* co-expressed genes in islets isolated from healthy donors and patients with metabolic diseases. ND—non-diabetic (healthy donors), PreD—pre-diabetic, T2D—type 2 diabetes, T3cD—type 3c patients. *—$p < 0.05$, **—$p < 0.01$, ****—$p < 0.001$

Considering that co-expressed genes are implicated in the same functions [174], we estimate the functional similarity of genes co-expressed with *TAAR1* in healthy controls and patients with prediabetic conditions or diabetes. GO biologic process terms semantic similarity was used to quantify the functional association of *TAAR1* co-expressed ($r > 0.5$, $p < 0.05$) genes. The semantic similarity was estimated by the Wang coefficient [175] by applying the GOSemSim [176] R package version 2.26.1 in each gene cluster. Then, the Wang coefficient values were compared between study groups by a Wilcoxon signed rank test. A random 400-gene set was included in the comparison.

We found that co-expressed genes in healthy donors have significantly higher ($p < 0.001$) functional relationships compared to prediabetic, T2D, and T3cD groups (Figure 5b). Also, both in healthy donors and pre-diabetic patients, the functional similarity of *TAAR1* co-expressed genes was higher ($p < 0.001$) than in the random gene set. The dramatic loss of functional similarity between *TAAR1* co-expressed genes was identified in pancreatic islets of T2D patients, and, especially, in pancreatogenic T3cD patients. This loss of functional links may be related to the disturbance of TAAR1-mediated biologic processes with diabetic-associated islets' damage.

8.3. RNA Sequencing of Single Cells and Cell Fractions Demonstrates Taar Expression in all Main Islet Cell Populations, Especially in δ-Cells

Analysis of single-cell RNA sequencing (scRNA-seq) data and data obtained by RNAseq in islet cell fractions confirms that *TAAR1* expression is observed in the δ-cells of the pancreatic islets (Table 3). According to the dataset GSE81547, TAAR1 is expressed in 7% of the human pancreatic δ cell population.

Table 3. Single-cell RNA seq and cell fractions transcriptome datasets included in the analysis.

Accession	Title	Cells	*n*	TAAR1	TAAR2	TAAR5	TAAR6	TAAR8	TAAR9
GSE57973	Age-dependent gene expression changes in human islets	α-cells	10	0–0.2 *	ND	0–0.07	0–0.08	0–0.08	0–0.04
		β-cells	10	0–1.14	0–0.4	0–0.96	0–0.97	0–0.2	0–0.04
GSE81547	Single-cell transcriptome analysis of human pancreas reveals transcriptional signatures of aging and somatic mutation patterns.	α-cells	998	0–2.25	0	0	0	0–111	0–0.79
		β-cells	348	0	0	0	0	0	0
		δ-cells	83	0–173	0	0	0	0	0
GSE80780	Human islets contain four distinct subtypes of β cells	β-cells	32	0–0.26	0–0.06	0–0.6	0	0–0.06	0–0.08
GSE86469	Single-cell transcriptomics defines human islet cell signatures and reveals cell-type-specific expression changes in type 2 diabetes (single-cell)	α/β/δ/ stellate cells	239/263/ 25/19/18	0	0	0	0	0	0
		PP	18	0–0.8	0	0	0	0	0
GSE73727	Single-cell transcriptomics reveals unique features of human pancreatic islet cell subtypes	α-cells	18	0.1–0.54	0.3–1.6	0.1–0.54	0.1–0.54	0.1–0.54	0.1–0.54
		β-cells	12	0.08–0.46	0.24–1.37	0.08–0.46	0.08–0.46	0.08–0.46	0.08–0.46
		δ-cells	2	0.18–0.24	0.55–0.71	0.19–0.23	0.18–0.024	0.18–0.024	0.18–0.24
		PP cells	9	0.1–0.58	0.3–1.74	0.1–0.58	0.01–0.58	0.01–0.58	0.01–0.58
GSE106148	α Cell Function and Gene Expression Are Compromised in Type 1 Diabetes	α-cells	5	0–1	0–1	0	0	0	0

* All expression values are represented in CPM.

TAAR1 mRNA was identified in other pancreatic cell populations, including α-cells. Analysis of other scRNA-sec datasets and transcriptomic datasets of cell fractions (GSE86469, GSE73727, GSE57973) also revealed *TAAR1* expression in β-cells and PP cells of pancreatic islets, and duct cells belonging to the exocrine pancreas.

Thus, the expression of *TAAR1* in islets is associated with δ-cells, while the expression of other TAARs in the endocrine pancreas cells seems to be sporadic. δ-cells are SST-producing cells. In human islets, the SST-secreting δ-cells make up 1–5% of the total islet cell population [177]. These cells express receptors that can sense appetite-regulating hormones leptin (LEPR) and ghrelin (GHSR), the growth factor neuregulin 4 (ERBB4), and the DA (D2R). The function of δ-cells is the integration of complex metabolic signals via different signaling pathways and regulation of insulin and glucagon synthesis in β- and α-cells, respectively [25]. It is possible that TAAR1 activation in δ cells causes the effect that then becomes amplified downstream.

9. Review Limitations

There are several limitations in this review that need to be acknowledged. Current knowledge of the trace amine and trace amine-associated function regulatory functions remains fragmental and needs further studies. The agonists of TAAR2, TAAR6, TAAR8, and TAAR9 are not well defined, and high-specific agonists are not available for these receptors. TAARs are considered rheostats of GPCR-mediated monoamine signaling, but their impact on the other GPCR-mediated pathways has not been studied. Thus, it is difficult to identify the biological base of some identified trends.

As the literature data for TAAR expression in pancreatic islet and islet cell populations are limited, we add the information mined from the GEO NCBI repository. It should be noted that TAARs are expressed at low levels, thus only transcriptomic datasets that were generated with suitable sequencing depth are suitable for the estimation of these receptors' expression. Therefore, the lack of TAAR expression in some samples may be because of false-negative results related to the insufficient sequencing depth. The RNA sequencing data from different groups cannot be completely standardized and combined for statistical analysis. We can only demonstrate some similar trends that are reproduced in different datasets and find out that the result is congruent and replicable.

The results generated by the systems biology methods like the semantic similarity analysis and GO terms or KEGG pathway enrichment analysis need further confirmation by the laboratory studies.

10. Conclusions

Currently, the concept of TAs and their receptors is gradually expanding outside their function in the olfactory system. TAAR1 primarily was shown outside the olfactory system, and this receptor has already been identified as the target of therapeutic agents for schizophrenia, depression, and drug addiction. Its role in the pathogenesis of other diseases, including MetS, continues to be studied. The pattern of TAAR expression allows suspecting the involvement of the TA system in the MetS pathogenesis. Other TAARs were also found in endocrine glands that control energy metabolism, including the hypothalamus, thyroid, and pancreatic islets. In addition, TAARs are important endogenous modulators of other neurotransmitter systems, in particular, dopaminergic transmission, which can be disturbed in MetS. Also, the antagonistic relationship between DA and insulin has been described. A deficiency in DA neurotransmission could cause MetS or its individual symptoms, including weight gain, insulin resistance, hypertension, and glucose intolerance. Meanwhile, insulin deficiency may trigger the development of depression and eating behavior disorders. DA signaling in the endocrine pancreas regulates insulin secretion and maintains glucose homeostasis. TAAR1 ligands in physiological concentration negatively regulate the DA system, which may draw attention to these compounds as putative therapeutic agents for MetS prevention or treatment. However, this hypothesis is contradicted by the effect of some natural TAAR1 agonists, which can elicit insulin resistance. Such results may be explained by the difference in exposure, dose, or the complex interaction of the TA system with other regulatory pathways.

By using bioinformatics approaches, we observed TAAR expression in pancreatic islets. Our in silico analysis demonstrates that TAARs, especially TAAR1, may be involved in the complex regulation of GSIS in the response to chemical stimuli, which is destroyed in type 2 diabetes and prediabetic metabolic disorders. Furthermore, TAARs were identified not only in insulin-producing β-cells but also in α- and δ-cells, which are involved in the GSIS regulation. The effect of TAAR ligands on these cells, especially on the δ-cells, has not yet been investigated. These facts, along with the reported effects of TAAR1 agonists, should be considered in future research to explore if MetS management with TAAR-targeting compounds is possible.

Author Contributions: Conceptualization, R.R.G.; methodology, A.N.V.; validation, A.N.V.; formal analysis, A.N.V.; investigation, T.S.S. and A.N.V.; resources, R.R.G.; data curation, A.N.V.; writing—original draft preparation, T.S.S., A.N.V., R.N.R. and C.S.; writing—review and editing, A.N.V., R.N.R., K.V.L. and C.S.; visualization, A.N.V., K.V.L.; software, A.N.V.; supervision, R.R.G.; project administration, R.R.G.; funding acquisition, R.R.G. All authors have read and agreed to the published version of the manuscript.

Funding: This research was funded by the Russian Science Foundation (19-75-30008-P).

Data Availability Statement: All datasets are available in the GEO database (https://www.ncbi.nlm.nih.gov/geo/, accessed on 15 September 2023).

Conflicts of Interest: The authors declare no conflict of interest.

References

1. Gainetdinov, R.R.; Hoener, M.C.; Berry, M.D. Trace Amines and Their Receptors. *Pharmacol. Rev.* **2018**, *70*, 549–620. [CrossRef] [PubMed]
2. Berry, M.D.; Gainetdinov, R.R.; Hoener, M.C.; Shahid, M. Pharmacology of Human Trace Amine-Associated Receptors: Therapeutic Opportunities and Challenges. *Pharmacol. Ther.* **2017**, *180*, 161–180. [CrossRef] [PubMed]
3. Borowsky, B.; Adham, N.; Jones, K.A.; Raddatz, R.; Artymyshyn, R.; Ogozalek, K.L.; Durkin, M.M.; Lakhlani, P.P.; Bonini, J.A.; Pathirana, S.; et al. Trace Amines: Identification of a Family of Mammalian G Protein-Coupled Receptors. *Proc. Natl. Acad. Sci. USA* **2001**, *98*, 8966–8971. [CrossRef] [PubMed]
4. Bunzow, J.R.; Sonders, M.S.; Arttamangkul, S.; Harrison, L.M.; Zhang, G.; Quigley, D.I.; Darland, T.; Suchland, K.L.; Pasumamula, S.; Kennedy, J.L.; et al. Amphetamine, 3,4-Methylenedioxymethamphetamine, Lysergic Acid Diethylamide, and Metabolites of the Catecholamine Neurotransmitters Are Agonists of a Rat Trace Amine Receptor. *Mol. Pharmacol.* **2001**, *60*, 1181–1188. [CrossRef] [PubMed]
5. Espinoza, S.; Sukhanov, I.; Efimova, E.V.; Kozlova, A.; Antonova, K.A.; Illiano, P.; Leo, D.; Merkulyeva, N.; Kalinina, D.; Musienko, P.; et al. Trace Amine-Associated Receptor 5 Provides Olfactory Input Into Limbic Brain Areas and Modulates Emotional Behaviors and Serotonin Transmission. *Front. Mol. Neurosci.* **2020**, *13*, 18. [CrossRef] [PubMed]
6. Samson, S.L.; Garber, A.J. Metabolic Syndrome. *Endocrinol. Metab. Clin. North Am.* **2014**, *43*, 1–23. [CrossRef] [PubMed]
7. Alberti, K.G.; Zimmet, P.Z. Definition, Diagnosis and Classification of Diabetes Mellitus and Its Complications. Part 1: Diagnosis and Classification of Diabetes Mellitus Provisional Report of a WHO Consultation. *Diabet. Med.* **1998**, *15*, 539–553. [CrossRef]
8. Eberly, L.E.; Prineas, R.; Cohen, J.D.; Vazquez, G.; Zhi, X.; Neaton, J.D.; Kuller, L.H.; Multiple Risk Factor Intervention Trial Research Group. Metabolic Syndrome: Risk Factor Distribution and 18-Year Mortality in the Multiple Risk Factor Intervention Trial. *Diabetes Care* **2006**, *29*, 123–130. [CrossRef]
9. de Toro-Martín, J.; Arsenault, B.J.; Després, J.-P.; Vohl, M.-C. Precision Nutrition: A Review of Personalized Nutritional Approaches for the Prevention and Management of Metabolic Syndrome. *Nutrients* **2017**, *9*, 913. [CrossRef]
10. Aguilar-Salinas, C.A.; Viveros-Ruiz, T. Recent Advances in Managing/Understanding the Metabolic Syndrome. *F1000Reserch* **2019**, *8*, F1000. [CrossRef]
11. Björntorp, P.; Rosmond, R. Neuroendocrine Abnormalities in Visceral Obesity. *Int. J. Obes. Relat. Metab. Disord.* **2000**, *24* (Suppl. S2), S80–S85. [CrossRef] [PubMed]
12. Sun, X.; Luquet, S.; Small, D.M. DRD2: Bridging the Genome and Ingestive Behavior. *Trends Cogn. Sci.* **2017**, *21*, 372–384. [CrossRef] [PubMed]
13. Brunerova, L.; Potockova, J.; Horacek, J.; Suchy, J.; Andel, M. Central Dopaminergic Activity Influences Metabolic Parameters in Healthy Men. *Neuroendocrinology* **2013**, *97*, 132–138. [CrossRef] [PubMed]
14. Revel, F.G.; Meyer, C.A.; Bradaia, A.; Jeanneau, K.; Calcagno, E.; André, C.B.; Haenggi, M.; Miss, M.-T.; Galley, G.; Norcross, R.D.; et al. Brain-Specific Overexpression of Trace Amine-Associated Receptor 1 Alters Monoaminergic Neurotransmission and Decreases Sensitivity to Amphetamine. *Neuropsychopharmacology* **2012**, *37*, 2580–2592. [CrossRef] [PubMed]
15. Freyberg, Z.; Saavedra, J.M. Trace Amines and Trace Amine-Associated Receptors: A New Frontier in Cell Signaling. *Cell Mol. Neurobiol.* **2020**, *40*, 189–190. [CrossRef] [PubMed]
16. Dinter, J.; Mühlhaus, J.; Jacobi, S.F.; Wienchol, C.L.; Cöster, M.; Meister, J.; Hoefig, C.S.; Müller, A.; Köhrle, J.; Grüters, A.; et al. 3-Iodothyronamine Differentially Modulates α-2A-Adrenergic Receptor-Mediated Signaling. *J. Mol. Endocrinol.* **2015**, *54*, 205–216. [CrossRef] [PubMed]
17. Revel, F.G.; Moreau, J.-L.; Gainetdinov, R.R.; Bradaia, A.; Sotnikova, T.D.; Mory, R.; Durkin, S.; Zbinden, K.G.; Norcross, R.; Meyer, C.A.; et al. TAAR1 Activation Modulates Monoaminergic Neurotransmission, Preventing Hyperdopaminergic and Hypoglutamatergic Activity. *Proc. Natl. Acad. Sci. USA* **2011**, *108*, 8485–8490. [CrossRef] [PubMed]
18. Straznicky, N.E.; Grima, M.T.; Lambert, E.A.; Sari, C.I.; Eikelis, N.; Nestel, P.J.; Phillips, S.E.; Hering, D.; Karapanagiotidis, S.; Dixon, J.B.; et al. Arterial Norepinephrine Concentration Is Inversely and Independently Associated With Insulin Clearance in Obese Individuals With Metabolic Syndrome. *J. Clin. Endocrinol. Metab.* **2015**, *100*, 1544–1550. [CrossRef]

19. Straznicky, N.E.; Grima, M.T.; Sari, C.I.; Eikelis, N.; Lambert, E.A.; Nestel, P.J.; Esler, M.D.; Dixon, J.B.; Chopra, R.; Tilbrook, A.J.; et al. Neuroadrenergic Dysfunction Along the Diabetes Continuum: A Comparative Study in Obese Metabolic Syndrome Subjects. *Diabetes* **2012**, *61*, 2506–2516. [CrossRef]
20. Halder, I.; Muldoon, M.F.; Ferrell, R.E.; Manuck, S.B. Serotonin Receptor 2A (HTR2A) Gene Polymorphisms Are Associated with Blood Pressure, Central Adiposity, and the Metabolic Syndrome. *Metab. Syndr. Relat. Disord.* **2007**, *5*, 323–330. [CrossRef]
21. Priest, C.; Tontonoz, P. Inter-Organ Cross-Talk in Metabolic Syndrome. *Nat. Metab.* **2019**, *1*, 1177–1188. [CrossRef] [PubMed]
22. Nolan, C.J.; Prentki, M. Insulin Resistance and Insulin Hypersecretion in the Metabolic Syndrome and Type 2 Diabetes: Time for a Conceptual Framework Shift. *Diabetes Vasc. Dis. Res.* **2019**, *16*, 118–127. [CrossRef] [PubMed]
23. Chen, Y.; Hong, F.; Chen, H.; Fan, R.-F.; Zhang, X.-L.; Zhang, Y.; Zhu, J.-X. Distinctive Expression and Cellular Distribution of Dopamine Receptors in the Pancreatic Islets of Rats. *Cell Tissue Res.* **2014**, *357*, 597–606. [CrossRef] [PubMed]
24. Di Cairano, E.S.; Moretti, S.; Marciani, P.; Sacchi, V.F.; Castagna, M.; Davalli, A.; Folli, F.; Perego, C. Neurotransmitters and Neuropeptides: New Players in the Control of Islet of Langerhans' Cell Mass and Function. *J. Cell Physiol.* **2016**, *231*, 756–767. [CrossRef] [PubMed]
25. Lawlor, N.; George, J.; Bolisetty, M.; Kursawe, R.; Sun, L.; Sivakamasundari, V.; Kycia, I.; Robson, P.; Stitzel, M.L. Single-Cell Transcriptomes Identify Human Islet Cell Signatures and Reveal Cell-Type-Specific Expression Changes in Type 2 Diabetes. *Genome Res.* **2017**, *27*, 208–222. [CrossRef] [PubMed]
26. Noguchi, G.M.; Huising, M.O. Integrating the Inputs That Shape Pancreatic Islet Hormone Release. *Nat. Metab.* **2019**, *1*, 1189–1201. [CrossRef] [PubMed]
27. Cataldo Bascuñan, L.R.; Lyons, C.; Bennet, H.; Artner, I.; Fex, M. Serotonergic Regulation of Insulin Secretion. *Acta Physiol.* **2019**, *225*, e13101. [CrossRef] [PubMed]
28. Paulmann, N.; Grohmann, M.; Voigt, J.-P.; Bert, B.; Vowinckel, J.; Bader, M.; Skelin, M.; Jevsek, M.; Fink, H.; Rupnik, M.; et al. Intracellular Serotonin Modulates Insulin Secretion from Pancreatic Beta-Cells by Protein Serotonylation. *PLoS Biol.* **2009**, *7*, e1000229. [CrossRef]
29. Kassi, E.; Pervanidou, P.; Kaltsas, G.; Chrousos, G. Metabolic Syndrome: Definitions and Controversies. *BMC Med.* **2011**, *9*, 48. [CrossRef]
30. Alberti, K.G.M.M.; Zimmet, P.; Shaw, J. Metabolic Syndrome–a New World-Wide Definition. A Consensus Statement from the International Diabetes Federation. *Diabet. Med.* **2006**, *23*, 469–480. [CrossRef]
31. Monzavi, R.; Dreimane, D.; Geffner, M.E.; Braun, S.; Conrad, B.; Klier, M.; Kaufman, F.R. Improvement in Risk Factors for Metabolic Syndrome and Insulin Resistance in Overweight Youth Who Are Treated with Lifestyle Intervention. *Pediatrics* **2006**, *117*, e1111–e1118. [CrossRef] [PubMed]
32. Kleinridders, A.; Ferris, H.A.; Cai, W.; Kahn, C.R. Insulin Action in Brain Regulates Systemic Metabolism and Brain Function. *Diabetes* **2014**, *63*, 2232–2243. [CrossRef]
33. Tokarz, V.L.; MacDonald, P.E.; Klip, A. The Cell Biology of Systemic Insulin Function. *J. Cell Biol.* **2018**, *217*, 2273–2289. [CrossRef] [PubMed]
34. Belfiore, A.; Frasca, F.; Pandini, G.; Sciacca, L.; Vigneri, R. Insulin Receptor Isoforms and Insulin Receptor/Insulin-like Growth Factor Receptor Hybrids in Physiology and Disease. *Endocr. Rev.* **2009**, *30*, 586–623. [CrossRef] [PubMed]
35. Bekerman, M. *Cellular Signaling in Health and Disease*; Springer: Berlin/Heidelberg, Germany, 2009.
36. Mackenzie, R.W.; Elliott, B.T. Akt/PKB Activation and Insulin Signaling: A Novel Insulin Signaling Pathway in the Treatment of Type 2 Diabetes. *Diabetes Metab. Syndr. Obes.* **2014**, *7*, 55–64. [CrossRef] [PubMed]
37. Hill, M.A.; Yang, Y.; Zhang, L.; Sun, Z.; Jia, G.; Parrish, A.R.; Sowers, J.R. Insulin Resistance, Cardiovascular Stiffening and Cardiovascular Disease. *Metabolism* **2021**, *119*, 154766. [CrossRef] [PubMed]
38. Milstein, J.L.; Ferris, H.A. The Brain as an Insulin-Sensitive Metabolic Organ. *Mol. Metab.* **2021**, *52*, 101234. [CrossRef]
39. McCracken, E.; Monaghan, M.; Sreenivasan, S. Pathophysiology of the Metabolic Syndrome. *Clin. Dermatol.* **2018**, *36*, 14–20. [CrossRef]
40. Rorsman, P.; Braun, M. Regulation of Insulin Secretion in Human Pancreatic Islets. *Annu. Rev. Physiol.* **2013**, *75*, 155–179. [CrossRef]
41. Amisten, S.; Salehi, A.; Rorsman, P.; Jones, P.M.; Persaud, S.J. An Atlas and Functional Analysis of G-Protein Coupled Receptors in Human Islets of Langerhans. *Pharmacol. Ther.* **2013**, *139*, 359–391. [CrossRef]
42. Bucolo, C.; Leggio, G.M.; Drago, F.; Salomone, S. Dopamine Outside the Brain: The Eye, Cardiovascular System and Endocrine Pancreas. *Pharmacol. Ther.* **2019**, *203*, 107392. [CrossRef] [PubMed]
43. Ustione, A.; Piston, D.W.; Harris, P.E. Minireview: Dopaminergic Regulation of Insulin Secretion from the Pancreatic Islet. *Mol. Endocrinol.* **2013**, *27*, 1198–1207. [CrossRef] [PubMed]
44. Ohara-Imaizumi, M.; Kim, H.; Yoshida, M.; Fujiwara, T.; Aoyagi, K.; Toyofuku, Y.; Nakamichi, Y.; Nishiwaki, C.; Okamura, T.; Uchida, T.; et al. Serotonin Regulates Glucose-Stimulated Insulin Secretion from Pancreatic β Cells during Pregnancy. *Proc. Natl. Acad. Sci. USA* **2013**, *110*, 19420–19425. [CrossRef]
45. Huising, M.O. Paracrine Regulation of Insulin Secretion. *Diabetologia* **2020**, *63*, 2057–2063. [CrossRef] [PubMed]
46. Braun, M.; Ramracheya, R.; Rorsman, P. Autocrine Regulation of Insulin Secretion. *Diabetes Obes. Metab.* **2012**, *14* (Suppl. S3), 143–151. [CrossRef] [PubMed]
47. Chandra, R.; Liddle, R.A. Modulation of Pancreatic Exocrine and Endocrine Secretion. *Curr. Opin. Gastroenterol.* **2013**, *29*, 517–522. [CrossRef] [PubMed]

48. Nagata, M.; Yokooji, T.; Nakai, T.; Miura, Y.; Tomita, T.; Taogoshi, T.; Sugimoto, Y.; Matsuo, H. Blockade of Multiple Monoamines Receptors Reduce Insulin Secretion from Pancreatic β-Cells. *Sci. Rep.* **2019**, *9*, 16438. [CrossRef]
49. Wei, H.; Zapata, R.C.; Lopez-Valencia, M.; Aslanoglou, D.; Farino, Z.J.; Benner, V.; Osborn, O.; Freyberg, Z.; McCarthy, M.J. Dopamine D2 Receptor Signaling Modulates Pancreatic Beta Cell Circadian Rhythms. *Psychoneuroendocrinology* **2020**, *113*, 104551. [CrossRef]
50. Cabrera, O.; Berman, D.M.; Kenyon, N.S.; Ricordi, C.; Berggren, P.-O.; Caicedo, A. The Unique Cytoarchitecture of Human Pancreatic Islets Has Implications for Islet Cell Function. *Proc. Natl. Acad. Sci. USA* **2006**, *103*, 2334–2339. [CrossRef]
51. Hartig, S.M.; Cox, A.R. Paracrine Signaling in Islet Function and Survival. *J. Mol. Med.* **2020**, *98*, 451–467. [CrossRef]
52. Underland, L.J.; Mark, E.R.; Katikaneni, R.; Heptulla, R. The Impact of Dopamine on Insulin Secretion in Healthy Controls. *Indian J. Crit. Care Med.* **2018**, *22*, 209–213. [CrossRef] [PubMed]
53. Han, X.; Li, B.; Ye, X.; Mulatibieke, T.; Wu, J.; Dai, J.; Wu, D.; Ni, J.; Zhang, R.; Xue, J.; et al. Dopamine D2 Receptor Signalling Controls Inflammation in Acute Pancreatitis via a PP2A-Dependent Akt/NF-KB Signalling Pathway. *Br. J. Pharmacol.* **2017**, *174*, 4751–4770. [CrossRef] [PubMed]
54. Feng, X.; Yan, J.; Li, G.; Liu, J.; Fan, R.; Li, S.; Zheng, L.; Zhang, Y.; Zhu, J. Source of Dopamine in Gastric Juice and Luminal Dopamine-induced Duodenal Bicarbonate Secretion via Apical Dopamine D2 Receptors. *Br. J. Pharmacol.* **2020**, *177*, 3258–3272. [CrossRef] [PubMed]
55. Mezey, E.; Eisenhofer, G.; Harta, G.; Hansson, S.; Gould, L.; Hunyady, B.; Hoffman, B.J. A Novel Nonneuronal Catecholaminergic System: Exocrine Pancreas Synthesizes and Releases Dopamine. *Proc. Natl. Acad. Sci. USA* **1996**, *93*, 10377–10382. [CrossRef] [PubMed]
56. Li, G.-W.; Li, J.; Feng, X.-Y.; Chen, H.; Chen, Y.; Liu, J.-H.; Zhang, Y.; Hong, F.; Zhu, J.-X. Pancreatic Acinar Cells Utilize Tyrosine to Synthesize L-Dihydroxyphenylalanine. *Exp. Biol. Med.* **2021**, *246*, 2533–2542. [CrossRef]
57. Ganic, E.; Johansson, J.K.; Bennet, H.; Fex, M.; Artner, I. Islet-Specific Monoamine Oxidase A and B Expression Depends on MafA Transcriptional Activity and Is Compromised in Type 2 Diabetes. *Biochem. Biophys. Res. Commun.* **2015**, *468*, 629–635. [CrossRef] [PubMed]
58. Liu, M.; Ren, L.; Zhong, X.; Ding, Y.; Liu, T.; Liu, Z.; Yang, X.; Cui, L.; Yang, L.; Fan, Y.; et al. D2-Like Receptors Mediate Dopamine-Inhibited Insulin Secretion via Ion Channels in Rat Pancreatic β-Cells. *Front. Endocrinol.* **2020**, *11*, 152. [CrossRef] [PubMed]
59. Rubí, B.; Ljubicic, S.; Pournourmohammadi, S.; Carobbio, S.; Armanet, M.; Bartley, C.; Maechler, P. Dopamine D2-like Receptors Are Expressed in Pancreatic Beta Cells and Mediate Inhibition of Insulin Secretion. *J. Biol. Chem.* **2005**, *280*, 36824–36832. [CrossRef]
60. Garcia Barrado, M.J.; Iglesias Osma, M.C.; Blanco, E.J.; Carretero Hernández, M.; Sánchez Robledo, V.; Catalano Iniesta, L.; Carrero, S.; Carretero, J. Dopamine Modulates Insulin Release and Is Involved in the Survival of Rat Pancreatic Beta Cells. *PLoS ONE* **2015**, *10*, e0123197. [CrossRef]
61. Butini, S.; Nikolic, K.; Kassel, S.; Brückmann, H.; Filipic, S.; Agbaba, D.; Gemma, S.; Brogi, S.; Brindisi, M.; Campiani, G.; et al. Polypharmacology of Dopamine Receptor Ligands. *Prog. Neurobiol.* **2016**, *142*, 68–103. [CrossRef]
62. Yung, K.K.; Bolam, J.P. Localization of Dopamine D1 and D2 Receptors in the Rat Neostriatum: Synaptic Interaction with Glutamate- and GABA-Containing Axonal Terminals. *Synapse* **2000**, *38*, 413–420. [CrossRef] [PubMed]
63. Aslanoglou, D.; Bertera, S.; Sánchez-Soto, M.; Benjamin Free, R.; Lee, J.; Zong, W.; Xue, X.; Shrestha, S.; Brissova, M.; Logan, R.W.; et al. Dopamine Regulates Pancreatic Glucagon and Insulin Secretion via Adrenergic and Dopaminergic Receptors. *Transl. Psychiatry* **2021**, *11*, 59. [CrossRef] [PubMed]
64. Sakano, D.; Choi, S.; Kataoka, M.; Shiraki, N.; Uesugi, M.; Kume, K.; Kume, S. Dopamine D2 Receptor-Mediated Regulation of Pancreatic β Cell Mass. *Stem Cell Rep.* **2016**, *7*, 95–109. [CrossRef] [PubMed]
65. Garcia-Tornadú, I.; Perez-Millan, M.I.; Recouvreux, V.; Ramirez, M.C.; Luque, G.; Risso, G.S.; Ornstein, A.M.; Cristina, C.; Diaz-Torga, G.; Becu-Villalobos, D. New Insights into the Endocrine and Metabolic Roles of Dopamine D2 Receptors Gained from the Drd2 Mouse. *Neuroendocrinology* **2010**, *92*, 207–214. [CrossRef] [PubMed]
66. Pijl, H. Reduced Dopaminergic Tone in Hypothalamic Neural Circuits: Expression of a "Thrifty" Genotype Underlying the Metabolic Syndrome? *Eur. J. Pharmacol.* **2003**, *480*, 125–131. [CrossRef] [PubMed]
67. DeFronzo, R.A. Bromocriptine: A Sympatholytic, D2-Dopamine Agonist for the Treatment of Type 2 Diabetes. *Diabetes Care* **2011**, *34*, 789–794. [CrossRef] [PubMed]
68. Beaulieu, J.-M.; Gainetdinov, R.R. The Physiology, Signaling, and Pharmacology of Dopamine Receptors. *Pharmacol. Rev.* **2011**, *63*, 182–217. [CrossRef]
69. Mishra, A.; Singh, S.; Shukla, S. Physiological and Functional Basis of Dopamine Receptors and Their Role in Neurogenesis: Possible Implication for Parkinson's Disease. *J. Exp. Neurosci.* **2018**, *12*, 1179069518779829. [CrossRef]
70. Seeman, P.; Van Tol, H.H. Dopamine Receptor Pharmacology. *Trends Pharmacol. Sci.* **1994**, *15*, 264–270. [CrossRef]
71. Schultz, W. Predictive Reward Signal of Dopamine Neurons. *J. Neurophysiol.* **1998**, *80*, 1–27. [CrossRef]
72. Taber, K.H.; Black, D.N.; Porrino, L.J.; Hurley, R.A. Neuroanatomy of Dopamine: Reward and Addiction. *J. Neuropsychiatry Clin. Neurosci.* **2012**, *24*, 1–4. [CrossRef] [PubMed]
73. Baik, J.-H. Dopamine Signaling in Food Addiction: Role of Dopamine D2 Receptors. *BMB Rep.* **2013**, *46*, 519–526. [CrossRef]
74. Baldo, B.A.; Sadeghian, K.; Basso, A.M.; Kelley, A.E. Effects of Selective Dopamine D1 or D2 Receptor Blockade within Nucleus Accumbens Subregions on Ingestive Behavior and Associated Motor Activity. *Behav. Brain Res.* **2002**, *137*, 165–177. [CrossRef] [PubMed]

75. Teegarden, S.L.; Nestler, E.J.; Bale, T.L. Delta FosB-Mediated Alterations in Dopamine Signaling Are Normalized by a Palatable High-Fat Diet. *Biol. Psychiatry* **2008**, *64*, 941–950. [CrossRef] [PubMed]
76. Halpern, C.H.; Tekriwal, A.; Santollo, J.; Keating, J.G.; Wolf, J.A.; Daniels, D.; Bale, T.L. Amelioration of Binge Eating by Nucleus Accumbens Shell Deep Brain Stimulation in Mice Involves D2 Receptor Modulation. *J. Neurosci.* **2013**, *33*, 7122–7129. [CrossRef] [PubMed]
77. Johnson, P.M.; Kenny, P.J. Dopamine D2 Receptors in Addiction-like Reward Dysfunction and Compulsive Eating in Obese Rats. *Nat. Neurosci.* **2010**, *13*, 635–641. [CrossRef] [PubMed]
78. Rubí, B.; Maechler, P. Minireview: New Roles for Peripheral Dopamine on Metabolic Control and Tumor Growth: Let's Seek the Balance. *Endocrinology* **2010**, *151*, 5570–5581. [CrossRef]
79. Freeman, M.E.; Kanyicska, B.; Lerant, A.; Nagy, G. Prolactin: Structure, Function, and Regulation of Secretion. *Physiol. Rev.* **2000**, *80*, 1523–1631. [CrossRef]
80. Labouesse, M.A.; Sartori, A.M.; Weinmann, O.; Simpson, E.H.; Kellendonk, C.; Weber-Stadlbauer, U. Striatal Dopamine 2 Receptor Upregulation during Development Predisposes to Diet-Induced Obesity by Reducing Energy Output in Mice. *Proc. Natl. Acad. Sci. USA* **2018**, *115*, 10493–10498. [CrossRef]
81. Zhang, Z.; Hao, C.-J.; Li, C.-G.; Zang, D.-J.; Zhao, J.; Li, X.-N.; Wei, A.-H.; Wei, Z.-B.; Yang, L.; He, X.; et al. Mutation of SLC35D3 Causes Metabolic Syndrome by Impairing Dopamine Signaling in Striatal D1 Neurons. *PLoS Genet.* **2014**, *10*, e1004124. [CrossRef]
82. Wu, C.; Garamszegi, S.P.; Xie, X.; Mash, D.C. Altered Dopamine Synaptic Markers in Postmortem Brain of Obese Subjects. *Front. Hum. Neurosci.* **2017**, *11*, 386. [CrossRef]
83. Kleinridders, A.; Pothos, E.N. Impact of Brain Insulin Signaling on Dopamine Function, Food Intake, Reward, and Emotional Behavior. *Curr. Nutr. Rep.* **2019**, *8*, 83–91. [CrossRef] [PubMed]
84. Taniguchi, C.M.; Emanuelli, B.; Kahn, C.R. Critical Nodes in Signalling Pathways: Insights into Insulin Action. *Nat. Rev. Mol. Cell Biol.* **2006**, *7*, 85–96. [CrossRef] [PubMed]
85. Ding, J.; Liu, Y.; Sullivan, D.A. Effects of Insulin and High Glucose on Human Meibomian Gland Epithelial Cells. *Investig. Ophthalmol. Vis. Sci.* **2015**, *56*, 7814–7820. [CrossRef] [PubMed]
86. Beaulieu, J.-M.; Sotnikova, T.D.; Marion, S.; Lefkowitz, R.J.; Gainetdinov, R.R.; Caron, M.G. An Akt/Beta-Arrestin 2/PP2A Signaling Complex Mediates Dopaminergic Neurotransmission and Behavior. *Cell* **2005**, *122*, 261–273. [CrossRef] [PubMed]
87. Owens, W.A.; Williams, J.M.; Saunders, C.; Avison, M.J.; Galli, A.; Daws, L.C. Rescue of Dopamine Transporter Function in Hypoinsulinemic Rats by a D2 Receptor-ERK-Dependent Mechanism. *J. Neurosci.* **2012**, *32*, 2637–2647. [CrossRef] [PubMed]
88. Kleinridders, A.; Cai, W.; Cappellucci, L.; Ghazarian, A.; Collins, W.R.; Vienberg, S.G.; Pothos, E.N.; Kahn, C.R. Insulin Resistance in Brain Alters Dopamine Turnover and Causes Behavioral Disorders. *Proc. Natl. Acad. Sci. USA* **2015**, *112*, 3463–3468. [CrossRef] [PubMed]
89. Luo, S.; Ezrokhi, M.; Cominos, N.; Tsai, T.-H.; Stoelzel, C.R.; Trubitsyna, Y.; Cincotta, A.H. Experimental Dopaminergic Neuron Lesion at the Area of the Biological Clock Pacemaker, Suprachiasmatic Nuclei (SCN) Induces Metabolic Syndrome in Rats. *Diabetol. Metab. Syndr.* **2021**, *13*, 11. [CrossRef]
90. Farino, Z.J.; Morgenstern, T.J.; Maffei, A.; Quick, M.; De Solis, A.J.; Wiriyasermkul, P.; Freyberg, R.J.; Aslanoglou, D.; Sorisio, D.; Inbar, B.P.; et al. New Roles for Dopamine D2 and D3 Receptors in Pancreatic Beta Cell Insulin Secretion. *Mol. Psychiatry* **2020**, *25*, 2070–2085. [CrossRef]
91. Freudenthal, R.; Moncrieff, J. "A Landmark in Psychiatric Progress"? The Role of Evidence in the Rise and Fall of Insulin Coma Therapy. *Hist. Psychiatry* **2022**, *33*, 65–78. [CrossRef]
92. Mansur, R.B.; Fries, G.R.; Subramaniapillai, M.; Frangou, S.; De Felice, F.G.; Rasgon, N.; McEwen, B.; Brietzke, E.; McIntyre, R.S. Expression of Dopamine Signaling Genes in the Post-Mortem Brain of Individuals with Mental Illnesses Is Moderated by Body Mass Index and Mediated by Insulin Signaling Genes. *J. Psychiatr. Res.* **2018**, *107*, 128–135. [CrossRef] [PubMed]
93. Lamos, E.M.; Levitt, D.L.; Munir, K.M. A Review of Dopamine Agonist Therapy in Type 2 Diabetes and Effects on Cardio-Metabolic Parameters. *Prim. Care Diabetes* **2016**, *10*, 60–65. [CrossRef] [PubMed]
94. Tavares, G.; Marques, D.; Barra, C.; Rosendo-Silva, D.; Costa, A.; Rodrigues, T.; Gasparini, P.; Melo, B.F.; Sacramento, J.F.; Seiça, R.; et al. Dopamine D2 Receptor Agonist, Bromocriptine, Remodels Adipose Tissue Dopaminergic Signalling and Upregulates Catabolic Pathways, Improving Metabolic Profile in Type 2 Diabetes. *Mol. Metab.* **2021**, *51*, 101241. [CrossRef] [PubMed]
95. Ahrén, B. Autonomic Regulation of Islet Hormone Secretion–Implications for Health and Disease. *Diabetologia* **2000**, *43*, 393–410. [CrossRef] [PubMed]
96. Rosengren, A.H.; Jokubka, R.; Tojjar, D.; Granhall, C.; Hansson, O.; Li, D.-Q.; Nagaraj, V.; Reinbothe, T.M.; Tuncel, J.; Eliasson, L.; et al. Overexpression of Alpha2A-Adrenergic Receptors Contributes to Type 2 Diabetes. *Science* **2010**, *327*, 217–220. [CrossRef] [PubMed]
97. Santulli, G.; Lombardi, A.; Sorriento, D.; Anastasio, A.; Del Giudice, C.; Formisano, P.; Béguinot, F.; Trimarco, B.; Miele, C.; Iaccarino, G. Age-Related Impairment in Insulin Release. *Diabetes* **2012**, *61*, 692–701. [CrossRef]
98. Costa, V.M.; Grando, L.G.R.; Milandri, E.; Nardi, J.; Teixeira, P.; Mladěnka, P.; Remião, F. On behalf of The OEMONOM Natural Sympathomimetic Drugs: From Pharmacology to Toxicology. *Biomolecules* **2022**, *12*, 1793. [CrossRef]
99. Teixeira, G.P.; Faria, R.X. Influence of Purinergic Signaling on Glucose Transporters: A Possible Mechanism against Insulin Resistance? *Eur. J. Pharmacol.* **2021**, *892*, 173743. [CrossRef]

100. Hoffstedt, J.; Wahrenberg, H.; Thörne, A.; Lönnqvist, F. The Metabolic Syndrome Is Related to B3-Adrenoceptor Sensitivity in Visceral Adipose Tissue. *Diabetologia* **1996**, *39*, 838–860. [CrossRef]
101. Martin, A.M.; Young, R.L.; Leong, L.; Rogers, G.B.; Spencer, N.J.; Jessup, C.F.; Keating, D.J. The Diverse Metabolic Roles of Peripheral Serotonin. *Endocrinology* **2017**, *158*, 1049–1063. [CrossRef]
102. Kim, K.; Oh, C.-M.; Ohara-Imaizumi, M.; Park, S.; Namkung, J.; Yadav, V.K.; Tamarina, N.A.; Roe, M.W.; Philipson, L.H.; Karsenty, G.; et al. Functional Role of Serotonin in Insulin Secretion in a Diet-Induced Insulin-Resistant State. *Endocrinology* **2015**, *156*, 444–452. [CrossRef] [PubMed]
103. Ohta, Y.; Kosaka, Y.; Kishimoto, N.; Wang, J.; Smith, S.B.; Honig, G.; Kim, H.; Gasa, R.M.; Neubauer, N.; Liou, A.; et al. Convergence of the Insulin and Serotonin Programs in the Pancreatic β-Cell. *Diabetes* **2011**, *60*, 3208–3216. [CrossRef] [PubMed]
104. Georgescu, T.; Lyons, D.; Heisler, L.K. Role of Serotonin in Body Weight, Insulin Secretion and Glycaemic Control. *J. Neuroendocr.* **2021**, *33*, e12960. [CrossRef] [PubMed]
105. Guenette, M.D.; Giacca, A.; Hahn, M.; Teo, C.; Lam, L.; Chintoh, A.; Arenovich, T.; Remington, G. Atypical Antipsychotics and Effects of Adrenergic and Serotonergic Receptor Binding on Insulin Secretion In-Vivo: An Animal Model. *Schizophr. Res.* **2013**, *146*, 162–169. [CrossRef] [PubMed]
106. Bennet, H.; Mollet, I.G.; Balhuizen, A.; Medina, A.; Nagorny, C.; Bagge, A.; Fadista, J.; Ottosson-Laakso, E.; Vikman, P.; Dekker-Nitert, M.; et al. Serotonin (5-HT) Receptor 2b Activation Augments Glucose-Stimulated Insulin Secretion in Human and Mouse Islets of Langerhans. *Diabetologia* **2016**, *59*, 744–754. [CrossRef] [PubMed]
107. Xie, K.-Y.; Chien, S.-J.; Tan, B.C.-M.; Chen, Y.-W. RNA Editing of 5-HT2C R Impairs Insulin Secretion of Pancreatic Beta Cells via Altered Store-Operated Calcium Entry. *FASEB J.* **2021**, *35*, e21929. [CrossRef] [PubMed]
108. Gilles, M.; Wilke, A.; Kopf, D.; Nonell, A.; Lehnert, H.; Deuschle, M. Antagonism of the Serotonin (5-HT)-2 Receptor and Insulin Sensitivity: Implications for Atypical Antipsychotics. *Psychosom. Med.* **2005**, *67*, 748–751. [CrossRef] [PubMed]
109. Burchett, S.A.; Hicks, T.P. The Mysterious Trace Amines: Protean Neuromodulators of Synaptic Transmission in Mammalian Brain. *Prog. Neurobiol.* **2006**, *79*, 223–246. [CrossRef]
110. Lindemann, L.; Ebeling, M.; Kratochwil, N.A.; Bunzow, J.R.; Grandy, D.K.; Hoener, M.C. Trace Amine-Associated Receptors Form Structurally and Functionally Distinct Subfamilies of Novel G Protein-Coupled Receptors. *Genomics* **2005**, *85*, 372–385. [CrossRef]
111. Martin, F.J.; Amode, M.R.; Aneja, A.; Austine-Orimoloye, O.; Azov, A.G.; Barnes, I.; Becker, A.; Bennett, R.; Berry, A.; Bhai, J.; et al. Ensembl 2023. *Nucleic Acids Res.* **2023**, *51*, D933–D941. [CrossRef]
112. Khan, M.Z.; Nawaz, W. The Emerging Roles of Human Trace Amines and Human Trace Amine-Associated Receptors (HTAARs) in Central Nervous System. *Biomed. Pharmacother.* **2016**, *83*, 439–449. [CrossRef] [PubMed]
113. Liberles, S.D. Trace Amine-Associated Receptors Are Olfactory Receptors in Vertebrates. *Ann. N. Y. Acad. Sci.* **2009**, *1170*, 168–172. [CrossRef] [PubMed]
114. Liberles, S.D.; Buck, L.B. A Second Class of Chemosensory Receptors in the Olfactory Epithelium. *Nature* **2006**, *442*, 645–650. [CrossRef] [PubMed]
115. Efimova, E.V.; Kozlova, A.A.; Razenkova, V.; Katolikova, N.V.; Antonova, K.A.; Sotnikova, T.D.; Merkulyeva, N.S.; Veshchitskii, A.S.; Kalinina, D.S.; Korzhevskii, D.E.; et al. Increased Dopamine Transmission and Adult Neurogenesis in Trace Amine-Associated Receptor 5 (TAAR5) Knockout Mice. *Neuropharmacology* **2021**, *182*, 108373. [CrossRef]
116. Efimova, E.V.; Kuvarzin, S.R.; Mor, M.S.; Katolikova, N.V.; Shemiakova, T.S.; Razenkova, V.; Ptukha, M.; Kozlova, A.A.; Murtazina, R.Z.; Smirnova, D.; et al. Trace Amine-Associated Receptor 2 Is Expressed in the Limbic Brain Areas and Is Involved in Dopamine Regulation and Adult Neurogenesis. *Front. Behav. Neurosci.* **2022**, *16*, 847410. [CrossRef]
117. Vaganova, A.N.; Murtazina, R.Z.; Shemyakova, T.S.; Prjibelski, A.D.; Katolikova, N.V.; Gainetdinov, R.R. Pattern of TAAR5 Expression in the Human Brain Based on Transcriptome Datasets Analysis. *Int. J. Mol. Sci.* **2021**, *22*, 8802. [CrossRef]
118. Katolikova, N.V.; Vaganova, A.N.; Efimova, E.V.; Gainetdinov, R.R. Expression of Trace Amine-Associated Receptors in the Murine and Human Hippocampus Based on Public Transcriptomic Data. *Cells* **2022**, *11*, 1813. [CrossRef]
119. Vaganova, A.N.; Katolikova, N.V.; Murtazina, R.Z.; Kuvarzin, S.R.; Gainetdinov, R.R. Public Transcriptomic Data Meta-Analysis Demonstrates TAAR6 Expression in the Mental Disorder-Related Brain Areas in Human and Mouse Brain. *Biomolecules* **2022**, *12*, 1259. [CrossRef]
120. Ito, J.; Ito, M.; Nambu, H.; Fujikawa, T.; Tanaka, K.; Iwaasa, H.; Tokita, S. Anatomical and Histological Profiling of Orphan G-Protein-Coupled Receptor Expression in Gastrointestinal Tract of C57BL/6J Mice. *Cell Tissue Res.* **2009**, *338*, 257–269. [CrossRef]
121. Zucchi, R.; Chiellini, G.; Scanlan, T.S.; Grandy, D.K. Trace Amine-Associated Receptors and Their Ligands. *Br. J. Pharmacol.* **2006**, *149*, 967–978. [CrossRef]
122. Raab, S.; Wang, H.; Uhles, S.; Cole, N.; Alvarez-Sanchez, R.; Künnecke, B.; Ullmer, C.; Matile, H.; Bedoucha, M.; Norcross, R.D.; et al. Incretin-like Effects of Small Molecule Trace Amine-Associated Receptor 1 Agonists. *Mol. Metab.* **2016**, *5*, 47–56. [CrossRef] [PubMed]
123. Miller, G.M.; Verrico, C.D.; Jassen, A.; Konar, M.; Yang, H.; Panas, H.; Bahn, M.; Johnson, R.; Madras, B.K. Primate Trace Amine Receptor 1 Modulation by the Dopamine Transporter. *J. Pharmacol. Exp. Ther.* **2005**, *313*, 983–994. [CrossRef] [PubMed]
124. Salahpour, A.; Espinoza, S.; Masri, B.; Lam, V.; Barak, L.S.; Gainetdinov, R.R. BRET Biosensors to Study GPCR Biology, Pharmacology, and Signal Transduction. *Front. Endocrinol.* **2012**, *3*, 105. [CrossRef] [PubMed]
125. Espinoza, S.; Salahpour, A.; Masri, B.; Sotnikova, T.D.; Messa, M.; Barak, L.S.; Caron, M.G.; Gainetdinov, R.R. Functional Interaction between Trace Amine-Associated Receptor 1 and Dopamine D2 Receptor. *Mol. Pharmacol.* **2011**, *80*, 416–425. [CrossRef] [PubMed]

126. Espinoza, S.; Ghisi, V.; Emanuele, M.; Leo, D.; Sukhanov, I.; Sotnikova, T.D.; Chieregatti, E.; Gainetdinov, R.R. Postsynaptic D2 Dopamine Receptor Supersensitivity in the Striatum of Mice Lacking TAAR1. *Neuropharmacology* **2015**, *93*, 308–313. [CrossRef]
127. Harmeier, A.; Obermueller, S.; Meyer, C.A.; Revel, F.G.; Buchy, D.; Chaboz, S.; Dernick, G.; Wettstein, J.G.; Iglesias, A.; Rolink, A.; et al. Trace Amine-Associated Receptor 1 Activation Silences GSK3β Signaling of TAAR1 and D2R Heteromers. *Eur. Neuropsychopharmacol.* **2015**, *25*, 2049–2061. [CrossRef]
128. Bradaia, A.; Trube, G.; Stalder, H.; Norcross, R.D.; Ozmen, L.; Wettstein, J.G.; Pinard, A.; Buchy, D.; Gassmann, M.; Hoener, M.C.; et al. The Selective Antagonist EPPTB Reveals TAAR1-Mediated Regulatory Mechanisms in Dopaminergic Neurons of the Mesolimbic System. *Proc. Natl. Acad. Sci. USA* **2009**, *106*, 20081–20086. [CrossRef]
129. Geracitano, R.; Federici, M.; Prisco, S.; Bernardi, G.; Mercuri, N.B. Inhibitory Effects of Trace Amines on Rat Midbrain Dopaminergic Neurons. *Neuropharmacology* **2004**, *46*, 807–814. [CrossRef]
130. Leo, D.; Mus, L.; Espinoza, S.; Hoener, M.C.; Sotnikova, T.D.; Gainetdinov, R.R. Taar1-Mediated Modulation of Presynaptic Dopaminergic Neurotransmission: Role of D2 Dopamine Autoreceptors. *Neuropharmacology* **2014**, *81*, 283–291. [CrossRef]
131. Pretorius, L.; Smith, C. The Trace Aminergic System: A Gender-Sensitive Therapeutic Target for IBS? *J. Biomed. Sci.* **2020**, *27*, 95. [CrossRef]
132. Schwartz, M.D.; Canales, J.J.; Zucchi, R.; Espinoza, S.; Sukhanov, I.; Gainetdinov, R.R. Trace Amine-Associated Receptor 1: A Multimodal Therapeutic Target for Neuropsychiatric Diseases. *Expert. Opin. Ther. Targets* **2018**, *22*, 513–526. [CrossRef] [PubMed]
133. Regard, J.B.; Sato, I.T.; Coughlin, S.R. Anatomical Profiling of G Protein-Coupled Receptor Expression. *Cell* **2008**, *135*, 561–571. [CrossRef] [PubMed]
134. Fan, F.S. Ractopamine Residues in Meat Might Reduce the Risk of Type 2 Diabetes. *Nutr. Health* **2023**, *29*, 5–7. [CrossRef] [PubMed]
135. Ferragud, A.; Howell, A.D.; Moore, C.F.; Ta, T.L.; Hoener, M.C.; Sabino, V.; Cottone, P. The Trace Amine-Associated Receptor 1 Agonist RO5256390 Blocks Compulsive, Binge-like Eating in Rats. *Neuropsychopharmacology* **2017**, *42*, 1458–1470. [CrossRef] [PubMed]
136. Adriaenssens, A.; Lam, B.Y.H.; Billing, L.; Skeffington, K.; Sewing, S.; Reimann, F.; Gribble, F. A Transcriptome-Led Exploration of Molecular Mechanisms Regulating Somatostatin-Producing D-Cells in the Gastric Epithelium. *Endocrinology* **2015**, *156*, 3924–3936. [CrossRef] [PubMed]
137. Shimada, M.; Date, Y.; Mondal, M.S.; Toshinai, K.; Shimbara, T.; Fukunaga, K.; Murakami, N.; Miyazato, M.; Kangawa, K.; Yoshimatsu, H.; et al. Somatostatin Suppresses Ghrelin Secretion from the Rat Stomach. *Biochem. Biophys. Res. Commun.* **2003**, *302*, 520–525. [CrossRef] [PubMed]
138. Apryatin, S.A.; Zhukov, I.S.; Zolotoverkhaya, E.A.; Kuvarzin, S.R.; Khunagov, T.A.; Ushmugina, S.V.; Klimenko, V.M. Protein Metabolism Changes and Alterations in Behavior of Trace Amine-Associated Receptor 1 Knockout Mice Fed a High-Fructose Diet. *Neurol. Int.* **2023**, *15*, 339–351. [CrossRef] [PubMed]
139. Murtazina, R.Z.; Zhukov, I.S.; Korenkova, O.M.; Popova, E.A.; Kuvarzin, S.R.; Efimova, E.V.; Kubarskaya, L.G.; Batotsyrenova, E.G.; Zolotoverkhaya, E.A.; Vaganova, A.N.; et al. Genetic Deletion of Trace-Amine Associated Receptor 9 (TAAR9) in Rats Leads to Decreased Blood Cholesterol Levels. *Int. J. Mol. Sci.* **2021**, *22*, 2942. [CrossRef]
140. Bandyopadhyay, D.; Qureshi, A.; Ghosh, S.; Ashish, K.; Heise, L.R.; Hajra, A.; Ghosh, R.K. Safety and Efficacy of Extremely Low LDL-Cholesterol Levels and Its Prospects in Hyperlipidemia Management. *J. Lipids* **2018**, *2018*, 8598054. [CrossRef]
141. Mühlhaus, J.; Dinter, J.; Jyrch, S.; Teumer, A.; Jacobi, S.F.; Homuth, G.; Kühnen, P.; Wiegand, S.; Grüters, A.; Völzke, H.; et al. Investigation of Naturally Occurring Single-Nucleotide Variants in Human TAAR1. *Front. Pharmacol.* **2017**, *8*, 807. [CrossRef]
142. Zhai, L.; Xiao, H.; Lin, C.; Wong, H.L.X.; Lam, Y.Y.; Gong, M.; Wu, G.; Ning, Z.; Huang, C.; Zhang, Y.; et al. Gut Microbiota-Derived Tryptamine and Phenethylamine Impair Insulin Sensitivity in Metabolic Syndrome and Irritable Bowel Syndrome. *Nat. Commun.* **2023**, *14*, 4986. [CrossRef]
143. Carpéné, C.; Viana, P.; Fontaine, J.; Laurell, H.; Grolleau, J.-L. Multiple Direct Effects of the Dietary Protoalkaloid N-Methyltyramine in Human Adipocytes. *Nutrients* **2022**, *14*, 3118. [CrossRef] [PubMed]
144. Lillich, F.F.; Imig, J.D.; Proschak, E. Multi-Target Approaches in Metabolic Syndrome. *Front. Pharmacol.* **2020**, *11*, 554961. [CrossRef] [PubMed]
145. Tremmel, E.; Hofmann, S.; Kuhn, C.; Heidegger, H.; Heublein, S.; Hermelink, K.; Wuerstlein, R.; Harbeck, N.; Mayr, D.; Mahner, S.; et al. Thyronamine Regulation of TAAR1 Expression in Breast Cancer Cells and Investigation of Its Influence on Viability and Migration. *Breast Cancer* **2019**, *11*, 87–97. [CrossRef]
146. Vattai, A.; Akyol, E.; Kuhn, C.; Hofmann, S.; Heidegger, H.; von Koch, F.; Hermelink, K.; Wuerstlein, R.; Harbeck, N.; Mayr, D.; et al. Increased Trace Amine-Associated Receptor 1 (TAAR1) Expression Is Associated with a Positive Survival Rate in Patients with Breast Cancer. *J. Cancer Res. Clin. Oncol.* **2017**, *143*, 1637–1647. [CrossRef] [PubMed]
147. Revel, F.G.; Moreau, J.-L.; Pouzet, B.; Mory, R.; Bradaia, A.; Buchy, D.; Metzler, V.; Chaboz, S.; Groebke Zbinden, K.; Galley, G.; et al. A New Perspective for Schizophrenia: TAAR1 Agonists Reveal Antipsychotic- and Antidepressant-like Activity, Improve Cognition and Control Body Weight. *Mol. Psychiatry* **2013**, *18*, 543–556. [CrossRef] [PubMed]
148. Koblan, K.S.; Hopkins, S.; Kent, J.; Cheng, H.; Goldman, R.; Loebel, A. 138 Efficacy and Safety of SEP-363856, a Novel Psychotropic Agent with a Non-D2 Mechanism of Action, in the Treatment of Schizophrenia. *CNS Spectr.* **2020**, *25*, 287–288. [CrossRef]

149. Correll, C.U.; Koblan, K.S.; Hopkins, S.C.; Li, Y.; Heather Dworak Null; Goldman, R.; Loebel, A. Safety and Effectiveness of Ulotaront (SEP-363856) in Schizophrenia: Results of a 6-Month, Open-Label Extension Study. *NPJ Schizophr.* **2021**, *7*, 63. [CrossRef] [PubMed]
150. Achtyes, E.D.; Hopkins, S.C.; Dedic, N.; Dworak, H.; Zeni, C.; Koblan, K. Ulotaront: Review of Preliminary Evidence for the Efficacy and Safety of a TAAR1 Agonist in Schizophrenia. *Eur. Arch. Psychiatry Clin. Neurosci.* **2023**, *273*, 1543–1556. [CrossRef]
151. Cripps, M.J.; Bagnati, M.; Jones, T.A.; Ogunkolade, B.W.; Sayers, S.R.; Caton, P.W.; Hanna, K.; Billacura, M.P.; Fair, K.; Nelson, C.; et al. Identification of a Subset of Trace Amine-Associated Receptors and Ligands as Potential Modulators of Insulin Secretion. *Biochem. Pharmacol.* **2020**, *171*, 113685. [CrossRef]
152. Michael, E.S.; Covic, L.; Kuliopulos, A. Trace Amine-Associated Receptor 1 (TAAR1) Promotes Anti-Diabetic Signaling in Insulin-Secreting Cells. *J. Biol. Chem.* **2019**, *294*, 4401–4411. [CrossRef]
153. Tan, E.S.; Miyakawa, M.; Bunzow, J.R.; Grandy, D.K.; Scanlan, T.S. Exploring the Structure−Activity Relationship of the Ethylamine Portion of 3-Iodothyronamine for Rat and Mouse Trace Amine-Associated Receptor 1. *J. Med. Chem.* **2007**, *50*, 2787–2798. [CrossRef] [PubMed]
154. Homuth, G.; Lietzow, J.; Schanze, N.; Golchert, J.; Köhrle, J. Endocrine, Metabolic and Pharmacological Effects of Thyronamines (TAM), Thyroacetic Acids (TA) and Thyroid Hormone Metabolites (THM)–Evidence from in Vitro, Cellular, Experimental Animal and Human Studies. *Exp. Clin. Endocrinol. Diabetes* **2020**, *128*, 401–413. [CrossRef] [PubMed]
155. Rutigliano, G.; Bandini, L.; Sestito, S.; Chiellini, G. 3-Iodothyronamine and Derivatives: New Allies Against Metabolic Syndrome? *Int. J. Mol. Sci.* **2020**, *21*, 2005. [CrossRef] [PubMed]
156. Manni, M.E.; De Siena, G.; Saba, A.; Marchini, M.; Landucci, E.; Gerace, E.; Zazzeri, M.; Musilli, C.; Pellegrini-Giampietro, D.; Matucci, R.; et al. Pharmacological Effects of 3-Iodothyronamine (T1AM) in Mice Include Facilitation of Memory Acquisition and Retention and Reduction of Pain Threshold. *Br. J. Pharmacol.* **2013**, *168*, 354–362. [CrossRef] [PubMed]
157. Regard, J.B.; Kataoka, H.; Cano, D.A.; Camerer, E.; Yin, L.; Zheng, Y.-W.; Scanlan, T.S.; Hebrok, M.; Coughlin, S.R. Probing Cell Type–Specific Functions of Gi in Vivo Identifies GPCR Regulators of Insulin Secretion. *J. Clin. Investig.* **2007**, *117*, 4034–4043. [CrossRef] [PubMed]
158. Tamayo, A.; Gonçalves, L.M.; Rodriguez-Diaz, R.; Pereira, E.; Canales, M.; Caicedo, A.; Almaça, J. Pericyte Control of Blood Flow in Intraocular Islet Grafts Impacts Glucose Homeostasis in Mice. *Diabetes* **2022**, *71*, 1679–1693. [CrossRef]
159. Barrett, T.; Wilhite, S.E.; Ledoux, P.; Evangelista, C.; Kim, I.F.; Tomashevsky, M.; Marshall, K.A.; Phillippy, K.H.; Sherman, P.M.; Holko, M.; et al. NCBI GEO: Archive for Functional Genomics Data Sets–Update. *Nucleic Acids Res.* **2013**, *41*, D991–D995. [CrossRef]
160. Robinson, M.D.; McCarthy, D.J.; Smyth, G.K. EdgeR: A Bioconductor Package for Differential Expression Analysis of Digital Gene Expression Data. *Bioinformatics* **2010**, *26*, 139–140. [CrossRef]
161. Wu, T.; Hu, E.; Xu, S.; Chen, M.; Guo, P.; Dai, Z.; Feng, T.; Zhou, L.; Tang, W.; Zhan, L.; et al. ClusterProfiler 4.0: A Universal Enrichment Tool for Interpreting Omics Data. *Innovation* **2021**, *2*, 100141. [CrossRef]
162. Zhang, S.; Li, L.; Li, H. Role of Ectopic Olfactory Receptors in Glucose and Lipid Metabolism. *Br. J. Pharmacol.* **2021**, *178*, 4792–4807. [CrossRef] [PubMed]
163. Yang, Z.; Cheng, J.; Shang, P.; Sun, J.-P.; Yu, X. Emerging Roles of Olfactory Receptors in Glucose Metabolism. *Trends Cell Biol.* **2022**, *33*, 463–476. [CrossRef] [PubMed]
164. Leem, J.; Shim, H.; Cho, H.; Park, J.-H. Octanoic Acid Potentiates Glucose-Stimulated Insulin Secretion and Expression of Glucokinase through the Olfactory Receptor in Pancreatic β-Cells. *Biochem. Biophys. Res. Commun.* **2018**, *503*, 278–284. [CrossRef] [PubMed]
165. Seiron, P.; Stenwall, A.; Hedin, A.; Granlund, L.; Esguerra, J.L.S.; Volkov, P.; Renström, E.; Korsgren, O.; Lundberg, M.; Skog, O. Transcriptional Analysis of Islets of Langerhans from Organ Donors of Different Ages. *PLoS ONE* **2021**, *16*, e0247888. [CrossRef]
166. Cheng, J.; Yang, Z.; Ge, X.-Y.; Gao, M.-X.; Meng, R.; Xu, X.; Zhang, Y.-Q.; Li, R.-Z.; Lin, J.-Y.; Tian, Z.-M.; et al. Autonomous Sensing of the Insulin Peptide by an Olfactory G Protein-Coupled Receptor Modulates Glucose Metabolism. *Cell Metab.* **2022**, *34*, 240–255. [CrossRef] [PubMed]
167. Oh, J.H.; Han, Y.E.; Bao, Y.R.; Kang, C.W.; Koo, J.; Ku, C.R.; Cho, Y.H.; Lee, E.J. Olfactory Marker Protein Regulation of Glucagon Secretion in Hyperglycemia. *Exp. Mol. Med.* **2022**, *54*, 1502–1510. [CrossRef]
168. Park, J.-H.; Song, D.-K. Sweet Taste Receptors as a Tool for an Amplifying Pathway of Glucose-Stimulated Insulin Secretion in Pancreatic β Cells. *Pflug. Arch.* **2019**, *471*, 655–657. [CrossRef] [PubMed]
169. Kojima, I.; Nakagawa, Y. The Role of the Sweet Taste Receptor in Enteroendocrine Cells and Pancreatic β-Cells. *Diabetes Metab. J.* **2011**, *35*, 451–457. [CrossRef]
170. Oya, M.; Suzuki, H.; Watanabe, Y.; Sato, M.; Tsuboi, T. Amino Acid Taste Receptor Regulates Insulin Secretion in Pancreatic β-Cell Line MIN6 Cells. *Genes. Cells* **2011**, *16*, 608–616. [CrossRef]
171. Serrano, J.; Meshram, N.N.; Soundarapandian, M.M.; Smith, K.R.; Mason, C.; Brown, I.S.; Tyrberg, B.; Kyriazis, G.A. Saccharin Stimulates Insulin Secretion Dependent on Sweet Taste Receptor-Induced Activation of PLC Signaling Axis. *Biomedicines* **2022**, *10*, 120. [CrossRef]
172. Calvo, S.S.-C.; Egan, J.M. The Endocrinology of Taste Receptors. *Nat. Rev. Endocrinol.* **2015**, *11*, 213–227. [CrossRef]
173. Kyriazis, G.A.; Soundarapandian, M.M.; Tyrberg, B. Sweet Taste Receptor Signaling in Beta Cells Mediates Fructose-Induced Potentiation of Glucose-Stimulated Insulin Secretion. *Proc. Natl. Acad. Sci. USA* **2012**, *109*, E524–E532. [CrossRef]

174. Eisen, M.B.; Spellman, P.T.; Brown, P.O.; Botstein, D. Cluster Analysis and Display of Genome-Wide Expression Patterns. *Proc. Natl. Acad. Sci. USA* **1998**, *95*, 14863–14868. [CrossRef]
175. Wang, J.Z.; Du, Z.; Payattakool, R.; Yu, P.S.; Chen, C.-F. A New Method to Measure the Semantic Similarity of GO Terms. *Bioinformatics* **2007**, *23*, 1274–1281. [CrossRef]
176. Yu, G.; Li, F.; Qin, Y.; Bo, X.; Wu, Y.; Wang, S. GOSemSim: An R Package for Measuring Semantic Similarity among GO Terms and Gene Products. *Bioinformatics* **2010**, *26*, 976–978. [CrossRef]
177. Arrojo e Drigo, R.; Jacob, S.; García-Prieto, C.F.; Zheng, X.; Fukuda, M.; Nhu, H.T.T.; Stelmashenko, O.; Peçanha, F.L.M.; Rodriguez-Diaz, R.; Bushong, E.; et al. Structural Basis for Delta Cell Paracrine Regulation in Pancreatic Islets. *Nat. Commun.* **2019**, *10*, 3700. [CrossRef] [PubMed]

MDPI AG
Grosspeteranlage 5
4052 Basel
Switzerland
Tel.: +41 61 683 77 34

Biomolecules Editorial Office
E-mail: biomolecules@mdpi.com
www.mdpi.com/journal/biomolecules

www.ingramcontent.com/pod-product-compliance
Lightning Source LLC
Chambersburg PA
CBHW061848210926
51850CB00006B/1007
9783725870561